The Multidimensionality of Hermeneutic Phenomenology

CONTRIBUTIONS TO PHENOMENOLOGY

IN COOPERATION WITH THE CENTER FOR ADVANCED RESEARCH IN PHENOMENOLOGY

Volume 70

Scope

The purpose of the series is to serve as a vehicle for the pursuit of phenomenological research across a broad spectrum, including cross-over developments with other fields of inquiry such as the social sciences and cognitive science. Since its establishment in 1987, *Contributions to Phenomenology* has published nearly 60 titles on diverse themes of phenomenological philosophy. In addition to welcoming monographs and collections of papers in established areas of scholarship, the series encourages original work in phenomenology. The breadth and depth of the Series reflects the rich and varied significance of phenomenological thinking for seminal questions of human inquiry as well as the increasingly international reach of phenomenological research.

For further volumes:
http://www.springer.com/series/5811

Babette Babich • Dimitri Ginev
Editors

The Multidimensionality of Hermeneutic Phenomenology

Springer

Editors
Babette Babich
Fordham University
New York, NY, USA

Dimitri Ginev
St. Kliment Ohridski University
Sofia, Bulgaria

ISSN 0923-9545
ISBN 978-3-319-01706-8 ISBN 978-3-319-01707-5 (eBook)
DOI 10.1007/978-3-319-01707-5
Springer Cham Heidelberg New York Dordrecht London

Library of Congress Control Number: 2013957066

Printed on acid-free paper

Springer is part of Springer Science+Business Media (www.springer.com)

Foreword

The Universality of Hermeneutics in Joseph Kockelmans's Version of Hermeneutic Phenomenology

In an autobiographical sketch, Joseph Kockelmans (2008) reflects on his *Denkweg* in a manner that allows him to delineate the profile of his version of hermeneutic phenomenology. Based essentially on this sketch, I should like in what follows to bring into focus three principal moments of his "journey into phenomenological philosophy" that allude to his idea of the universality of interpretation in all culturally specified modes of being-in-the-world. I will call these moments respectively (a) the phenomenological reformulation of the Greek episteme; (b) the integration of the ontological difference in the theory of scientific truth; and (c) the historicity of objectifying thematization.

There is in Professor Kockelmans's works from the 1950s a gradual transition from Nikolai Hartmann's theory of the ontological modalities and categories (addressed in its capacity to serve as a prerequisite for reconstructing the ontological assumptions of basic scientific theories) to a kind of hermeneutic ontology. This transition is especially palpable in his reading of Hartmann's "Philosophy of Nature." In Hartmann's categorial metaphysics of knowledge *Dasein* and *Sosein* (as ways of being) are subordinated to the modes and spheres of being. The transition was by no means accomplished via a direct borrowing of Heidegger's concept of *Dasein*. It is rather the idea that the very metaphysics of knowledge should seek to make sense of the ontological categories by having recourse to the interrelations of *Dasein* and *Sosein* within the scope of scientific knowledge. A true "Philosophy of Nature" cannot avoid addressing the revealing of nature's being in these interrelations.

Professor Kockelmans's subsequent transformation of Hartmann's concept of *Dasein* in terms of ek-sistence as a pre-categorial way of being opened the avenue to hermeneutic phenomenology. The constitution of meaning is the "facticity" which the theory of categories presupposes, being unable at the same time to reflect upon it. Yet important motifs of a categorial metaphysics of knowledge were retained in the new philosophical project. These motifs precisely informed the desire for a rehabilitation of the Greek episteme within the ontological framework. Still in his Dutch period, Professor Kockelmans adopted the view that philosophy is neither a meta-scientific world-view nor can it be "naturalized" by recasting its problematic in scientific terms and languages. The constitution of meaning in human ek-sistence is the subject which philosophy has to address. Philosophy can master this task by developing a kind of hermeneutic ontology that leaves enough room for epistemological investigations. It is the rehabilitation of the Greek episteme that provides the chance for reconciling such investigations with the ontological search for meaning constitution and truth as un-concealment.

But what kind of epistemology does this rehabilitation imply? An answer to that question is to be found in Professor Kockelmans's long-standing critical encounter and dialogue with the post-empiricist philosophy of science. Roughly speaking, in this dialogue he was after an epistemology that is capable to complement hermeneutic phenomenology in a manner that can bridge the analytic of meaning constitution with a theory of epistemic-thematic articulation of various kinds of objects. Obviously, such a theory has little to do with the established (in the analytic philosophy) concept of epistemology as a normative theory about "justified true beliefs." By exploring the leeway released by the combination of the Greek episteme with the Greek phronēsis, Professor Kockelmans unfolded in diverse directions the claim that there is a horizonal understanding at the root of all specific forms of articulated knowing. It is this understanding that is a subject shared by both, hermeneutic phenomenology and the kind of epistemology which he looked for. Reflecting on horizonal understanding provides the access to both the transcendence of the world, i.e. to what is at issue in the ontology of the potentiality-for-being, and to the ongoing fore-structuring of knowing by contextualized epistemic practices and procedures. Kockelmans (1993, p. 101) made the case that horizonal understanding "has in itself the eksistential structure of being a projection." Because of this structure it acquires epistemological relevance.

Understanding as "grasping by anticipation"—so the argument goes—fore-structures the formation of each epistemic-thematizing attitude toward the world. (Professor Kockelmans was preoccupied in the first place with the triads of fore-having, fore-sight, and fore-conception that characterize the kinds of scientific thematization qua objectification of the world.) The "anticipatory sighting" of what gets constituted by epistemic practices assures the passage from hermeneutic ontology to hermeneutically pertinent epistemology. Horizonal understanding is at once a constitutive ontological phenomenon and (via its interpretative specification) the fore-structure of each kind of knowing (including the knowing achieved by procedures of idealization in the natural sciences). In his long-standing elaborations on the "being of knowing," Joseph Kockelmans gained deservedly the reputation of the philosopher who in the most profound manner succeeded in demonstrating the hermeneutic-phenomenological unity of (non-metaphysical) ontology and (non-representationalist) epistemology.

In working out the variety of epistemology which takes into account the "being of knowing," Professor Kockelmans dedicated serious efforts to criticizing the holist epistemological strategies (offered by Lakatos, Kuhn, Stegmüller, Hübner and several others) for their reflexive deficits and characteristic failures to make intelligible the fore-structuring of (the production of) scientific knowledge. Yet the focus on this fore-structuring did not promote a search for a radicalization of the intrinsic hermeneutic tendencies (as these have been most clearly exhibited in Mary Hesse's work) in the post-positivist historicism. It was rather a criticism aimed at a retrieval of what has gotten lost in the post-empiricist turn. In this regard, Professor Kockelmans undertook an original rendering of logical empiricism's problematic of meaningfulness with the intent to "repeat" this problematic in a hermeneutic-phenomenological framework. At stake was the eradication of the empiricist foreclosing of any approach to theory-observation interrelations that might take into consideration the interpretative contextualization of scientific thematization—the contextualization which is to be strongly distinguished from the contextual-epistemic interpretability implied by the post-empiricist thesis of theory-ladenness. Against logical empiricism Professor Kockelmans made the point that the interpretative nature of scientific thematization cannot be recast in terms of logical semantics. Finally, the hermeneutic approach to the fore-structuring of scientific knowing (as an approach that mediates between ontology and epistemology) calls into question basic assumptions shared by all parties in the realism-debate. The procedural-empirical laying bare of formally symmetric structures that unite measurable and quantifiable entities as domains of research proved to be a shared doctrine of constructive empiricism (as a particular position in this debate) and Kockelmans's program for a hermeneutic phenomenology of the natural sciences.

On constructive empiricism, not the relationship of correspondence but that of constructive co-interpretability of theoretical models and data models (as deliverances of experimental and observational experience) is at the heart of scientific enterprise. The epistemological counterpart of the relationship of co-interpretability is the empirical adequacy of a theory (in van Fraassen's technical sense). The greatest merit of constructive empiricism is the overcoming of the static subject-object relation's epistemology. The hermeneutic circle involved in the mathematical saving of phenomena, on which van Fraassen insists in his earlier work, strongly bears resemblance to the circularity of the horizonal understanding's epistemic specification within the objectifying research of the natural sciences. Yet the constructive empiricist skips the possibility to reflect on the hermeneutic circle of saving phenomena in a manner that would allow her to reinstitute the problematic of scientific truth by means of transcendental arguments. By amending her conception through such arguments, the constructive empiricist would be able to arrive at a concept of truth beyond the technical discussion of theory's empirical adequacy, avoiding at the same time making concessions to scientific (and structuralist) realism. More generally, since constructive empiricism offers only a subtle and cogent "description of what from an empiricist point of view it means to be an empirical scientist" (Kockelmans 1993, p. 138), this doctrine ignores the transcendental dimension of scientific objectification whose approaching reinstitutes the problematic of scientific truth against the background of the ontological difference. Reflecting on the constructive-hermeneutic circularity of models and data (appearances) should open the way from epistemology of science's empirical adequacy to the ontological specification of what scientific truth is.

The way in which Professor Kockelmans puts in his earlier work the fore-structuring of scientific knowledge first leads to the second principal moment in his philosophical journey. This moment gets expressed by a principal thesis to be found in several papers of him in the 1970s: With regard to the kinds of fore-structuring of thematizing knowledge that mediates between the ontological disclosure of scientific domains and the epistemic organization of scientific research, three basic hermeneutic situations of scientific thematization are to be distinguished related accordingly to the objectification through mathematical projection, phenomenological description of profiles that remain invariant within manifolds of variations (as this is shown in particular by phenomenological psychology), and interpretative-dialogical reflexivity in making sense of cultural phenomena. Though these three kinds of scientific thematization correspond to a certain extent to three types of scientific disciplines (objectifying [empirical] sciences, descriptive-phenomenological social sciences, and interpretative human studies), Professor Kockelmans has good reasons to insist that he is dealing with hermeneutic situations of doing research and not with institutionalized disciplines. Each of the situations may in principle take place in every scientific discipline. Thus, the second moment in his philosophical journey is the triple specification of the research processes' interpretative nature with regard to three kinds of science's basic hermeneutic situations.

Let me stress some important consequences that Professor Kockelmans drew from the way in which he spelled out the concept of hermeneutic situation of scientific thematization. The first one is the argumentation against the strategy of shifting essentialism from science's cognitive structures to invariants (groups of symmetries) of pre-scientific perception. Perception, however elementary it could be, is always already in a (pre-scientific or scientific) hermeneutic situation. In other words, there is no perception that precedes the constitution of meaning. All perceptive acts are contextualized by meaning-constituting practices. A paradigmatic alternative to the hermeneutic-contextual view of perception is suggested by various structuralist doctrines. Thus, Cassirer's gestalt-psychological view (expressed for the first time as early as in the conception of the *symbolische Prägnanz* from the 1920s, and clearly formulated in his celebrated paper "The Concept of Group and the Theory of Perception") restores the spirit of epistemological essentialism on a pre-scientific level by emphasizing structural invariants in the sensory flux of perception. Cassirer tried to advocate the view that it is not (only) the formal structure of scientific knowledge, but also the "structure of perception" that remains invariant/symmetric with respect to a group of transformations. On Kockelmans's argument, since symmetries of perceptual spaces and perceptual objects inevitably take place in a context, it is the meaning-constituting contextualization (and not the symmetries) that has to be taken as a point of departure of epistemological analyses within the scope of hermeneutic philosophy.

Another consequence from the scrutiny of the concept of hermeneutic situation is the new argumentation against the hypostatization of mathematical essences.

A domain of scientific research gets disclosed not through the projection of a mathematical formalism. A domain's being-disclosed is always in a hermeneutic situation in which practices and procedures of idealization and the related to them ongoing projection of formal structures come into being. The formation of basic mathematical formalism is always interpretatively contextualized.

The hermeneutic situation in which the regimes of epistemic practices get established and a particular domain of knowing gets disclosed is not outside the reality of being-in-the-world. It belongs to that reality which becomes at once revealed and concealed by being disclosed in a hermeneutic situation. This observation has a substantive implication for the specificity of scientific truth: The characteristic way of revealment/concealment defined by the hermeneutic situation of a scientific thematization is the *ontological truth* of that thematization. In stressing this kind of truth that is ignored by the analytical philosophers of science, Joseph Kockelmans does not go on to get rid of the epistemic (correspondent, coherentist, consensualist, pragmatic, instrumentalist, and so on) kinds of scientific truth. Yet he argued that the *ontic truth* (either of particular scientific propositions and statements or of holist conceptual frameworks like those of scientific theories) is to be circumscribed in semantic and epistemological terms only when one manages to determine the ontological truth of the basic scientific thematization (for instance, the thematization by means of which the domains of classical physics are disclosed). The rationale for this claim is that the formulation of all epistemological/semantic criteria for truth as well as the carrying out of all formal and non-formal procedures of verification take place in a reality that is always already disclosed by a scientific thematization. The ontological truth of the latter stipulates the conditions of possibility of the ontic truth within scientific knowing. The truth of formal invariants and groups of symmetries "shows itself" also in a hermeneutic situation of thematization. This is why it is also only a kind of ontic truth.

The third moment in Joseph Kockelmans's philosophical journey is his conception of the "critical studies in the history of science." His hermeneutic vision of science's historical dynamics opposes the post-empiricist division between internal and external history of science. The treatment of the historical horizon of scientific thematization resists any relegation in the competence of one of the two types of historiography. Within this horizon there is a constant interplay of practices belonging to various discursive formations. To be sure, one has to distinguish clearly between two cases. For the sake of brevity, think on bacteriology and quantum electrodynamics as typical manifestations of these cases. Bacteriology became disclosed within a heterogeneous discursive formation that involves non-scientific practices and administrative policies as well as clinical activities and research practices of physiology, classical immunology, cytology, zoology, and chemistry. The objectifying thematization and delineation of relevant objects of inquiry had been "prepared" by meanings of various kinds constituted by this discursive formation. Accordingly, the research articulation of the domain of bacteriology "found" in the period of its inception "ready-made entities" already distinguished by hygienic, clinical, and biological meanings. As Bruno Latour in particular shows, entities like contagiousness, miasma, aetiologic agents, different kinds of microorganisms, "model organisms" and so on circulated with important functions in spaces of political power. The initial objectifying thematization in bacteriology transformed these entities into scientific objects. Thus, the founding hermeneutic situation in bacteriology involves the task of scientification of "life-world's entities." This task is completely alien to the inception of quantum electrodynamics. The domain got disclosed by recasting of objects constituted entirely by research practices of older domains. There was no "provocation" from external problems arising out of non-scientific social practices. The founding hermeneutic situation involved the task of "enfranchising" of already existing scientific objects. Accordingly, the objectifying thematization was determined by research practices entitled to accomplish the recasting in question—computations based on perturbation theory, conceptual practices of overcoming incompatibilities between special relativity and quantum mechanics, formal practices of searching for covariant formulations of experimental results, etc.

Yet regardless of the way in which the domain had been historically disclosed—so Professor Kockelmans's basic argument goes—the hermeneutic situation of thematization makes the constitution of meaningful scientific objects a function solely of research practices. In other words, once disclosed, a scientific domain is characterized by a research process that projects its own horizon of possibilities. This is also the horizon of relevant problematization within the everydayness of scientific research. Once brought into play in a characteristic hermeneutic situation, the research process is dependent only on the possibilities projected by the practices of this process.

On Kockelmans's conception, the "rational reconstruction" of science's historical dynamics is a hermeneutic task. This does not mean that social-pragmatic interests have no impact on the research process. They certainly make enormous impact. Yet this impact gets refracted by the horizon of research possibilities. The very refraction provides a protection against cognitive deformations of scientific research caused by external pressure on the research process. It is a protection that is again of hermeneutic nature: Within the hermeneutic circle of the constitution of meaning and meaningful objects in scientific research, the external aims and interests get "translated" in possibilities of doing research that are proper to the articulation of the respective scientific domain.

Hermeneutic phenomenology of the natural sciences seems to be both a highly esoteric and a too exotic initiative. Yet it is of prime importance for everybody who champions the post-metaphysical universalizing of hermeneutics. Without approaching the interpretative nature of the natural sciences, philosophical hermeneutics would be essentially restricted. Without doing this it would have had to refrain from laying claim to the conceptually most sophisticated form of culture. Professor Kockelmans dedicated a great deal of his work to the removal of this restriction imposed for several historical reasons on philosophical hermeneutics. In his final work he concentrated his efforts on supplementing the natural sciences' hermeneutic ontology with various approaches developed in methodical hermeneutics (Kockelmans 2002). At issue are the formation of textual traditions and the effective-historical series of contextualization of classical texts in the history of physical disciplines. This was an additional contribution of his to the post-metaphysical universalizing of hermeneutics.

Sofia, Bulgaria Dimitri Ginev

References

Kockelmans, Joseph J. 1993. *Ideas for a hermeneutic phenomenology of the natural sciences*. Dordrecht: Kluwer Academic Publishers.

Kockelmans, Joseph J. 2002. *Ideas for a hermeneutic phenomenology of the natural sciences. Vol. II: On the importance of methodical hermeneutics for a hermeneutic phenomenology of the natural sciences*. Dordrecht: Kluwer Academic Publishers.

Kockelmans, Joseph J. 2008. My journey into hermeneutic phenomenology of the natural sciences. In *Aspekte der phänomenologischen Theorie der Wissenschaft*, ed. D. Ginev, 99–113. Würzburg: Königshausen und Neumann.

Contents

Introduction

Babette Babich

The Multidimensionality of Hermeneutic Phenomenology: From Philology through Science and Technology to Theology

Studies of hermeneutics have historically invoked horizons and numbered dimensions[1] and hermeneutic phenomenology is inherently multidimensional. In part this is due to the different disciplines to be reviewed, such as the essential connection between hermeneutics and philology, attesting to the relevance of Nietzsche as well as Heidegger.[2] In addition to the hermeneutic tradition of classical philology, there is theology and law and there is the historical and specifically methodic legacy of Wilhelm Dilthey. Hence Joseph J. Kockelman's 2003 *Ideas for a Hermeneutic Phenomenology of the Natural Sciences* invokes "The Importance of Methodical Hermeneutics."[3] With this description, echoing the contributions of his friend and long-time colleague, Thomas Seebohm, Kockelmans himself relates Dilthey to Boeckh and thus to the classic tradition of hermeneutics including Gadamer.[4] Hence speaking of methodical hermeneutics, what Kockelmans (and to

[1] See E. D. Hirsch, Jr. (1972) in addition to the collection edited by Günter Figal and Hans-Helmuth Gander (2005) as well as an earlier collection featuring both legal and literary contributions, Winfried Hassemer (1984), in addition to Ronald Bontekoe's overview (1996), etc.

[2] See here the contributions to Helmut Flaschar, Karlfried Gründer and Axel E.-A. Horstmann (1979). See too for a discussion with reference to Gadamer as well as Husserl and Heidegger, István Fehér (1999) or (2001).

[3] Kockelmans (2003). See for a discussion of Boeckh and Dilthey, Otto Friedrich Bollnow's (1982) as well as Thomas M. Seebohm's monograph (2004) in addition to Seebohm's (1984). See too in connection with Boeckh's teacher, Schleiermacher, E. D. Hirsch, Jr. (1975). In connection with Nietzsche, see Whitman (1986) as well as Poschl (1979) and more recently Christian Benne (2005).

[4] Kockelmans, Joseph (2003). See for a discussion of Boeckh and Dilthey, Otto Friedrich Bollnow's (1982) and, adding, methodical hermeneutics, Thomas M. Seebohm's (2004) monograph as well as his (1984) essay on Boeckh and Dilthey and see too in connection with Boeckh's teacher, Schleiermacher, E. D. Hirsch, Jr. (1975). Giovanni Leghissa also includes a discussion of Boeckh in his contribution to the current volume.

be sure what Seehohm) understood as hermeneutic phenomenology comprised the full scope of the scholarly and 'scientific' traditions of classical philology just where classical philology subsumes under its aegis not only archaeology but the disciplinary breadth of aesthetics and history as well as philosophy and theology. In this methodical fashion, classical philology—like Husserl's famous phenomenological call to the 'things themselves'—refers to nothing less than the 'words' themselves.

Although even otherwise hermeneutically sensitive scholars routinely limit their conception of Nietzsche to his supposed proclamation of the death of God,[5] such a limitation can obscure Nietzsche's explicitly hermeneutic philology. Indeed, what can be seen to be Nietzsche's hermeneutic phenomenology is clearly expressed in his philological study of ancient Greek lyric and tragedy out of what he called "the spirit of music."[6] Thus philologically, i.e., attuned to the words themselves, Nietzsche there undertook to 'hear' Greek lyric and tragic poetry, hearing 'with his eyes' as he described the philological task in question. For Nietzsche, *aesthetics*, defined *as a science* corresponded to the rigorously methodical and thus *scientific* question of his own discipline of ancient or classical philology. And he had posed this question even before his first book inasmuch as the *critical* perspective Nietzsche urges beginning with his inaugural lecture in Basel is also the reason he concludes that lecture with a conversion of Seneca's dictum: philology is to become philosophically critical which is also to say that philology must be set on the path of a critical science.

In this sense, we can begin to comprehend Nietzsche's otherwise difficult to understand self-critique (or self-defense), as he claims that in his first book, *The Birth of Tragedy out of the Spirit of Music*, he found himself grappling with what we may here describe as the 'multidimensional' problem of the multidimensionality of science itself: "something frightful and dangerous, a problem with horns … in any case, a new problem … science considered for the first time as problematic, as questionable."[7] Asking how science, as such, is possible *qua* science (which is what I have elsewhere described as a critical philosophy of science),[8] Nietzsche was in this sense the first to propose a hermeneutics of science.

Nietzsche would go on to address physics itself, characterizing the natural scientist's 'interpretation' of nature as a "lack of philology,"[9] invoking his own scientific expertise or authority ("speaking as an old philologist"), to accuse natural scientists

[5] See Adriann T. Peperzak's contribution in the essays below in addition to Kockelmans' own (1983).

[6] See Babich (2005) as well as Christophe Corbier (2009) and see the final chapters of Babich (2013) for more discussion and further references. Damir Barbarič (2005) explores the question of hearing in an effort to differentiate Heidegger's rhetorically attuned hermeneutics from Gadamer's hermeneutics but he does not raise the question Nietzsche does in terms of the music of words, that is to say of the sounding of the text.

[7] Friedrich Nietzsche (1980a).

[8] Thus see Babich (2010a, 2009).

[9] Nietzsche, *Jenseits von Gut und Böse*, §22; Nietzsche (1980), Vol. 5, 37.

of misinterpreting their interpretations, that is to say: of forgetting that their interpretations corresponded to "interpretation rather than text."[10]

In my own Nietzsche-indebted overview of different approaches to continental philosophy of science—including philosophies of science other than the traditional preoccupation with physics that characterizes mainstream or analytic philosophy of science—I discuss both philology and method, echoing Karl Jaspers' along with Karl Reinhardt's additional reflections, in order to argue for the clear multidimensionality of the philosophy of science itself: "The Case for -P Philosophies of Science, where P = Physics."[11]

Kockelmans alludes to Nietzsche's famous reflections on science as interpretation in *Beyond Good and Evil*, focusing in his case on the issue of text *qua* text.[12] To be sure, Kockelmans' own concern was methodical hermeneutics, and like others, Kockelman's does not speak as Nietzsche speaks of 'text,' i.e., as a metaphor for the object as such but conventionally, i.e., with respect to the traditions of scientific interpretation. For Nietzsche however, as for Heidegger, the 'text' when it comes to natural science will be its objects, or else, as Patrick Heelan has also offered several hermeneutic studies of these, its instruments, its 'readable' technologies.

The relevance of hermeneutics and science in particular must be foregrounded as it is central to the current collection but also given the sometimes peripheral presence of such approaches in mainstream histories and philosophies of science. Although one can also explore this peripherality in terms of the analytic or mainstream tendency to distinguish the history of science (and its more traditionally text-based or historiographically hermeneutic orientation) from the philosophy of science (and its traditional orientation to theory and experiment), one can also, as noted above, trace this back to the old distinction, in Dilthey's formulation, nature we explain, the life of the mind we understand: *Die Natur erklären wir, das Seelenleben verstehen wir.*[13] This distinction has been decisive, especially for what would become today's analytic and even expressly logical positivist philosophies of science (e.g., von Wright's 1971 *Explanation and Understanding*).[14] To this day we continue to contrast the natural and the human sciences, whereby the natural sciences dominate our ideal notion of science as science. Hence physics is the pre-eminent or archetypical science (the "-P-sciences" mentioned above are no less natural sciences but include the philosophy of chemistry as well as the earth sciences including geology, as well

[10] Ibid.

[11] See Babich (2010a), 359ff. On Nietzsche and Reinhardt and history, see Wolfgang Müller-Lauter (1999) and see too for a discussion of Löwith and history, Rodolphe Gasché's essay, "The Remainders of Faith: On Karl Löwith's Conception of Secularization" in the present collection below.

[12] See Kockelmans (2003), ix. Kockelmans here refers to Paul van Tongeren's (2000).

[13] Wilhelm Dilthey (1916–1967), Vol. VII, 144 See further, Dilthey (1991). Sabine Müller, a philosophical physicist includes Dilthey along with an explicit reference to hermeneutics in her (2004) but even where Dilthey is not mentioned by name he remains influential—rather in the Hegelian spirit that is marked by a disinclination to draw connections to other authors.

[14] Georg Henrik von Wright (1971).

as biology).[15] The human sciences, by contrast, include history and literary studies as well as art history and theology but they also traditionally include the more quantifiably promising disciplines of psychology, sociology, ethnography and political science in addition to other so-called social sciences. Thus in his 1930s Nietzsche lectures, Heidegger highlights the academic tendency to connect the arts and the sciences, less a conjunction than a contest, an agonistic tension nicely expressed in Rorty's pragmatic bon mot as "physics envy."[16] Rorty's phrase captures the relation to the natural sciences particularly asserted by analytic philosophy, evident in the conflict that has in the interim peaked (without for that being fully resolved) under the rubric of the so-called science wars[17] but also in the ongoing debates on the relevance or irrelevance of philosophy (as expressed from the point of view of physicists like Stephen Hawking),[18] where what counts as philosophy excludes hermeneutic and phenomenological kinds and is pretty much defined as Paul or Patricia Churchland define it, i.e., as good will advocates for brain scans or as dedicated, in P.M.S. Hacker's more pithy phrase, to "singing the Hallelujah chorus for the sciences."[19] Indeed, Hawking's and other scientist's complaints would seem to make it plain that the scientists see themselves as perfectly capable of bandleading on their own behalf.[20]

Dilthey's contrast between explication and understanding is a clear one and articulates an importantly hermeneutic truth when it comes to the relation between subject and subject in the human sciences. This recurs in Gadamer's existential emphasis in his reminder that we always understand otherwise, when we understand, inasmuch as, in this very Diltheyan sense, understanding is always understanding another—an other, any other's—understanding. But despite its clarity and correctness (as Heidegger distinguishes ontic truth), Nietzsche challenges that although we may give our science the name of "'Explanation'… it is 'description' that distinguishes us from older stages of knowledge and science. Our descriptions

[15] See my above cited: "Towards a Critical Philosophy of Science" for this distinction and for extensive references to the philosophy of chemistry, including Eric Scerri as well as Jaap van Brakel—whose work appears in a different context in the present collection—as well as the philosophy of geology, including the work of Rom Harré and Bob Frodeman (and this collection features some of Frodeman's work), in addition to the philosophy of biology (and to which Dimitri Ginev's contribution in this collection also belongs) including the complex case examples of Haeckel and Franz Moewus as well as Rupert Sheldrake, Lynn Margulis and the molecular cancer researcher and AIDs epidemiologist, Peter Duesberg.

[16] Richard Rorty (1994). See for further references, Babich (2010b).

[17] The science wars were instigated by disgruntled thinkers on the side of physics and traditionally positivistic philosophy of science. See for complete references and a hermeneutic account Babich (2002b). See also the introduction to the same volume: Babich (2002c).

[18] Hawking has been saying this for some time—and it is complemented by his ambition to be heard as a philosophically as well as scientifically in his *A Brief History of Time.* See for one account in the popular press: Matt Warman (2011). For this, see Stephen Hawking and Leonard Mlodinow (2010).

[19] In interview with James Garvey (2010). For a measured discussion, see Maxwell R. Bennett and Peter M. S. Hacker (2003).

[20] This I emphasize in an interview: Babich (2011).

are better—we do not explain any more than our predecessors."[21] As Nietzsche goes on to reflect:

> How could we possibly explain anything? We operate only with things that do not exist: lines, planes, bodies, atoms, divisible time spans, divisible spaces. How should explanations be at all possible when we first turn everything into an image, our image![22]

Nietzsche later observes that "It is perhaps just dawning on five or six minds that physics, too, is only an interpretation and exegesis of the world (to suit us, if I may say so!) and not a world-explanation."[23] Explanation turns out to be little more than a redescription of the unfamiliar in familiar terms, whereby the unknown is able to be 'taken' as known, *as if* known—a point not lost on the neo-Kantian philosopher Hans Vaihinger.[24]

In addition to Nietzsche's hermeneutic and phenomenological thinking,[25] the range of approaches to hermeneutic phenomenology including but not limited to the philosophy of science characterizes the breadth of not only Martin Heidegger in his writing on science and technology but also Maurice Merleau-Ponty, in addition to a range of philosophers of science cutting across the contemporary analytic-continental divide, where some are patently analytically minded and others more traditionally, or classically, continentally framed. The term can be applied, arguably—by which I mean descriptively—to many thinkers and scholars including, anthropologists and sociologists and even poets of science, like Gaston Bachelard, theorists and historians of science, such as Günther Abel, Karl-Otto Apel, Babette Babich, Gaston Bachelard, Nancy Cartwright, Peter Caws, Bob Crease, Martin Eger, Jacques Ellul, Paul Feyerabend, Dagfinn Føllesdal, Dieter Freundlieb, Steve Fuller, Carl F. Gethmann, Ronald Giere, Dimitri Ginev, Trish Glazebrook, Ian Hacking, Lee Hardy, Patrick Heelan, Kurt Hübner, Peter Janich, Pierre Kerszberg, Ted Kisiel, Joseph J. Kockelmans, Bruno Latour, Hans Lenk, Reinhard Löw, Alfred Nordmann, Gerard Radnitzky, Joseph Rouse, Thomas Seebohm, Michel Serres, Isabel Stengers, Bas C. van Fraassen, and Carl Friedrich von Weizsäcker, John Ziman, among many others. Although hardly to be reduced to any one tradition, if only to the extent that each of the above names—and many more could be added—represent sometimes opposed

[21] Nietzsche, *Die fröhliche Wissenschaft*, §112; (1980), Vol. 3.

[22] Ibid.

[23] Nietzsche, *Jenseits von Gut und Böse*, §14; (1980), Vol. 5.

[24] See Vaihinger (1924). I discuss Vaihinger and Nietzsche together with the philosopher of chemistry and early interpreter of Nietzsche and science, Alwin Mittasch in Babich (1994). For a related discussion but particularly with refernce to Robert Julius Mayer, see Günter Abel (1998).

[25] Gadamer had already written about Nietzsche and hermeneutics some time ago along with Paul Ricoeur and Gianni Vattimo, in addition, of course, to almost everyone who has ever written on Nietzsche and interpretation. And anyone concerned with Nietzsche and science was perforce reflecting upon yet another dimension of hermeneutic phenomenology, to wit Vaihinger as well as Mittasch but also Walter del Negro and Reinhardt Löw, Jean Granier, Friedrich Kaulbach, Wolfgang Müller-Lauter and others. Several collections have appeared drawing out the lines of Nietzsche and phenomenology, most recently and most comprehensively, Élodie Boubil and Christine Daigle (2012).

but always distinct philosophical approaches in their own right, along with their own specializations, the breadth of these hermeneutic and phenomenological approaches to the history and philosophy of science is highlighted to an astonishing degree in Kockelmans' several approaches to the philosophy of science beginning with a concern with the history and philosophy of mathematics[26] and physics[27] and, as he himself emphasizes,[28] with Husserl.[29] On his own account of this and after his initial work in the philosophy of mathematical physics, Kockelmans' intellectual development works through Merleau-Ponty[30] as well as Heidegger's philosophical reflections on science in *Being and Time* and throughout his later writings (including Heidegger's reflections on art),[31] before Kockelmans goes on to offer his own overview in his two-volume study, the first volume published in 1993 and the second volume almost a decade later in 2002: *Ideas for a Hermeneutic Phenomenology of the Natural Sciences*.[32]

Heidegger had argued that reflective or meditative thinking or philosophy is questioning—meaning that it both presupposes and that it entails—questioning. In this questioning hermeneutic sense, the Heidegger of 1929/1930 is able to contend that "all science is perhaps only a servant with respect to philosophy."[33] The same spirit of this early suggestion can be heard in the later Heidegger's provocative dictum on science in his *What is Called Thinking* that, and above all: 'science does not think.'[34]

In this sense, the Gadamerian hermeneutic philosopher, Jean Grondin, seemingly argues that continental philosophy is hermeneutics—as it were—all the way down.[35] But traditional practitioners of hermeneutic philosophy have tended to keep themselves well clear of the broad themes of philosophy, especially epistemology but above all philosophy of science, emphasizing as students of hermeneutics tend to do, a focus on text or literary traditions of the same. The result can lead to misprisions in the classic debates over the years between Gadamer and Habermas or the debate specifically relevant to the current context, between Patrick Aidan Heelan and

[26] Joseph J. Kockelmans (1953).

[27] Kockelmans (1958, 1962).

[28] See for this emphasis: Kockelmans (1993), ix ff.

[29] See for example, Kockelmans (in Dutch) (1964), (in English as 1967) as well as his (1987). See too Kockelmans' monograph on Husserl which begins with a reprint and translation of Husserl's 1928 article on "Phenomenology" in the Encyclopedia Britannica: Kockelmans (1994) as well as Kockelmans (1970).

[30] Joseph J. Kockelmans (1970) as well as (1964).

[31] Kockelmans (1985).

[32] Kockelmans (1993, 2002).

[33] Heidegger 1995, 5. The focus on questioning is the meaning of critique, foregrounded as essential in Kant and post-Kantian thought in Nicholas Rescher's contribution to the current volume. See also Richard Tieszen (2005) who emphasizes the importance for Gödel of this likewise Husserlian emphasis on the role of philosophy.

[34] Heidegger (1968), 8ff. For discussion see Jean-Michel Salanskis (1995), Babich (2003). Ginev (1997).

[35] See Jean Grondin (2000).

György Markus.[36] Markus takes the literary scholar's conventional understanding of the hermeneutic tradition as his point of departure, invoking the "cultural organization of the Author-Text-Reader relation."[37] Markus then goes on to insist that when it comes to the philosophy of science, meaning the natural sciences, "writings explicitly addressed to such an undertaking are very rare."[38] But this circular insistence exemplifies what Nietzsche called the acoustic (*ceteris paribus*: cognitive) illusion, that where one hears (or can conceptualize) nothing, there is nothing. Thus if we have not bothered to read widely—and many of us, even many of the more scholarly among us, do not bother—we assume that what we have read exhausts the extant texts, which then allows us to go on to say that such approaches are either nonexistent or rare. The tendency is self-confirming and convenient. Coupled with the tendency scholars have to focus on just a few names at the tip of the fashionable disciplinary iceberg, the attention deficit disorder Nietzsche called a 'lack of philology' continues to this day.

But the problem is worse than a lack of research. For a hermeneutics of natural science goes beyond the texts themselves in the very phenomenological direction of the things themselves, indeed towards a hermeneutic phenomenology of natural science. Thus in order to contend in 1987 that there is "No Hermeneutics of Natural Sciences," Markus was required not only to overlook Heidegger himself—who offers a precisely hermeneutic account of the natural sciences and specifically naming physics as such and thereby amplifying Husserl's phenomenological project for the sciences in Heidegger's 1927 *Being and Time*—but also, and more expressly, Heelan's 1965 monograph on Heisenberg's philosophy of science,[39] as well as Kockelmans' 1966 monograph on the philosophy of physical science (leaving out Kockelmans' earlier Dutch language studies),[40] in addition to, among excluded others, Heidegger's successor in Freiburg after the war, the Hungarian philosopher of science, Wilhelm Szilasi who published a very hermeneutical minded study of science (in Heidegger's spirit) in 1945 and so on.[41]

The history of the philosophy of science itself, in the meetings of the American Philosophy of Science Association and in its publications, which in the 1960s received the work of Kockelmans as it also received Heelan's contributions with an openness that was as striking as it would prove to be short-lived, has yet to be written, but any account would need to review the changing rubrics that must rule the reference to hermeneutics.[42] Thus Kockelmans pointed out that he himself originally spoke of "existential phenomenology" and only later came to speak of the

[36] See here Gyorgy Markus's patently circular (1987), as well as Heelan's patient rejoinder (1989). Largely engaging Markus, see Dimitri Ginev (1997). See yet more broadly, Heelan (1998).

[37] Markus (1987), 5.

[38] Ibid., pp. 5–6.

[39] Heelan (1965).

[40] See Kockelmans (1966), which in turn was a translation of an earlier text written in Dutch (1962).

[41] Wilhelm Szilasi (1945) as well as Szilasi (1961).

[42] Heelan's own biographical reflections, (for a beginning and a chronological review of Kockelmans' as indeed of Heelan's own publications in this matter) can also be revealing.

same as "hermeneutic phenomenology"[43] and Heelan too would experiment with terms like context-dependence and interpretation.

Another part of the problem may well be traceable to my own teacher, Hans-Georg Gadamer, who maintained, perhaps because his own father was a well-known professor of chemistry, a certain distance from the sciences, and who, when he did engage the sciences in his long life, did little to supersede the effects of this same distance. Thus Gadamer's *Reason in the Age of Science* repeated Dilthey without going beyond him.[44] More troublesome was the conventional distinction lent to studies of the social sciences (already burdened by the old fact/value distinction as sciences of 'spirit' in a German context) by authors who did not really introduce hermeneutics at all into books that were nonetheless so titled, such as Zygmunt Bauman's *Hermeneutics and Social Science* which was rather more of a primer for anthropological sociology than anything else.[45] By contrast, of course, Kockelmans always sought to include both phenomenology and hermeneutics in his own discussions of the social sciences.[46]

Nevertheless there are significant signs that things are changing. I read Alfred Nordmann's "Getting the Causal Story Right" as an important step in such a direction interior to mainstream philosophy of science, beyond the continued damage done by Markus's limitation of hermeneutics to the "interpretive encounter of a reader with a text"[47] rather than and as Kockelmans himself had read Heidegger's hermeneutic transformation of the phenomenological return to Husserl's things themselves in the schemes that Heidegger contended made up the region or delimited an individual science *qua* science, or as Merleau-Ponty saw this as

[43] This is also to be seen in the original title for the largest society for the study of continental philosophy in North America, the Society for Phenomenology and Existential Philosophy. Some years ago there was talk of changing the name of the society to reflect not only hermeneutics but other significant trends. Similar emphases also can be seen in the leading journal for continental philosophy which was originally called *Man and World* and is now called, obviously enough, *The Continental Philosophy Review*.

[44] Gadamer (1981). But of course all of this collection is about showing the precise relevance of Gadamer's thinking to science as exemplified, just for one example by an essay featuring medical and nursing professionals among the collective authors: Nancy J. Moules et al. (2013).

[45] Simon Glynn's concluding essay, "The Hermeneutics of God, the Universe, and Everything", offers an exception to this claim. See Zygmunt Bauman's (1978) also avoided significant engagement with Gadamer, reading hermeneutics to be sure as a literary tradition. I should also note that although Richard Bernstein's study of pragmatism and hermeneutics invokes science in the title of his book, Bernstein does not in fact speak to philosophy of science. Similarly, the rhetoric of science can fail to engage the broad tradition of hermeneutics as can be seen by more rather than less conventional studies such in evidence in monographs and collections such as Allan G. Gross and William M. Keith (1997). By contrast and although also analytically inclined Chrysostomos Mantzavinos (2005) offers a systematic approach to what may count, very provisionally, as a new beginning.

[46] Thus see in particular Kockelmans' important essay (1975) as well as his (1976), his (1978) and (1979) essays.

[47] Alfred Nordmann's (2008).

informing sense-perception including measurement, as Heelan would also argue in both Heidegger's and Husserl's sense, as well as theory.

If Heidegger could call Husserlian phenomenology the *Urwissenschaft* in 1919,[48] his signal contribution was his articulation of an explicitly hermeneutic phenomenology. Thus if the Heidegger of 1925, almost in the very same terms that Husserl uses, refers to the "crisis of philosophy as science," he reflects in the same spirit—and indeed one that will recur almost verbatim in the early section of *Being and Time*—that all "sciences and groups of sciences are undergoing a great revolution of a productive kind that has opened up new modes of questioning, new possibilities, and new horizons."[49] Heidegger goes on to detail the theory of relativity in physics along with the crisis of foundations in mathematics, to which one must add quantum mechanics along with the movement against mechanistic thinking in the biological sciences. For Heidegger, what is at issue is the *constitution* of modern technological and mathematizable (measurable, calculable, model-oriented) science, conceived in both the Husserlian phenomenological sense and the mechanically explicit sense of standardized manufacture and institutional technology.[50]

But it would be difficult to characterize Joseph Kockelmans' approach better than Ted Kisiel has where Kisiel also has recourse to the above-mentioned distinctions and contrasts to do so:

> Contrary to Patrick Heelan and me, Joe K's hermeneutic approach to the philosophy of science consistently follows an (to me unremarkable) approach sketched out by MH in SZ 363 of "thematizing objectification" with math physics as its ultimate model, which via mathematical projection abstracts and demarcates a domain of objects, which it regards as Nature, for research by way of formalization and other such theoretical systematizations. All this summarized in his book (Kluwer, 1993) entitled *Ideas for a Hermeneutic Phenomenology of the Natural Sciences*.[51]

To this extent, and as we may, following Kisiel here, review Kockelmans' own philosophical trajectory in the philosophy of science, Heidegger himself also followed and complemented Husserl's own approach to science. In the same way, as Kockelmans has also foregrounded this conjunction, both Heidegger and Husserl significantly regarded phenomenology as an approach needed for any philosophy of science that might come forth as such.[52] But in the same spirit, and this is where

48 Heidegger (2000), 3, 11ff . See further Ted Kisiel (2002), 17ff.

49 Heidegger (2002), 148.

50 This is a complex point, and later the same Heidegger who will foreground *Gelassenheit*, suggests in the 1930s that the trajectory of modern technology may be described as a "humanism" — reading humanism here as Nietzsche speaks of the human, all too human. See for this reading of the *Beiträge* of the 1930s and 1940s, Babich (2012a) as well as my own essay included in the collection below on Heidegger's 1949/1950 lectures as well as, for a critical account relevant to our own times, Babich (2012–2013).

51 Ted Kisiel, email to the author. 12:00 AM, 11 June 2013.

52 Indeed although the great majority of the contributions show the dominant influence of analytic philosophy, the contributions to Carlo Ierna, Hanne Jacobs, and Filip Mattens (2010) illustrate this point as does R. L. Tieszen (1989) as well as Ginev (1997) and the contributions to Babich (2002a) as well as Glazebrook (2012).

many readers of Heidegger's philosophy of science will tend to shy away, recognizing this as a critical reservation, Heidegger also opposes sense-directed reflection [*Besinnung*] to the rational, calculative project of Western technologically articulated and advancing science. Thus Heidegger discusses the relation between science and philosophy in *Being and Time*, noting as he does there that philosophical logic can either 'limp along' after the sciences,[53] or else it can leap ahead, as a literally "productive logic."[54] For Heidegger this generative logic that leaps ahead "into some area of Being, discloses it for the first time, in the constitution of its Being, and, after thus arriving at the structures within it, makes these available to the positive sciences as transparent assignments for their inquiry."[55]

Reflecting on the "future" of hermeneutic philosophy, Otto Pöggeler, along with Bas C. van Fraassen, a contributor to Tim Stapleton's edited Festschrift in Kockelmans honor,[56] could observe that no possibility that is not adequately anticipated or sufficiently met can come to be. What is then lacking is not a failure of possibility with respect to what has or what might come to pass but a deficiency in the prerequisite or condition for the possibility of matching such a possibility in advance and as *point de départ* in the present time.[57]

Kockelmans' own *Ideas for a Hermeneutic Phenomenology of the Natural Sciences*[58] begins with a sober recollection of the breadth of his background in his introduction to this collection, going back to his fairly patently hermeneutic 1958 study on *Time and Space*.[59] Although it bears directly on the issue at hand, i.e., although it is precisely relevant to the multifarious dimensionality or dimensionalities of hermeneutic phenomenology of science precisely qua philosophy of science, there is here no way to detail the history of reception and lack of reception, i.e., to explicate the antecedents and consequents of what are (or what become) received viewpoints vs. the unreceived viewpoints that collectively make up the hermeneutic constellation of what is routinely included within and what is excluded from what is called philosophy of science. Some of this is due to what is widely condemned as scientism or analytic philosophy's 'physics envy' quoted from Rorty above.[60] Other elements are doubtless due to a related trend on the part of analytic philosophy to bar from its ranks anything, anyone and indeed any themes that might compromise analytic philosophy's ongoing effort to be taken as the sole arbiter of science and reason—even in place of scientists as such—but may also be accounted to the

[53] Heidegger (1962), 31.

[54] Ibid., 30.

[55] Ibid., 31.

[56] See the contributions to Timothy Stapleton (1994).

[57] See here Otto Pöggeler (1994).

[58] Kockelmans (1993).

[59] Cited as: "*Time and Space: The Meaning of Einstein's Relativity Theory for a Phenomenological Philosophy of Nature* (Haarlem: Bohn, 1958)" in Footnote 2 of Kockelmans, 1993: "Preface", ix. Original (1958).

[60] Rorty (1994) and see, again, for further references and discussion, Babich (2010b).

extreme rigor of hermeneutic phenomenology which from the start conceived its own approach as scientific, and of the very first rank.

It is in this fashion that Heidegger reflects on the reflexive contradiction of the claim that "there is no absolute certainty."[61] Like Nietzsche's claim that there is no truth (only interpretation), Heidegger does not dispute the argument countering that this claim advances "a claim to absolute certainty that there is no absolute certainty." Nevertheless and just as Nietzsche does not dispute but much rather encourages the critic who observes that the claim that 'everything is interpretation' is itself an interpretation, the issue for philosophical and logical reflection is exactly, as Heidegger points out, that "this apparently unshakable argument nevertheless carries no weight."[62] At issue is the lived dynamic of philosophy or "freedom" for Heidegger, a freedom which also corresponds to an "innermost ambiguity,"[63] the same ambiguity that appears in Nietzsche's writings as "change" or "becoming." It is because of the "turbulent" freedom of philosophizing, as human beings must philosophize, that everything that belongs to the human condition "belongs just as essentially to the truth of philosophy."[64] Hence and in a Nietzschean (and indeed Avenarius-cum-Machian) moment reflecting on the economy of knowledge, Heidegger observes that "No knower necessarily stands so close to the verge of error at every moment as the one who philosophizes."[65]

For Heidegger—and this reflects the overall spirit of the present collection on the multidimensionality of hermeneutics—philosophy is called upon to think on science. But Heidegger also contends not only that science is infamously innocent of thought but in what we may now see to be an echo of Nietzsche's remarks on physics and interpretation in *Beyond Good and Evil*, Heidegger also writes that "Physics as physics can make no assertions about physics."[66] To this extent—and this is why hermeneutics cannot be dispensed with, perhaps particularly when it comes the natural sciences—Heidegger's objections are, logically, formal ones. As Ted Kisiel explains Heidegger's gnomic pronouncement on error: "In order to reflect on any science, it is necessary to transcend that science and adopt a transcendental vantage point, to put it in Kantian terms."[67] For Heidegger, a scientist philosophizes, with all the risks of the same, as a philosopher and not as a scientist when reflecting on the foundations of his own discipline.

When Kockelmans concludes the first volume of his *Hermeneutic Phenomenology of Natural Science* by reflecting on the same foundations with respect to the history and philosophy of science, his point concerns the conceptual framework of science

[61] Heidegger (2001), 18, cf. 17.

[62] Heidegger, *The Fundamental Concepts of Metaphysics*, 18.

[63] Ibid., 19.

[64] Ibid.

[65] Ibid.

[66] Heidegger (1977), here 176.

[67] Kisiel (1970), 167–183, here 170.

as this itself "essentially depends on its mathematical character."[68] In this sense Kockelmans stresses, as Hilbert had already argued as necessary point of departure for mathematics as a science, the foundational point Heidegger makes above, that "mathematics is not a means to express a rationality that is already there" but much rather that mathematics "constitutes the rationality of our description of the observed phenomena."[69] The essays to follow exemplify this rigor and above all they testify to the multidimensionality of hermeneutic phenomenology not only in the philosophy of science but also for the philosophy of technology as well as metaphysics and epistemology, and including aesthetics, as well as explorations of the history of philosophy and theology.

Plan of the Text

In his lead essay, "A Paradox of Cognition" in the first section, *Cognition, Bio-Hermeneutics, and Lifeworld*, Nicholas Rescher offers a reflection on the classical ironic circumstance that finds us increasingly aware of the limitations of our knowledge the more we know. Rescher takes his point of departure from Kant's observation that every answer to our questions provides new materials for the development of further questions. As knowledge expands, the lineaments of our ignorance are brought even more clearly into sight. Questioning is thus an earmark of hermeneutic phenomenology. In his essay to follow, Dimitri Ginev turns to a case study drawn from vectorial biochemistry in his "The Articulation of a Scientific Domain from the Viewpoint of Hermeneutic Phenomenology: The Case of Vectorial Metabolism." Ginev's case study involves both theoretical objects related to anisotropic processes of trans-membrane transport and objects of inquiry contextually ready to hand within a configuration of scientific practices, especially including the hermeneutic fore-structure of scientific research in terms not only of scientific practices but also hermeneutic and horizontal possibilities as well as spaces of representation in addition to readable technologies.

The next essays take up the social sciences. Gregor Schiemann in his contribution, "One Cognitive Style Among Others: Towards a Phenomenology of the Lifeworld and of Other Experiences," addresses the work of Alfred Schütz in the phenomenology of the social sciences. Schiemann emphasizes Schütz's pluralist theory of experience. Speaking not only on cognitive styles but of the lifeworld as a world of perception as Husserl expressed it but also of the layer-model of the lifeworld developed by Schütz and Thomas Luckmann, Schiemann shows that "lifeworld" does not denote a category that encompasses culture or nature but refers to a delimited action-space and goes on to deploy Schütz's criterion-catalogue to characterize both experimental science and subjectivity. Then, in his essay to follow, "Steps Toward a Postfoundational Phenomenology," Giovanni Leghissa explores the problem of

[68] Kockelmans (1993), 281.

[69] Ibid.

historicity together with the paradoxes of foundation for the sake of a more comprehensive inquiry into the concept of lifeworld. Drawing upon Husserl and Blumenberg, Leghissa explores the relationship between history and the lifeworld as well as the paradoxes contained in the *Krisis.*

The concluding essays in this first section turn to practical hermeneutic dimensions in the natural sciences, including the philosophy of geology as well as measurement. Robert Frodeman discusses "Hermeneutics in the Field: The Philosophy of Geology," arguing that geological reasoning provides a rich and realistic account of both the power and limitations of scientific reasoning. Frodeman shows that geological reasoning highlights the hermeneutic and historical nature of reasoning, scientific or otherwise, in addition to the neglected kinship between reasoning in the sciences and the humanities. To conclude this first section, Robert Crease examines measurement as an 'emblematic technology' in his essay, "The Metroscape: Phenomenology of Measurement." Reading measurement to develop and extend Heidegger's concept of *Gestell*, Crease argues that measurement is more than one tool among others, such as rulers, scales, and other instruments, measurement is a fluid and correlated network that is smoothly and intimately integrated into the world and its shape. This essay proposes the concept of *metroscape* to develop and extend Heidegger's concept of *Gestell.*

The second of the four sections in this collection, "Hermeneutic and Phenomenological Philosophy of Science and Technology" leads with an essay by Patrick Aidan Heelan, "Consciousness, Quantum Physics, and Hermeneutical Phenomenology" who begins with a powerful metaphor comparing Friedrich Schleiermacher's 'hermeneutic' transformation of Kant's anthropology (in order to include then-newly discovered peoples that Captain Cook had discovered in the South Sea Islands) to Kockelmans effort to update Kant's notion of natural science to include the phenomenological lifeworld syntheses of classical, relativity, and quantum physics. In this hermeneutical move, the 'observer' is 'embodied consciousness' and 'measure-numbers' represent 'observable presence.' For Heelan, the quantum notion of an "observable" introduces into the discursive language of physics the common sense lifeworld notion of "contextuality" as Heelan himself had earlier developed the notion of a context-dependent logic. In the next essay, Michael Stölzner begins by noting that usual treatments of Nietzsche's thesis of eternal recurrence tend to highlight its ethical or anthropological rather than its more scientific aspects. Stölzner reviews Oskar Becker's 1936 effort to defend the scientific and logical basis of Nietzsche's writings, noting that although Becker endorses Abel Rey's *Le retour éternel et la philosophie de la physique* (1927), he neglects the work of the mathematician Felix Hausdorff, particularly his *Das Chaos in kosmischer Auslese* (published in 1898 under the pseudonym Paul Mongré). For Stölzner, Becker's argument rests upon the constructivist standpoint in the foundations of mathematics and the Heideggerian underpinning of it by the temporality of mathematical thought that he had already given in his 1927 *Mathematische Existenz.* From this set-theoretical and cosmological perspective, the rest of the contributions in this section take up Heidegger and technology, beginning with Theodore Kisiel's essay "Heidegger and Our Twenty-first Century Experience of *Ge-Stell,*" where he proposes an etymological translation of *Ge-Stell*, Heidegger's word for the essence of modern technology, from its Greek and Latin roots as "syn-thetic com-posit[ion] ing." For Kisiel, the virtue of such a compound translation shows that Heidegger's Ge-Stell presciently portends our twenty-first century experience of what Kisiel calls "the internetted WorldWideWeb," with its virtual infinity of 'websites' in 'cyberspace,' but also Global Positioning Systems, interlocking air traffic control grids, world-embracing weather maps, the 24/7 world news coverage of cable TV networks like CNN, etc.,—all of which are structured by the complex programming based on the computerized and ultimately simple Leibnizian binary-digital logic generating an infinite number of combinations of the posit (1) and non-posit (0). Kisiel argues that the sharp contrast between the global time-space technologically foreshortened into instantaneity and simultaneity and the radically local time-space of our situated historical existence illuminates nothing less than the temporal-spatial tension between *Ge-Stell* and *Da-Sein* and Kisiel accordingly seeks to bring them together in contemporaneous compatibility. Babette Babich's essay, "Constellating Technology: Heidegger's *Die Gefahr/The Danger*," revisits the original 1949 lectures to the Club of Bremen arguing that a hermeneutic not only of Heidegger's reflections on technology but the context in which he offered his lecture series can offer insight into some of the more controversial passages in these texts. Like Kisiel, Babich adverts to today's media context, particularly the ecology of modern technicized consciousness (underlining that we are still in need of a greater integration of Heidegger's thinking and critical theory), as well as the increasing real-world ecological pressures of our own day to rethink, once again, the related notions of event [*Ereignis*] and ownedness [*Eigentlichkeit*]. Lin Ma and Jaap van Brakel, in their jointly authored essay, "Heidegger's Thinking on the 'Same' of Science and Technology," begin by noting that as opposed to the common view that modern technology derives from modern science, Heidegger presents a reverse picture in which science originated in the essence of technology, wherein Being speaks. Ma and van Brakel contend that in this sense Heidegger speaks of the Same [*das Selbe*] of science and technology as ultimately grounded in the history of Being. From 1938 to the end of his life in 1976, Heidegger constantly explored the question concerning the relation of science and technology and kept himself well-informed of both traditional and new types of technology and science, including quantum physics, nuclear technology, and biophysics. Ma and van Brakel argue however that one cannot ascribe to the Heidegger the view that these new developments originate a new Epoch of Being. In his concluding contribution to the first half of this collection, "Logos and the Essence of Technology," Holger Schmid contends that current convictions that nature is not 'nature' but social construction corresponds to the self-accomplishment of metaphysical Platonism, thereby opening a common hermeneutic horizon for two articles of Heideggerian doctrine: namely, that technology has a non-technological 'essence' and that the final outbreak of the 'principle of reason' follows an incubation period of more than two millennia. What thus unfolds for Schmid is the philosophic history of the word 'logos': not speech, as Heidegger rightly urges, but 'laying.' In this fashion, and including important references to Wilhelm von Humboldt on language and, more subtly, to Friedrich Georg Jünger on technology, Schmid continues to argue that today's technoscientific worldview increasingly determines the way reality is perceived, privileging the framework of the natural as opposed to the human sciences.

The second half of the collection begins with the section *Philosophical Truth, Hermeneutic Aesthetics, and History of Philosophy*. Graeme Nicholson, in his lead essay here, "On the Manifold Meaning of Truth in Aristotle," makes the case that when Aristotle treats true and false statements in his logical treatises, what is demonstrated is that truth and falsity are the pre-supposed, non-discursive grounding for statements themselves. Nicholson goes on to note that it is even more salient that Aristotle's ethical treatises show that intellectual virtues are constituted by truth whereas the *Metaphysics* shows that truth in thinking is sustained by the truth of being. As Nicholson argues, these diverse studies can be connected to one another by way of the Greek term for truth, *aletheia*, as Heidegger has treated it. Jeff Malpas' essay, "The Twofold Character of Truth: Heidegger, Davidson, Tugendhat," continues Nicholson's focus on the concept of truth as *aletheia*, or 'unconcealment.' Malpas differs from Nicholson's analysis in that he places his emphasis on Tugendhat's influential criticism of Heidegger's identification of truth with *aletheia* together with Donald Davidson's account. Malpas seeks to show why it remains the case that *aletheia* is to be understood as a mode of truth, arguing that this involves understanding a certain transcendental-topological structure as pertaining to *aletheia*, thereby understanding truth as standing in an essential relation to place or *topos* constituting the ground for genuine questioning or critique. In his essay, "What can Philosophy of Science Learn from Hermeneutics—What Can Hermeneutics Learn From Philosophy of Science?" Jan Faye challenges the traditional supposition that hermeneutics and phenomenology were the dominant positions in the philosophy of the humanities, whereby the validity of these constitutive acts of meaning depended on the historical situation of the interpreter and of the object of interpretation. Although agreeing with the hermeneutic-phenomenological tradition, Faye proposes a view of interpretation and understanding resting on the idea that human cognition is a natural phenomenon. Thus Faye argues that objective understanding exists in the humanities in the sense whereby the validity of an interpretation, like an explanation in the sciences, is independent of the interpreter's historical situation. As the concluding contribution to this section, Enrico Berti's "The Classical Notion of Person and its Criticism by Modern Philosophy" illustrates the definition of person given by Boethius as "an individual substance of a rational nature," and as derived from Aristotle. Berti explores the criticisms of this notion formulated by both modern and contemporary philosophers from David Hume to Derek Parfit and details the rediscovery of the classical notion of person, or of its Aristotelian elements, by Saul Kripke, David Wiggins, Paul Ricoeur, and Martha C. Nussbaum.

The concluding section, *Hermeneutic Science and First Philosophy, Theology, Hermetics and the Universe*, begins with a contribution that recollects the purview of the collection as a whole. In his essay, "Philosophie des sciences et philosophie première", Pierre Kerszberg argues that ever since the institution of Galilean science, the mathematical science of nature has wanted to surmount the deceptive appearances of everyday experience. Yet reference to familiar experience is insurmountable even for contemporary theory. Kerszberg, thus undertakes the project of first philosophy in terms of the horizon of a *mathesis universalis* in order to explore the possibilities of an epistemology that eliminates both the fantasy of absolute control of what is as well as the skepticism that inevitably follows the frustration of the same fantasy. For Kerszberg, Kant's transcendental phenomenology opens a path to such, including the contributions of modern and contemporary science to invent kinds of evidence that would engage anew the gestures of the body translated into the spaces of thought. Adriaan T. Peperzak's "A Re-Reading of Heidegger's "Phenomenology and Theology" in dialogue with Kockelmans own engagement with theology raises the question of the status of both science and theology, motivated by critical questions concerning his basic statements about the presence and absence of certain relations between faith and philosophy. Perperzak invokes traditional theological debates as well as a reflection on Franz Overbeck, usually noted in connection with Nietzsche but who was important for many contemporary debates on theology.

In the penultimate article in this collection, "The Remainders of Faith: On Karl Löwith's Conception of Secularization," Rodolphe Gasché's essay explores Löwith's notion of secularization. Gasché argues that this notion presupposes a conception of faith found only in the religions of the Book. For Gasché, Löwith's analyses of history, no matter whether eschatological or progressive, are adumbrated against the background of the Greek experience of the physical cosmos as this is characterized by cyclical time. The final contribution by Simon Glynn, "The Hermeneutics of God, the Universe, and Everything," offers a comprehensively global perspective on hermeneutic interpretation as a means of clarifying and resolving apparent incoherencies and contradictions within the scriptures as well as legal, classical, and other texts. Explicating such wide-ranging application within these diverse fields of human inquiry, Glynn concludes, along with Heidegger, that hermeneutic interpretation is central to all epistemological understanding, as it is to human existence.

Acknowledgments

Permission from Bildagentur dpa to reprint the photograph included in Babette Babich, "Constellating Technology: Heidegger's *Die Gefahr/The Danger*" is gratefully acknowledged.

Robert Frodeman's contribution has appeared in earlier variations as Frodeman, "Geological reasoning: Geology as an interpretive and historical science," *GSA Bulletin 107* (1995): 960–968 as well as Frodeman, *Geologic: Breaking Ground between Philosophy and the Earth Sciences* (Albany: State University of New York Press, 2003). An earlier version of Enrico Berti's "The Classical Notion of Person and its Criticism by Modern Philosophy" was initially published under the title of "The Classical Notion of Person in Today's Philosophical Debate" in the collection edited by Edmond Malinvaud and Mary Ann Glendon, *Conceptualization of the*

Person in Social Sciences, Proceedings of the Eleventh Plenary Session of the Pontifical Academy of Social Sciences, Vatican City (2006): 63–77 and was republished in part with the title "The Classical Notion of Person and its Criticism by Modern Philosophy" in *International Academy for Philosophy, News and Views*, New Series, Vol.1, No 1, 22 (Spring 2009): 9–19. The first instauration of Rodolphe Gasché's essay was published in *Divinatio. Studia Culturologica Series*, Sofia: MSHS, vol. 28 (2008): 27–50. An earlier version of Holger Schmid's essay was published in the *Proceedings of the 35th Meeting of the Heidegger Circle. On Heidegger's 1976 Letter to the 10th Meeting of the Heidegger Circle in Chicago* (New York: Fordham University, 2001): 101–114.

The co-editor of this collection, Dimitri Ginev, conceived the idea for this collection in honor of the memory of one of the pioneers of hermeneutic phenomenology, Joseph J. Kockelmans. Both editors together worked to bring the contributors here included in the present volume. This volume is both a tribute to Joseph Kockelmans as well as an invitation to read—and to re-read—Kockelmans' work.

As the corresponding and working coeditor, the present author thanks the reviewers of the volume for their helpful suggestions and Anita van der Linden-Rachmat as well as the series editor, Dermot Moran. In addition to my gratitude to Joseph Kockelmans, Jr. and Veronique Foti, I am also grateful to Tim Stapleton, Patrick Aidan Heelan and Dimitri Ginev. William J. Richardson, S.J., Richard Cobb-Stevens, and David B. Allison also provided helpful discussions. But my deepest tribute of thanks is owed to Ted Kisiel, who worked with Joe Kockelmans in Pittsburgh and who has inspired me in the years since I first encountered Ted in my reading of the collection he edited with Joe Kockelmans and again when I met him in person, for the first time, in Berlin and ever since.

References

Abel, Günter. 1998. *Nietzsche: Die Dynamik der Willen zur Macht und die ewige Wiederkehr.* Berlin: de Gruyter.

Babich, Babette. 1994. *Nietzsche's philosophy of science: Reflecting science on the ground of art and life*. Albany: State University of New York Press.

Babich, Babette, ed. 2002a. *Hermeneutic philosophy of science, Van Gogh's eyes, and god: Essays in honor of Patrick A. Heelan, S.J.* Dordrecht: Kluwer.

Babich, Babette. 2002b. Sokal's hermeneutic hoax: Physics and the new inquisition. In: *Hermeneutic philosophy of science Van Gogh's eyes, and god,* ed. B. Babich, 67–78. Dordrecht: Kluwer.

Babich, Babette. 2002c. The fortunes of incommensurability: Thoughtstyles, paradigms, and Patrick A. Heelan's hermeneutic of science. In: *Hermeneutic philosophy of science, Van Gogh's eyes, and god,* ed. B. Babich, 1–18. Dordrecht: Kluwer.

Babich, Babette. 2003. On the analytic-continental divide in philosophy: Nietzsche's lying truth, Heidegger's speaking language, and philosophy. In *A house divided: Comparing analytic and continental philosophy,* ed. Carlos G. Prado, 63–103. Amherst: Humanity Books.

Babich, Babette. 2005. The science of words or philology: Music in *The birth of tragedy* and the alchemy of love in *The gay science*. In *Revista di estetica.* n.s. 28. XLV, ed. Tiziana Andina, 47–78. Turin: Rosenberg & Sellier.

Babich, Babette. 2006. Heideggers 'Beiträge' zwischen politische Kritik und die Frage nach der Technik. Trans. Harald, Seubert with the author. In *Eugenik und die Zukunft,* ed. Stefan Sorgner, H. James Birx, and Nikolaus Knoepffler, 43–69. Freiburg: Karl Alber Verlag.

Babich, Babette. 2009. Nietzsche's philology and Nietzsche's science: On the 'problem of science' and *'fröhliche Wissenschaft.'* In *Metaphilology: Histories and languages of philology,* ed. Pascale, Hummel, 155–201. Paris: Philologicum.

Babich, Babette. 2010a. Towards a critical philosophy of science: Continental beginnings and bugbears, whigs and waterbears. *International Journal of the Philosophy of Science*. 24(4): 343–391.

Babich, Babette. 2010b. *ex aliquo nihil*: Nietzsche on science, anarchy, and democratic nihilism. *American Catholic Philosophical Quarterly*. 84(2): 231–255.

Babich, Babette. 2011. *An impoverishment of philosophy*, 37–71. Purlieu: Philosophy and the University.

Babich, Babette. 2012a. Heidegger's philosophy of science and the critique of calculation: Reflective questioning, *Gelassenheit*, life. In: *Heidegger on science,* ed. Patricia Glazebrook, 159–192. Albany: State University of New York Press.

Babich, Babette. 2012b. Heideggers Wille zur Macht: Die Beiträge lesen im Ruckblick auf Nietzsche, Wissenschaft und Technik. In *Heidegger und Nietzsche,* ed. Babich, Denker, and Zaborowski (trans: Harald Seubert), 283–321. Amsterdam: Rodopi Verlag.

Babich, Babette. 2012c. Philosophische Figuren, Frauen und Liebe: Zu Nietzsche und Lou. In *Nietzsche Forschung*, ed. Renate, Reschke, 113–139. Berlin: Akademie Verlag.

Babich, Babette. 2012–2013. *O, Superman!* or being towards transhumanism: Martin Heidegger, Günther Anders, and Media Aesthetics. *Divinatio*. 36(Autumn/Winter): 83–99.

Babich, Babette. 2013. *The Hallelujah effect. Philosophical reflections on music, performance practice and technology*. Surrey: Ashgate.

Barbarič, Damir. 2005. Hörendes Denken. In: *"Dimensionen des Hermeneutischen": Heidegger und Gadamer,* ed. Figal, Günter and Hans-Helmuth Gander, 37–58. Frankfurt: Klostermann.

Bauman, Zygmunt. 1978. *Hermeneutics and social science*. London: Hutchinson.

Benne, Christian. 2005. *Nietzsche und die Historisch-kritische Philologie*. Berlin: de Gruyter.

Bennett, Maxwell R., and Peter M.S. Hacker. 2003. *Philosophical foundations of neuroscience.* Oxford: Wiley.

Betti, Emilio. 1967. *Allgemeine Auslegungslehre als Methodik der Geisteswissenschaften.* Tübingen: Mohr.

Bernstein, Richard. 1983. *Beyond objectivism and relativism: Science, hermeneutics, and praxis.* Philadelphia: University of Pennsylvania Press.

Bollnow, Otto Friedrich. 1982. *Studien zur Hermeneutik Band I: Zur Philosophie der Geisteswissenschaften*. Freiburg: Alber.

Bontekoe, Ronald. 1996. *Dimensions of the hermeneutic circle*. Amherst: Humanity Books.

Boubil, Élodie and Christine Daigle, eds. 2012. *Nietzsche and phenomenology power, life, subjectivity*. Bloomington: Indiana University Press.

Corbier, Christophe. 2009. Alogia et eurythmie chez Nietzsche. *Nietzsche-Studien*. 38: 1–38. For more discussion and further references.

Dilthey, Wilhelm. 1916–1967. *Gesammelte Schriften*. Göttingen: Vandenhoeck and Ruprecht.

Dilthey, Wilhelm. 1991. *Introduction to the human sciences*. Princeton: Princeton University Press.

Duesberg, Peter H., D. Mandrioli, A. McCormack, J.M. Nicholson, D. Rasnick, C. Fiala, C. Koehnlein, H.H. Bauer, M. Ruggiero. 2011. AIDS since 1984: No evidence for a new, viral epidemic—not even in Africa. *Italian Journal of Anatomy and Embryology*. 116(2): 73–92.

Fehér, István. 1999. Philologie und Hermeneutik: Verständnis der Sache, Verständnis des Textes. *Berliner Beiträge zur Hungarologie (Schriftenreihe des Seminars für Hungarologie an der Humboldt-Universität zu Berlin)*. 11: 11–25.

Fehér, István. 2001. Hermeneutics and philology: 'Understanding the matter,' 'understanding the text.'" *Continental Philosophy Review*. XXXIV(3): 269–285.

Figal, Günter and Hans-Helmuth Gander, eds. 2005. *"Dimensionen des Hermeneutischen": Heidegger und Gadamer*. Frankfurt: Klostermann.

Fine, Arthur. 1993. "Fictionalism." *Midwest Studies in Philosophy*. 18(1): 1–18.

Flaschar, Helmut, Karlfried Gründer, and Axel, E.-A. Horstmann, eds. 1979. *Philologie und Hermeneutik im 19. Jahrhundert: zur Geschichte und Methodologie der Geisteswissenschaften*. Göttingen: Vandenhoeck & Ruprecht.

Ginev, Dimitri. 1995. *Die Mehrdimensionalität geisteswissenschaftlicher Erfahrung*. Essen: Die Blaue Eule.

Ginev, Dimitri. 1997. *A passage to hermeneutic philosophy of science*. Amsterdam: Rodopi.

Ginev, Dimitri. 1997. *Essays in the hermeneutics of science*. Aldershot: Ashgate.

Ginev, Dimitri. 2011. *The tenets of cognitive existentialism*. Athens: Ohio University Press.

Glazebrook, Trish, ed. 2012. *Heidegger on science*. Albany: State University of New York Press.

Grondin, Jean. 2000. Continental or hermeneutical philosophy: The tragedies of understanding in the analytic and continental perspective. In: *Interrogating the tradition: Hermeneutics and the history of philosophy,* ed. John Sallis, and Charles Scott, 75–83. Albany: State University of New York Press.

Gross, Allan G., and William M. Keith, eds. 1997. *Rhetorical hermeneutics: Invention and interpretation in the age of science*. Albany: State University of New York Press.

Ierna, Carlo, Hanne Jacobs, and Filip Mattens, eds. 2010. *Philosophy, phenomenology, sciences: Essays in commemoration of Edmund Husserl*. Frankfurt am Main: Springer.

Hacker, Peter M.S. with James Garvey. 2010. Hacker's challenge. *The Philosopher's Magazine,* October 25.

Hassemer, Winfried. ed. 1984. *Dimensionen der Hermeneutik: Arthur Kaufmann zum 60. Geburtstag*. Heidelberg: R.v. Decker and C. F. Müller.

Heelan, Patrick Aidan 1983. *Space-perception and the philosophy of science.* Berkeley: University of California Press.

Heelan, Patrick Aidan. 1975. Heisenberg and radical theoretic change. *Zeitschrift fur allgemeine Wissenschaftstheorie.* 6: 113–138.

Heelan, Patrick Aidan. 1970. Complementarity, context dependence, and quantum logic. *Foundations of Physics*. 1: 95–110.

Heelan, Patrick Aidan. 1965. *Quantum mechanics and objectivity: The physical philosophy of Werner Heisenberg.* The Hague: Martinus Nijhoff.

Heelan, Patrick Aidan. 1989. There is a hermeneutic philosophy of natural science: Rejoinder to Markus. *Science in Context*. 3: 469–480.

Heelan, Patrick Aidan. 1998. Scope of hermeneutics in the philosophy of natural science. *Studies in the History and Philosophy of Science*. 29: 273–98.

Heelan, Patrick Aidan. 2011. Le petit philosoph. In: *In search of the whole: Twelve essays on faith and academic life,* ed. John C. Haughey, 129–146. Washington, DC: Georgetown University Press.

Heidegger, Martin. 1927. *Sein und Zeit*. New York: Harper & Row.

Heidegger, Martin. 1962. *Being and time*. Trans. John Macquarrie, and Edward Robinson. New York: Harper & Row.

Heidegger, Martin. 1968. *What is Called Thinking*. Trans. J. Glenn Gray. New York: Harper.

Heidegger, Martin. 1977. Science and reflection. In *The guestion concerning technology and other essays* (trans: William Lovitt), 155–182. New York: Harper.

Heidegger, Martin. 2001. *The fundamental concepts of metaphysics: World, finitude, solitude*. Trans. William McNeill, and Nicholas Walker. Bloomington: Indiana University Press.

Heidegger, Martin. 2000. *Towards the definition of philosophy*. Trans. Ted Sadler. London: Continuum.

Heidegger, Martin. 2002. Wilhelm Dilthey's research and the struggle for a historical worldview. Trans. Charles Bambach. In Heidegger, M., *Supplements: From the earliest essays to "Being and time" and beyond*, ed. John van Buren. Albany: State University of New York Press.

Hirsch, Jr., E. D. 1972. Three dimensions of hermeneutics. *New Literary History*. 3(2), *On Interpretation: I* (Winter): 245–261.

Hirsch, Jr., E. D. 1975. Current issues in theory of interpretation. *The Journal of Religion*. 55(3): 298–312.

Hübner, Kurt. 1979. *Kritik der wissenschaftlichen Vernunft*. Alber: Freiburg/München.

Hawking, Stephen, and Leonard Mlodinow. 2010. *The grand design*. New York: Bantam.

Kisiel, Theodore. 1970. Science phenomenology and the thinking of being. In *Phenomenology and the natural sciences: Essays and translations*, eds. J. Kockelmans and T. Kisiel, 167–183. Evanston: Northwestern University Press.

Kisiel, Theodore. 2002. *Heidegger's way of thought*. London: Continuum.

Kockelmans, Joseph. 1953. *Philosophy of mathematics in the middle ages* (in Dutch). Langemark: Vonksteen.

Kockelmans, Joseph. 1958. *Tijd en ruimte : de opvattingen van Albert Einstein over de absolute betrekkingen volgens tijd en ruimte tussen voorvallen in de werkelijkheid en haar betekenis voor de wijsbegeerte der anorganische natuur* [*Time and space: The meaning of Einstein's relativity theory for a phenomenological philosophy of nature*]. Haarlem: Bohn.

Kockelmans, Joseph. 1962. *Phaenomenologie en Natuurwetenschap: een inleiding in de wijsbegeerte der natuurwetenschappen* [*Phenomenology and physics*]. Haarlem: Erven F. Bohn.

Kockelmans, Joseph. 1964. *Husserl's phenomenological psychology* [in Dutch]. The Hague and Tielt: Lannoo.

Kockelmans, Joseph. 1964. Merleau-Ponty on space and space-perception. *Review of Existential Psychology and Psychiatry*. 4: 69–105.

Kockelmans, Joseph. 1966. *Phenomenology and physical science: An introduction to the philosophy of physical science*. Pittsburgh: Duquesne.

Kockelmans, Joseph. 1967. *Edmund Husserl's phenomenological psychology. A historico-critical study*. Pittsburgh: Duquesne University Press.

Kockelmans, Joseph. 1970. The mathematization of nature in Husserl's last publication. In: *Phenomenology and the natural sciences essays and translations*, ed. Kockelmans, and Theodore J. Kisiel, 45–67. Evanston: Northwestern University Press.

Kockelmans, Joseph. 1970. Merleau-Ponty on space perception and space. In: *Phenomenology and the natural sciences*, eds. Joseph J. Kockelmans, and Theodore J. Kisiel, 274–311. Evanston: Northwestern University Press.

Kockelmans, Joseph. 1975. Toward an interpretative or hermeneutic social science. *Graduate Faculty Philosophy Journal*. 5: 73–96.

Kockelmans, Joseph. 1976. Hermeneutic phenomenology and the science of history. *Phänomenologische Forschungen*. 2: 130–179.

Kockelmans, Joseph. 1978. Reflections on social theory. *Human Studies: A Journal for Philosophy and the Social Sciences*. 1: 1–15.

Kockelmans, Joseph. 1979. Deskriptive oder interpretierende Phänomenologie in Schütz' Konzeption der Sozialwissenschaft. In: *Alfred Schütz und die Idee des Alltags in den Sozialwissenschaften*, eds. Walter M. Sprondel, and Richard Grathoff, 26–42. Stuttgart: Enke.

Kockelmans, Joseph. 1985. *Heidegger on art and art works*. Dordrecht: Martinus Nijhoff.

Kockelmans, Joseph. 1983. The challenge of Nietzsche's 'god is dead'. In: *The great year of Zarathustra (1881–1981)*, ed. David Goicoechea, 63–80. Washington, DC: University Press of America.

Kockelmans, Joseph. 1986. *A companion to Martin Heidegger's "being and time."* Washington, DC: Center for Advanced Research in Phenomenology and University Press of America.

Kockelmans, Joseph. 1987. Husserl's original view on phenomenological psychology. *Phenomenological Psychology Phaenomenologica*. 10(3): 3–29.

Kockelmans, Joseph. 1993. *Ideas for a hermeneutic phenomenology of the natural sciences*. Dordrecht: Kluwer.

Kockelmans, Joseph. 1994. *Edmund Husserl's phenomenology*. West Lafayette: Purdue University Press.

Kockelmans, Joseph. 2003. *Ideas for a hermeneutic phenomenology of the natural sciences. Volume II: On the importance of methodical hermeneutics for a hermeneutic phenomenology of the natural sciences*. Dordrecht: Kluwer Academic Publishers.

Mantzavinos, Chrysostomos. 2005. *Naturalistic hermeneutics*. Cambridge: Cambridge University Press.
Margulis, Lynn. et al., 2009. Spirochete round bodies. Syphilis, lyme disease & AIDS: Resurgence of 'the great imitator'? *Symbiosis*. 47: 51–58.
Markus, Gyorgy. (1987). Why is there no hermeneutics of natural sciences. *Science in Context*. 1/1(1987): 5–51.
Mittasch, Alwin. 1952. *Nietzsche als Naturphilosoph*. Stuttgart: Alfred Kroner Verlag.
Müller, Sabine. 2004. *Programm für eine neue Wissenschaftstheorie*. Würzburg: Königshausen & Neumann.
Müller-Lauter, Wolfgang. 2000. *Nietzsche: His philosophy of contradictions and the contradictions of his philosophy*. Trans. D. Parent. Bloomington: University of Illinois Press.
Nancy J. Moules, David W. Jardine, Graham P. McCaffrey, Christopher B. Brown. (2013). Isn't all of oncology hermeneutic? *Journal of Applied Hermeneutics* 1–10.
Nietzsche, Friedrich. 1980a. Versuch einer Selbstkritik. *Die Geburt der Tragödie*. In: *KSA: Kristische Studien-Ausgabe,* eds. Giorgio Colli, and Mazzini Montinari, Vol. 1. Berlin: de Gruyter.
Nietzsche, Friedrich. 1980b. *Jenseits von Gut und Böse*. KSA 5.
Nietzsche, Friedrich. 1980c. *Die fröhliche Wissenschaft*. KSA 3.
Nordmann, Alfred. 2008. Getting the causal story right: Hermeneutic moments in Nancy Cartwright's philosophy of science. In: *Nancy Cartwright's philosophy of science,* eds. Luc Bovens, Carl Hoefer, Stephan Hartmann, 369–388. New York: Routledge.
Pöggeler, Otto. 1994. The future of hermeneutic philosophy. In: *The question of hermeneutics,* ed. Timothy Stapleton, 17–35. Dordrecht: Kluwer.
Poschl, Viktor. 1979. Nietzsche und die klassische Philologie. In: *Philologie und Hermeneutik im 19. Jahrhundert*, eds. Flaschar, Gründer and Horstmann, 141–155. Göttingen: Vandenhoeck & Ruprecht.
Rorty, Richard, 1994. A tale of two disciplines. *Callaloo*. 17/2: 575–585.
Salanskis, Jean-Michel. 1995. Die Wissenschaft denkt nicht. *Tekhnema*. 2: 60–85.
Seebohm, Thomas M. 2004. *Hermeneutics: Method and methodology*. Frankfurt am Main: Springer.
Seebohm, Thomas M. 1984. Boeckh and Dilthey: The development of methodical hermeneutics. *Man and World*. 17: 325–346.
Sheldrake, Rupert. 2012. *The science delusion: Freeing the spirit of enquiry*. London: Coronet.
Stapleton, Timothy, ed., 1994. *The question of hermeneutics*. Kluwer: Dordrecht.
Szilasi, Wilhelm. 1945. *Wissenschaft als Philosophie*. Zürich: Europa Verlag.
Szilasi, Wilhelm. 1961. *Philosophie und Naturwissenschaft*. Bern: Francke.
Tieszen, Richard. 2005. *Phenomenology, logic, and the philosophy of mathematics*. Cambridge: Cambridge University Press.
Tieszen, Richard. 1989. *Mathematical intuition: Phenomenology and mathematical knowledge*. Dordrecht: Kluwer.
Warman, Matt. 2011. Stephen Hawking tells google 'philosophy is dead.' *The Telegraph*. May 17.
Whitman, James Q. 1986. Nietzsche in the magisterial tradition of German classical philology. *Journal of the History of Ideas*. 47: 453–468.
Vaihinger, Hans. 1924. *The philosophy of "As If."* Trans. C. K. Ogden. London: Kegan Paul.
Van Fraassen, Bas. 1994. *The question of hermeneutics,* ed. Stapleton, Timothy. Kluwer: Dordrecht.
Van Brackel, Jaap. 1999. On the neglect of the philosophy of chemistry. *Foundations of Chemistry* 1: 111–174.
van Tongeren, Paul. 2000. *An introduction to Nietzsche's philosophy*. West Lafayette: Purdue University Press.
von Wright, Georg Henrik. 1971. *Explanation and understanding*. Ithaca: Cornell University Press.

Part I
Science, Cognition, Hermeneutics, and Lifeworld

A Paradox of Cognition

Nicholas Rescher

Abstract The paper deals with variations on the theme of the ironic circumstance that the more we know the ampler our realization of the extent of our ignorance. For, as Kant already observed, every answer to our questions provides new materials for the development of further questions. The expansion of our knowledge thus brings the lineaments of our ignorance even more clearly into sight.

Knowledge narrows the range of acceptable possibility. If I know nothing about the matter, it is possible, as far as I am concerned, that a lion is on the mat—or that the mat is unoccupied. But if I know that the cat is on the mat, then these possibilities are eliminated.

However, while knowledge narrows the range of possibility, it expands the range of appropriate questions. For all questions hinge on presuppositions, and the more one knows, the more presuppositions are at one's disposal. Thus once I know that the cat is on the mat, then all sorts of new questions pop up on the agenda: Why is the cat on the mat? What age is the cat on the mat? And so on. So the more one knows, the larger the range of what one can meaningfully ask and wonder about.

The coming to be and passing away of questions is a phenomenon that can be mooted on this basis. A question *arises* if it then can meaningfully be posed because all its presuppositions are then taken to be true. And a question *dissolves* if one or another of its previously accepted presuppositions is no longer deemed acceptable. Any state of science will remove certain questions from the agenda and dismiss them as inappropriate. Newtonian dynamics dismissed the Aristotelian question "What cause is operative to keep a body in movement (with a uniform velocity in a straight line) once the impressed force that set it into motion has ceased to operate?" Modern quantum theory does not allow us to ask the classical "What caused

N. Rescher (✉)
Department of Philosophy, University of Pittsburgh, Pittsburgh, PA 15260, USA
e-mail: rescher@pitt.edu

B. Babich and D. Ginev (eds.), *The Multidimensionality of Hermeneutic Phenomenology*, Contributions to Phenomenology 70, DOI 10.1007/978-3-319-01707-5_1,

this atom on californium to disintegrate after exactly 32.53 days, rather than, say, a day or two later?" Scientific questions should thus be regarded as arising in an *historical* setting. They arise at some juncture and not at others; they can be born and then die away.

A change of mind about the appropriate answer to some question will unravel the entire body of questions that presupposed this earlier answer. For if we change our mind regarding the correct answer to one member of a chain of questions, then the whole of a subsequent course of questioning may well collapse. If we abandon the luminiferous aether as a vehicle for electromagnetic radiation, then we lose at one stroke the whole host of questions about its composition, structure, mode of operation, origin, and so on. The course of erotetic change is no less dramatic than that of cognitive change.

Epistemic change over time accordingly relates not only to what is "*known*" but also to what can be *asked*. Newly secured information opens up new questions. And when the epistemic status of a presupposition changes from acceptance to abandonment or rejection, we witness the disappearance of various old ones through dissolution. Questions regarding the *modus operandi* of phlogiston, the behavior of caloric fluid, the structure of the luminiferous aether, and the character of faster-than-light transmissions are all questions that have become lost to modern science because they involve presuppositions that have been abandoned.

The second of those aforementioned modes of erotetic discovery is particularly significant. The phenomenon of the ever-continuing "birth" of new questions was first emphasized by Immanuel Kant, who in his classic *Critique of Pure Reason* depicted the development of natural science in terms of a continually evolving cycle of questions and answers, where, "*every answer given on principles of experience begets a fresh question, which likewise requires its answer* and thereby clearly shows the insufficiency of all scientific modes of explanation to satisfy reason."[1] This claim suggests the following Principle of Question Propagation—Kant's Principle, as we shall call it: "The answering of our factual (scientific) questions always paves the way to further as yet unanswered questions."

There is a fundamentally relational pathway to new knowledge. Suppose I know that item No. 1 has the property *F* but not *G*, while No. 2 has *G* but not *F*. Over and above these two individual facts, a whole host of *relational* issues now arises. Why do the two differ in point of *F-G*? How did those come about? Must it be so? Is there a connection here of such a sort that if No. 1 changes in this regard, No. 2 must do so as well. And so on. Plural facts invariably pose relational issues and open the way to further knowledge. Combining facts engenders new questions. When physicists postulate a new phenomenon they naturally want to know its character and modus operandi. When chemists synthesize a new substance they naturally want to know how it interacts with the old ones.

This circumstance has ominous cognitive implications. New answers breed new questions. And the more the merrier. The ironic fact is that the more one knows the greater the arena of one's recognizable ignorance becomes.

---

[1] Kant (1911), Sect. 57; 352.

The process at work here is that new facts are generated (both substantively and cognitively) by interrelating (conjoining, coordinating, combining) old ones. In this way it will always be possible to extrude from n facts at the least n^2 additional ones. But now if—under steady-state conditions regarding inquiry—the body of known fact grows linearly while the body of fact involucrated therein grows (at least) exponentially, then it is clear that the ratio of accessed to accessible fact will always diminish. With the development of knowledge, the manifold of undeniable ignorance grows ever larger. As we know more, the range of what we cannot but acknowledge as unknown grows ever larger. For the ironic fact of it is that as our determinate knowledge grows, the range of our determinable ignorance grows ever faster since every determination opens a doorway to further detail.

To be sure, various cognitive resources can countervail against our ignorance. One of these is generalization. For knowledge can be either generic or specific. It is one thing "*to know that all lions have manes*" (Kx (∀y) ($Ly \supset My$)), and something quite different "*to know of every lion that it has a mane*" (∀y) Kx ($Ly \supset My$). Specific knowledge of universal facts is generally inaccessible to finite knowers. But generalization will often alternate the deficiencies of knowledge.

Then too, approximation can also help here. If asked about the present population of Los Angeles I could not claim exact knowledge of the answer. But I would unhesitatingly say that it is:

- A great many
- Roughly ten million
- More than five million and less than fifty

Many questions that we cannot answer—strictly speaking—exactly become answerable once approximation is admitted.

This line of consideration indicates the cognitive value of detail and precision. For what we usually understand by knowledge is precise knowledge and by answers to questions we mean exact answers. The growth of knowledge is not betokened by the range of questions that we can answer correctly, but by the range of questions that we can answer with precise detail!

Clearly if we relax these conditions/requirements, the range of our "knowledge" could be vastly expanded. The situation stands as per the following diagram which illustrates the reciprocal complementarity on the volume and precision of our knowledge (Fig. 1).

We frequently have recourse to this circumstance. All too commonly we settle for imprecise answers to difficult questions. The extent of our ignorance is then hidden away in a cloud of unknowing.

So, what can we say about the substance of our ignorance? A crucial consideration here is the unmeetability of the challenge: "Give me an example of a fact you do not know." One is obviously stymied at this point. For one cannot coherently claim in one and the same breath that something is a fact and that one does not know it to be so. We can know *that* there are facts we do not know individually or collectively. But we cannot identify specifically an individually *what* they are.

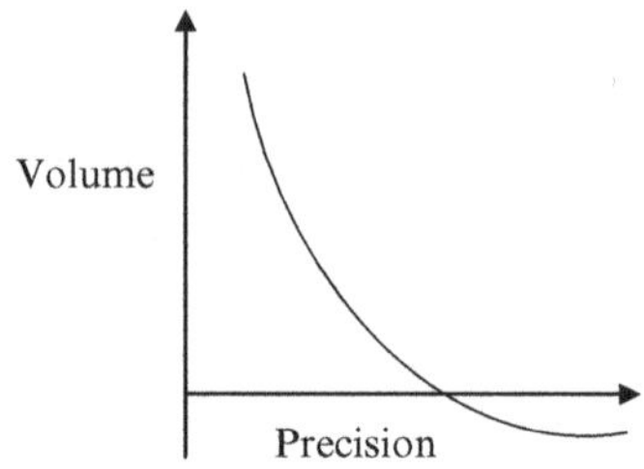

Fig. 1 Volume and precision of knowledge

The geography of our ignorance cannot be mapped with exactitude; its boundaries cannot be pinpointed.

At this point the difference between knowledge and wisdom becomes critical. For wisdom requires us to grasp that the cognitive life extends beyond the limits of knowledge, and that we do not adequately honor the priceless value of our knowledge if we fail to acknowledge that it also has its limits and limitations.

Reference

Kant, Immanuel. 1911. *Prolegomena to any future metaphysics that will be able to present itself as a science prolegomena zu einer jeden künftigen Metaphysik, die als Wissenschaft wird auftreten können*. In: *Kant's Gesammelte Schriften,* Vol. 4 of the edition of the Koeniglach Preussischen Akademie. der Wissenschaften, 255–383. Berlin: G. Reimer.

The Articulation of a Scientific Domain from the Viewpoint of Hermeneutic Phenomenology: The Case of Vectorial Metabolism

Dimitri Ginev

Abstract By making use of an approach stemming from hermeneutic phenomenology, this paper explores the constitution of meaningful objects in the domain of vectorial biochemistry. At stake are both theoretical objects related to anisotropic processes of trans-membrane transport and objects of inquiry contextually ready to hand within a configuration of scientific practices. The concept of the hermeneutic fore-structure of scientific research is discussed. The domain's formal, conceptual and experimental articulation gets fore-structured in the horizon of possibilities projected by the interrelated practices. The appropriation of possibilities constitutes meaningful objects. Some basic trends of the domain's articulation are addressed through analyzing three aspects of interpretative fore-structuring.

1 The Concept of Hermeneutic Fore-Structure of Scientific Research

There is a mythical moment involved in science's objectifying thematization. This seemingly extravagant claim has been most successfully advocated not by Paul Feyerabend or Kurt Hübner. Its true champion is Joseph Kockelmans. In trying to elucidate this qualification, let me start out with a disclaimer. Kockelmans's position has nothing to do with a kind of New Age mythologization of scientific theories aiming at a new world-view. His program for a hermeneutic phenomenology of science is the most irreconcilable antagonist to all distorted and confused holistic *Weltanschauungen*. Kockelmans is a decisive opponent to the view that mythical forms of understanding and mythical narratives might serve the function of counterparts of scientific reasoning. What he advocates is rather an elaborated version of

D. Ginev (✉)
St. Kliment Ohridski University, Sofia, Bulgaria
e-mail: dimiginev@abv.bg

B. Babich and D. Ginev (eds.), *The Multidimensionality of Hermeneutic Phenomenology*, Contributions to Phenomenology 70, DOI 10.1007/978-3-319-01707-5_2,

the conception that the mythical mode of making the world meaningful is part and parcel of the scientific attitude itself, insofar as this mode seems to be implied in the formulation of basic theoretical assumptions (Kockelmans 1993, pp. 164–169).

The mythical moment in objectifying thematization is to be understood by having recourse to Kockelmans's interpretation of "man's mythical mode of Being"—an interpretation that unfolds Heidegger's ideas developed in the seminal review of the second volume of Cassirer's *The Philosophy of Symbolic Forms*. In this mode of Being (which is closest to the characteristics of the primordial being-in-world) the mythical bringing into being a totality of meaning embraces primary horizons of world's understanding, initial forms of world's discursive articulation, and kinds of primordial attunement to the world.[1] Accordingly, what gets disclosed in the mythical mode of Being is *the world as a totality of meaning*. On Kockelmans's conception, all basic theoretical assumptions in science presuppose tacitly such a disclosure of the world. Otherwise, the objectifying projection of the world would be impossible. This disclosure institutes the primary horizon of scientific thematization. By implication, it plays the role of a fore-structure of the constitution of scientific objects and the articulation of domains of scientific research. What takes place in scientific research through the mythical disclosure of the world is a unity of fore-having, fore-sight, and fore-conception of how a meaningful domain is to be articulated in objects of inquiry. Once involved in the objectifying projection (and the related theoretical assumptions), it becomes a unity that constantly operates in the research process. I call this unity a hermeneutic fore-structuring of scientific objectification.[2] Due to the changeability of fore-having, fore-sight, and fore-conception in the constitution of meaningful objects, the hermeneutic fore-structuring constantly contextualizes in an intrinsic manner the research process.

---

[1] Kockelmans strongly distinguishes his Heideggerian interpretation of the mythical understanding from the established mythological paradigms in cultural anthropology and the phenomenology of religion. The disclosure of the world's totality of meaning within the mythical mode of Being does not demand a narrative that relates the time of the historical events to the pre-historical time of origins, or a narrative that is capable to found rites. In fact, Kockelmans's conception is closer to Bultmann's program of demythologization which aims at revealing the primary mythical horizon of religious imagery. It is the existential interpretation of the Holy Script that reveals this horizon.

[2] In analyzing the way in which Heidegger spells out the expression "to let be" with regard to scientific discovery and the formulation of scientific laws, John Haugeland reaches the conclusion that "it is little odd to say that the law of gravity was not true before Newton discovered it… And there remains the question of what to say about Einstein's discovery that there is (was and will be) no force of gravity—just curved space-time. Does this mean that, through Newton, his laws became true, but only for a while?" (Haugeland 2007, 100) In my view, the way in which Haugeland formulates and tries to address and settle such questions indicates a naturalist treatment of Heidegger's claim that there is (scientific) truth only so long and so far as Dasein is. Because of the situated transcendence of Dasein-as-epistemic-subject (and the finitude of all ontic knowledge) the theoretical formulations known as scientific laws are always already hermeneutically fore-structured. It is this hermeneutic fore-structuring that Haugeland does not take into consideration in his interpretation of Heidegger's claim. An additional shortcoming of this interpretation is the avoidance of the distinction between ontic/epistemic and ontological truth. On the alternative interpretation I am going to subscribe in this essay, each scientific law is true within a characteristic hermeneutic situation that reveals and conceals the objectified world in a specific manner.

In the remainder I will be preoccupied with the ongoing hermeneutic fore-structuring of the formal, conceptual, and experimental articulation of a scientific domain. Following Kockelmans's ideas, I will treat hermeneutic phenomenology as a philosophical approach to the constitution of meaningful objects within settings of interrelated practices. The hermeneutic philosophy of science applies this approach to the dynamics of scientific practices in the process of articulating a domain of research.

To begin with, the notions of understanding and interpretation get a special construal in hermeneutic phenomenology. They are no longer treated as particular cognitive procedures. Understanding and interpretation are rather attributed to the ongoing becoming of the reality of what gets articulated. Accordingly, these notions undergo an ontological reformulation. Understanding is the horizon whose projection discloses entities in their possible usability. More specifically, an entity within-the-world is projected upon the totality of its possible involvements in contexts of configured practices. Thus considered, entities have meaning within the projected horizon of understanding of the contexts in which they are ready to hand for possible manipulations. It is the reality of the entities contextually ready to hand that is in a state of ongoing interpretative articulation. The latter consists in a constant appropriation and actualization of projected possibilities for manipulating the entities that are already understood in their contextual usability. This is the paradigm of the constitutional analysis of meaning suggested by hermeneutic phenomenology.[3] It admits that the interpretative articulation of what gets understood proceeds through appropriation of possibilities whose actualizing constitutes meaning. In scientific research the articulation concerns the domain's structure and its particular objects of inquiry.

The research process is at any given moment situated in a configuration of practices of formulating hypotheses; introducing appropriate mathematical idealizations; constructing systems of differential equations, phase diagrams, mathematical plots, and other formal tools; developing experimental systems; replicating experiments; measuring parameters of such systems and constituting data-models of measurements; creating and calibrating instruments; accomplishing computer simulations; elaborating on the empirical reading of theoretical concepts; etc. The outcome of performing the concerted and configured practices is a step in the domain's articulation.

For the sake of illustration consider the following configuration of practices in cytology. At stake in the research process are issues of the investigation of the mitochondrial cytochrome systems (for instance, the cristae membrane of mammalian mitochondria). Such a system consists of a class of colored proteins that

---

[3] In its classical formulation this paradigm is to be found in Heidegger (1927/1962, 188–203). The basic concepts here are the "fore-structure of understanding" and the "as-structure of interpretation." The interplay between the fore-structure qua horizon of possibilities and the as-structure qua ongoing articulation of meaningful units takes on the form of hermeneutic circle. The hermeneutic-phenomenological analysis of a particular interrelatedness of practices (like the practices in a domain of scientific research) requires an interpretation of the way in which the hermeneutic circle operates in articulating the domain disclosed by that interrelatedness of practices.

play crucial roles in oxidative processes and energy transfer during cell metabolism and cellular respiration. A cytochrome system is an electron carrier. In serving this function a mitochondrial cytochrome system is responsible for the coupling between respiratory mechanism and proton transport.[4] The configuration of scientific practices in the early 1960s was devoted to the identification of electron-carrying and hydrogen-carrying components across the mitochondrial membranes. This configuration includes practices of developing conceptual models that relate the enzyme-mediated chemical group (or electron) transfer to the catalysis of solute transport mediated by the mobility of specific molecules; elaborating on new mathematical plots about enzyme kinetics; creating controlled experimental systems for observing the role the protons transfer plays by enzymes like the ironsulphur-flavoprotein dehydrogenases; measurements of substrates concentrations in the proton-translocating redox process; spectrophotometric measurements; measurements of trans-cellular electric currents; cytomechanical investigations of mitochondria membrane structure; observations of patterns formation of fluctuations arising within a homogeneous cellular region; constructing data-models for proton- and electron-conducting pathways; and many others. Each of these practices was distinguished by a particular space of representing what is under investigation.

In their interrelatedness the configured practices project a leeway of possibilities for doing research by manipulating contextually various entities. Yet the same practices appropriate and actualize the possibilities. Each configuration of practices in the research process projects and partially appropriates its own leeway of possibilities. In the practices' performance certain possibilities remain necessarily not actualized. They get either forgotten (pushed aside) or recast in a new context of inquiry brought into being by another configuration of practices. Furthermore, the appropriation of possibilities by performing practices generates new possibilities that transcend the leeway projected by the current configuration. This leads to the emergence of a new configuration of practices, and accordingly, to opening a new leeway of possibilities. Thus, the research process is at any moment at once situated in a configuration and transcended by possibilities that cannot be appropriated and actualized by this configuration. Because of this "situated transcendence" the research process is constantly moving in an open horizon of possibilities.

On this account, not only the research process but also the formal, conceptual, and experimental articulation of a domain is in a state of situated transcendence. This is why the domain's articulation is "always already" projected upon possibilities. A scientific domain's "being as actual presence" is always a derivative from the potentiality-for-being of this domain.

The way of performing each particular practice in a certain configuration entails a kind of "reading something." Thus, one is reading instruments, experimental

[4] With regard to the catalysis of solute transport of specific molecules across lipid membranes this conclusion is drawn for the first time by Davson and Danielli (1943), pp. 72–79. This classical work is also the pioneering study of the thermodynamic aspects of the vectorial processes of cellular physiology, in particular processes running against the concentration gradient.

systems, semantic models of theoretical concepts, differential equations, diagrams, and so on. In other words, a "readable technology" (Heelan 1983, 1998) is indispensably embedded in each particular scientific practice. The three dimensions (structural, conceptual, and experimental) of the domain's articulation which takes place in a configuration of practices are functions of reading that appropriates possibilities through its formal, theoretical, and empirical technologies. The configuration in cytology from the early 1960s I am discussing managed to articulate a formal apparatus for describing and calculating the effects of the electric membrane potential difference and the pH difference across a certain cytochrome complex on the electronic spin and redox states of the haem iron centers (Quagliariello et al. 1976; Mitchell 1976). This was a step in the domain's structural/formal articulation. New idealizations and new kinds of conceptualizing the topological arrangement of various cytochromes as represented by equations about the dynamics of trans-membrane flows contributed to the domain's conceptual articulation. Experiments aimed at verifying predictions of working hypotheses regarding the coupling between proton/electron transport and respiratory mechanism (for instance, experiments with varying the electric membrane potential, the pH membrane potential, and the total protonic membrane potential) provided new data-models that articulated the domain empirically.

Let me return to the way in which hermeneutic phenomenology is applied in this paper. On this philosophical approach, there is a nexus of understanding and interpretation that gets revealed in terms of the triad of fore-having, fore-sight, and fore-conception of "something that becomes meaningful as something." The orientation toward entity's meaning as an outcome of configured practices is a fore-having. The expectation of visualizing that meaning (within a scientific space of representation) is a fore-sight. In the articulation of a scientific domain the most important is the visualization of the meaning of the domain's theoretical objects. Accordingly, the visualizability refers to the possible spaces of the trans-empirical entities' visual representation by means of things that are immediately ready to hand. The fore-sight accomplishes the primary specification of what has been taken into one's fore-having. Finally, the anticipation of possible further contextualization of entity's meaning is a fore-conception. Whenever something is interpreted as something in a context of practices, the triad of fore-having, fore-sight, and fore-conception characterizes interpretation of entities that are ready-to-hand in the performance of practices (Heidegger 1927/1962, 191).[5] The unity of these three moments does make sense in two ways which correspond to two basic hermeneutic dimensions of scientific research. I will accordingly convey the dimensions to the concepts of "hermeneutic fore-structure" and "characteristic hermeneutic situation." Both of them refer at once to the contextualization of readable technologies, the contextual appropriation of possibilities for doing

[5] Fore-having, fore-sight, and fore-conception are three relationships between fore-structure of understanding and as-structure of interpretation. In other words, they are three characteristics of the hermeneutic circle which operates in the articulation of a domain disclosed by an interrelatedness of practices. See also Kockelmans (1986).

research, and the constitution of contextually meaningful objects of inquiry. In this paper I will be entirely preoccupied with the concept of hermeneutic fore-structure.

Against the background of the foregoing considerations the concept of the hermeneutic fore-structure of scientific research comes to the fore as a result of the efforts to specify the fore-having, fore-sight, and fore-conception with regard to the appropriation of projected possibilities by making use of readable technologies. Those who are involved in the research process taking place in a given domain understand the domain in the first place as possibilities (of structural, conceptual, and experimental articulation) whose ongoing appropriation comes to pass in scientific spaces of representation. Implementing readable technologies creates at the same time such spaces in which objects of inquiry become constituted. On another formulation, one appropriates possibilities for doing research by employing readable technologies that bring into play series of deferring and displacing one another spaces of representation within a configuration of scientific practices. By making use of the expression "deferring spaces" I mean that the outcome of a particular representation is always already in another space of representation. Thus, for instance, the outcome of a given experiment gets meaning in the space of constructing data-models of measurements; represented as diagrams these models are already in a space of representation provided by a theory's mathematical formalism; the calculations based on theoretical models developed through this formalism get meaning through devising a new kind of measurements that become possible through the construction and calibration of new instruments; the results of instrumentation get meaning in the space of representing the empirical contents of theoretical concepts; etc. Put differently, what becomes deferred in any space of the chain of deferring and displacing one another spaces of representation is the constituted meaning of the objects of inquiry. The meaning is dispersed/deferred in the totality of readable technologies and spaces of representation of the configuration of practices.

Hans-Jörg Rheinberger (1997, p. 108) observes that "scientific objects come into existence by comparing, displacing, marginalizing, hybridizing, and grafting different representations with, from, against, and upon each other."[6] One articulates the domain of research by actualizing possibilities of making readable what circulates in the spaces of representation. In this regard, interpretation is appropriation of possibilities by means of readable technologies, whereby the actualized possibilities within spaces of representation articulate the domain of research. Now, any particular space does not represent isolated objects but objects entangled in structures that are informed by mathematical idealizations. Like the objects, the (formal) structures are in statu nascendi during the research process. The domain's interpretative articulation consists in the constitution of objects of inquiry that (formally) depend on the structures' symmetry groups. A space of representation visualizes contextual objects of inquiry embedded in changeable structures.

[6] Rheinberger commits this claim also to a "dialectic of fact and artifact" taking place in the constitution of the scientific research's "epistemic things".

To sum up, the hermeneutic fore-structure of scientific research refers to having, seeing, and grasping in advance of what gets meaningfully constituted in the domain's structural, conceptual, and empirical articulation. (For reasons that will be clarified later, I intentionally refrain from taking a position here with regard to the debate on whether structures have an ontological priority over objects, or vice versa.) Understanding (the domain of research as a horizon of possibilities) and interpretation (that articulates meaningful objects-embedded-in-structures through readable technologies in spaces of representation) are in constant interplay during the domain's potentially infinite articulation. It is this interplay that I treat as an ongoing fore-structuring of a domain's (three-dimensional) articulation. In the remainder the fore-structuring of the domain of vectorial metabolism (as it came into being through research practices introduced in the first place for approaching protonmotive systems) will be scrutinized. No doubt, the configuration of practices in cytology being considered so far has much to do with the formation of this domain.

The changes of the domains' architectonic in biology leading to the emancipation of new domains are usually attributed to the rise of new hypotheses. Doubtless, Mitchell's chemiosmotic hypothesis which will be discussed later provoked such a change that eventually led to the formation of the domain of vectorial metabolism. This is why some authors are speaking of a "chemiosmotic revolution" as the point of departure for exploration of molecular parts and functional wholes (Harold 1991, 348; Weber 1991, 581). Douglas Allchin (1997, 5) makes convincingly the case that in the ox-phos controversy (between the believers in chemiosmotic hypothesis and the believers in the existence of high-energy intermediates of oxidative phosphorylation) one of the main point of disagreement was about the boundaries of the domain of cellular bioenergetics. This disagreement contributed to the emancipation of the domain of vectorial metabolism. In fact, however, the new domain was disclosed within a broad configuration of scientific practices through which the classical bag-of-enzymes biochemistry and membrane biochemistry were recast in terms of transport. The practices of topological osmochemistry (not to mention the practices devoted to revealing patterns of morphogenesis) were by no means a simple continuation of the chemiosmotic hypothesis. Much more correct is the conclusion that the formulation of hypotheses regarding the osmotic trans-membrane transport (like Mitchell's hypothesis and Lehninger's 1960 hypothesis that there is in mitochondrial membrane a geometrical structure that controls the regime of components' interactions) was a particular practice in the configuration being mentioned. I will try to address the interpretative fore-structuring of scientific research in the genesis of the domain of vectorial metabolism. In so doing, my concern will be the contextual constitution of meaningful objects of inquiry through reading/representing spatially oriented metabolic reactions. The standard story about this domain goes as follows.

In 1961 Peter Mitchell introduced in the domain of enzymology and bioenergetics a theoretical scenario for the coupling of respiration and ATP synthesis. (More specifically, this was a scenario of coupling between redox reactions and phosphorylation or dehydration reactions.) It is designed to provide explanation of the energetic resources for the phosphorylation step from ADT to ATP. On Mitchell's account, the coupling of respiration and ATP synthesis is mediated by an electrochemical gradient of protons across the mitochondrial membrane. The emphasis was placed on the causal link between the flow of electrons through the respiratory chain enzymes and the translocation of protons across the inner mitochondrial membrane. This account was deliberately forged as an alternative to the chemical mechanism of the oxidative phosphorylation which appeals not to anisotropic flaws but to enzyme-bound chemical compounds (Griffiths 1965). The chemical approach was advocated at that time by such authorities in biochemistry like Fritz Lipmann and Bill (E.C.) Slater.[7] Allchin (1996, 32) nicely summarized the wrong assumption of those biochemists who laid stress on the possibility to identify stable high-energy compounds in the phosphorylation of ATP: "Chemists wanted to isolate and identify a set of high-energy intermediate molecules, but found the task unduly difficult. They began to suspect that the compounds might be tightly bound to membrane proteins. That, at least, could account for their persistent failures."

In trying to find a way beyond the wrong assumption and the persistent failures, Mitchell admitted that the membrane plays a "topological" role in oxidative phosphorylation. Accordingly, the removal of a molecule of water in the synthesis ATP from ADP follows two opposite direction: The hydrogen ion (proton) and the hydroxyl ion are removed to the opposite side of the membrane. The osmotic potential provides the energy for ATP synthesis. The respiratory chain in its turn drives the flow of ions in the opposite direction. The reactions of this chain act also vectorially. On this account, the enzyme ATPase that phosphorylates ADP to ATP is responsible for the return of the protons, thereby enabling the energy currency of the cell. In the perspective of the standard story, it was Mitchell's chemiosmotic theory alone that initiated the research work in the domain of "vectorial metabolism." The orientation of further research work was indicated by Mitchell's interpretation of oxidative phosphorylation—the mechanisms of coupling membrane-transport systems and metabolic systems. Due to this orientation the domain of vectorial metabolism became situated between biochemistry and physiology. The initial structure of chemiosmotic hypothesis (and later theory) combined the physiology of the mitochondrial membrane with the chemistry of moving ions across the membrane. Let me now move on to a more detailed historical synopsis.

[7] There is a long history of sociological, historical, and philosophical reconstructions (case studies) of Mitchell's chemiosmotic hypothesis and the subsequent ox-phos controversy. At the beginning of this history is Gilbert and Mulkay's (1984) study designed in terms of a sociological discursive analysis. The significant value of this study is due to the way in which the authors make use of interviews with 34 participants in the ox-phos controversy. The historical reconstruction of Allchin (1997) presents not only the genesis of Mitchell's hypothesis, but also a quite vivid picture of the development of the chemical theory. Weber's (2002a, b) nice philosophical reconstructions of Mitchell's research program and its experimental verifications (a reconstruction that takes into consideration also the problematic of incommensurability) try to evaluate the positions in the ox-phos controversy in terms of normative epistemology, whereas Prebble (2001) investigates in a highly original manner the role of the "personal knowledge and tacit knowing" (in the sense of Michael Polanyi's conception) in the formation of Mitchell's philosophical and scientific views.

2 The Domain of Vectorial Metabolism

The ideas of vectorial biochemical processes that subsequently lead to models of pathways of morphogenesis are to be traced back to the pioneering studies of Elmer Lund from the mid 1920s. He devised experimental practices for investigating trans-membrane vectorial reactions. Lund managed to discover patterns of directive force (attributed to cell functions) in the formation of new structures. At that time, however, there was not in biochemistry a stable configuration of practices devoted to the investigation of the topological arrangements of coupled metabolic reactions that can generate morphogenetic processes. The plausibility of such a configuration was rather adumbrated by some negative research results. Thus, in 1930 Vladimir Engelhardt described for the first time the mechanism of oxidative phosphorylation as a link between the oxidation-reduction reactions and the synthesis of ATP. Within the existing interrelatedness of scientific practices the link had had to be discovered as chemical intermediates. Albeit several proposals for chemical mechanism of substrate phosphorylation were formulated, the long search (running from the model for substrate-level phosphorylation of succinyl coenzyme A in the early 1950s to the late 1960s when more than dozen mechanisms of chemical transfer were under scrutiny) for "scalar" chemical intermediates that are capable to transfer energy from molecule to molecule failed to find them. Biochemistry still had to wait for a configuration of research practices through which the spatial organization of the metabolic processes could have been objectified as a reality of specific entities whose interpretation would have gone hand in hand with the appropriation of new possibilities for doing research.

The rise of this configuration proceeded step by step. Due to Linus Pauling's work in the early 1950s a trend of research took shape that paid much attention to the enzymes catalyzing group transfer by lowering the free energy of the transition state. By the same time Fritz Lipmann advanced the view that the group-potential gradient gets transmitted between coupled reactions. Lipmann's attention was concentrated on phosphate derivatives arising in the process of group activation through phosphoryl transfer from ATP. The original idea of the anisotropic nature of metabolism in biochemistry stemmed from the work of Henrik Lundegårdh (1945) who suggested that cytochrome pigments might provide an electron-conducting pathway across plant cell membranes so that oxygen could be reduced on the one side while hydrogenated substrates were oxidized on the other side.[8] Lundegårdh generalized cytochrome-catalyzed electron translocation by introducing the concept of enzyme-catalyzed group translocation. A particular result of his work (that made him a

[8] Mitchell indicates that Malcolm Dixon in his 1941 lecture course at Cambridge had presented for the first time the notion of group transfer as a spatial migration of a donor group to an acceptor group. For Mitchell, however, Dixon had been not committed to the view of a vectorial mechanism of group translocation. Nevertheless, the very vectorial representation of diagrams of enzymes had been of great importance for Mitchell's orientation toward vectorial chemistry. (See Prebble and Weber 2003, 35)

forerunner to Mitchell) was the observation that the uptake of ions was driven by the vectorial flow of hydrogen ions across the plasmalemma (Larkum 2003).

In addressing the question as to which are the prime movers in metabolically-coupled translocation reactions, the search for patterns of spatially extended chemiosmotic proton circulation acquired currency. An important event on the way to the inception of research work in vectorial metabolism was the introduction of spectrophotometric measurements that suggested the existence of anisotropic protonmotive forces. Another experimental practice invented by Albert Claude (1946) showed the osmotic stabilization of mitochondria by the organic compound sucrose. This finding was later re-contextualized in the articulation of vectorial biochemistry. Though not directly related to the research of ions translocation across membranes, Waddington's formal models about how gene regulatory products could generate developmental phenomena are also to be mentioned in this historical context. Waddington's ideas got new actuality in the domain of vectorial metabolism and morphogenesis in the late 1980s when it was realized that the relationship of genes to cell form is not like that of genes to proteins.

By the end of the 1950s a configuration of practices pertinent to the topological arrangements of intracellular metabolic and physiological processes came into being. Within this configuration it was realized that the distinction between vectorial chemistry and vectorial physiology is a matter of degree. There is no difference in principle between reading (through experiments and instruments) the topology of enzyme catalyzed reactions and the topology of physiological processes. A paradigmatic example became a muscle contraction whose directionality in cellular space expresses a joint chemical-physiological spatiality. The configuration brought into play contexts of theoretical and instrumental/experimental reading in which new research objects got constituted. Their total interconnectedness made plausible the existence of a research domain *sui generis* that might call into question traditional borderlines between biochemistry and physiology. This was a domain in which the search for proton circuits promised the discovery of patterns of vectorial intracellular processes. The configuration involved practices of experimentation with inserting enzymes into a membrane in such a manner that the reaction pathway crosses the barrier, measurements that have to establish the translocation of a chemical group across the membrane, construction of theoretical models of group transfer reactions that proceed along spatial trajectories within the protein molecules, observation of how enzymes become integrated in larger structural complexes that determine the directionality of chemical processes, and devising data-models of energy transduction.

A joint research work of Peter Mitchell and Jennifer Moyle in the late 1950 proved that enzyme systems are the conductors of membrane transport and that metabolic energy is converted to osmotic work by the formation of covalent links. More specifically, they (Mitchell and Moyle 1958) were able to demonstrate a transfer of phosphate group vectorially through an enzyme. The idea of metabolic directionality was introduced with respect to the different directions the two substrates are approaching the active site of an enzyme from in order for the transfer to take place. This initial idea of a vectorial metabolism placed emphasis on the membranes of mitochondria which supposedly act as chemiosmotic links between the media that they separate. In investigating the group translocation as a form of trans-membrane transport, Mitchell and Moyle managed to integrate the metabolic inter-conversions and fluxes across the various cellular membranes into a single mechanism. This was a decisive step in disclosing the domain of vectorial metabolism as an autonomous domain of scientific research whose focus is the mechanisms of coordination of transport with metabolism. Parallel to Mitchell and Moyle's research work Robertson carried out investigations of gastric acid secretion which were also guided by anisotropic spatial model of reactions. He assumed that the act of secretion depends on the separation of positive and negative charge in the electron transport system.

At this point I have to mention again the role of the chemiosmotic hypothesis in instituting the new domain. In its original form (Mitchell 1961) this hypothesis stated that the coupling between oxidoreduction and phosphorylation systems could be due to the channeling of translocation of protons which could play the part of the donor/acceptor intermediate between the oxidase and ATPase systems.[9] This hypothesis required a complex experimental verification that included investigations of mitochondria membrane impermeability to protons, the "proton-pump" created by the respiratory chain, the rate of proton translocation during oxidoreduction, the magnitude of the total protonmotive force across the membrane, and the identification of proton-coupled porter systems in the membrane that regulate the internal pH, thereby maintaining osmotic stability. In fact however, Mitchell's (1962) ambition was not restricted to this verification, but to developing a unifying framework for addressing the processes that underlie metabolism, transport and morphogenesis. With the rise of chemiosmotic hypothesis the "vectorial mystery" of the cellular space became the new domain's kernel problematic. By the late 1950s it was a matter of belief in this mystery; a belief that abstractly denied that there are unidentified molecules with a high-energy bond (as this was postulated in Slater's 1953 "chemical hypothesis"). The further theoretical conceptualization and experimental verification of the chemiosmotic hypothesis transformed the abstract belief into assumptions integrated in new research everydayness. It was the everydayness of performing research practices devoted to deciphering that structural integrity of membrane barrier which separate in a special way inside and outside compartments,

[9] Initially, it was a hypothesis that mitochondria ejected protons and that proton gradient drives ATP synthesis. It had been introduced as an explanatory mechanism of the coupling of electron transport to phosphorylation. What was challenged by this hypothesis was the view that transport and metabolism were essentially separate processes. As an experimentally verified theory the chemiosmotic one concerns the process of energy conservation in mitochondrial and bacterial respiration and in photosynthesis. Its main cellular object of inquiry is the inner membrane where is the site of the enzyme ATPase for ATP synthesis. It is important to be indicated that chemiosmotic theory was regarded by Mitchell as a special case of nowadays is called "vectorial metabolism." Yet Prebble (2001), 447 is right when stressing that "the notion of vectorial metabolism is given its most fulsome treatment almost concurrently with promulgation of the chemiosmotic theory." The linkage between the coupled reactions catalyzed by two osmoenzymes depends on the electrochemical potential gradient generated by one reaction and consumed by the other.

thereby creating orientation of enzyme catalyzed reactions. At the same time, the alternative research everydayness in which biochemists were looking for high-energy intermediates of oxidative phosphorylation became significantly enriched with new practices of constructing theoretical models, experimentation and inventing novel techniques of measurement. Nonetheless, this kind of normal science did not obtain positive results. Allchin (1997, 17) goes on to summarize the fate of this failed research work in two ways: "(a) researcher could not generalize their findings from one set of experiments to a wider domain or scope of phenomena, or (b) results interpreted as fitting within one domain or causal network were late found to fit in a different domain. In the former, prospective facts were discarded; in the latter, the facts were displaced."

Mitchell changed radically the research work (for several authors this was a revolutionary paradigm shift in biochemistry)[10] by assuming that the oxidation-reduction reactions were so arranged in the membrane that they were coupled to transfer of protons outwards across the membrane. Thus, the mechanism of electrochemical gradient got introduced according to which the oxidation-reduction reactions are mediated by a proton gradient and its accompanying membrane potential. By implication, the intermediate between the oxidation in the respiratory chain and the ATPase would have had no longer to be conceived of as a specific chemical substance. Though the chemiosmotic hypothesis was conceptually coherent and complete (especially after Mitchell 1967 enriched it with the concept of "solute porters" which facilitate the solute between the two aqueous areas), several historical case studies demonstrate that one is not able to single out a particular historical moment at which the ox-phos controversy could be counted as resolved in favor of the chemiosmotic theory. As Marcel Weber (2002a) shows, it was the change of the situation (including the emergence of new research practices) by 1975 which brought decisive arguments for the acceptance of Mitchell's hypothesis. He analyses the reasons why several experiments in the 1960s and the early 1970s (including the earlier experimental search for reconstituting oxidative phosphorylation in Racker's laboratory) failed to resolve the controversy (or failed to play the role of a crucial experiment), and reaches the conclusion that while the role of the membrane was highly ambiguous around 1970 and impeded an independent test that can resolve the controversy, the final reconstitution experiments (creating cellular artifacts from isolated components) in conjunction with new knowledge about the membrane functions provided sufficient base for ruling out the chemical hypothesis and accepting the chemiosmotic one (Weber 2002a, 43–45). In another historical case study John Prebble (2001) lays particular emphasis upon the research community's "implicit knowledge" which resisted the search for designing appropriate experiments for checking up whether mitochondria ejected protons, whether a proton gradient would drive ATP synthesis, and whether the mitochondrial inner membrane was impermeable to protons. More specifically, the community's

---

[10] See, in particular Prebble and Weber (2003, p. 3), Weber (2002b), Mulkay and Gilbert (1984), Skulachev (1987). This view is supported even by Slater (2003), one of Mitchell's major opponents.

implicit knowledge prevented (or at least restricted the leeway of) the elaboration on research scenarios uniting chemical with physiological mechanisms.

In the mid 1960s Mitchell extended his original theory by assuming that the electron-carrying arm of the redox loop could include a photoelectric reaction so that the potential of this reaction could be transformed into protonmotive force. In the context of this extension Mitchell launched the view that the conformational mobility of the translocation catalysts is associated with their normal group-translocation and solute-translocation functions. This view was consonant with several principal new ideas in biochemistry and enzymology from the early 1960s like Koshland's induced-fit interpretation of enzyme kinetic data, the hypothesis that the activation energy for group transfer may be lowered by the balancing of stress–strain relationships, and Perutz's elaborations on conformational changes in haemoglobin molecules during oxygenation. These ideas paved the way to the cognitive institutionalization of the domain of vectorial metabolism (Mitchell 1972). The research work in this domain was concentrated on effective translocations of protons. "Mitchell bade us look upon cell growth as a grand symphony of transport" (Harold 1991, 365). Morphogenesis is a function of vectorial metabolism. Configurations of scientific practices were designed for studying reactions leading to chemiosmotic coupling. Two possible types of such coupling were recognized—one taking place on molecular-level enzyme associations, and the other applying to associations of enzymes in subcellular vesicles.

In fact, Mitchell came to the scenario of "chemiosmotic coupling in energy transduction" (so the technical expression he proposed) in following a particular scientific practice—the one of drawing analogies. In his case, the analogy was between the osmotic translocation reactions (for instance, the coupling of phosphate translocation against arsenate translocation) and the enzyme catalysed group-transfer reactions (Mitchell 1972). There was nothing unusual in pursuing this thoroughly conventional scientific practice. Mitchell's scenario was by no means an exotic hypothesis since the "primary chemical coupling" was known as early as the 1930s when a research group reported on coupling in group-transfer reactions in studying the role of the NAD coenzyme. Nonetheless, until the mid 1960s the "chemiosmotic mechanism" of energy generation implied by Mitchell's scenario remained without resonance in scientific community. This mechanism brings into play theoretical objects like "the transmembrane electrochemical potential powering the enzyme of the mitochondrial ATPase," "the anisotropic enzymes requiring two aqueous phases separated by a membrane," "the phosphorylation reactions driven by proton-motive force," and most of all "the electrochemically based vectorial metabolism" (i.e. metabolism whose catalytic systems are distinguished by a spatial orientation of the reactions' dynamics).

It was the change of the configuration of practices that made possible the spaces of representation/reading of vectorial metabolism of respiration. Based on the assumption that protons might be directly involved in the oxidative phosphorylation, several laboratories undertook experiments on proton gradients. In addition, more precise methodic of establishing the location of phosphorylating enzymes was introduced. Later on the practice of measuring protons translocation in respiring mitochondria came on the scene. As Marcel Weber (2002b) makes it clear, thanks to this practice several predictions made by Mitchell's initial theory were confirmed. Though many of the experimental results obtained explanations also in the framework of the chemical theory, the confirmation through independent tests provided a rationale for legitimizing (at least some) of the theoretical objects in whose existence the supporters of the chemiosmotic theory believed. However, much more important was the new hermeneutic fore-structure provoked by the (relatively small) change of the configuration of practices. New possibilities for doing research became projected, including possibilities whose appropriation was in line with the assumptions of the anisotropic flows of energy and the vectorial metabolism. The theoretical objects envisaged by chemiosmotic theory were significantly specified and "inscribed" on the horizon of new possibilities. Also a "conversion to the belief in Mitchell's objects" took place. Whole laboratories launched research programs inspired by the belief in the existence of these objects. As a result, several research groups started to articulate the domain of enzymological bioenergetics in accordance with possibilities for instantiating the chemiosmotic mechanism via data-models which was one the ways leading to the emancipation of vectorial biochemistry. Although accepted by the most members of oxidative phosphrylation research community, there are still influential opponents of chemiosmotic theory. Thus, for instance, Mitchell's hypothesis is accused for not having acquired a status of a genuine theory. More specifically, the critical point is that the depiction of group translocation has never been documented in scientific literature. Saier (2000) makes the case that Mitchell's hypothesis can be transformed into a theory if the mechanisms of trans-membrane group transport can be recast in terms of "molecular phylogeny." The constitution of three-dimensional structures of integral membrane transport proteins qua specific objects of inquiry provides new possibilities for studying the evolutionary histories of transport proteins.

In the second half of the 1970s and the 1980s the domain's research everydayness was dominated by studies of the ways in which chemical and osmotic forces are balanced against one another. Very important research objects became the proton-coupled reversible ATPases. It was established that these enzymes consist of two parts, one located in membrane that has no enzymatic function and another that catalyzes ATP hydrolysis. The research work in Racker's laboratory demonstrated that the membrane-located part acts as a proton-conducting channel when the catalytic part is detached. In this context, the chemiosmotic theory underwent a transformation through a revision of the claim that hydrolysis of ATP is reversed by a current of protons driven though the catalytic part of the enzyme. On the new version, it is the membrane-located part that acts as a reversible proton pump being conformationally coupled to the enzymatic part (Mitchell 1991, 330–333). Another significant trend of research in the 1980s that essentially articulated further the domain was the experimentation with uncouplers of oxidative phosophrylation (inhibitors of ATP synthesis that prevent the coupling in a manner that the energy produced by the redox cannot be used for phosphorylation). The uncouplers do not affect the activities of electron flow and ATPase, but ATP synthesis cannot occur. In particular, studies in protonophoric uncouplers (such ones that allow protons to cross lipid bi-layers) revealed mechanisms of induction of biological activities by modification of the state of a specific membrane protein. Some results helped one in extending the domain of vectorial biochemistry toward studies in molecular and cellular morphogenesis.

The search for patterns of morphogenesis on the level of cellular physiology became a "natural continuation" of the inquiry into topology of enzyme catalyzed reactions. Thus, the problematic of the spatiality of growth, morphogenesis, and self-organization is intimately related to the domain of vectorial metabolism. The contemporary studies in vectorial metabolism are dealing with spontaneous emergence of regular spatial patterns. The notion of morphogenetic field consisting of diffusible molecules acquired a special importance. The spatial and temporal features assure the passages between molecular biology and physiology. The ions play the role of cellular morphogenes (i.e. genes dedicated to the task of cell shaping). A promising trend of inquiry in the domain's articulation is the investigation of how ionic currents localize morphogenetic events (Goodwin and Trainor 1985). The latter are mediated by diverse enzyme activities. (Harold 1990) Experimental practices with "morphogenetic mutants" whose aim is to establish the patterns of transition from molecular morphogensis to cellular growth and morphogensis. To be sure, Prigogine's dissipative structures (and more generally, the doctrine that thermodynamic consequences of metabolism are responsible for spatial organization of physiological processes) provide patterns and models in this regard. However, there is no configuration of scientific practices that constitutes objects of inquiry through which nonlinear, autocatalytic chemical reactions might be connected to cellular morphogenesis. The integration of the paradigm of self-organization at chemical level in the domain of vectorial biochemistry remains a desideratum. The same conclusion is to be drawn with regard to the study of the genetic specification of biochemical topology.

3 The Hermeneutic Fore-Structure of the Research Domain of Vectorial Metabolism

The preceding presentation provides, I hope, a summarized picture of the main lines of structural, conceptual, and experimental articulation of the domain of vectorial biochemistry. My aim now is to address the hermeneutic fore-structure of the research process in that domain. In line with my foregoing considerations, the talk of a new hermeneutic fore-structure is to be comprehended as a new specification (thanks to the readable technologies and the spaces of representation) of the fore-having, fore-sight, and fore-conception in the articulation of an emerging domain of scientific research.

The fore-having is the orientation within (what seems to be) a familiar horizon of understanding the domain's contextualized entities. What Mitchell's research work from the early 1960s managed to change was precisely the horizon of understanding of the biochemical processes' specificity. Those who were affected by the new way of looking at the problematic of cellular bioenergetics began to lose their "feeling of familiarity" in dealing with what is ready to hand in the "chemical" configuration of scientific practices that became displaced and replaced (at least in certain laboratories) by practices constituting research objects of protonmotive forces and transmembrane transport systems. As a result of Mitchell's research work, the horizon of possibilities for investigating the relevance of topologically arranged osmotic processes and ionic currents to the cellular organization became a horizon of research everydayness. Brought into being was the understanding of chains of biochemical reactions in the perspective of making them meaningful entities that take place in elaborately organized systems and organized complexities. Furthermore, the research process's horizon of understanding was projected as a horizon of possibilities for investigating the anisotropic protein structure as providing the chemical foundations upon which rests the spatial organization of biological activities. In this new horizon, the spatio-temporal organization was conceived of as the biological activity itself.[11] Briefly, the fore-having of the research process consists in one's understanding upon all possibilities one can appropriate in treating the biological activity in anisotropic terms.

This horizon of understanding called into question the strong dominance of analytical methods and attitudes in biochemistry. The possibilities of exploring growth and morphogenesis cannot be appropriated by means of these methods. (Even the simplest scenarios of morphogenesis studies like that of the spontaneous association of macromolecules into structures of higher order cannot be devised by traditional analytical methods.) The understanding of the "biochemical world" as the world of proton circuits and groups translocation was initiated not only by projecting the new horizon but by inventing spaces of representation of the actualized possibilities as well. In appropriating the possibilities within the emerging configuration of practices, the readable technologies provided not only readings of particular items (experimental results, measurements, calculations, etc.). They manage in their synergy to create a whole "text of vectorial metabolism" whose "units" are the domain's meaningful objects.

Yet this text which (since the domain's inception) has remained always open to further revisions and extensions in the horizon of possibilities for doing research was never given as a pure presence. The text has existed through the contextualized readable technologies which constantly re-create it within the configuration of practices. On this account, the readable technologies are not reproducing meaning already produced before (or independent of) their implementation. Reading by

[11] Hermeneutic phenomenology of scientific research is to be distinguished from the studies of the hermeneutic pre-understandings of individual scientists. Prebble (2001) provides an excellent example of such a study in which at issue are Mitchell's philosophical pre-understandings as tacit knowledge. The author manages to show how the philosophical concept of fluctoid underlies as a tacit element the development of Mitchell's formulation of the chemiosmotic hypothesis and its verification as a theory. The hermeneutic fore-structure of scientific research has a trans-subjective status, and accordingly cannot be regarded as belonging to the personal knowledge. Fore-having, fore-sight, and fore-conception are not to be attributed to the tacit (including collective) knowledge as well.

means of these technologies is an ongoing textualization that always transcends the formation of stable textual structure. In accordance with the paradigm of constitutional analysis advanced so far, the textualization constitutes meaning by appropriating the possibilities projected by the configuration of practices. Stressing the priority of textualization over textual structure amounts to insisting on the inevitable hermeneutic fore-structuring of textual structure as this occurs in the constitution of meaningful/readable entities. The text gets continuously re-created by changing the meaning of objects of inquiry already constituted and by incorporating in it new meaningful units. *Per definitionem*, the total interconnectedness of the objects of inquiry constituted within a configuration of practices is a "text" that exists in the variability of its contextualization and its ongoing textualization.[12]

The new *fore-having* of the biochemical entities in the formation of vectorial metabolism domain was informed by a configuration of scientific practices that brought into play readable technologies of the organization of processes in the cellular space. More specifically, this was the idea that the directionality of all cellular processes is ultimately built upon molecular anisotropy. Granted that the aqueous environment of the cell cytoplasm is an isotropic medium, the fore-having of the research process oriented the work within the new configuration of practices toward identifying the processes which can give rise to a vectorial reaction. The success in this regard hinged on the way in which the hypothetical object of enzyme-catalyzed group translocation will be represented by visualizable research objects which are contextually ready-to-hand in various research situations. Different kinds of phosphoryl group and substrates served in Mitchell and Moyle's work the function of such objects. Quite indicative for the new fore-having is F. Harold's (1991, 348–349) observation that "protein molecules are regular and shapely bodies, not blobs. They come with clefts, cavities and sometimes channels, undergo ordered changes of conformation and possess intrinsic asymmetry or polarity. This is a commonplace today, colourfully illustrated in every textbook, but in the fifties it was a speculative and pregnant notion."

However novel this fore-having (orientation) of the research process in the early 1960s had been it was still consonant with established prejudices (here I am using

[12] Another aspect of the text's openness is to be spelled out in terms of Rheinberger's conception where the semiotic aspect of constituting "epistemic things" in scientific research is presented as constant creation in the research process of inscriptions, traces, and graphematic articulations. In this perspective, textualazing amounts to creating not a codified text, but a textual whole of traces—production of traces within the deferring spaces of representation engendered by the configuration. Thus, the text is intimately related to what Rheinberger calls a tracing game. For him, in scientific research there is a permanent replacement of any presumed signified by signifier (Rheinberger 1997, 104). What is read in a certain space of representation takes its meaning not from things that are beyond the research process but from what is read in other spaces of representation. Within a configuration of scientific practices as readable technologies the potentially endless production of traces becomes foreseeable. It is this fore-sight that makes the totality of what gets read in a configuration a relatively homogeneous (but by no means a structurally codified) text. Following further Rheinberger's characterization, it is a text without "referent" and without assignable "origins." Furthermore, it is a text that is doomed to be radically recast and re-contextualized in new configurations of scientific practices.

this notion in the sense of hermeneutic philosophy as a pre-judgment) incorporated in central doctrines of cytology and biochemistry. Thus, a case in point is the "topological idea" that a semi-permeable membrane serves as a selective barrier between the cytoplasm and the environment, thereby transporting actively substances into the cell. A further specification of the primary horizon of understanding as a *fore-sight* led to an alternative (to that of traditional biochemistry) "normal science" (research everydayness). The new fore-sight set the stage for a normal-scientific research guided by the intention to single out conditions under which scalar reactions (like the most enzyme reactions) would become vectorial with regard to the reactants arrived on the one side of the membrane and the products left on the other. Along the normal-scientific orientation of biochemical inquiry toward revealing important phenomena related to the topology of metabolic reactions the expectation came to the fore of obtaining visualized representations of the vectorial properties of the catalysts of the respiratory chain inlaid in the lipid membrane. One expected to disclose patterns and mechanisms of the chemiosmotic reactions. The normal scientific everydayness within the new configuration of practices was distinguished by the search for electron- and proton-conducting pathways across the membrane. Initially, it was confined to simple electrostatic interpretation of the electric potential difference caused by the migration of ions. Later on, the domain's normal scientific work brought into being more sophisticated models of topological arrangements of coupled reactions. Generally speaking, in generating the fore-sight (the search for a visualization of what is admitted to have a contextual meaning) the fore-having in this case concerns the way of "having in advance" the coupling between proton translocation and electron transfer in key metabolic reactions.

The new *fore-sight* consisted in the ways of making visible (theoretical) objects projected on the new horizon of understanding (Ginev 2008).[13] Fore-sight is the specification of this horizon through visualizing such objects. Roughly speaking, these were objects presupposed in approaching enzyme-catalyzed group transfer reactions in terms of topological translocations of ions. The visualizability refers in the first place to the translocation of a chemical group or of the reaction product(s) across the membrane. The most decisive step in visualizing vectorial metabolism's theoretical objects was the emergence in the 1970s of experimental practices of producing "cellular chimeras" (a mosaic of components stemming from different organisms that is entitled to demonstrate how the cell production of ATP comes into play). Douglas Allchin (1996, 34–36) who is the author of the philosophically most interesting historical reconstruction of the "experimental and conceptual chimeras" in biochemistry, draws the attention to the fact that it was the failure to provide *in vitro* visualization of oxidative phosophorylation (as this is the case with the experimental visualization of the critic acid cycle, for instance) that brought into being the creation of "surrogate reality" in the attempts at identifying hypothetical objects. The practices of production of "chimeras" (composed by fragments stemming

[13] A theoretical object is projected upon a horizon of research possibilities when the ongoing actualization of these possibilities in various contexts of research contributes to visualizing contextually the properties ascribed to the object by the respective theory.

from evolutionary divergent groups like bacteria, plants and animals) were designed to proving experimentally a visualized vindication of the chemiosmotic hypothesis. (The artificial chimeric composition had had to have to demonstrate how mitochondrial vesicle transfers energy necessary for generating ATP.) At stake was the experimental visualization of how cytochromes create a proton membrane gradient. This was a visualization that broke the borderline not only between *in vivo* and *in vitro* experiments, but between the factual and the artificial too. Eventually, each of the theoretical elements of the chemiosmotic hypothesis was visualized by the artificial chimeric mitochondrial vesicles. Yet this visualization would have been impossible without the fore-sight operating in the constitution of vectorial metabolism's objects.

There is a proto-normative function of the fore-sight that poses requirements to the ways of reading/representing the specific theoretical objects: Given that the visualization of the theoretical objects takes place in spaces of representation, the objects' treatment within the configuration of practices has to incorporate the topological arrangements of electron-carrying and hydrogen-carrying catalytic components in the spaces of representations that visualize the objects. It is hard to imagine how whatever visualization might be addressed to Mitchell's chemiosmotic theory which Leslie Orgel (1999) many years after the domain's institutionalization still qualified as a deeply counter-intuitive idea. The ongoing visualization (of this idea that is in conflict with the commonsensical visualizability) took place not when the hypothesis's validity was demonstrated in the energy transduction in oxidative and photosynthetic phosphorylation but when the hypothetical theoretical objects involved in the investigation of the coupling between chemical transformation and transport in biochemistry became integrated in deferring spaces of representation. More specifically, the visualization required spaces of representation that united scalar chemical entities and vectorial physical and physiological processes. It is the dualist nature of these hypothetical objects that demands visual representations of scalar-vectorial unities.[14] In recapitulating his work Mitchell (1991, 298) writes that a main preoccupation of him had been to foster "a self-consistent research process based on simple principles" that may have scalar attributes and osmotic transport attributes.

As early as the late 1960s there were various possible trajectories of a vectorial biochemical process' visualization taking place in the deferring spaces of representation that circulate within the configuration of practices. To reiterate, due to the ongoing deferring the outcome of a particular representation is always already in another space of representation. By the end of the sixties the domain of vectorial

[14]Actually, the interpretations of the dualist nature led to articulation of objects of inquiry that transcended the scope of the domain of vectorial biochemistry. Allchin (1997, 22) is absolutely right when reaching the conclusion that however strong the domain of vectorial biochemistry became emancipated, "Mitchell's alternative theory and experimental gestalt did not wholly eclipse all aspects of the chemical hypothesis or its domain. Many of the findings that initially led biochemists to search for the intermediates were 'composted' into other areas of research practice or domains that are not addressed by the chmiosmotic model."

metabolism was characterized by an ongoing interpenetration of such spaces of representation as equations representing types of transfer mechanisms, flow diagrams of equations representing the transpositions of ubiquinone couples, diagrams representing the topology of proton-motive forces, electron spin resonance measurements representing values of quantifiable variables, spectrophotometric measurements representing stability constant of certain anions, data-models representing various types of cytochrome system of mitochondria, invention of instruments capable of measuring/representing the miniscule electric fields generated by single cells in the surrounding medium, measurements of potential differences in the nanovolt range, searching for patterns of trans-cellular electric currents providing clues to the mechanisms of generation of spatial order, experiments with cytochromes in rat liver and pigeon heart mitochondria representing transport systems, experiments designed to invalidate the loop models of the classical cytochrome system, experiments with antimycin inhibition, and computer simulations of spontaneous generation of spatial patterns from reaction schemes.

Within the whole circularity of deferring and displacing each other spaces the vectorial metabolism was represented as transport systems that are linked through the thermodynamic parameters of substrate concentration and electrical potential, slow diffusion of ions through cytoplasmic space, transducting proteins of bacterial chemiotaxis, microtubules and microfilaments as polarized structures due to intrinsic asymmetries, kinds of molecular polarity conferring physiological direction, patterns of enzyme kinetics that take into account molecular anisotropy, coupled activities of osmoenzymes, sodium fluxes that support ATP synthesis, the role of potential gradient as driving force for the uptake of amino acids, a kind of ATPase that expels potassium ions in exchange for hydrogen ions, complementarity between molecular asymmetries and cellular vectors, and so on. The total interconnectedness of these meaningful (hypothetical, theoretical, formal, and experimental) objects of inquiry (as projected upon a horizon of possibilities for their further manipulability) was the "original text" of vecorial metabolism which was contextualized in diverse ways by the readable technologies and spaces of representation within the initial configuration of practices.

The *fore-conception* of the domain's research process is the anticipation of further possible contextualization of the "original text." It is this anticipation that informs changes of current agenda in the research process due to implementation of new readable technologies in new spaces of representation. Such changes imply revisions and extensions of the whole text which gets read in a new context. The fore-conception refers to anticipating integral re-contextualization of the text. It is the fore-grasping of possibilities that will be actualized in an essentially new context. A typical example for an early extension of the vectorial metabolism's original text is the actualization of possibilities to distinguish between primary and secondary transport systems. By means of this distinction a more resilient view about the balance between anisotropic and isotropic trans-membrane transport was gained. The former was attributed to actions catalyzed by osmoenzymes that create transport systems whose mechanism is based on covalent bonds. Such a transport system is characterized by a mutual reinforcement of chemical reactions and osmotic processes. By contrast, the secondary transport systems do not involve covalent bonds and mediate translocation of solutes isotropically (Mitchell 1967). In the context of this distinction the "original text" was recast with regard to ascertaining the way in which membranes act both as barriers and as links. One got a clearer view about why the conventionally scalar process of group transfer becomes the vectorial process of group translocation. Furthermore, new vectorial pathways were involved in the "original text." Mitchell's (1975) introduction of the model for Q-cycle (a mechanism whereby protons due to oxidation of ubiquinol to ubiquinone move across the membrane) provides another illustration for an extension of the "original text".

From the viewpoint of hermeneutic phenomenology, scrutinizing the fore-conception reveals the situated transcendence of the text (being at once situated in readable technologies and spaces of representation and transcended by an open hrorizon of still not appropriated possibilities). It is this dual status of what gets read that enables the text's openness to further readings and re-contextualization. The fore-conception makes possible that direction of the domain's research process which enables the transition from vectorial biochemistry to molecular and cellular morphogenesis. It operates in this regard as anticipations of text's extensions that reflect the transition. I would like to conclude this paper by indicating two prominent examples of such extensions.

The first one is the contextualization of vectorial metabolism's original text within practices of studying systems of biochemical reactions as dissipative structures by means of appropriate mathematical models and computer simulations. At stake are stable patterns of chemical reactions that arise in systems far from thermodynamic equilibrium. A second contextualization (starting with Wolpert 1969) is achieved through practices of exploring diffusible morphogen gradients and positional information that include transplant experiments and experiments on intercalation. Practices of studying morphogenetic fields (like endogenous electric fields, Nuccitelli 1988) provide a third example of contextualization in which spatial order is paralleled with a timeframe, and the whole "chronotope" of morphogenesis is related to new conceptions of the cytoskeleton's functions. However, the skepticism this kind of contextualizing the original text of vectorial biochemistry provokes is not to be ignored. Thus, Davies (2005, 8) cogently argues that "the idea of morphogenetic gradients and fields is very useful for providing a high-level view of events, but it cannot be a part of a molecular-level explanation because a gradient is, by definition, non-local and cannot be sensed directly by a single molecule." Several authors champion the view that the quest for mechanisms of morphogenesis should be based on the assumption that accounts of the shape changes at the scales of cells and tissues are to be given in terms of events that take place only at the scale of individual molecules.

The continuous reading of the vectorial metabolism's original text is carried out through readable technologies and spaces of representation that (in contrast to the initial, one-dimensional entities) constitutes three-dimensional contextually meaningful objects of inquiry. Cases in point are the so-called "morphogenetic mutants" which are defective in the cell division cycle. They are used in experiments in which growth is arrested at a particular morphological stage. Some of them generate grossly aberrant shapes. Quite pertinent to the passage from vectorial metabolism to cellular morphogenesis are those mutants which are defective in enzymes (like adenylate cyclase) that catalyze reactions at metabolic branch points (Harold 1990). Yet the identification of "morphogenes" that are dedicated to the task of cell shaping is still a trend of research in the domain being discussed as well as in the developmental biology without significant successes. Sure for the moment is only that molecular morphogenesis is directly gene-controlled.

New possibilities for bridging the gap between vectorial biochemistry and topological physiology have emerged. This gap is still the domain's basic unsolved problem. Perhaps, the domain is waiting for a new configuration of scientific practices in which a text will be constituted that will unite the topology of the biochemical autocatalytic reactions systems characterized by self-organization with cytological mechanisms of the ways in which cells can translate commands to make shape into that shape itself. These mechanisms which operate on physiological level belong supposedly to the repertoire of morphogenesis. Indications that the new configuration will arise provide the interrelatedness of experimental and theoretical practices of studying phenomena of vectorial metabolism in terms of adaptive self-organization. A configuration of practices that will be capable to appropriate the possibilities projected by it, and to read and represent the objects of inquiry it constitutes still does not guarantee that it will bridge the gap I mentioned. Yet in revealing the mechanisms in question, it will integrate in a unitary text supramolecular complexes and morphogenetic processes in individual cells. To be sure, this is an indispensable step in the transition from vectorial biochemistry to topological physiology. Undertaking this step promises one to give meaning to the concept of emergence of a topological arrangement that runs through molecular, cellular, and supracellular-physiological level.

References

Allchin, D. 1996. Cellular and theoretical chimeras: Piecing together how cells process energy. *Studies in History and Philosophy of Science* 27: 31–41.

Allchin, D. 1997. A twentieth-century phlogiston: Constructing error and differentiating domains. *Perspectives on Science* 5: 81–127.

Claude, A. 1946. Fragmentation of mammalian liver cells by differential centrifugation: II. Experimental procedures and results. *Journal of Experimental Medicine* 84: 61–89.

Davies, J.A. 2005. *Mechanisms of morphogenesis*. Burlington/San Diego/London: Elsevier.

Davson, H., and J.F. Danielli. 1943. *The permeability of natural membranes*. Cambridge: University Press.

Gilbert, G.N., and M. Mulkay. 1984. *Opening Pandora's box. A sociological analysis of scientists' discourse*. Cambridge: Cambridge University Press.

Ginev, D. 2008. Cognitive existentialism. *The Jerusalem Philosophical Quarterly* 57: 227–242.

Goodwin, B.C., and L.E.H. Trainor. 1985. Tip and whorl morphogenesis in *Acetabularia* by calcium-regulated strain-fields. *Journal of Theoretical Biology* 117: 79–106.

Griffiths, D.E. 1965. Oxidative phosphorylation. *Essays in Biochemistry* 1: 91–120.

Harold, F.M. 1990. To shape a cell: An inquiry into the causes of morphogenesis in microorganisms. *Microbiological Reviews* 54: 381–431.
Harold, F.M. 1991. Biochemical topology: From vectorial metabolism to morphogenesis. *Bioscience Reports* 6: 347–385.
Haugeland, J. 2007. Letting Be. In *Transcendental heidegger*, eds. S. Crowell and J. Malpas, 93–103. Stanford: Stanford University Press.
Heelan, P. 1983. Natural science as hermeneutic of instrumentation. *Philosophy of Science* 50: 181–204.
Heelan, P. 1998. Scope of hermeneutics in the philosophy of natural science. *Studies in History and Philosophy of Science* 29: 273–298.
Heidegger, Martin. 1927/1962. *Being and Time*. Trans. J. Macquarrie, and E. Robinson. San Francisco: Harper & Row.
Kockelmans, Joseph. 1986. On the hermeneutic dimension of the natural sciences. *Etudes Phénoménologique* 3: 33–81.
Kockelmans, Joseph. 1993. *Ideas for a hermeneutic phenomenology of the natural sciences*. Dordrecht: Kluwer.
Larkum, A.W.D. 2003. Contributions of Henrik Lundegårdh. *Photosynthesis Research* 76: 105–110.
Lehninger, A. 1960. Oxidative phosphorylation in submitochondrial systems. *Federation Proceedings* 19: 952–962.
Lundegårdh, H. 1945. Absorption, transport and exudation of inorganic ions by the roots. *Arkiv för Botanik* 32: 1–7.
Mitchell, P. 1961. Biological transport phenomena and the spatially anisotropic characteristics of enzyme systems causing a vector component of metabolism. In *Membrane transport and metabolism*, ed. A. Kleinzeller and A. Kotyk, 22–34. New York: Academic.
Mitchell, P. 1962. Metabolism, transport and morphogenesis: Which drives which? *Journal of General Microbiology* 29: 25–37.
Mitchell, P. 1967. Translocations through natural membranes. *Advances in Enzymology and Related Areas of Molecular Biology* 29: 33–85.
Mitchell, P. 1972. Chemiosmotic coupling in energy transduction: A logical development of biochemical knowledge. *Journal of Bioenergetics* 3: 5–24.
Mitchell, P. 1975. The protonmotive Q cycle: A general formulation. *FEBS Letters* 59: 137–139.
Mitchell, P. 1976. Possible molecular mechanisms of the protonmotive function of cytochrome systems. *Journal of Theoretical Biology* 62: 327–367.
Mitchell, P. 1991. Foundations of vectorial metabolism and osmochemistry. *Bioscience Reports* 11: 297–346.
Mitchell, P., and J. Moyle. 1958. Group translocation: A consequence of enzyme-catalysed group transfer. *Nature* 182: 372–373.
Nuccitelli, R. 1988. Ionic currents in morphogensesis. *Experientia* 44: 657–666.
Orgel, L.E. 1999. Are you serious, Dr. Mitchell? *Nature* 402: 17.
Prebble, J. 2001. The philosophical origins of Mitchell's chemiosmotic concepts. *Journal of the History of Biology* 34: 433–460.
Prebble, J., and B. Weber. 2003. *Wandering in the gardens of the mind. Peter Mitchell and the making of glynn*. Oxford/New York: Oxford University Press.
Quagliariello, E., S. Papa, F. Palmieri, E.C. Slater, and N. Siliprandi (eds.). 1976. *Electron transfer chains and oxidative phosophorylation*. Amsterdam: North-Holland.
Rheinberger, Hans-Jörg. 1997. *Toward a history of epistemic things: Synthesizing proteins in the test tube*. Stanford: Stanford University Press.
Saier, M.H. 2000. Vectorial metabolism and the evolution of transport systems. *Journal of Bacteriology* 182: 5029–5035.
Skulachev, V. 1987. Introduction to the Russian Edition of G. N. Gilbert and M. Mulkay's *Opening Pandora's box*. Moscow: Progress. 9–11. (In Russian)
Slater, E.C. 1953. Mechanism of phosphorylation in the respiratory chain. *Nature* 172: 975–978.

Slater, E.C. 2003. Metabolic gardening. *Nature* 422: 816–817.

Weber, B. 1991. Glynn and the conceptual development of the chemiosmotic theory: A retrospective and prospective view. *Bioscience Reports* 11: 577–617.

Weber, M. 2002a. Theory testing in experimental biology: The chemiosmotic mechanism of ATP synthesis. *Studies in History and Philosophy of Biological and Biomedical Sciences* 33: 29–52.

Weber, M. 2002b. Incommensurability and theory comparison in experimental biology. *Biology and Philosophy* 17: 155–169.

Wolpert, L. 1969. Positional information and the spatial pattern of cellular differentiation. *Journal of Theoretical Biology* 25: 1–47.

One Cognitive Style Among Others: Towards a Phenomenology of the Lifeworld and of Other Experiences

Gregor Schiemann

Abstract In his pioneering sociological theory, which makes phenomenological concepts fruitful for the social sciences, Alfred Schütz has laid foundations for a characterization of an manifold of distinct domains of experience. My aim here is to further develop this pluralist theory of experience by buttressing and extending the elements of diversity that it includes, and by eliminating or minimizing lingering imbalances among the domains of experience. After a critical discussion of the criterion-catalogue Schütz develops for the purpose of characterizing different cognitive styles, I move on to examine its application to one special style, the lifeworld. I appeal, on the one hand, to Husserl's characterization of the lifeworld as a world of perception, and on the other hand to the layer-model of the lifeworld developed by Schütz and Thomas Luckmann. A consequence of this approach is that the lifeworld appears as a socially definable context that is detached from other experiences but on an equal footing with them with respect to their claim of validity. The term "lifeworld" does not denote a category that encompasses culture or nature but refers to a delimited action-space. Finally, I draw upon Schütz's criterion-catalogue to characterize two domains of experience outside of the lifeworld, which play a central role for the process of differentiation of experience in modernity and for the phenomenological analysis of types of experience: experimental science and subjectivity.

G. Schiemann (✉)
Philosophisches Seminar & IZWT, Bergische Universität Wuppertal,
Gaußstr. 20, D-42119 Wuppertal, Germany
e-mail: schiemann@uni-wuppertal.de

B. Babich and D. Ginev (eds.), *The Multidimensionality of Hermeneutic Phenomenology*, Contributions to Phenomenology 70, DOI 10.1007/978-3-319-01707-5_3,

1 Introduction

In his pioneering sociological theory, which makes phenomenological concepts fruitful for the social sciences, Alfred Schütz has laid the foundations for a characterization of a *manifold of distinct domains of experience*.[1] Alongside science and the lifeworld, one can distinguish among others the domains of dreams, fantasies, religious practices, children's play and insanity.[2] Schütz characterizes these different domains by attributing to each its own distinctive "cognitive style" [*Erkenntnisstil*]. His understanding of "cognitive" [*erkenntnismäßig*] is expressed by his view that knowledge is experience that we take to be real.[3] "Cognitive," here, does not refer to justified belief but, on the contrary, to beliefs that are assumed not to require justification because the reality of their objects is taken to be sufficiently firm.[4]

My aim here is to further develop the pluralist theory of experience by buttressing and extending the elements of diversity that it includes, and by eliminating or minimizing lingering imbalances among the domains of experience. One of the achievements of Schütz's social phenomenology is that the lifeworld, which remains the most socially important non-scientific domain of experience, has been established as an object of enquiry for sociology.[5] Following Edmund Husserl's lifeworld analysis, he ascribes a foundational status to this domain, which is an ineliminable determinant of all other experiences—partly just as a presupposition, partly as a fundamental, universal structural property.

Schütz justifies the *privileging of the status of the lifeworld* by appealing to its character as a world of working. "Working" [*Wirken*] refers to all bodily movements performed to carry out "action in the outer world, based upon a project,"[6] which are centered around objects lying within the perceptual field of an individual. The core of the lifeworld includes the telephone that an individual uses, for example, as well as the voice of her interlocutor, but not the interlocutor herself. Cognitive styles outside of the lifeworld, in contrast, are primarily characterized by forms of action that are directed not toward objects of perception but toward past impressions, present products of imagination, or abstract entities. They rest upon different background presuppositions than the lifeworld, follow from divergent social relations, and stand at a certain distance from the demands of everyday life.

In granting a special status to the lifeworld, Schütz invests his conception with an absolute claim to validity, for his conception of the lifeworld refers to an historically

[1] On Schütz's sociology, see Natanson (1970); List and Srubar (1988); Embree (1999).

[2] Schütz (1971), 266.

[3] Ibid., 265.

[4] The foundation for this approach are spelled out in "On Multiple Realities" (1945), and "Symbol, Reality, and Society" (1955), in: Schütz (1973), 207 ff. and 287 ff.

[5] On Schütz's theory of the lifeworld, cf. Grathoff and Sprondel (1979); Srubar (1988); Endress, Psathas and Nasu (2004).

[6] Schütz (1971), Vol. 1, 243.

and culturally invariant structure, without which human life and its various modes of experience would be unimaginable.[7] One potential advantage of assuming such a universal matrix is that it could conceptually do justice to the relatively inalterable basic conditions of human life that may obtain.

If, on the other hand, one conceives of the cognitive styles as products of a typically modern *process of differentiation of experiences*, it is no longer possible to regard the lifeworld as an unqualified point of reference. Contemporary non-lifeworld cognitive styles are—I would submit—no longer as integrated within the lifeworld as they once were. Instead, they have become autonomous and partially professionalized, and attained an influence as such upon the lifeworld. Artistic, religious, scientific and other modes of experience outside of the lifeworld make the lifeworld appear to be unreal. Along with the lifeworld, they form a pluralist structure in which different domains border on each other and determine individuals' lives with specific intensities and durations. As a rule, individuals are not simultaneously in different domains of experience. Action in the lifeworld does not leave space for scientific work; dream-worlds extinguish waking consciousness; religious practice is not normally acquainted with the open-endedness of fantasy etc. The switch from one domain to another, which can already take place several times during one single day, implies the possibility of boundary experiences.

Correcting Schütz's conception of cognitive styles makes it possible to distinguish among various autonomous domains of experience. The dissolution of the foundational status of the lifeworld leads to a manifold of *equally justified contexts*, which represent the plurality of experiences that is characteristic of modernity. The non-lifeworld experiences mentioned by Schütz can be supplemented by further cognitive styles belonging, for example, to professional domains, to more finely differentiated scientific domains of experience, and to the public domain as a generally accessible sphere of societal communication, the objects of which are, to speak with Kant, "whatever necessarily interests everyone,"[8] or subjectivity as an experience in which a subject's attention is directed toward its own conscious events or conscious states.

Schütz developed a catalogue of criteria for characterizing different cognitive styles. There are only a few, easily corrected, places where the catalogue reflects Schütz's privileging of the lifeworld. Section 1 will be devoted to presenting a modified version of the catalogue and illustrating it with examples. Then I take a closer look at its application to the lifeworld. I will be drawing, on the one hand, upon Husserl's definition of the lifeworld as a world of perception, and on the other hand on Schütz's and Thomas Luckmann's leveled model of the lifeworld. Consequently, the lifeworld appears as a socially bounded context that is distinct from other, equally valid, experiential domains. The term "lifeworld," then, does not refer to a category that encompasses the entirety of nature and culture, but to a restricted space of action in (next section). Finally, I will use Schütz's catalogue of criteria to characterize two

---

[7] Cf. Heller (1986), 154.

[8] Kant (1900 ff.), Vol. V, B 868.

non-lifeworldly domains of experience that play an prominent role in the modern process of differentiation of experience and for the phenomenological analysis of types of experience: experimental natural science and subjectivity (Sect. 3).

2 Schütz's Conception of Cognitive Styles

A "certain set of our experiences" qualifies as "a finite province of meaning" according to Schütz, "if all of them show a specific cognitive style and are—*with respect to this style*—not only consistent in themselves but also compatible with one another."[9] Schütz can refer to a domain of meaning as "finite" because the condition of compatibility does not apply among cognitive styles: "that which is compatible with the province of meaning P be also compatible within the province of meaning Q."[10] For Schütz, it is by virtue of the lifeworld that communication among domains of meaning is nevertheless possible.[11]

The expression "cognitive style" makes it clear that Schütz is not so much interested in the content as in the form of knowledge. The term *"cognitive"* does not refer only to true beliefs but also to "valid as well as the invalidated [experiences],"[12] insofar as they exhibit the same style and are thus held with conviction to be real. The exclusive restriction to domains of meaning, however, invests the concept of knowledge with a statical character that underestimates the significance of *boundary experiences*, which alone enable the transformation of the conditions for the constitution of meaning. If one presupposes a division among domains of meaning, experiences arising from an overlap of cognitive styles can result either from the transgression or from the dissolution of borders. Schütz describes such experiences with a terminology that is also suitable for the transition between incommensurable worlds. Thus, the shift between domains of meaning does not occur gradually but suddenly, and is connected with an emotional shudder that Schütz compares with "Kierkegaard's experience of the 'instand' as the leap."[13] Among other things, the "shock of falling asleep as the leap into the world of dreams; the inner transformation we endure if the curtain in the theater rises […]; the radical change in our attitude if, before a painting, we permit our visual field to be limited by what is within the frame […], our quandary, relaxing into laughter, if, in listening to a joke, we are for a short time ready to accept the fictitious world of the jest as a reality."[14]

[9] Schütz (1971), Vol. 1, 264.

[10] Ibid., 397 (267).

[11] Ibid., 296ff., 392 and 395.

[12] Ibid., 265.

[13] Ibid., 266 (267).

[14] Ibid., 266 (397f.).

In order to make it possible to characterize conceptually "at least some of the features"[15] of the cognitive styles specific to domains of meaning, Schütz formulated his *catalogue of criteria*. As already noted, I will be making some modifications to the catalogue that result from criticism of the foundational status Schütz attributed to the lifeworld. In its social-phenomenological orientation, the catalogue presupposes an egological structure that rests upon subjective experience (which Schütz does not invest with a specific cognitive style), and then integrates phenomena of intersubjectivity. The catalogue of criteria includes six features, the concise definitions of which are supplemented by Schütz with examples to illustrate their relationship with the domains of meaning.[16]

1. The foundational criterion, modeled upon Bergson's "*attention à la vie*," is the so-called "*tension of consciousness*,"[17] which is understood as an "orienting toward and attending to life."[18] Schütz distinguishes various intensities of attention, as well as various resultant degrees of reality. Since it is "the meaning of experience and not the ontological structure of the objects which constitutes reality," reality is dependant upon the attentional structures of the conscious.[19] Following Bergson, Schütz ascribes to the lifeworld the highest degree of attention, the "wide-awakeness,"[20] and to the dream-world the lowest degree, a merely "passive attention."[21] By characterizing "wide-awakeness" with reference to "normal [...] adults",[22] Schütz distinguishes the lifeworld from fantasy worlds—which he also regards as characteristic for children—and from worlds of (pathological) insanity. I will be adopting this point of reference, along with the attention toward external objects, as a criterion for the lifeworld's intuitive orientation, and will be using Husserl's concept of perception in order to fine-tune it to this orientation.

   In order to revoke the privileged status of the lifeworld, it is necessary to abandon the ranking of attentional intensities as a means of individuating cognitive styles. The principally equal justification of cognitive styles implies that they have equivalent possibilities for validity claims upon reality. By no means, however, does this undermine the necessity of the criterion of tension of consciousness. For all cognitive styles remain "names [...] of one and the same consciousness, and it is the same life [...], which is attended to in different modifications."[23]

[15] Ibid., 265.

[16] This presentation of the criterion catalogue follows Schiemann (2002), 86ff.

[17] Schütz (1971), Vol. 1, 243ff. and 265, in the original not emphasized, as for all further characterizations of the criteria.

[18] Ibid., 243.

[19] Ibid., 264 (393).

[20] Ibid., 265.

[21] Ibid., 244.

[22] Schütz and Luckmann (1979), 47.

[23] Schütz (1971), Vol. 1, 297.

Taking part in the "stream of consciousness" is thus a necessary condition for cognitive styles that could not otherwise determine the experiential space of *one* individual with varying intensities and duration.

2. The next criterion is concerned with different contents of attention, and captures the "*prevalent form of spontaneity*" within a cognitive domain.[24] Unlike the tension of consciousness, which refers to an internal attitude, it marks out a relation between the subject's working and the experiences she takes to be real. The distinctions among forms of spontaneity arise from the qualitative experiential differences between thinking and conducting [*Verhalten*], and between acting [*Handlung*] and not acting. Among the forms of thinking action, for example, fantasizing differs from contemplation by virtue of the absence of any intention to make the fantasy real.[25] Moreover, dreaming lacks a structuring of action, as the subject does not have the freedom to direct the flow of events.[26] Hence, dreaming is neither an experience of being conducting nor one of performing actions. The paradigm of a subject performing actions is the lifeworld. As already noted, its *prevalent* spontaneity is, for Schütz, working [*Wirken*].

   Following Oswald Schwemmer, I would like to further restrict the spontaneity *prevalent* of the lifeworld by pointing to its *unprofessionality*. Lifeworldly action can "just as well be performed by and expected of others as from us," since we "attribute competency in principle to everyone."[27] With this further characteristic, the scope of lifeworldly action becomes far narrower, and the lifeworld loses its status as universal mediator for communication among the cognitive domains. This characterization can probably be connected with the following, negative, criterion of the "epoché."

3. Although the word "*epoché*" is borrowed from Husserl, Schütz uses it in a different way, namely to refer to the bracketing or suspending of particular aspects of reality.[28] Schütz adds a hierarchization, similar to that of the tensions of consciousness, when he claims that it is only in the lifeworld that all doubts about reality are suspended and it is accepted as self-evident.[29] Nevertheless, this ranking does not lead to the privileging of the lifeworld, as the doubts about the validity of reality in other domains of experience apply only to particular phenomena. To formulate it as a paradox: the general epoché of the lifeworld, which is characterized by a complete absence of doubt, is the specifically lifeworldly epoché. For other domains of meaning, Schütz himself introduces examples of specific epochés: thus, dreaming experience is not concerned with the validity of

[24] Ibid., 265.

[25] Ibid., 270.

[26] Ibid., 277.

[27] Schwemmer (1987), 207.

[28] Schütz (1971), Vol. 1, 263 and 265.

[29] Ibid., 265 and 268.

"certain logical axioms"[30]; scientific reflection puts aside, among other aspects, the bodily existence of the researcher "as psycho-physical human being".[31]

4. While the characteristics mentioned so far have to do with the subjective constitution of objects and meaning, the next one introduces the specific "*form of sociality*"[32] as a further definitional feature. Roughly speaking, it distinguishes the domains of meaning that are experience exclusively alone (e.g., dreams, contemplation) from those that can be experience only together (e.g., lifeworld) also from those that can be experienced alone or together (e.g. fantasy, religion and, according to Schütz, science). On Schütz's view, as I will elucidate further below, the familiar/well-known essence of lifeworldly social forms results from a minimalization of the typification of social relations.
5. Possibilities for the production of complex relational networks within and between domains of meaning are opened up by the two final criteria: "time-perspectives" and "*experiencing one's self*."[33] The criterion of different "time-perspectives" is closely tied to the aforementioned criterion of sociality. It refers to the objective time of the world that is beyond control, to biological times (bodily rhythms, seasons, etc.), to individual biographical time, to the subjective time of the stream of consciousness with internal duration, and to the social time of the calendar and standardized intersubjective time.[34]
6. The criterion of "*experiencing one's self*,"[35] which I merely mention here, postulates the emergence of patterns of personal identification that are specific to different domains of meaning and that are dependent upon validity elements to reality.

3 The Lifeworld as a Restricted Perceptual World

Three of the characteristics formulated by Schütz can be conceived as *necessary criteria* for the lifeworld's cognitive style: "wide-awakeness" as a specific *tension of consciousness*; the *prevalent form of spontaneity* characterized by unprofessional action that is directed toward objects within the perceptual field of an individual; and the specific *epoché* that places the validity of reality beyond doubt.[36] The lifeworldly *form of sociality* discussed by Schütz, namely familiar/well-known

[30] Ibid., 279.

[31] Ibid., 286.

[32] Ibid., 265.

[33] Ibid., 265.

[34] Schütz and Luckmann (1979), Vol. 1, 73.

[35] Schütz (1971), Vol. 1, 265.

[36] The concept of a context of lifeworldly experience, which draws upon Husserl and Schütz, has been elucidated further in Schiemann (2005), chapter 1.1.2. Along with a given background knowledge, the three criteria form a sufficient condition for lifeworldly experience, cf. ibid. Chapter 1.1.2, paragraph 3.1.2.

intersubjectivity, does not in my view count among the necessary criteria, since individuals (such as Robinson Crusoe) can also live alone in a lifeworld. Insofar as an individual has a perceptual world bound to its body, that perceptual world determines the familiarity and intuitiveness of the things that can be altered by direct (unprofessional) actions. What an individual of course conceives as a uniform world constitutes a *subjective lifeworld*, around which are concentrically arranged spheres of what can only be experienced indirectly. There are as many subjective lifeworlds as there are normal adult individuals, and as many *shared lifeworlds* as made possible by the integration of different subjective worlds into unified worlds in social action spaces.

In order to give a more precise characterization of the lifeworldly *tension of consciousness*, the next step will be to base the specific *epoché* upon Husserl's notion of perception. Lifeworldly attention has to do primarily with perception of the outer world of bodies that is not symbolically mediated. This characterization will make it possible for me to apply Schütz's and Luckmann's layered model for specifying the *prevalent* forms of *spontaneity*, time-*perspective*, and *sociality*.

The absence of—or freedom from—doubt leads to the emergence of a core of unshakable certainties, the objects of which are not any particular contents but the validity of the world. This self-evident world-belief is what Husserl refers to as a "*general thesis of the natural attitude*" [*Generalthesis der natürlichen Einstellung*].[37] The lifeworld is unquestioningly accepted as a unity and as existence that is independent of the act of knowing. Husserl's general thesis is to a certain extent within the Cartesian tradition, insofar as the given world that corresponds to this natural attitude is primarily conceived as being given to intuition. Where reflection about experience or knowledge does not occur, it is the testimony of perception that dominates. The lifeworld is the "world of perception," because it is a world that is taken as self-evident in its being.[38] "Perception," refers exclusively to the "mode of self-presence" of an appearing object, which Husserl calls the "original mode of intuition," [*Urmodus der Anschauung*] in contrast to the recollective or anticipatory intuition of a currently absent object.[39] This "originally giving" [*originär gebende*] experience is oriented toward "mere bodiliness."[40] "Through sight, touch, audition, etc., bodily things are *simply there for me in* the different ways appropriate to the various senses and in some particular spatial distribution."[41] In this sense, the lifeworld encompasses the things within the visible surroundings of the subject that are currently present and point as signs to an Other. In a broader sense, it also extends to things that are invisible, occluded or absent at the moment but are present in memory in a "conscious sense" [*bewußtseinsmäßig*].[42]

[37] Husserl (1977), 63f.—also emphasized in the original.

[38] Husserl (1976), 49f., 171; Husserl (1968), 58f.

[39] Husserl (1976), 107.

[40] Husserl (1948), 54.

[41] Husserl (1977), 57.

[42] Ibid.

Husserl applies the general thesis not only to the "world of things" [*Sachenwelt*]. The same world is there "in the same immediacy as *world of values, world of goods, practical world*."[43] In the lifeworld, bodies do not appear independently of their cultural, social and practical evaluations—and neither, in turn, do these evaluations attain an autonomous existence: "In order for something to be given as usable, beautiful, terrible, frightful, attractive or whatever else, it must be somehow present to the senses."[44] The perception of meaning presupposes the perception of bodies, which is why the lifeworld remains, even from a socio-cultural perspective, primarily a world of perception.[45]

In subscribing to the assumption of bodily primacy, I conceptualize the lifeworld as a cognitive style that is founded upon the achievements of outer perception. Schütz and Luckmann have used the term *"stratifications"* [*Aufschichtungen*] to refer to the structures of such a domain of experience. I will be using this model in order to specify the circle of necessary conditions for a *restricted social context*. In so doing, I will be pursuing the strategy of limiting the scope of the claims to validity of the definitions I am borrowing from Schütz and Luckmann in order to make space for consideration of experientially different contexts. The *spatial* stratification specifies the *prevalent* spontaneity of direct action that is characteristic of the lifeworld and which takes on foremost significance in a world of perception. Schütz and Luckmann, who characterize the lifeworld through socially mediated as well as through immediate experience, follow the same ranking. The *temporal* and *social* stratifications confirm the structural features of the spatial stratification and provide additional definitions.

3.1 Spatial Stratification

The centerpoint of the spatial structure is the locus of the bodily presence of the lifeworldly subject. The scope of its direct actions spans the so-called "primary zone of efficacy" [*primäre Wirkzone*]. This zone contains all the things and people that the subject can influence solely through bodily movement (within a present that spans a certain time interval)—including the use of technical devices (tools, means of locomotion, etc.). A "secondary zone of efficacy" [*sekundäre Wirkzone*] includes the use of technical devices to alter bodies outside of the primary zone. The primary zone is entirely, and the secondary zone partially, within the sphere of perceptible things. The "*world currently within reach*" [*Welt der aktuellen Reichweite*], which is made up of these two zones, can be divided according to sensory modalities (what is visible in the distance can only seldom be smelled, whereas what can be smelled generally has a broader spatial range than what can be touched or tasted, etc.).

[43] Ibid., 59—emphasized also in the original.

[44] Husserl (1948), 53.

[45] Ibid., 55.

It borders upon the "*world potentially within reach* [*Welt der potentiellen Reichweite*]," which includes objects that are not currently present but can be in the future.[46]

The notion of spatial stratification can be illustrated with the example of a student's domestic workspace (which may be sufficiently similar to Husserl's rather inadmissibly tidy lifeworld): her primary zone of efficacy includes the computer she was just using a moment ago as well as the part of the speaker and door opener she is presently using. The secondary zone of efficacy contains visible and invisible things: the door-opening mechanism itself (invisible), the front door (partially visible) and the person outside who just rang the doorbell (invisible). Along with the inventory of the primary zone of efficacy, the world that is currently within the student's reach also contains perceptible objects outside of the action-radius—including, for example, objects that her gaze falls upon outside the window (houses across the street, distant stretches of landscape). It does not however include the person who has not yet entered, although this person may be part of the world potentially within reach. None of these worlds contains invisible components of mechanical-electronic devices, such as keyboards and speakers, unless the user has an interest in their function and would thus draw them into the world potentially or even currently within her reach. As far as modern technical devices are concerned, the lifeworld is a world of button-pushing. It is only concerned with relevant in- and outputs, and does not normally extend beyond the surface of technical apparatuses.

3.2 Temporal Stratification

The temporal structures of the lifeworld that Schütz and Luckmann investigate are exceedingly complex, and can be contrasted with other closed domains of meaning by virtue of the flowing together of otherwise separate forms of time.[47] The spatial stratification in fact already implies the basic structure of the temporal stratification: the present occurs primarily in the world that is currently within reach; future events and (shared) experiences make up the world that is potentially within reach.

The state of the world beyond its potentially reach effects a subject in a different way temporally than spatially. Whatever is spatially out of reach tends to escape her interest and also her knowledge. Lifeworlds, in their essential locality, are only marginally involved with distant events. Of the events that are temporally out of reach, however, there are two that have unconditional significance for individual subjects: one's own *birth* and *death*. One cannot have any direct lifeworldly contact with these events.[48] One knows about them only through the accounts of others, and assumes that one's own birth and death must be similar to those of others. Of course,

[46] Schütz and Luckmann (1979), Vol. 1, 63ff.

[47] Ibid., 73ff.

[48] Ibid., 74f.

the mediacy of these experiences is not symmetrical. One's own birth is far more removed from one's experience than one's death, which announces its approach in bodily aging and dying.

For Husserl, Schütz and Luckmann, the transcendence of birth and death is the basis of the strict *generational orientation* of a lifeworldly context, in which much older and much younger people appear only at the margins. The zenith of efficacy of one's predecessors, which is manifest in cultural achievements, belongs to the past, while the coming generation participates in adult life only partially.

3.3 Social Stratification

According to Schütz's and Luckmann's starting premise, each individual subject assumes unquestioningly "that other men exist."[49] The social stratification deals with a differentiation of social experience according to *degrees of anonymity*: anonymity stands in, so to speak, an inversely proportional relation to the immediacy of experience, while no social experience is so immediate that it could do entirely without the application of anonymous definitions. "Anonymity" refers to typologies that are applied to people, i.e. the formation and use of types for identifying people ("our mailman Mr. Martin," "an enemy of the people," etc.).[50] It concerns aspects of actions less in the primary zone of efficacy than in the world potentially within reach.

For Schütz and Luckmann, the reach of the socially stratified lifeworld fades off into indeterminacy. They impute not only the weakest but even the strongest forms of mediacy of social experience to the lifeworld, which retains its midpoint but extends to the entire social and cultural world. Such radical openness corresponds, on the one hand, to the horizonedness [*Horizonthaftigkeit*] of lifeworldly experience, on the other hand it precludes a classification of social experience in which lifeworldly experience stands alongside other kinds of experience. In contrast to Schütz and Luckmann, I would therefore like to apply the criterion of anonymity in such as way as *to highlight the part of the social world that is already restricted by the spatial stratification*. Anonymity should deliver definitions that make it possible to distinguish at least roughly between people with whom one shares a lifeworld and people who appear in one's lifeworld but do not properly belong to it. In carrying out such a demarcation, it is possible to build upon Schütz's and Luckmann's conception of the lifeworld as a world of perception. Experience in which anonymity is as minimal as possible can only be had, according to Schütz and Luckmann, in "face-to-face situation," i.e. through bodily presence. The term for one subject's attentively turning to the other is the "*thou-orientation*," and the other who is experienced as an equal person is a "fellow being" [*Mitmensch*].

[49] Ibid., 87.

[50] On lifeworldly typologies, cf. Schiemann (2005), chapter 1.1.2, paragraph 3.1.2.

Whenever this orientation is mutual, the social relationship can be called a "*we-relationship*."[51]

We-relationships can be classified according to the degree of anonymity. The more concrete, detailed and presumably also persevering an unmediated social relationship is, the less important typologies become in the "thou-orientation."[52] The alter ego may appear as a singular individual that cannot be captured adequately by any typology at all. Higher degrees of anonymity arise through the successive replacement of immediate social experience by typologies. In the extreme case, typologies can fully take the place of social experience. Since we-relationships depend upon a turning of attention for which typologies can provide no substitute, they are progressively destroyed by increasing anonymity. It is in we-relationships with minimal degrees of anonymity that I see the essential social feature of the shared lifeworld.

3.4 A Proto-Concluding Characterization

As a world of perception, the subjective and the shared lifeworlds have a *centric nature*. The shared lifeworld, which has been the focus here, appears to individuals as a familiar social space in which they linger sometimes with more attention and sometimes with less, but which they can also sometimes leave and to which they can again return. One knows the objects and people in one's lifeworld on their own terms. That is the way in which they remain in one's memory when they are no longer present or when one is outside of one's lifeworld (while dreaming or fantasizing, while being in public, at work, etc.). Other spaces of experience are layered upon the lifeworld in varying degrees of familiarity, and partially overlap with it. Persons, which enter the lifeworldly domain coming out of these spaces of experience, remain distincted from the lifeworld at first. Since everything that belongs to the lifeworld must be perceptible, its temporal modus is essentially the present. The past is the source the experiences come that make perception possible and orientate action, and it is toward the future that desires, expectations and action plans are directed.

As an equal cognitive style alongside others, the lifeworld does not lose its *foundational significance for the conceptualization of processes of modernization*. Increasing professionalization can only be understood against the backdrop of the existing unprofessional spaces of experience, which find their very core in the re-defined lifeworld. Moreover, the integration of the lifeworld into a manifold of differing domains of meaning offers the possibilities for conceiving of the lifeworld as a cultural phenomenon that arises only within a plurality of historically changing lifeworlds. The cultural-historical process of differentiation of experience leads to a multiplication of non-lifeworldly cognitive styles.

[51] Schütz and Luckmann (1979), Vol. 1, 90f.

[52] Ibid., 95ff.

4 Examples of Non-Lifeworldly Cognitive Styles

Using examples of two non-lifeworldly experiences—experimental natural science and subjectivity—I would now like to discuss the application of Schütz's criterion catalogue to the demarcation of different cognitive styles. Schütz and Luckmann never mention these two experiential contexts as closed domains of meaning. Only in a few places do they speak—very generally—about *science* and about the "scientific contemplation" that they impute to science.[53] Without being able to give any convincing justification, Schütz contrasts the latter with the lifeworld. Moreover, his understanding of scientific contemplation has less in common with cognitive styles found in the sciences than with Husserl's concept of subjective experience.[54]

Since Schütz and Luckmann do not consider *subjectivity* as a closed domain of meaning, they overlook the constitutive significance that this cognitive style has for modern self-understanding. In his persuasive study of modern forms of identity, Charles Taylor refers to what he calls the three "sources of the self": alongside the "voice of nature" and the "affirmation of ordinary life" that is bound up with the lifeworld, there is also the "inwardness" that characterizes the demarcation of reason from the outside world as well as its formation as an autonomous site of knowledge and action.[55] The mainstream discourse on subjectivity in early modernity got underway with a ritual turning away from the lifeworld, which opened up newfound space for loneliness.[56] Subjectivity soon became the subject of methodological analyses, which subsequently influenced introspective procedures in psychology (Wundt, James, Titchener and the Würzburg school with Külpe) as well as phenomenological method of reduction (Husserl).[57] Finally, it has, as a space of sensory qualities, experienced an unanticipated renaissance in the past few decades in the philosophy of mind.[58]

4.1 Experimental Natural Science

Schütz's concept of a cognitive style offers the possibility of a classification within the plurality of modern modes of experience in the first approximation. Thus, the uniform characterization of the experience that is characteristic for experimental science is also only sensible by virtue of the contrast to other cognitive styles.

---

[53] Ibid., 48f., 356ff.; Schütz (1971), Vol. 1, 281ff.

[54] Cf. Schiemann (2002).

[55] Taylor (1992), 109ff., 207ff. I use the concept of modernity here in a sense that encompasses modernity in its contempary period; cf. Schiemann (2009).

[56] Cf. Bürger (1998).

[57] Varela and Shear (1999).

[58] Smith and Thomasson (2005); Kriegel and Williford (2006).

It cannot claim to do justice to the diversity of scientific methods, domains and conceptions. A conceptual demarcation of experimental science is of crucial importance. These disciplines have been a—perhaps the—motor driving the *scientification and technologization* that is characteristic of modernity.

Experimental procedures in the natural sciences present a context of purposive-rational action, which aims to state or to create phenomena, and serves to develop, evaluate or criticize knowledge of objectifiable reality.[59] It is found in just about all natural sciences and can be invoked as their defining feature. The term "*phenomenon*" here refers to events, processes or states that are expected or that have been demonstrated to admit of conceptual description, to appear regularly under the appropriate conditions, and to accessible to theoretical explanation. In a broader sense, it can also (or alternatively) be used to refer to regularity itself. Thus, the experimental creation of phenomena need not impose any particular conception upon them. Countless phenomena have indeed been discovered in experiments that were not devised with that aim. In general, it can be said that the minimization of parameters and variables that is necessary for the systematic investigation of phenomena necessitates a *demarcation* of the objects of the experiment from their surroundings. In order to fulfill this condition optimally, experiments are often conducted in technically manipulable apparatuses. Experimental phenomena are perceived or observed. "*Perception*" refers to the acquisition of sensory impressions from external bodies, such as also are encountered in the lifeworld. "*Observation*," on the other hand, is the empirical referral to theoretical entities (electrons, atoms, genes, black holes, etc.) that is mediated by the apparatus. Many scientific experiments have to do with properties that are too small, too large, too distant or too transient for perception. If observations can be made with the help of or through measurement devices, sensory perception may be limited to reading information from dials or other presentations of results, i.e. interchangeable representations of data of theoretical entities. Observations, however, are also mediated by the senses insofar as they arise through actions. The ongoing automatization of experimental techniques has however made forms of observation possible that are only peripherally or not at all dependent upon human perception.

The aforementioned condition of demarcation from the surroundings allows for a spatio-temporal localization of experiments. Its intended claim to validity is, however, not local but universal. Scientific knowledge is supposed to be testable under reproducible conditions, and stakes a claim upon *unrestricted intersubjective validity*. This claim to universality is matched by the boundary expanding structure of research organizations. The bearers of experimental science are not single individuals but the worldwide, networked "scientific communities" of the various disciplines. The knowledge they produce has only a relative objectivity, though, insofar as experimental science—like science in general—is essentially a *socio-cultural endeavour*. Hence, scientists follow changeable and replaceable "thought styles" (Ludwik Fleck) in their beliefs, or they organize their work within

[59] For overviews of recent literature on experimentation, see Heidelberger and Steinle (1988); Radder (2002).

changeable and replaceable "paradigms" (Thomas S. Kuhn). Under these historically contingent conditions, the predominant rationality of the experimental sciences has proven itself to be *instrumental*. It answers, roughly speaking, partly to the inner dynamics of acquisition of knowledge through problem-solving, partly to the external societal factors that support or hinder the development of methods and investigations of object-domains.

This admittedly rather rudimentary description of experimental science can be re-formulated and made more precise with the terminology of Schütz's criterion catalogue. Experimental science shares the "wide-awakeness" of acting subjects that is characteristic of the lifeworld and inscribed in the first criterion of *tension of consciousness*. But this attentional intensity can in the context of experiments be directed to things that function as signs for results that cannot themselves be accessed perceptually, as the discussion of observation showed. It follows that the spatial structure of experimental design does not necessarily point back to perceptual conditions. Rather, it varies with institutional, methodological and object-dependent requirements. The spectrum can span from a narrowly bounded locality up to an earth encompassing global level. Modern communication systems make it possible to carry out experiments, if necessary, in spatially very disparate locations, and to evaluate the data somewhere else altogether. The second criterion, too, is relatively independent of the reach of an individual's perceptual field: the *prevalent spontaneity* of experimental action is directed toward the design, the execution and the evaluation of experiments, as well as to the development and application of the attendant conceptual and theoretical constructions. Successful experiments demand a high-degree of professionalism in practical and systematic respects, which is normally guided by thought styles or paradigms. As an example of the specific *epoché*, the third criterion, Schütz mentions the bracketing of the bodily existence of the researcher as a "psycho-physical human being" (see above). The *form of sociality*, the fourth criterion, describes the organization in "scientific communities." If the experimental praxis also requires an ability that may resemble lifeworldly forms of action, the claim to universality, on the other hand, requires the interchangeability of acting subjects. Finally, the definitions discussed so far, and the limits of their applicability to the diversity of varieties of experimental science, are reflected in the final criterion of specific time-*perspectives*. A common element could be portrayed more closely with Hans Blumenberg's concept of "world-time" as "core of all conceivable chronologies" and contrasted with the "life-time" typical of everyday praxis.[60]

4.2 Subjectivity

Subjective experience is experience in which a subject's *attention is directed toward her own conscious events or states* in experiencing or reflecting upon them. Husserl's transcendental epoché is one example of subjective experience, as are the forms of affective involvement (*Betroffenheit*) and of self-consciousness with

[60] Blumenberg (1986).

self-attribution described by Hermann Schmitz.[61] People to whom subjective experience is ascribed have *privileged access* to that experience. Their reports have the status of uncorrectability, and are given in the first-person singular. The features of the corresponding cognitive style are related in a partially complementary fashion to those of the lifeworld.[62]

The direction of the *tension of consciousness* is, so to speak, contrary to that of the lifeworld: in the lifeworld, attention is directed toward external objects of perception, whereas, in the case of subjectivity, it is directed to one's own conscious states (sensations, perceptions, thoughts, attitudes, moods, etc.), which are withdrawn from their everyday, practical meaning. Whereas the locus of bodily presence is the immovable midpoint of the spatial structure of the lifeworld, it can be left aside in subjective experience. Thoughts that are not tied up with perception and sensations are free of the necessity of a spatial localization. It is possible to go to a particular location merely in thought, or to be lost in thought and forget one's actual physical location. While the *prevalent spontaneity* of a person in the lifeworld aims toward specific interventions in an external world that is shared with fellow human beings, in subjective experience it is conducted as an invisible act within the inner world of consciousness, and does not primarily have an agentive but, rather, an experiential or a reflective character. Professionalism, which need not be present in the lifeworld, can indeed be formed in subjectivity. Individual awareness of one's own inwardness does not presuppose any specific competency, because the possibility of its realization is an integral component of modern self-understanding. But, on the other hand, the methods of introspection and phenomenological reduction, which are available for systematic mental self-exploration, must be acquired through learning and practice. The subjective *epoché* does not bracket out doubts about the reality of external objects but the practical goals that are prominently bound up with them in other worlds. The questions that arise within a subjective cognitive style are not concerned with properties of perceived objects that are relevant for action but with the experiential qualities, thoughts or propositional attitudes that are connected with perception, how their mode of presentation depends upon spatio-temporal position, etc. With respect to its *form of sociality*, subjective experience is not communalized but essentially lonesome experience, and not limitlessly shareable with other people. Time-*perspectives*, finally, focus upon dimensions of inner time-consciousness that are not at issue within the lifeworld.

5 Conclusion

In order to illustrate Schütz's conception of cognitive styles, I have discussed three domains of meaning that are of significance for *modern processes of differentiation* of experience: the lifeworld can be seen as the core of unprofessionalized spaces

[61] Schmitz (1990).

[62] The definitions belonging to a subjective context of experience were developed in Schiemann (2005), Chap. 1.2.2, and are discussed critically in relation to Schütz's concept of "scientific contemplation" in Schiemann (2002).

of experience, experimental natural science is to be regarded as a central motor driving scientification and technologization, subjectivity is foundational for human self-understanding today.

The three areas, however, also have particular relevance for *phenomenological analysis*. Traditionally, the phenomenology of the *lifeworld* coming from Husserl has been accorded a foundational role in two senses: it has been thought to present the foundation of human experience in general and of scientific knowledge in particular. Against this characterization, which conflicts with the currently predominant plurality of experience, I have drawn upon Schütz in order to establish not only the lifeworld but also *experimental science* as autonomous contexts.

Husserl was right in pointing out that the analysis of the lifeworld must avail itself of a perspective external to the lifeworld. This insight must be generalized in two respects. Every analysis of a domain of experience requires a reference point that does not lie in the domain. The plurality of experience delivers just such reference points by bringing out the differences among various cognitive styles. *Subjectivity*—the decisive domain of meaning for the phenomenological analysis of the lifeworld—is among them. Subjectivity is no more a privileged context than the lifeworld. It is not the "subjectivity that creates all world-validities with their contents and in all pre-scientific and scientific ways."[63] Rather, it corresponds to a cognitive style that other experiences do not require exclusively for their validity. It owes its existence to an attitude that can be taken by anyone, with or without practice, and which can be more or less beneficial.[64] Subjectivity is in this sense a generally accessible cognitive style, but one that can be experienced only individually and which is typical for modern self-understanding.

References

Blumenberg, Hans. 1986. *Lebenszeit und Weltzeit*. Frankfurt am Main: Suhrkamp.

Bürger, Peter. 1998. *Das Verschwinden des Subjekts: eine Geschichte der Subjektivität von Montaigne bis Barthes*. Frankfurt am Main: Suhrkamp.

Embree, Lester (ed.). 1999. *Schützian social science*. Dordrecht: Kluwer.

Endress, Martin, George Psathas, and Hishashi Nasu (eds.). 2004. *Explorations of the life-world: Continuing dialogues with Alfred Schütz*. Dordrecht: Kluwer.

Grathoff, Richard, and Walter M. Sprondel (eds.). 1979. *Alfred Schütz und die Idee des Alltags in den Sozialwissenschaften*. Stuttgart: Ferdinand Enke.

Heidelberger, Michael, and Friedrich Steinle (eds.). 1988. *Experimental essays – Versuche zum Experiment*. Baden-Baden: Nomos Verlagsgesellschaft.

Heller, Agnes. 1986. The sociology of everyday life. In *Sociology*, vol. 2, ed. Ulf Himmelstrand. London: Sage.

[63] Husserl (1976), 70.

[64] Husserl compares the change in attitude that occurs in the transition from the natural to the transcendental attitude with the change in attentional directedness that "normal people" make when they move from lifeworldly occupations ("father") to their jobs ("shoemaker") or to politics ("citizen") (ibid., 139ff., 154), i.e. when they cross the boundaries separating domains of experience.

Husserl, Edmund. 1948. *Erfahrung und Urteil. Untersuchungen zur Geneologie der Logik.* Hamburg: Claassen & Goverst.

Husserl, Edmund. 1968. In *Husserliana* IX: *Phänomenologische Psychologie. Vorlesungen Sommersemester 1925*, ed. Walter Biemel. The Hague: Nijhoff.

Husserl, Edmund. 1976. In *Husserliana* VI: *Die Krise der europäischen Wissenschaften und die transzendentale Phänomenologie. Eine Einführung in die phänomenologische Philosophie*, ed. Walter Biemel. The Hague: Nijhoff.

Husserl, Edmund. 1977. In *Husserliana* III/1: *Ideen zu einer reinen Phänomenologie und phänomenologischen Philosophie, Erstes Buch: Allgemeine Einführung in die reine Phänomenologie, 1. Halbband*, ed. Karl Schuhmann. The Hague: Nijhoff.

Kant, Immanuel. 1900ff. In *Kant's gesammelte Schriften*, ed. Königlich Preußischen (later the Deutschen) Akademie der Wissenschaften (zu Berlin) (Berlin 1900ff). Vol. V.

Kriegel, Uriah, and Kenneth Williford (eds.). 2006. *Self-representational approaches to consciousness*. Cambridge, MA: MIT Press.

List, Elisabeth, and Ilja Srubar (eds.). 1988. *Alfred Schütz: Neue Beiträge zur Rezeption seines Werkes*. Amsterdam: Rodopi.

Natanson, Maurice (ed.). 1970. *Phenomenology and social reality*. The Hague: Nijhoff.

Radder, Hans (ed.). 2002. *The philosophy of scientific experimentation*. Pittsburgh: University of Pittsburgh Press.

Schiemann, Gregor. 2002. Rationalität und Erfahrung. Ansatz einer Neubeschreibung von Alfred Schütz' Konzeption der, Erkenntnisstile. In *Zugänge zur Rationalität der Zukunft*, ed. Nicole C. Karafyllis and J.C. Schmidt. Stuttgart: Metzler.

Schiemann, Gregor. 2005. *Natur, Technik, Geist. Kontexte der Natur nach Aristoteles und Desccartes in lebensweltlicher und subjektiver Erfahrung*. Berlin: de Gruyter.

Schiemann, Gregor. 2009. *Hermann von Helmholtz's mechanism: The loss of certainty. A study on the transition from classical to modern philosophy of nature*. Dordrecht: Springer.

Schmitz, Hermann. 1990. *Der unerschöpfliche Gegenstand*. Bonn: Bouvier.

Schütz, Alfred. 1971. *Gesammelte Aufsätze*, vol. 1. The Hague: Nijhoff.

Schütz, Alfred. 1973. *Collected Papers*, vol. I: The problem of social reality. The Hague: Nijhoff.

Schütz, Alfred, and Luckmann Thomas. 1979. *Strukturen der Lebenswelt*, vol. 1. Frankfurt am Main: Suhrkamp.

Schwemmer, Oswald. 1987. *Handlung und Struktur. Zur Wissenschaftstheorie der Kulturwissenschaften*. Frankfurt am Main: Suhrkamp.

Smith, D.W., and A.L. Thomasson (eds.). 2005. *Phenomenology and philosophy of mind*. Oxford/New York: Clarendon Press/Oxford University Press.

Srubar, Ilja. 1988. *Kosmion. Die Genese der pragmatischen Lebensweltheorie von Alfred Schütz und ihr anthropologischer Hintergrund*. Frankfurt am Main: Suhrkamp.

Taylor, Charles. 1992. *Sources of the self: The making of the modern identity*. Cambridge: Cambridge University Press.

Varela Frances J., and J. Shear (eds.) 1999. The view from within: First-person approaches to the study of consciousness. *Journal of Consciousness Studies* 6(2–3): 1–14.

The Infinite Science of the Lifeworld: Steps Toward a Postfoundational Phenomenology

Giovanni Leghissa

Abstract This essay considers the problem of historicity together with the paradoxes of foundation. The concept of lifeworld has been studied mainly in order to clarify two orders of problems, the first concerning the relationship between history and the lifeworld, the second concerning the paradoxes contained in the *Krisis* regarding the possibility to attain a new foundation of transcendental phenomenology. Both issues must be taken into consideration before we question the possibility of a science of the lifeworld.

1 The Problem of Historicity and the Paradoxes of Foundation

The concept of lifeworld has been studied mainly in order to clarify two orders of problems, the first concerning the relationship between history and the lifeworld, the second concerning the paradoxes contained in the *Krisis* regarding the possibility to attain a new foundation of transcendental phenomenology. Both issues must be taken into consideration before we question the possibility of a science of the lifeworld.

On the one hand, starting from Ricoeur's and Landgrebe's pioneering essays,[1] much attention has been paid to the problem of historicity. The aim was not to show how Husserl, a philosopher with strong interests in mathematics, succeeded in structuring a coherent philosophy of history at the end of his career, after decades

[1] Ricoeur (1949), Landgrebe (1963, 1968, 1982).

G. Leghissa (✉)
Dipartimento di Filosofia e Scienza dell'educazione, University of Turin,
Via Sant'Ottavio, 20, Torino 10124, Italy
e-mail: giovanni.leghissa@unito.it

B. Babich and D. Ginev (eds.), *The Multidimensionality of Hermeneutic Phenomenology*, Contributions to Phenomenology 70, DOI 10.1007/978-3-319-01707-5_4,

spent in a deep and, to a certain extent, ambiguous dialog with Dilthey's historicism.[2] Much more interesting was the explanation of the strategies adopted by Husserl in developing a transcendental historicity, which can be seen as a consequence—and a deepening—of some aspects of Husserl's philosophy developed during the Twenties, that is, time consciousness, passive synthesis and intersubjectivity. It could be disappointing not to find in the late Husserl's philosophy a fulfilled project culminating in a philosophy of history, the scope of which should be the unfolding of all the intentional ties binding together human knowledge of the world and human self-positioning in the world itself. Nevertheless, we must be content with the result accomplished by Husserl—and with the well grounded supposition that this result was precisely what Husserl himself aimed at: the theory of the lifeworld has not been conceived to found a new concept of historicity, it is rather the tool needed by phenomenology in order to achieve a solid point of departure that makes a phenomenologically oriented reflection on both the natural and the scientific attitude possible.[3] (I will return to this point later).

On the other hand, it is the paradoxical structure that characterizes Husserl's path towards transcendental phenomenology that has drawn much attention. This paradoxical structure seems to be no obstacle, but rather a rhetoric necessity within the argument set out in the *Krisis*; more precisely, it seems to be intrinsically related to the impossibility of maintaining a distinction between the transcendental and the empirical subject (Hua VI 1954, 182–185).[4] What we discover after having accomplished the reduction, is nothing but the interweaving of ourselves and the general structure of the world, the latter considered as the sedimentation [*Sedimentierung*] of the never-ending human activity of sensemaking. Surely, according to what constitutes the main achievement of Husserl's philosophy since the *Prolegomena*, all sensemaking relies on logical structures, the validity of which does not depend on their being accomplished within a singular act of judgement. However, the kind of intentional analysis developed by phenomenology makes it necessary to make evident not only how their instantiation—for example within a singular act of judgement—constitutes the condition of possibility for the world to be known by any subject whatsoever, but also how the subject itself plays a role in constituting the way in which pieces of the world appear in time. The analysis of constitution shows that it is not possible to consider the constituting subject as if it were an external moment of the process of constitution itself. The phenomenological domain described by the term subjectivity is the result of a peculiar way of looking at the process of constitution, thanks to which we can point out the role played by the subject; but, as a result of the phenomenological analysis, what reveals itself to be peculiar to the domain described this way is the fact that the subject which constitutes the objects that appear in the flow of consciousness and confers them with identity and stability, at the same time constitutes itself and its own persistence as an identical pole of the subjective processes.

---

[2] Ströker (1987), 160–186.

[3] Carr (1974), Ströker (1987), 139–159.

[4] Ströker (1987), 115–138.

From this interconnection of constituting and being constituted stems the fact that it becomes more and more difficult to maintain the purity that should characterize the transcendental sphere of the ego—which is another way to articulate the aforementioned difficulty of maintaining the distinction between the transcendental and the empirical. This state of affairs becomes more evident as soon as we consider that the ego envisaged by phenomenological analysis is not simply the individual self, but rather a pole belonging to a more extended intentional process. This extension encompasses not only the intentional life of an individual consciousness, but also everything that links this individual consciousness to previous acts of constitution, accomplished by other subjects. Every present intentional act of mine is surrounded by a living horizon within which a complex set of relationships is still operating, and precisely these relationships form the context outside which no individual belief could be meaningful: every singular belief is thus interwoven not only with my past experience, but also with a wider texture of beliefs constituting the shared meaning of the world (Hua VI 1954, 152, 169). Therefore, the whole generative intersubjectivity, meant as the alternation of human generations within the course of history, cannot be detached from the idiosyncratic styles of perceiving and conceiving the world that are peculiar to each subject (Hua XI 1966, 218f).

In stressing the importance of the intersubjective structure of the lifeworld, Husserl underlines the fact that there is no access to any shared and common experience without our being affected bodily by other subjects: the mere presence of others, their actual being here, within my own *Umwelt*, or their having been present as a member of a historical community I still belong to, constitute the founding moment of any actual experience. In this way, phenomenology extends the reach of its own analysis to fields where the boundary of the egological sphere reveals not the mark of a closure, but rather what allows exteriority to affect the inner structure of the subject. Intersubjectivity, otherness, corporeality build up together a unit of themes the articulation of which shows how much dimension we could qualify as anthropological matters for the self-fashioning of the transcendental sphere.[5]

Thus, both Husserl and his interpreters have to cope with the uncomfortable paradox contained in the fact that the subject, called—or, better, summoned—to guarantee for the validity of every act of constitution, depends for its own constitution on what belongs to a sphere of exteriority affected by a relentless contingency. The efforts made by Husserl to sidestep this difficulty could be interpreted as the expression of the desire to save both the transcendental role played by the subject and the awareness that it is impossible to bestow on the transcendental ego the properties required by the agent supposed to guarantee for what classical philosophy called *Letztbegründung*. Regardless of Husserl's own intentions, one could say that phenomenology has simply substituted any form of *Letztbegründung* with the infinite operativity (*Fungierung*) of intersubjective life, or, which amounts to the same, with the uninterrupted work of constitution carried out by the human community. If one wants, on the contrary, to take into account the intentions that

[5] Zahavi (2003), 79–140, stresses the importance of the connection between time, corporeality, and intersubjectivity in order to understand the meaning of the life-world.

animated the Husserlian project, the main effort should consist in clarifying the reason why a philosophical account of reality should have its point of departure in the subject, as it is the case within the phenomenological version of the transcendental argument, and why precisely the argumentative structure of such an account marks the difference between phenomenology and other forms of discourse claiming for their part to shape through linguistic categories a persuasive description of human being in the world.

However, what is at stake here is not a defence of Husserl's attempts to maintain the 'purity' of the transcendental sphere. Much more important is the question concerning the benefit we can obtain in accepting the transformation that the transcendental stance undergoes within the *Krisis*. And, again, the way we deal with this transformation is important not to measure our fidelity to the Husserlian project of a transcendental phenomenology, but to give us the possibility of saving the epistemic goal pursued by phenomenology *even in absence of a transcendental foundation*.

This sounds like a strong claim and therefore requires a more extensive explanation. As we have seen, no matter whether we take into consideration the problem of history, or the question of what is needed to accomplish a transcendental foundation, we encounter the following questions: is it still possible to speak of a foundation if the transcendental subjectivity that should sustain the burden of it coincides with the process of constitution itself? Are we forced to see in the Husserlian solution nothing but a doomed attempt to elude the *fact* that the transcendental domain coincides with the domain of history—a domain affected by an irreducible finiteness? An answer should not be attempted too hastily. There are good reasons, in fact, for questioning any attempt to let phenomenology flow into hermeneutics too quickly, which would be precisely the result of an assumption that transcendentality and historicity coincide. Even if we feel uncomfortable with the term 'transcendental,' we must recognize that the rhetorical function it exerts cannot be renounced too easily: the transcendental stance is plainly the philosophical stance meant as the possibility of a critical attitude towards reality, as the uninterrupted work of questioning every realm of the world with no reference to an agency supposed to transcend the world itself. In Husserlian terms, it is the attitude we attain whenever we keep the world at a distance in order to reflect on our being engaged in the world, or, better, on our being engaged in that given network of relationships that make possible the validity of the common knowledge of the world (Hua VI 1954, 153).

Furthermore, forcing Husserl's late philosophy into any form of historicism yields little benefit in philosophical terms. The battle fought by Husserl against any form of psychologism—with historicism as one of its forms—belongs to the perennial effort made by philosophy to state that concepts like 'truth,' not to mention the concept of 'concept,' are *not* parts of a whole called 'nature,' or the 'empirical world' if you prefer. It could be more appropriate, then, to find the solution for the questions posed above precisely in turning the paradox into an epistemic resource.[6] As we shall see, what is aimed at in this way is not the validation of Husserl's efforts

[6] Even if with different aims, a similar solution has been suggested in Dodd (2004).

to safeguard the transcendental character of phenomenology at any rate. Much more important is the yield obtained by investing in a philosophical project the generative element of which allows for the renouncing of any foundational stance.[7] Postfoundationalism means here not only to assume that phenomenology achieves its significance as a philosophical project only through its capability of establishing *only* the conditions of possibility of validity, and by far *not* the conditions of possibility of experience; it also means that the possibility of a transcendental questioning depends on historical factors, an exhaustive analysis of which is, in principle, impossible. The coincidence between empirical and transcendental ego seen by Husserl as the paradox of transcendental phenomenology could be thus interpreted as the clue of the fact that a transcendental questioning remains a *possibility*, which each subject can always put into operation, yet without being able to justify the *necessity* of accomplishing it.

Truly, there is no phenomenological self-positioning in the world if the subject avoids carrying out the *epoché*, as it can be stated, for example, by looking at the sense of the controversy between Husserl and Heidegger during the late Twenties. In other words, a theoretical necessity, deeply rooted in the phenomenological mode of thought, brings Husserl to establish the *epoché* as the doorway to phenomenology. But this necessity, I argue, is largely rhetorical in nature and related to the difficulties phenomenology meets when it has to articulate, at the argumentative level, its own paradoxical position. As Husserl himself states, carrying out the *epoché* is an exercise (Hua VI 1954, 140; Hua VII 1959a, 279f; Hua VIII 1959b, 11f), a *praxis* which can not be justified only in theoretical terms; or, better, it is a performance carried out by a subject that was not otherwise able to manage its own position as a transcendental agent in the awareness that this position does not mean exteriority with respect to experience. Considered this way, the *epoché* loses its foundational character, but it maintains its role within a transcendental argument, the core of which is to be displaced differently from the traditional way of understanding it, that is, it is to be repeated and re-shaped as a figure of speech which performs the accomplishment of that stance which we can at best define through the adjective 'critical.'

The core of the argument I am putting forth here consists of the aforementioned fact that the position of the subject exercising the critique cannot be considered as the point of departure that allows access to an absolute exteriority. On the one hand, Husserl's phenomenology pursued as its deepest aim the destruction of any form of ingenuous realism, that is, of the idea that reality can be approached with immediacy—or, if we prefer to say the same in more poetic terms, phenomenology could be seen as the rebellion against the tyranny of the sense data. According to Husserl, this tyranny must be fought principally in the name of a methodological clarification of what informs the scientific construction of reality. On the other hand, the decision to bring scientific constructs back to the lifeworld in order to show that they depend

[7] Even if oriented towards issues that differ from the one discussed here, the question of a postfoundational attitude within the phenomenological project is also raised and discussed in Mensch (2001) and Drummond (1990).

on intersubjectivity is not motivated by the goal to substitute scientific realism with that form of realism which accompanies our ordinary experience of the world. The latter, in fact, knows its own way to conceal the role played by intersubjectivity as a presupposition for any human sensemaking, and therefore is not structurally different from the scientific attitude, if we look at both from the perspective of phenomenology. The problem is that the phenomenologist finds herself in a position that is rooted in the field of intesubjectivity as well: a position that allows looking at the processes of constitution from outside this field simply does not exist. This is the reason why critique pursued by phenomenology finds its ultimate motivation in an 'existential' choice, of which it is not possible to give an account thoroughly articulated in purely conceptual terms (Hua VI 1954, 60).

Along with the path just traced, where we suggested that the relationship between transcendentality and historicity should be interpreted as the starting point for the articulation of a postfoundational position, it becomes possible to put the question concerning the science of the lifeworld on a basis that should reach beyond the impasses met by Husserl. If we are not afraid of considering the coincidence between transcendental and empirical as the normal status of the phenomenological investigation whenever the question of foundation is at stake, and, furthermore, if we are ready to accept that anthropology is the final destiny of phenomenology, then to put in practice a science of the lifeworld should not mean to describe the structures of a realm the existence of which is needed to establish the system of all that exists; it could rather mean to describe the possibility of invariants within given forms of experiencing the world. In this sense, I will argue—and this constitutes the main goal of the present essay—that the place the humanities occupy within the encyclopaedia is the only possible site where we can find the accomplishment of a science of the lifeworld.

2 Positing the Lifeworld, or How to Supplement the Transcendental Position

In order to pursue this aim, it should be shown what characteristics could be attributed to the lifeworld. The first thing to be kept in mind is that the lifeworld is not a place you can go through, you can inhabit, or where you can smell aromas or touch objects. It is not the world of everyday life.[8] Of the latter it is possible to have a descriptive science, able to provide us with the main tools needed to build up a theory of social action. A theory of social action, to put it briefly, has to explain why human groups, or communities, act in a particular way under circumstances that are given each time. Of course, a social theory which does not confine itself to a simple collection of case studies can provide good—or bad—explanations for the reason why a motivation is not precisely a form of causation, that is, it can frame

[8] Grathoff (1989), 91–121.

its own conceptuality in such a way that it is possible to distinguish the realm of human actions from the realm of natural events. Nevertheless, a social theory is not forced to give any account of the constitutive operations that led to its own establishment. In other words, there is no need for it to practice a second order reflection on its own tools and operations. If this is the case, we are looking at a social theory that reveals an influence either by Husserlian phenomenology,[9] or by another philosophical stream.[10] Precisely the absence of such a second order reflection makes a given social theory not very different from a natural science, which is characterized, according to Husserl, by the claim of being able to account for the deep structure of reality without exposing the subjective operations that brought about the shaping of its own theoretical edifice. What the phenomenologist wants to provide, instead, is a theory of intersubjectivity, the function of which is to explain how the social construction of reality works—including that peculiar form of social construction we usually call 'scientific knowledge.' The point here is that Husserl needs to create a 'place' (a 'stage,' I would even say) where the intersubjective construction of reality is made evident—is made visible in absolute purity. Without stressing this 'purity,' Husserl would hardly be able to attribute a transcendental character to intersubjectivity itself. This is also the reason why Husserl emphasizes so much the evident character of the objects met in the lifeworld—an emphasis which renders the lifeworld a 'place' where ordinary events use to happen almost regularly, where we expect to meet a given set of objects and not others. In other words, in the lifeworld we are expected to encounter all the possible objects, whereas it is the inner legality governing this totality that determines the way we encounter them (Hua VI 1954, 142f).

No need to say that the lifeworld is a 'world' where even the philosopher would be out of place.[11] Philosophy is a very peculiar form of sensemaking, universal in its forms of expression. "Wonder" [*thaumazein*] was the name given by Greek philosophers to the attitude enabling us to question the mere 'givenness' of things, in order to grasp what makes the world to be what it is. A similar will to look at the things we find in the world without trusting the immediacy that characterizes tour experience and knowledge of them is surely to be found among all human cultural traditions. This ubiquity of the philosophical gaze at the world must be stressed in order to correct the conviction, expressed by Husserl, according to which philosophy is strictly tied with the cultural development of the Western cultural tradition. But, apart from any concern for the relationship between philosophy and the cultural context in which it arises, what must be put in evidence now is the fact that philosophy gives expression to the desire both to investigate the intimate nature of things, and to explore the possibility that things could be different from what they are. Strictly related to that desire is the awareness, which increases through methodical observation, that we cannot exclude the possibility that, sometimes,

---

[9] Luhmann as well as Bourdieu are the authors that could be mentioned in the present context (see in particular Bourdieu 1977 and Luhmann 1995).

[10] For example: Winch (1958) and Bloor (1976).

[11] Blumenberg (2010).

there could be a discrepancy between perception and belief, or between how we represent the world and how the world effectively is. In other words, it is peculiar to what we call philosophy to pose the question of truth. Further refined, this question gives rise to complex forms of knowledge, thanks to which, if properly institutionalized, it becomes possible not only to establish some statements supposed to be true about reality, but also to justify why they should be held for true. Yet if the lifeworld is the place where evidence characterizes the encounter with every object that can be found in it, then the philosophical questioning about the condition of possibility of truth remains pointless.

In the lifeworld the philosopher would know exactly what grounds those harmonious experiences that she, like any other human being, must presuppose in order to cope with any possible alteration of validity. Within our usual intercourse with objects it can occur—and in fact occurs—that harmony disappears and our experience ceases to be harmonious; both thanks to the corrections I can make, and thanks to communalization of what is perceivable, that is, thanks to the fact that the world I perceive is the same world that is perceived by others, whose communication of what they perceive completes what I perceive, it is always possible to re-establish a harmonious experience. But such a common understanding of the world, such a communalization reached through the uninterrupted re-agreement between what I perceive and what is, or has been perceived by others, is possible only because I 'know' that other human beings are related to same world: both the individual subject and others have a common horizon which encompasses the totality of all possible things to be met in the course of experience (Hua VI 1954, 167). What Husserl points at by describing the transcendental character of this horizon, is the fact that it operates as a presupposition within each form of human intercourse. Without having signed any contract, an agreement is always possible as regards the concordance of what is perceived within the world. This concordance (or 'harmony,' the English translation for *Einstimmigkeit*) is precisely the result of the awareness that the horizon is always operating as a presupposition held to be present both by myself and by others. But it would be misleading to think that this concordance, once achieved, coincides with the lifeworld; the lifeworld itself is to be found rather on the side of the horizon. In this sense, the lifeworld names the total system of multiplicities that makes possible the unity of human experience of the same world. This unity can even be missed on occasion, but what counts is the possibility of it—a possibility that, according to Husserl, accompanies every single act of perception. In itself, however, the horizon cannot become an object we can grasp or perceive (Hua VI 1954, 145f).[12] What can be perceived intuitively is, rather, the legal character of it, that is, the fact that the structure of the lifeworld is subject to the same inner logic found in every system of multiplicity.

However, if we state that the lifeworld is no place for philosophers, because within its realm there is no need for any philosophical investigation about the reason why the harmony of our experience can sometimes fail, what brought Husserl to establish the necessity of the life-world? Why did he choose to introduce

[12] Held (1991).

this concept? The answer is well known and sounds quite naive: because he needed a new point of departure for phenomenology. According to Husserl, this point of departure must maintain the purity that is proper to any realm supposed to play a transcendental role. Nevertheless, there is something that induces us to think that the lifeworld is more or less a phenomenologist's invention: both the average man and the scientist keep on living in the world and experiencing the world without any knowledge of the fact that their living in the world or experiencing it presupposes the lifeworld. Put like this, it does not sound as serious as it should. Yet, the invention I am talking about must meet the requirements that characterize the conceptual objects philosophers are used to dealing with. In fact, without the positioning (in the strong sense of *Setzung*) of the lifeworld, there would be nothing to justify the peculiar self-positioning of the philosopher in front of both the natural and the scientific attitude.[13] In this sense, I argue, the lifeworld is supposed to be the proxy of that transcendental position that should exist if the philosopher could hold it. We can at best explain this argument by turning it into the question about the origin of the lifeworld. At a first glance, the lifeworld, being the horizon of our experience, has no origin for its own. Not very differently from what happens with the law of logic investigated first in the *Prolegomena*, and then in *Formale und transzendentale Logik*, what we envisage when we meet the lifeworld is an object the property of which is to exist without any reference to space-time: omnitemporality [*Allzeitlichkeit*] is the expression used by Husserl to describe the way of existence peculiar to such logical entities. Nevertheless, Husserl did not withdraw from the task of phenomenologically investigating the fact that even logical forms and laws can be analysed in relation to their being intentioned by a subject. Precisely this investigation marked the gap between his position and Frege's from the beginning of their philosophical contention.[14] Furthermore, it is a radicalisation of Husserl's displacement of what defines the logical form within a theory of intentional consciousness that led him to add the genetic method to the previous static one. Thus, it is no surprise that the thematisation of the lifeworld, which made its appearance in correlation with the establishment of the genetic method, has been carried out by Husserl in such a way that the question about the origin of the lifeworld is far from being irrelevant.

But how can the lifeworld be originated on its part if its function is precisely to make every form of origination possible? It is important not to forget that the lifeworld is the result of a two-fold reduction accomplished in relation both to the natural attitude and to the scientific attitude.[15] Within the social exchange that characterizes our every day life, we meet other people and objects, or we deal with institutions; in none of these circumstances are we aware of the fact that every form

[13] Ströker (1987, p. 87f).

[14] Mohanty (1982), Willard (1984).

[15] What follows is an oversimplification, in the sense that I won't account for the steps necessary to pass from the *epoché* of the natural attitude to the reduction of the scientific one to the evidences we can seize in the life-world. For a more detailed account, see Dodd (2004), 175–206. As far as the peculiarity of the reduction of the scientific attitude is concerned, see also Kisiel (1970).

of human relationship, every meaningful experience of the external world, as well as every commitment we met to the various forms of sociability, formally or informally institutionalised, stems from our being involved in a broader web of intersubjective relationships (Hua VI 1954, 149). Precisely these relationships make it possible for our multifaceted exchange with the world to maintain a constant meaning. But they remain constantly concealed: we keep on acting individually as if we were detached from the broader context made up by intersubjectivity, a context that frames all our individual action. As far as the concealment of this universal frame is concerned, not far more different is the situation that characterizes the scientific attitude.[16] In the case we decide to investigate the world from a scientific perspective, we can then attain a comprehensive knowledge of the world from which we would be precluded if we were still involved in the natural attitude.[17] But each member of the scientific community works on her own research program, closely with other colleagues, without questioning the operations needed to attain scientific knowledge itself. In Husserl's terms, the scientific interest that motivates each member of the scientific community is such as to induce a forgiveness of the intersubjective operations that brought both to the establishment of science, meant as a peculiar way of approaching reality, and to the construction of singular scientific theories or doctrines, meant as given instantiation of what science is (Hua VI 1954, 134).[18] In order to establish the visibility of those intentional ties that connect both our every day experience of the world and the systematic knowledge of the world furnished by science to the web of intersubjective relations, it is thus necessary, according to Husserl, to perform the phenomenological reduction—or, better, a twofold one: the first in order to bracket the natural attitude, the second in order to disconnect the scientific one. As a consequence, the lifeworld is what remains, is what we can see operating as system of intersubjective relations presupposed both by the natural and by the scientific attitude. What must be stressed in the present context is the fact that the subject performing the reduction can only be the phenomenologically oriented philosopher, with 'philosopher' not referring to a particularly gifted person: the philosopher in question is simply someone who shares the same word in its facticity with others, and can, on occasion, decide to investigate the world scientifically, but presently makes up her mind to perform the reduction in order to disclose the operativity [*Fungierung*] of intersubjectivity. Truly, we have to do here with an individual position, which, in principle, everybody can partake

[16] Held (1991) has made the point very clear by defining the scientific attitude as a second-level natural attitude.

[17] As stated by several passages from his work, Husserl nor put in doubt the achievements of scientific knowledge, neither was willing to disrupt the idea that scientific knowledge is the only one giving us the possibility to access the 'true' world. Whether the Husserlian conception of the relationship between scientific knowledge and truth can still be maintained, is an issue we cannot address here. On the subject, see Hacking (1992).

[18] Fleck (1979) and Bourdieu (2001) not only move in the same direction as Husserl, but also show how productive a phenomenologically oriented sociology of knowledge could be; nevertheless, these contributions still find scarce recognition within Anglo-Saxon sociology of knowledge (even if Fleck's work was issued in 1935).

in—and such a possibility guarantees for the fact that this position maintains affinity with the universality supposed to characterize the transcendental subjectivity. But this affinity cannot efface the irreducible contingency affecting the position of the subject that wants to perform the reduction.

Thus, it seems to be justified to claim that the lifeworld is a sort of 'invention' by phenomenologists in order to make the phenomenological stance possible. The latter reveals to be one possibility, one among others, of looking at the world; other attitudes, which are motivated by other interests, can underlie different ways of looking at the world, which, of course, still remains the same world we all have in common as human beings. As a consequence, the contingent character of the motivation that underlies the phenomenological attitude thoroughly affects the structure of the lifeworld as well: if taken as the result of a peculiar way of looking at the role played by intersubjectivity in order to better understand how the uninterrupted work of sensemaking come to existence, the lifeworld seems to be nothing but the result of the individually chosen positional act that produces its visibility. What stems from this contingency is that the lifeworld is to be understood more as an archive of some invariant patterns than as the totality encompassing all what exists within human experience. If the lifeworld is the place where these invariants are to be found, and if phenomenology is to be the science of the lifeworld, we must now identify the peculiarities of the phenomenological inquiry with respect to other disciplines that claim, on their part, to be better appointed to carry out the same inquiry.

3 The Encyclopaedia of the Humanities as Infinite Description of the Lifeworld

As a result of the previous analysis, what we want to achieve now is the possibility of a science of the lifeworld that coincides with the definition of the anthropological bases needed to understand the most general pattern of human behaviour. This science could claim to be still understood as 'philosophy' because of the fact that it coincides neither with the natural nor with the scientific attitude. What it loses, nevertheless, is the adjective 'transcendental,' of which Husserl is so fond. Hans Blumenberg has spent a lot of philosophical energy showing that the obsession with the purity that should characterise the phenomenological discourse prevented Husserl from allowing phenomenology to turn itself into a philosophical anthropology.[19] If phenomenology remains a transcendental discourse, than purity is safeguarded. However, this does not offer a great advantage: a 'pure' science of the lifeworld seems to have a limited descriptive power; as Husserl knew, in fact, psychology (as well as history) can offer better accounts of how human beings act in the world. Furthermore, a 'pure' science of the lifeworld seems to have a limited prescriptive power as well: as we have already pointed out above, both the scientist

[19] Blumenberg (2006).

and the human beings who find their way about in the world by taking advantage from that shared knowledge we call 'common sense' are not aware of the evidence that informs the lifeworld, and nevertheless they keep on doing what they do seemingly very well (whereas 'very well' means 'in an adaptive manner,' or 'in a satisfying manner from an adaptive point of view'). The result of Blumenberg's analysis is that renouncing a transcendental stance is a benefit, and not a loss—a benefit I would describe as an injection of anthropological flesh into the skeleton of the lifeworld.

We cannot follow the whole argument made by Blumenberg in order to demonstrate both why Husserl was not able to accept an anthropological stance, and why an anthropological turn within phenomenology could be seen as an improvement; nevertheless, one point should not be overlooked, and this point can be seen as a development of Blumenberg's whole argument. The reason why phenomenology needs to be turned into anthropology, which obviously has as a result the loss of transcendentality, is tightly related with the transcendental stance itself. If the transcendental subject reflects upon its own self-positioning, and takes into account the finiteness that marks its position, the consequence that must be drawn is that the subject, precisely for transcendental reasons, is subdued to this finiteness, or, to put it differently, that the subject must ascribe to its own finiteness a transcendental character. The finiteness we are dealing with here is not a metaphorical designation for our being mortal, it is rather the main property of our being related to the network of the intentionally structured sensemaking processes that constitute the common world and are all together called 'intersubjectivity' by the phenomenologist: the subject can access this network only through a given number of entrances, each of which, in part, is shaped in conformity with the social and communicative competences achieved by the subject itself. Thus, this finiteness is the condition of possibility of our being in the world. Questioning the functioning of the relationship between the aforementioned entrances and competences can precisely constitute the task of a philosophical investigation of intersubjectivity. This investigation will maintain an affinity with the transcendental analysis of the lifeworld in order to define itself as philosophical, but it will at the same time spread through the whole complex of disciplines gathered together under the title of Humanities, and will therefore coincide, at least in part, with an analysis thoroughly anthropological in character.

The question is whether such an anthropologically oriented philosophy still maintains a relationship with the phenomenological project. By thematizing the lifeworld as an object we can describe and analyse, that is, as an object we can experience, the phenomenologist is not claiming to be able to put herself in a position that is external with respect to the lifeworld itself, as the latter is a totality that encompasses both the evidences presupposed by the subjects who act within the natural or the scientific attitude, and the phenomenologist's self-positioning that makes the coming-to-light of the lifeworld possible. A phenomenological analysis of the lifeworld is simply the result of a different way of looking at the lifeworld, a way that posits it as the ground [*Boden*] for all what is presupposed by any subject that looks at the world from a point of view different from the phenomenological one. We can adopt this solution because we have already stated that the lifeworld can emerge and become visible only if the phenomenologist's glance performs its emergence and its visibility. Should Husserl have formulated two different concepts, one for the lifeworld meant as the totality that encompasses both the subjects who experience the world and their reflection upon the world, and another one for the lifeworld meant as the ground of all validity?[20] If we answer in the affirmative, we run the risk of missing the productive character of the paradox enunciated by Husserl, namely the paradox of a twofold lifeworld, which splits up in order to make possible for the subject to reflect upon that totality in which it still remains included. And the productivity of this paradox can be shown precisely in the moment in which we deal with the necessity to submit the lifeworld to analysis.

Husserl himself can figure out the analysis we are talking about only in the form of an anthropological analysis. As is to be expected, he makes enormous efforts to avoid drawing all the consequences implied by such an analysis. However, it is not without significance that he speaks of invariants in order to define the object of this analysis. These invariants should constitute the object of investigation of an a priori anthropology, which is to be understood as an ontology of the world as well (Hua XXXIX 2008, 57). Without this a priori anthropology the various ways of sensemaking, differing from each other both historically and geographically could not be perceived as variations of the same world. The world that is supposed to remain the same is the world we are familiar with, it is the world that, thanks to its own presence and consistence shapes our habits, or, better, makes the emergence of habits possible. Now, the fact that the never changing structure of the world is strictly interwoven with the possibility for human habits to change both from time to time, and from region to region seems to be the presupposition Husserl needs to state the identity of an ontology of the world with an a priori anthropology. When he asks which is the main characteristic of mankind, that is, which is the peculiarity all human beings must share in order to understand themselves as human beings, the answer is: their historicity (Hua XXXIX 2008, 344). No reader of the *Krisis* will be surprised by this answer. What could be—if not surprising, perhaps a little disturbing—is the way Husserl depicts this historicity. What Husserl points at, in fact, is the rootedness of each individual in its community, where it was born, has acquired acquaintance with the world, and has gained the opportunity to turn the world itself into the mute horizon both of human experience in general, and of its own experience in particular. The same holds for human groups, no matter whether their dimension is small or big as in the case of a nation. Historicity, in this case, means to be rooted in a country that allows a strong form of identification—a country, therefore, which can be understood in terms of homeland. It is in our homeland that the world becomes familiar to us. In fact, there would be no familiarity with the world without that form of acquaintance with shared values and shared forms of life we can gain only when we participate in the common work that is necessary to guarantee the prosecution of the tradition we belong to. Thus, historicity coincides

[20]The problems concerning the plurivocity of Husserl's notion of life-world are discussed in Claesges (1972).

with the persistence of a generative tradition, with the power possessed by a tradition to live on, to reproduce itself and overwhelm the opposite power of time to destroy the traces human beings have left on the earth. Following this train of thought, we can find in the third appendix of the *Krisis*, namely in the text Fink titled *The Origin of Geometry*, a historicity that coincides with the capability to leave traces, whereas the vitality of a tradition consists of the capability to institutionalize the way in which these traces are both reproduced and interpreted (Hua VI 1954, 371).

The argumentative strategy Husserl adopts to describe the connection between the rootedness in a *Heimat* and the historicity that marks the essence of mankind could recall the similar tones we find in Heidegger's commentaries of Hölderlin's hymns "Germanien" and "Der Rhein."[21] But it would be misleading to follow the superficial resemblance of tones, even if the temptation to do so could seem appealing (especially if we consider that the text where Husserl speaks of human historicity is more or less coeval with Heidegger's lecture courses). Notwithstanding his insisting on the *völkisch* dimension that is taken for characteristic of any human rootedness in a country, Husserl is able to draw a connection between the feeling of belonging to one country and the human capability to cross boundaries and to perceive the whole of humanity as an extension of our homeland: even if the way I perceive the world is biased by the manner in which the human group I belong to has always perceived it, nevertheless what I perceive is the same world I share with the rest of humanity. Precisely the possibility to turn back to the unique world, meant as the source of all objectivity, allows me also to perceive the unity encompassing all different cultural traditions (Hua XXXIX 2008, 340). In this sense, we must recognize how deep Husserl's commitment to the tradition of the Enlightenment was, a statement that does not apply to Heidegger's philosophical position. At the same time, we must recognize how strong, even in Husserl's case, has been the temptation, which never ceased to haunt the European tradition of the Enlightenment, to identify the history of Europe with the most successful example of a unitary cultural tradition, which would have revealed itself capable of overwhelming its own internal differences (Hua XXXIX 2008, 349).[22]

Now, regardless the rhetoric of 'belonging' that affects Husserl's description of human historicity, what must be underlined here is the fact that Husserl's late reflections are able to provide a convincing account of the reason why historicity is to be considered as the ultimate horizon of human experience.[23] Above all, it must

[21] Heidegger (1999).

[22] Derrida (1991).

[23] It is worth taking notice that historicity, according to Husserl, constitutes even the ultimate horizon of animal life in general. If the way a subject can experience historicity depends on its rootedness in a territory, then an experience of the world that can be defined as historical cannot be denied with respect to animals. However, animals are not able to generate a tradition, which remains a peculiarity of human beings. On the other hand, the source of the human capability to generate a tradition, that is, to make sense of the experience of the world we all share as human beings, is deeply rooted in a biologically based characteristic, namely in the fact that we can produce signs by using the expressive potential of our corporeality (Hua XXXIX 2008, 344–346). Even if confined to a footnote, this clue of how complex Husserl's analysis of historicity is seems to me no less important than the main objective I want to pursue within the present essay.

be stressed that the horizon that is at stake here is to be referred not only to the human *Umwelt*, namely to all that surrounds human experience in both geographical and cultural terms (belonging to a territory, speaking a language, sharing a given set of historically determined values, and so on). If it were so, nothing but the empirical dimension of our being rooted in the world would be affected by historicity. Husserl seems also to be tempted to confer a historical character even to the transcendental dimension that makes possible both the emergence and the formation of objectivity. Indeed, he claims that every act of knowledge, every form of knowledge, every formation bearing in itself the result of an act of knowledge (*Gebilde* is the expression used by Husserl) is motivated. This means that knowledge, not differently from any other human activity, is part of a tradition, or, which is the same, is rooted in the unity of human history. Husserl makes this statement in a context introduced by a question concerning the absence of presuppositions that is supposed to characterise knowledge (Hua XXIX 1993, 343). It is a common-sense statement that the absence of presuppositions is precisely the main characteristic of any scientific undertaking. But here we are taught not only that knowledge does not occur in absence of presuppositions, but also that its own historicity is precisely that which knowledge presupposes at a deepest level. Husserl goes on with his argument as follows. Acts that confer a meaning on an object, and do so in a way that raises this meaning to the level of universal validity, cannot be detached from the historical horizon to which they belong, nor from the objects the validity of which they attempt to establish. Notwithstanding its being generated within the horizon of history, this established validity of objects does not cease to inform the complex of scientific knowledge. If we operate at the level of the latter, as scientists or so, we can be forgetful of the historical process that generated it. But if we continue to adhere to the phenomenological way of looking at scientific knowledge, we cannot overlook the dependence of acts conferring objectivity upon the broader context of intersubjectivity. Thus, the objectivity possessed by the objects of the lifeworld is thoroughly historical (Hua XXIX 1993, 347f).

Some important consequences can be drawn from the relationship between historicity and the self-positioning of phenomenology as a science of the lifeworld. On the one hand, phenomenology can turn itself into an ontology of the lifeworld—or, an a priori anthropology—without losing its transcendental character only if the task of this ontology consists of an investigation of the invariants that mark the human being in the world. Yet, if the main invariant, to which all other invariants are to be traced back, is historicity, then there is no place at all for an investigation that is supposed to differ essentially from the one carried out by Humanities. If we consider the research project that informs the humanities in general, we potentially gain a complete description of the different ways of inhabiting the world, a description that includes even those invariants that are to be found within every cultural tradition. This description may be a finite one at a given moment of its own internal development, but it is virtually infinite in the sense that the horizon within which it takes place is precisely the infinite horizon of human history. On the other hand, in the present context an important role is played by the relationship investigated by phenomenology between the realm of logical forms and the extent to which they suspend their peculiar onmitemporality to become part of the structure of meaning the subject needs in order to build up a coherent and consistent account of the world. If this relationship forms one of the most important issues of phenomenological investigation (if not the most important one), and if the result of this investigation, as we have seen above, brings us to acknowledge the insurmountable historicity of those processes that lead to the formation and establishment of objectivity, then it turns out to be inevitable to suppress any difference between phenomenology and a critical epistemology. The aim of the latter is precisely to make evident which historical, cultural, and political biases must be taken into consideration to explain the emergence and the sedimentation of any form of knowledge.[24]

Are the consequences just drawn above a strained interpretation of some isolated passages of Husserl's late reflection? Probably not, however, if we want to be consistent with respect to the main goal we are pursuing here, namely to explore the possibility of a postfoundational phenomenology, what we are concerned with should not be biased by this question. The interconnection between the empirical and the transcendental subject, which goes through Husserl's reflection on the lifeworld (Hua VI 1954, 190, 214, 268), is the point of departure we need to justify our attempt at moving from a 'pure' phenomenology to a phenomenologically-oriented stance that understands itself as a discourse that cannot be detached from the field occupied by the Humanities. The transcendental subject is the last instance to which the process of foundation must terminate; the final result of the latter is the discovery that every sense-formation [*Sinnbildung*] depends on intersubjectivity, and this is the reason why Husserl insists on emphasizing the fact that even the objectivity that characterizes scientific knowledge is subject-relative. A part of this discovery is the historical nature of the problems discussed by phenomenology as a science of the lifeworld (Hua VI 1954, 378). Husserl was surely close to claiming that only by taking seriously the historicity of the lifeworld itself would it have been possible to achieve the final scope promised by a radical phenomenological foundation. What he was not ready to recognize, however, was that this radical foundation should have been understood in terms of a thorough historicisation of subjectivity as well.[25] As a clue of the resistance offered by Husserl against this historicisation, we should look at the way in which he speaks, in some passages, of the invariants that are also constitutive of human experience of the world. Differently from the above quoted passages where the anthropological nature of these invariants has been stated very clearly, Husserl tries sometimes to define these invariants in opposition to what could be the result of the efforts made by a historian in order to

[24] A good example of what could be understood as a sound and convincing accomplishment of a critical epistemology can be found in Foucault's work (especially in Foucault 1972, where the interweaving of empirical and transcendental within the production of scientific discursivity has been made explicit as an object of investigation). It is also worth mentioning the relationship between Foucault's philosophy and the way in which Cavaillès took up and modified Husserlian phenomenology: in doing so, Cavaillès prepared the ground necessary to every further development along the path we are suggesting here (see Cavaillès 1947).

[25] Ströker (1993), 165–205.

describe the process of sense-sedimentation [*Sinnsedimentierung*]. Husserl gives indeed the impression that it is always possible to distinguish the empirical work of the historian from a transcendental analysis of the invariant structures of the historical world. The meaning of these structures could be caught and perceived with evidence thanks to a reflection on the historical material the aim of which is to purify it from its empirical character (Hua XXIX 1993, 241). Once again, the obsession with 'purity' seems to be Husserl's main concern.

However, for the argument we are putting forth here it is above all worth drawing our attention to how Husserl understands the method we need to attain the knowledge of the invariants that underlie any form of given experience of the world. Husserl simply refers to the already well-functioning method of free variation. The method has been available within phenomenology for a long time and has been applied whenever the old question of the universal needed to undergo a phenomenological investigation. As in the previous cases, Husserl attempts to preserve the purity of the essence attained thanks to the free variation; further, he does not seem to deviate from the conviction that the pure essence of the singular object starting from which the variation begins is nothing but what the object has always included as a constitutive part of its own object-like character. On one hand, we could reproach Husserl for having not discussed the difference (if any) between the free variation and the classical method of induction. On the other hand, we could observe that the method is caught in a vicious circle, in the sense that certain knowledge of the universal is already presupposed whenever we choose a given singular object and decide to 'extract' the essence it contains by applying the method of free variation. At any rate, we cannot really address here the objections the method of free variation could easily undergo. In order to do this, we should engage in a deeper discussion of the whole issue. In the present context, the only thing we should not overlook is the coincidence between the phenomenological method of free variation and the comparison between similar states of affairs that can be found within the praxis of the Humanities. The world that surrounds me refers to an infinite horizon that includes both events and processes that occurred in the past, and events and processes that are taking place now elsewhere. At the same time, I also know that it is always possible to refer to the general structure of the lifeworld (Hua VI 1954, 142). The latter, as we have seen above, makes both the cultural and the historical crossing possible. What I perceive along with those crossings can be recorded, measured, tested, analysed, accounted for: in other words, scientific knowledge of both historical and cultural differences is possible. The epistemic basis for any scientific recording of and accounting for cultural and historical data is given by my capability to put forth a continuum that begins with the already-known and moves to the unknown all the possible forms of historical existence. Husserl claims that this modalisation of my own horizon is not completely free, in the sense that it is still subdued to the spatial and temporal biases that make up every human experience of the world (Hua XXIX 1993, 63–65). We can easily agree with this claim without evoking once again the aforementioned difficulties related to the method of free variation. But there is a further step to be taken, namely to notice that a methodology based on the progressive extension of horizons has been constituting the core of any form of scientific knowledge since the humanities began to reflect on their own epistemic status.

A clear awareness of this issue can be found in the discussion about the reliability of our historical sources concerning ancient Rome. This discussion, which took place during the first part of the eighteenth century, came before the establishment of a self-confident historical discipline, but it is worth mentioning because it clearly shows that the historical consciousness, once raised, bears with itself the necessity to cope with questions of methodological nature. Some decades later, at the beginning of the nineteenth century, on German soil, which has been understood as the cradle of a rigorous philology since that moment, we encounter a Friedrich August Wolf, who was able to state very clearly the hypothetical nature of all assertions made within the historical reconstruction of the past—a hypothetical nature that does not imply a diminished rigour. But it is due to August Boeckh if we can better grasp the fact that any historical knowledge rests on the possibility to modify our own horizon until we reach a sound understanding of what makes up the peculiarity of other cultures and other historical ages. "*Erkenntnis des Erkannten*" ("knowing of the known") was the formula uttered by Boeckh in order to make clear the necessary relationship between the point of departure of scientific knowledge, that is, the living horizon within which I act as a subject, and the alien world that must be submitted to investigation. Not different from the modalisation of the horizon within which the subject of knowledge operates is the procedure adopted by the anthropologist. The anthropological discipline, born officially in the second part of the nineteenth century, is in fact based on a method that can be seen as an application of the philologist's method in a field where we cannot rely on written sources as far as the access to otherness is concerned.

The examples that could be mentioned here are innumerable—and if I do not go into details, it is not because the length of the present contribution would not allow it: more simply, it is the whole history of the Humanities during the modern age that should be taken into consideration here.[26] But the point that deserves our attention remains the aforementioned similarity between the methodology suggested by Husserl in order to capture the invariants of the lifeworld and the method effectively applied within the Humanities in order to achieve the necessary acquaintance with alien forms of life.

The last—and conclusive—point of our investigation concerns the question whether an autonomous place for a philosophical questioning can be maintained. The proposed postfoundational phenomenology could be read as a suggestion to merge any philosophical inquiry with the encyclopaedia of the Humanities. But there is a specific function that is still to be accomplished by philosophy.

[26] In Leghissa (2007), there is a more detailed account of the epistemic structure of the Humanities with special reference to classical philology, which has been the first discipline among the Humanities to develop the methodological awareness we are dealing with here.

As Husserl knew very well, scientific disciplines do not always bind together the theory they attempt to shape of a specific field with the epistemology the same theory rests on (Hua XVII 1974, 8, 32). To elucidate the various forms taken by the relation between the two would be precisely the task of a phenomenologically-oriented philosophy. Such a task would not coincide with a 'transcendental foundation' of the sciences of the lifeworld, as the phenomenologist's glance is entirely internal to the level where the description of the lifeworld takes place. It would rather mean to turn the phenomenological attitude from a 'reflection above' the validity of the world into a 'reflection within' the historical processes that inform the intesubjective constitution of the world. Put in this way, we can bestow a new meaning on Husserl's seemingly obscure remarks on the *einströmen*. Under this notion, which is present both in the *Krisis* (Hua VI 1954, 115, 213) and in the related texts (Hua XXIX 1993, 77–83), Husserl referred to the fact that phenomenology, as well as any other form of theory that brings in itself the awareness of the subjective-relative character of knowledge, 'flows into' the lifeworld it reflects on. Such a 'flowing into,' or *einströmen*, can be well understood as the form phenomenology assumes in the moment in which it decides to accompany the efforts made by human beings when they keep the world at a distance in order to gain a critical attitude towards it. Not different from other forms of critical theory, but perhaps better equipped than they are as far the exercise of distantiation is concerned, phenomenology can then present itself as a praxis, more precisely as that specific form of exercise that is required whenever we have to deal with the paradox generated by the intersubjective constitution of the world.

References

References to Husserl's works are enclosed in parentheses (round brackets) and embedded in the text. First comes "Hua," which is the abbreviation of the German Complete Edition (*Husserliana—Gesammelte Werke*), and then the volume number followed by the page number.

Hua VI. 1954. *Die Krisis der europäischen Wissenschaften und die transzendentale Phänomenologie*. Herausgegeben vonW. Biemel. Den Haag: Nijhoff.

Hua VII. 1959a. *Erste Philosophie (1923–1924). 1. Kritische Ideengeschichte*. Herausgegeben von R. Boehm. Den Haag: Nijhoff.

Hua VIII. 1959b. *Erste Philosophie (1923–1924). 2. Theorie der phänomenologischen Reduktion*. Herausgegeben von R. Boehm. Den Haag: Nijhoff.

Hua XI. 1966. *Analysen zur passiven Synthesis: Aus Vorlesungs- und Forschungsmanuskripten, 1918–1926*. Herausgegeben von M. Fleischer. Den Haag: Nijhoff.

Hua XVII. 1974. *Formale und transzendentale Logik: Versuch einer Kritik der logischen Vernunft*. Herausgegeben von P. Janssen. Den Haag: Nijhoff.

Hua XXIX. 1993. *Die Krisis der europäischen Wissenschaften und die transzendentale Phänomenologie. Ergänzungsband. Texte aus dem Nachlass 1934–1937*. Herausgegeben von R.N. Smid. Dordrecht-Boston-London: Kluwer.

Hua XXXIX. 2008. *Die Lebenswelt. Auslegungen der vorgegebenen Welt und ihrer Konstitution. Texte aus dem Nachlass (1916–1937)*. Herausgegeben von R. Howa. Dordrecht: Springer.

Further Literature

Bloor, D. 1976. *Knowledge and social imagery*. Chicago: The University of Chicago Press.

Blumenberg, H. 2006. *Beschreibung des Menschen. Aus dem Nachlaß herausgegeben von M. Sommer*. Frankfurt am Main: Suhkamp.

Blumenberg, H. 2010. *Theorie der Lebenswelt*. Frankfurt am Main: Suhrkamp.

Bourdieu, P. 1977. *Outline of a theory of practice*. Cambridge: Cambridge University Press.

Bourdieu, P. 2001. *Science de la science et reflexivité*. Paris: Editions Raison d'agir.

Carr, D. 1974. *Phenomenology and the problem of history: A study of Husserl's transcendental philosophy*. Evanston: Northwestern University Press.

Cavaillès, J. 1947. *Sur la logique et la théorie de la science*. Paris: Presses Universitaire de France.

Claesges, U. 1972. Zweideutigkeiten in Husserls Lebenswelt-Begriff. In *Perspektiven transzendentalphänomenologischer Forschung*, eds. U. Claesges and K. Held, 85–101. Den Haag: Nijhoff.

Derrida, J. 1991. *L'autre cap, suivi de la démocratie ajournée*. Paris: Minuit.

Dodd, J. 2004. *Crisis and reflection. An essay on Husserl's crisis of the European sciences*. Dordrecht: Kluwer.

Drummond, J.J. 1990. *Husserlian intentionality and non-foundational realism*. Dordrecht: Kluwer.

Fleck, L. 1979. In *Genesis and development of a scientific fact*, eds. Trenn, T.J. and R.K. Merton. Chicago/London: The University of Chicago Press.

Foucault, M. 1972. *The Archeology of Knowledge*. Trans. A.M. Sheridan Smith. New York: Pantheon Books.

Grathoff, R. 1989. *Milieu und Lebenswelt. Einführung in die phänomenologische Soziologie und die sozial-phänomenologische Forschung*. Frankfurt am Main: Suhrkamp.

Hacking, I. 1992. The self-vindication of the laboratory sciences. In *Science as practice and culture*, ed. A. Pickering, 29–64. Chicago/London: The University of Chicago Press.

Heidegger, M. 1999. In *Hölderlins Hymnen "Germanien" und "Der Rhein" (Wintersemester 1934/35)*, ed. S. Ziegler, (GA Bd. 39). Frankfurt am Main: Klostermann.

Held, K. 1991. Husserls neue Einführung in die Philosophie: der Begriff der Lebenswelt. In *Lebenswelt und Wissenschaft. Studien zum Verhältnis von Phänomenologie und Wissenschaftstheorie*, ed. C.F. Gethmann, 79–113. Bonn: Bouvier.

Kisiel, T.J. 1970. Phenomenology as the science of science. In *Phenomenology and the natural sciences*, eds. J.J. Kockelmans and T.J. Kisiel, 5–44. Evanston: Northwestern University Press.

Landgrebe, L. 1963. *Der Weg der Phänomenologie*. Gütersloh: Mohn.

Landgrebe, L. 1968. *Phänomenologie und Geschichte*. Gütersloh: Mohn.

Landgrebe, L. 1982. *Faktizität und Individuation. Studien zu den Grundfragen der Phänomenologie*. Hamburg: Meiner.

Leghissa, L. 2007. *Incorporare l'antico. Filologia classica e invenzione della modernità*. Milan: Mimesis.

Luhmann, N. 1995. *Social systems*. Stanford: Stanford University Press.

Mensch, J.R. 2001. *Postfoundational phenomenology: Husserlian reflection on presence and embodiment*. University Park: Pennsilvania State University Press.

Mohanty, J.N. 1982. *Husserl and Frege*. Bloomington: Indiana University Press.

Ricoeur, P. 1949. Husserl et le sens de l'histoire. In *Revue de Métaphysique et de Morale* 54: 280–316. Now in Ricoeur. 2004. *A l'école de la phénoménologie*, 19–64. Paris: Vrin.

Ströker, E. 1987. *Phänomenologische Studien*. Frankfurt am Main: Klostermann.

Ströker, E. 1993. *Husserl's transcendental phenomenology*. Stanford: Stanford University Press.

Willard, D. 1984. *Logic and the objectivity of knowledge. A study in Husserl's early philosophy*. Athens: Ohio University Press.

Winch, P. 1958. *The idea of a social science*. London: Routledge & Kegan Paul.

Zahavi, D. 2003. *Husserl's phenomenology*, 79–140. Stanford: Stanford University Press.

Hermeneutics in the Field: The Philosophy of Geology

Robert Frodeman

Abstract Geology has had a marginal place within the philosophy of science; its processes and results have not matched our traditional ideas concerning the nature and outcomes of scientific reasoning. This is a reflection of the fact that philosophy of science has been, with few exceptions, implicitly or explicitly the philosophy of physics, and more generally the philosophy of lab science. In actuality, geological reasoning provides a rich and realistic account of the power and limitations of scientific reasoning. It also highlights the hermeneutic and historical nature of reasoning, scientific or otherwise, and the neglected kinship between reasoning in the sciences and the humanities.

In what follows I argue for the importance of the field sciences as a model for understanding the nature of scientific reasoning. This is in opposition to the long-held *disciplinary* bias across the sciences, where truth has been defined in terms of that which can be walled off from outside forces. In seeking to shift our model of reasoning from what obtains the lab to what happens in the great world beyond we set aside some of the arrogance and presumption that has attached to scientific knowledge.[1] This is not done with the goal of refuting science or reducing scientific reasoning to simply one among many different outlooks or worldviews. Rather, I seek to promote an Aristotelian mean, where science is understood as a vibrant but far from unequivocal way of gaining knowledge about ourselves and the world. The upshot is that scientific reasoning comes to be seen as a humanistic enterprise,

[1] I have made similar arguments concerning the nature of philosophy, which also has been excessively disciplined, and which needs greater exposure to the field perspectives (e.g., Frodeman 2010, 2013).

R. Frodeman (✉)
CSID/Philosophy, University of North Texas, 1155 Union Circle #310920,
Denton TX 76203-0920, USA
e-mail: frodeman@unt.edu

B. Babich and D. Ginev (eds.), *The Multidimensionality of Hermeneutic Phenomenology*, Contributions to Phenomenology 70, DOI 10.1007/978-3-319-01707-5_5,

elevating our conception of the humanities at the same time we deflate some of our presumptions concerning science.

The great creative leap of Greek thought was to imagine the possibility of *logos*—that the world and its occurrences were not random. But order can take many forms. The Greeks came to focus upon a specific type of *logos*, directing the rational gaze upward, toward the stars, and inward, toward a pure mental order. They abstracted from the world as we find it, or rather saw this world, our world, as merely a dim reflection of another perfect world. The Greek search for *logos* thus became a quest for a very particular type of order—one that was distant, regular, immutable, and certain.

The motivation behind this is clear enough. Our knowledge of things in the sensible world is constantly changing, and thus always questionable. Temporality can be seen as the enemy of rationality, rendering every truth claim inconclusive and suspect. Thus the message above the portal to Plato's Academy: "Let no one deficient in geometry enter." Only in the realm of thought, and in the celestial sphere, a region thought to be beyond material corruption, would we find the proper conditions for truth.

This notion of rationality has placed us in a peculiar situation. It has given us an unprecedented control over the world; we have conquered many diseases, and now can satisfy our desires at the touch of a button. But it has also deformed our personal and social relations, prompting a culture-wide nihilism for those aspects of life (ethics, politics, metaphysics, aesthetics, and religion) that cannot be parameterized and controlled and which are immersed in time and contingency.

The effects of our traditional definition of *logos* upon geology—the discipline ostensibly concerned with understanding the Earth—are also clear. The limitations of what counts as "understanding the Earth" can be revealed, in part, by a comparison with medicine. What would we think of an "understanding of health" that wasn't concerned with actually making people healthy? For a normative element to knowledge is, or should be, as fundamental to the Earth sciences as it is to medicine. But rather than moving toward a grand synthesis, a geo-logos that would end in geo-ethics, geologists have been trained to search for lawlike generalizations.

These presumptions about rationality have influenced not only how the Earth sciences are constituted, but also their very origins. The study of the heavens has been pursued for over two millennia, while the systematic study of the Earth is barely 200 years old. The Earth has been thought to be beneath us, a subject too earthy to be worthy of serious attention. In contrast to the clarity of mathematics, or the mathematicized heavens, the Earth was inaccessible, impenetrable, and subject to sudden and unpredictable violence. And even before the discovery of geologic expanses of time, the vast expanse of the Earth, in both space and time, mocked the very idea of grasping it whole.

Except for a few early studies of mineralogy and metallurgy, the science of geology dates from Hutton's and Werner's investigations at the end of the eighteenth century. We have only recently (i.e., since the 1960s) gained an overall (if still incomplete) *logos* of the natural world, where an understanding of plate tectonics is combined with links to processes across land, ice, ocean, air, and biota. But our greatest task still lies before us: integrating Earth scientific knowledge and perspectives into our social, political, and spiritual lives. Grappling with the issues of global climate change, the loss of biodiversity, and the geologically immanent loss of natural resources requires the marriage of the Earth sciences and the humanities. By enlarging geology in this way we will gain a better purchase upon our environmental challenges.

Overall, geology has received little attention from the humanities (but see McPhee 1981, 1983, etc.). Contemporary philosophy has hardly recognized the field as a subject worthy of reflection.[2,3] There is no philosophy of geology or of the Earth sciences as there are philosophies of physics and of biology. The two main schools of contemporary philosophy, analytic and continental, are one in ignoring geology.[4] It has been assumed (few thought to argue the point) that examining the Earth sciences was unnecessary to understand the nature of science. Statements by philosophers on the status of geology sound a common refrain: "Geology is a science just like other sciences, for example physics or chemistry."[5]

Nothing shows this disregard of geology better than the lack of attention humanists have paid to the concept of geologic time. The discovery of deep, or geologic, time parallels in importance the widely acknowledged Copernican revolution in our conception of space.[6] The concept of time plays an especially prominent role within contemporary continental or European philosophy. Nevertheless, philosophers and historians have ignored the Huttonian and Wernerian revolution's decisive role in reshaping our sense of time. In fact, the typical conclusion drawn from the terrific span of geologic time is that it renders all our human efforts insignificant. Geologic time opposes human time, rather than encloses it, mocking our efforts (cf. Shelley's Ozymandias) rather than being seen as part of and ennobling them.

Insofar as it has been considered at all, geology has been viewed as a derivative science, consisting of a few rules of thumb (e.g., the principles of uniformity and superposition) that guide the use of mathematics and the application of the laws of chemistry and physics to geologic phenomena. Geology, it was thought, suffers from a host of problems that undercuts its claims to real knowledge: incomplete data, because of the gaps in and the poor resolution of the stratigraphic record; the lack of experimental control that is possible in the laboratory-based sciences; and the great spans of time required for geologic processes, making direct observation impossible.

The geologic community itself has been the main source of reflection on the philosophic aspects of geology. Gilbert's and Chamberlin's essays (e.g., 1886 and 1890), dating from the classic era of nineteenth century geology, embody the

[2] Cf. Laudan (1987).

[3] Exceptions to this general neglect include David B. Kitts (1977), W.V., Engelhardt and J. Zimmermann (1988 [1982]), Ronald Giere (1988), Oreskes et al. (1994), Frodeman (1995, 2003), Rom Harré (2000), Robert John Inkpen (2009), and Bechtel and Herschbach (2010).

[4] In addition, Babich (2010) features a section on "Philosophy of Geology or Modelling and Its Discontents," 362ff. in addition to Babich (2013) for a section "Grounding Physical Science: Geology and Deep Time," 271ff.

[5] Nelson Goodman (1967).

[6] But see Cervato and Frodeman (2012).

attitudes of natural philosophy. In the twentieth century, a few geologists have reflected upon the methodology underlying particular subfields of geology or have offered general accounts of geological research.[7] The work of Stephen Jay Gould is especially notable, but the writings of Niles Eldridge, Peter Ward, and Edward Wilson show that the tradition of natural philosophy isn't entirely extinct.[8] Nonetheless, Earth scientists continue to practice "the reverential reference"—treating physics as the paradigm of reasoning, trolling within physics or mathematics for an approach (relativity, quantum mechanics, fractals, complexity theory) that gives a patina of legitimization to their "softer" discipline.[9]

Are the Earth sciences best understood as merely applied and imprecise physics, vainly attempting to achieve the degree of resolution and predictability typical of (some parts of) physics? I offer a different view: geological reasoning exemplifies both the power and limitations of reasoning, scientific and otherwise.

According to some, there is no distinction to be made between analytic and continental philosophy.[10] There is only good philosophy—a claim usually made by analytic philosophers, who refuse to recognize the possibility that there could be distinctive approaches to what counts as the philosophic project. From a continental point of view, however, two points concerning science stand out: (1) whereas science offers us a powerful tool for the discovery of truth, science is not the only, or even necessarily the best way that humans come to know reality, and (2) the belief there is one distinctive scientific method is a myth. Science has neither primacy in the discovery of truth, nor the unity and cohesiveness of one identifiable method, nor the distance from ethical, epistemological and metaphysical commitments that analytic philosophy had claimed.

When we view the Earth sciences from the perspectives of continental philosophy, certain features that had been left in the shadows begin to show themselves. Consider first the perspectives of hermeneutics, one of the most characteristic tools of nineteenth and twentieth century continental philosophy. A text (by which is meant, typically, a literary work, but which I want to expand to include the outcrop) is a system of signs the meaning of which is not apparent, but must be deciphered. This deciphering takes place when we assign differing types or degrees of significance to the various elements making up the text. The status of this deciphered meaning has been the source of some dispute: in the nineteenth century some claimed that, when properly applied to a text, hermeneutic technique resulted in knowledge as objective as that of the natural sciences. In the twentieth century, however, hermeneuts have claimed that the deciphering of meaning always involves the subtle interplay of what is "objectively" there in the text with what presuppositions and expectations the reader brings to the text. Hermeneutics rejects the claim that facts can ever be completely independent of theory.

---

[7] E.g., Stanley A. Schumm (1991) and Derek V. Ager (1993).

[8] See Stephen Jay Gould (1997 [1987]), Niles Eldredge (1995), Peter D. Ward (1998), Edward O. Wilson (1998).

[9] Doreen Massey (1999).

[10] Leiter 2007.

In the twentieth century, hermeneutics moved from being a rather straightforward methodology of the *Geistwissenshaften* to a more general account of knowing. Hermeneutic philosophers such as Heidegger have argued that *all* human understanding is fundamentally interpretive. Not only books, but all of reality is a text to be read: rarely do we find completely objective data or information that is "purely given." How we perceive a thing is always shaped by how we conceive and act upon it with the sets of tools, concepts, expectations and values that we bring to it.

We are all familiar with the hermeneutical aspect of understanding, the shift in our awareness of an object when we approach it with a fresh set of concepts or expectations. It is an experience that happens regularly to students when they are first introduced to a subject. In an introductory art history course each class may begin with lights dimmed, the professor showing a slide of a famous work of art and giving the students a few minutes to consider it on their own. Typically—especially at the beginning of the semester—students will see nothing of any significance. Yet after a few minutes of lecture, the piece undergoes the most striking transformation. Aided by concepts introduced by the professor, the piece of art reveals itself for the first time. Like art history, with which it shares a strongly visual component, geology is a deeply hermeneutic science; the outcrop typically means nothing to the uninitiated until the geologist introduces concepts for seeing the rock.[11]

The claim that all human knowledge is fundamentally hermeneutic—that our perceptions are always to some degree structured by our conceptions—has portentous implications for our understanding of the nature of scientific knowledge, and for the relation between science and society at large. For it makes the question of human interests, personal, ethical and political, and metaphysical, intrinsic rather than extrinsic to the work of science. The theoretic assumptions that the scientist brings to his or her work—what counts as significant, and what research is worth doing—structure all that is examined, seen and reported to 1° or another. Contemporary hermeneutics claims that this mix of percept and concept is fundamental to all human understanding. For Merleau-Ponty, all understanding is a combination of eye and mind.[12]

Hermeneutics does not offer methodological principles analogous to how analytic philosophy understood the scientific method. The role of hermeneutics is not to develop a set of rules for proper interpretation, but to clarify the general conditions under which understanding takes place. Nevertheless there are three concepts that play a fundamental role in any hermeneutic process, including geological reasoning: the hermeneutic circle, the fore-structures of understanding, and the historical nature of knowledge.

Heidegger argued that understanding is fundamentally circular: when we try to comprehend something, we understand the meaning of its parts from their relation to the whole, and conceive the whole from an understanding of its parts.[13] So the

[11] Martin J. S. Rudwick (1976).

[12] Maurice Merleau-Ponty (1993).

[13] Martin Heidegger (1962).

meaning of this sentence is understood in terms of the entire essay, and vice versa. Similarly, our understanding of a rock outcrop is based upon our understanding of the individual bedding layers within it, which are in turn made sense of in terms of their relation to the entire outcrop. This back-and-forth process operates on all levels; wholes at one level of analysis become parts at another. Thus our understanding of a region's geology is based on our interpretation of the individual outcrops in that region, and vice versa; and our interpretation of an individual bed within an outcrop is based upon our understanding of the sediments and structures that make up that bed.

Circular reasoning is viewed as a vice. But Heidegger argued that this type of circularity is not only unavoidable; it is also the process through which understanding progresses. Understanding begins when we develop an intuition of the object's overall meaning. Without this initial conception, we would lack a criterion for judging the pertinence of a given piece of evidence. This provisional interpretation is called into question when details in the object or text don't jibe with our overall sense of things. This forces us to revise our interpretation of the whole as well as our interpretation of the other particulars. Comprehension deepens in this circular fashion, as we revise our conception of the whole by the new meaning suggested by the parts, and our understanding of the parts by our new understanding of the whole.

One consequence of the hermeneutic circle is that it puts to rest the claim that it is possible to approach an object in a neutral manner. Rather, we always come to our object of study with a set of prejudgments: an idea of what the problem is, what type of information we are looking for, and what will count as an answer. What keeps these prejudgments from slipping into dogmatism and prejudice—what makes science still possible as distinguished from ideology—is the fact that they are not blind. We remain open to correction, allowing the text or object to instruct us and suggest new meanings and approaches.

In *Being and Time* Heidegger identified three types of prejudgments. First are our *pre-conceptions*, the ideas and theories that we rely upon when thinking about an object. Concepts are not neutral tools; rather, through them we get hold of an object in a specific way, opening up certain possibilities while closing off others. "Liberal" and "conservative" structure our political conversations, just "ophiolite complexes" and "accretionary terranes" affect what we see in the field. These pre-conceptions include our initial definition of the object as well as the criteria used to identify the significant facts and the insignificant ones. Second is our *pre-sight,* our idea of our inquiry's presumed goal and our sense of what will qualify as an answer. Without at least a vague sense of what type of answer we are looking for, we would not recognize it when we find it. Third, we approach our object of study with a set of practices that Heidegger calls our *pre-having*. These are our culturally acquired sets of implements, skills and institutions. In field geology, implements include the geologist's hammer, 0.10 % HCl, a measuring tape, a hand lens, a Jacob's staff, pencil and paper and a Brunton compass. In the lab, there is another set of tools: purified chemicals, mass spectrometers, computers, and a scanning electron microscope. With a different set of tools, we might gather new data that would give us a different (possibly quite different) sense of the world.

Heidegger's "pre-having" also includes the various skills that the scientist learns in the field or the laboratory: map-making, measuring strike and dip, preparing samples, cleaning and preserving specimens, and even wielding a hammer properly to split the rock without destroying the fossils within. Just as crucial, and often ignored, are the social and political structures of science: professors, graduate students, research groups, and professional associations. Science is a social as well as a mental activity depending on having colleagues to bounce ideas off of, professional societies and journals to define hot topics and favored lines of research, and graduate students to help run the labs and collect samples.[14]

Hermeneutics also emphasizes is the historical nature of understanding. The claim here—distinct from the argument below—is that the particular prejudgments we start with have a lasting effect. Some assume that, no matter what assumptions or goals we begin with, the scientific method will eventually bring us to the same final understanding of objective reality. Hermeneutics claims that our original goals and assumptions result in our discovering certain facts rather than others, which in turn lead to new avenues of research and sets of facts—a point known in economics as 'path dependence.[15] As decisions get multiplied over the decades, bodies of scientific and political knowledge come to have strongly historical components.

A further feature of geologic reasoning is worth highlighting: its nature as a historical science. A historical science (which includes other disciplines such cosmology, paleontology, and anthropology) is defined by the role that historical explanation plays in its work. Explanations within the historical sciences involve the tools common to all sciences (e.g., the deductive-nomological model of explanation), but are also distinguished by three additional elements: the limited relevance of laboratory experiments, the problem of natural kinds, and the role of narrative.

To the degree that scientific research is based on laboratory experimentation, it is essentially non-historical. In principle, the particularities of space and time in principle play no role in the reasoning process. Not only is the space idealized, set up so that other researchers can recreate the experiment's identical conditions within their own laboratory; in a fundamental sense, history does not exist. Of course, time and history are inescapable parts of every instance of scientific research: a chemical reaction takes time to complete, and every chemical reaction is historical in that it has some feature, no matter how insignificant, that distinguishes it from every other reaction. But our *interest* in chemical reactions lies not in chronicling the specific historical conditions affecting a given reaction, but rather in abstracting a general or ideal truth about a class of chemical reactions. A particular chemical reaction is approached as an instance of a general law or principle, rather than as a part of the great irretrievable sweep of historical events.[16]

In the historical sciences, the specific causal circumstances surrounding the subject of investigation—what led up to it, and what issued from it—are the researcher's main concern. In geology, for instance, the goal is often not to identify

[14] Andrew Pickering (1992), Karin Knorr-Cetina (1999).

[15] Paul A. David (2000).

[16] Cf. Nancy Cartwright (1983).

general laws, but rather to chronicle the particular events that occurred at a given location (at an outcrop, for a region, or for the entire planet). Hypotheses are not testable in the way they are in the experimental sciences. Although the geologist may be able to duplicate the laboratory conditions of another's experiment (e.g., studying the nature of deformation through experiments with play-doh), the relationship of these experiments to the realities of the Earth's history (e.g., the formation of the Rocky Mountains) will always remain uncertain.

The crucial point here is that the historical sciences are distinguished by a different set of criteria for *what counts as an explanation*. To borrow and adapt an example from David Hull, when we ask why someone has died, we are not satisfied with the appeal to the law of nature that all organisms die, true as that is.[17] We are asking for an account of the particular circumstances surrounding that person's death. Similarly, in the Earth sciences we are largely interested in the specific histories of historical phenomena (a particular stream, a region such as the Western Interior Seaway, a trilobite species). We might identify general laws in geology that have explanatory power; but the weight of our interest lies elsewhere.

A second aspect of the historical sciences merits our attention. Historical entities present a unique challenge to the researcher; for how does we define our object of study? In some sciences, the objects appear as "natural kinds": for instance, the nucleus of an atom consists of neutrons and protons, a distinction well grounded in the very structure of the atom. But historical entities do not spring into being fully formed, nor do they remain unchanged until their destruction. For instance, in investigating the history of the Colorado River (which seems to have run in different directions at different times in its history), we first face the riddle of when it first became the 'Colorado River.'[18] The researcher of historical entities is faced with identifying the set of characteristics that define the particular individual, and with deciding how much change can occur before we have a new individual rather than simply a modification of the old.

Hayden White argues that the concept of a *central subject* allows us to construct historical explanations.[19] A central subject is the organizational identity that ties together disparate facts and incidents. In human history, a wide variety of entities can function as central subjects: individuals or social groups, corporate entities (for instance, nations), even concepts (the idea of progress). In the Earth sciences there is a similar range of historical individuals: the Animas River, the Rocky Mountains, the species *Mytiloides mytiloides*, and the Pleistocene. Central subjects provide the coherence needed to construct an intelligible narrative out of a seemingly disconnected set of objects or events. But since these subjects are not natural kinds, they can be defined in different ways.

Finally, the historical sciences are distinguished by the role that narrative plays in their accounts. In the experimental sciences, predictions are produced by combining general laws with a description of initial conditions. But the historical sciences

---

[17] Hull (1976).

[18] Ivo Lucchitta (1990).

[19] Hayden White (1963).

are not primarily in the business of making predictions: rather than explaining an event by subsuming it under a generalization, they make sense of it by integrating the event into the flow of a story.[20] To make sense of a river, an outcrop, or a political event is to show how it is part of, and contributes to, a larger narrative. In science, narrative is commonly ignored: it is seen as a mere literary form lacking the logical rigor and evidential support necessary for real truth claims. But this dismissal ignores the fact that narrative has its own distinctive logic—and begs the question of whether scientific explanation itself depends upon the logic of the story.

Continental philosophers have been prominent in arguing that scientific explanation and narrative understanding in fact complement one another—science providing facts that parameterize an issue, narrative providing the overall goal and moral purpose of research. In *Time and Narrative*, Paul Ricoeur claims that narrative is our most basic way of making sense of experience. In Ricoeur's view, scientific explanation itself depends on a preceding narrative: framing the scientific project and making sense of its results depends upon the place that this project occupies within one or more storylines. These storylines (e.g., the pursuit of fame or riches, the righting of a public or private wrong, the desire for truth, or the wish for a better common future) provide the essential contexts for science. For instance, the development and testing of global circulation models (GCMs) gains its rationale in terms of the story we tell ourselves about the possible dangers of global climate change. Such "Earth stories" as how much oil or copper do we have left? How likely is a catastrophic flood or volcanic eruption? What are the possible scenarios for our climate's future? make sense only when it is placed within the structure of a story.

Finally, it is worth noting that narratives are distinguished from scientific knowledge by the fact that the former have an inherent moral structure. Narratives look to the future, not in the scientific sense of making predictions, but in Aristotle's sense of being concerned with final causes. A story always expresses a moral vision of what the future should look like (in the case of dis-utopias, through dialectical reversal). Historians, philosophers, and littérateurs excel at creating and interpreting the stories used to frame the work of the sciences, bridging the chasm that separates science and society.

The Earth sciences only partially live up to the classic model of scientific reasoning. But rather than viewing itself as a lesser or derivative science, geological reasoning provides an outstanding model of another type of scientific reasoning based in the approaches of hermeneutics and the historical sciences. Geology is a preeminent example of a synthetic science, combining a variety of logical techniques to solve its problems. The geologist exemplifies Levi-Strauss' *bricoleur*, the thinker whose intellectual toolbox contains a variety of tools that he or she selects as is appropriate to the job at hand.[21]

[20] Naomi Oreskes (2000).

[21] I offer versions of this argument in Frodeman (1995) as well as Frodeman (2003).

References

Ager, Derek V. 1993. *The nature of the stratigraphical record*. Chichester: Wiley.

Babich, Babette. 2010. Towards a critical philosophy of science: Continental beginnings and bugbears, whigs and waterbears. *International Journal of the Philosophy of Science* 24(4): 343–391.

Babich, Babette. 2013. Early continental philosophy of science. In *The new century: Bergsonism, phenomenology, and responses to modern science*, ed. K. Ansell-Pearson and Alan Schrift, 263–286. Durham: Acumen.

Bechtel, W., and M. Herschbach. 2010. Philosophy of the cognitive sciences. In *Philosophy of the sciences*, ed. Fritz Allhoff, 237–261. Oxford: Blackwell.

Cartwright, Nancy. 1983. *How the laws of physics lie*. Oxford: Oxford University Press.

Cervato, Cinzia, and Robert Frodeman. 2012. The significance of geologic time. *Earth in mind, II, GSA special volume*.

Chamberlin, T.C. 1890. The method of multiple working hypotheses. *Science* 15: 92–96.

David, Paul A. 2000. Path dependence, its vritics and the quest for historical economics. In *Evolution and path dependence in economic ideas: Past and present*, ed. P. Garrouste and S. Ioannides. Cheltenham: Edward Elgar Publishing.

Eldredge, Niles. 1995. *Dominion*. New York: H. Holt.

Engelhardt, W.V., and J. Zimmermann. 1988 [1982]. *Theory of earth science*. New York: Cambridge University Press.

Frodeman, Robert. 1995. Geological reasoning: Geology as an interpretive and historical science. *GSA Bulletin* 107: 960–968.

Frodeman, Robert. 2003. *Geologic: Breaking ground between philosophy and the earth sciences*. Albany: State University of New York Press.

Frodeman, Robert. 2010. Experiments in field philosophy. *New York Times*. November 23.

Frodeman, Robert. 2013. Philosophy dedisciplined. *Synthese* 190(11): 1917–1936. July 2013.

Giere, Ronald N. 1988. *Explaining science: A cognitive approach*. Chicago: University of Chicago Press.

Gilbert, G.K. 1886. The inculcation of scientific method by example. *American Journal of Science* 31: 284–299. 3rd ser.

Goodman, Nelson. 1967. Uniformity and simplicity. In *Uniformity and simplicity: A symposium on the principle of the uniformity of nature*, New York: Geological society of America special paper, vol. 89, ed. C.C. Albritton Jr. et al., 93–99.

Gould, Stephen Jay. 1997. *Time's arrow, time's cycle*. Cambridge, MA: Harvard University Press.

Harré, Rom. 2000. Philosophy and geology. In *The Oxford companion to the Earth*, ed. P. Hancock and B.J. Skinner, 821–822. Oxford: Oxford University Press.

Heidegger, Martin. 1962. *Being and Time*. Trans. John, MacQuarrie, and Edward, Robinson. London: SCM Press.

Hull, D.L. 1976. Central subjects and historical narratives. *History and Theory* 14: 253–274.

Inkpen, Robert John. 2009. The philosophy of geology. In *A companion to the philosophy of history and historiography*, ed. A. Tucker, 318–329. Chichester: Wiley-Blackwell.

Kitts, David B. 1977. *The structure of geology*. Dallas: SMU Press.

Knorr-Cetina, Karin. 1999. *Epistemic cultures: How the sciences make knowledge*. Cambridge: Harvard University Press.

Lauden, Rachel. 1987. *From mineralogy to geology: The foundations of a science, 1650–1830*. Chicago: University of Chicago Press.

Leiter, Brian. 2007. *The future for philosophy*. Oxford: Oxford University Press.

Lucchitta, Ivo. 1990. The history of the Colorado river and the Grand Canyon. In *Grand Canyon geology*, ed. Stanley Beus and Michael Morales. Flagstaff: Museum of Northern Arizona Press.

Massey, Doreen. 1999. Space-time, science, and the relationship between physical geography and human geography. *Institute of British Geography* 24: 264.

McPhee, John. 1981. *Basin and range*. New York: Farrar, Straus and Giroux.

McPhee, John. 1983. *In suspect terrain*. New York: Farrar, Straus and Giroux.
Merleau-Ponty, Maurice. 1993. Eye and mind. In: *The Merleau-Ponty Aesthetics Reader,* ed. Galen A. Johnson (trans: Michael B. Smith). Evanston: Northwestern University Press.
Oreskes, Naomi. 2000. Why predict? Historical perspectives on prediction in Earth science. In *Prediction: Science, decision making, and the future of nature*, ed. Sarewitz Daniel, Roger A. Pielke Jr, and Radford Byerly Jr., 23–40. Washington, DC: Island Press.
Oreskes, Naomi, Kristin Shrader-Frechette, and Kenneth Belitz. 1994. Verification, validation, and confirmation of numerical models. *Science* 263(5147): 641–646. Feb 4 1994.
Pickering, Andrew (ed.). 1992. *Science as practice and culture*. Chicago: University of Chicago Press.
Rudwick, Martin J.S. 1976. The emergence of a visual language for geological science, 1760-1840. *History of Science* 14: 149–195.
Schumm, Stanley A. 1991. *To interpret the Earth*. New York: Cambridge University Press.
Ward, Peter D. 1998. *Time machines: Scientific explorations in deep time*. New York: Copernicus.
White, Hayden. 1963. The logic of historical narration. *Philosophy and History* 3: 4–14.
Wilson, Edward O. 1998. *Consilience: The unity of knowledge*. New York: Knopf.

The Metroscape: Phenomenology of Measurement

Robert P. Crease

Abstract Measurement is an emblematic technology: it is ubiquitous, transforms our experience, withdraws into the background, and is noticed primarily when it breaks down. Measurement is not merely one tool among others, belonging only to separate elements such as rulers, scales, and other instruments, but a fluid and correlated network that is smoothly and intimately integrated into the world and its shape. This paper proposes the concept of *metroscape* to develop and extend Heidegger's concept of *Gestell.*

Measurement is an emblematic technology: it is ubiquitous, transforms our experience of the world, withdraws into the background, and is noticed primarily when it breaks down. Once measuring is embodied, and metrology becomes a social institution open to issues of trust and distrust, it is no longer a neutral activity but tied to justice, the good, and human enrichment—with a dark side having to do with injustice, exploitation, and alienation. The story of measurement therefore encompasses more than the tale of how today's network of standards, instruments, and institutions came to be, but includes the changes that take place in measurement's meaning. Every age has a metrosophy, a shared cultural understanding of why we measure and what measuring delivers to us, an understanding that evolves over time and across cultures.[1] Metrosophy is more difficult to discuss than the material network of standards, instruments, and institutions. But it is an important feature without which any discussion of measurement is incomplete (Crease 2011).

Prior to the advent of the metric system, measuring systems arose from local resources and practices to serve local needs. Systems from different communities

[1] The principal person to promote the concept of metrosophy is Hans Vogel, who applied it to the Chinese context (1994). However, I am vastly expanding the scope of this term.

R.P. Crease (✉)
State University of New York at Stony Brook,
250 West 90th St. #P3B, New York, NY 10024, USA
e-mail: robert.crease@stonybrook.edu

B. Babich and D. Ginev (eds.), *The Multidimensionality of Hermeneutic Phenomenology*, Contributions to Phenomenology 70, DOI 10.1007/978-3-319-01707-5_6,

were as original and varied as their artworks, political systems and other forms of cultural life, and the views of the point and purpose of measurement equally diverse. The more important a society viewed some aspect of the environment—gold in West African cultures, salt in Mesoamerican communities, court ritual in China, distance in nomadic tribes, agriculture in pre-modern Europe—the finer and more elaborate measures of this aspect tended to be, and the more these measures were specified and regulated. Witold Kula's study of European measures goes so far as to claim that measures are key to the character and vitality of pre-modern European life.[2] Those who do not understand the use and abuse of European measures, according to Kula, cannot understand Europe itself.

Within a short time, historically speaking—about 200 years—virtually all such systems became consolidated into one *universal* system of measurement—the metric system, now the SI—adopted by virtually every country on the planet. Most historians of metrology seem to feel that metrology thereby evolved beyond metrosophy. Even Kula, that eminent and sensitive decipherer of the social logic of European medieval measures and how closely they integrated into human life, shared this view. True, in *Measures and Men* he often dropped remarks to the effect that the metric system is "sheer convention," has "no practical social meaning," lacks a connection to "social values," and has "no inherent social significance whatsoever." Nevertheless, he grudgingly admits to being an "admirer" of the metric system, which has brought about a "higher level of mutual understanding among people," and "taken us very far along the road of more effective and fruitful international understanding and cooperation"[3] The book's final sentence—"And in the end, a time will come when we shall all understand one another so well, so perfectly, that we shall have nothing further to say to one another"[4]—however ironic, makes it clear that Kula faults the modern world, not the modern system of measurement so perfectly adapted to its realities.

The great scholar has nodded. The modern system of measurement is not devoid of social meaning, of metrosophy. The thoroughgoing project of stripping the imprint of regions, products, and times from measures, of abstracting measures from each and every local context in order to make the world measurable, calculable, and universal for human beings and to put it at our disposal, has a deep social meaning indeed. Heidegger's famous concept of *Gestell* points to this meaning. This paper aims to show that this meaning can be further elaborated via the concept of *metroscape*, a concept to capture the idea that measurement is not merely one tool among others, belonging only to separate elements such as rulers, scales, and other instruments, but a fluid and correlated network that is smoothly and intimately integrated into the world and its shape.

The suffix "-scape" commonly refers to a kind of space (landscape, seascape, or cityscape) that is extended, produced by human interaction with nature, has a particular character, and shapes how human beings relate to nature and each other.

---

[2] Kula (1986).

[3] Kula (1986), 121.

[4] Ibid., 288.

In words like *soundscape* and *ethnoscape*, the suffix is applied to more virtual kinds of spaces with a similar impact. A "-scape" is neither simply material nor mental but both at once; it inhabits the world and its features and simultaneously the way we perceive and relate to this world. The modern metroscape is not the doing of the SI, which is a consequence rather than a cause of the metroscape.

A contrast between two images can serve to begin a discussion of the metroscape. Consider first of all da Vinci's famous and much copied and caricatured drawing known as the Vitruvian Man, for he evidently had that architect's passage about measures in mind. This Vitruvian Man displays how the proportions of the human body, and the units drawn from it, participate in an ideal of beauty. The Vitruvian Man shows us the organization of measures can have symbolic and spiritual significance. Now consider an image by early twentieth century industrial designer Henry Dreyfuss of a pair of archetypical human beings, "Joe" and "Josephine," whose line drawings and measurements—the fruit of decades of data collection and research—were intended to allow engineers to incorporate human form and behaviors into products and machinery. Joe and Josephine are the new Vitruvian Man, gendered into a couple and measured over their entire lifespans. Joe and Josephine are utterly devoid of the boldness, nobility, and beauty of da Vinci's Vitruvian Man. Their reason for being—why they were created and what they show us about the world—is not beauty and symmetry but efficiency. They do not help connect us humans with something beyond the world, but help engineers and the individual human beings for whom the engineers design to get a better grip on this one. The central feature of the modern metrosophy is efficiency.

The modern metroscape, illustrated by this contrast, involves a new relation between measurement and how we relate to the world and each other. This metroscape—which tends to hide but can be brought to light—shapes what we make, what we purchase, how we classify things, and what we consider real. It shapes products, workers, markets, and businesses, and reflects and reinforces social, political and economic currents. Busch and Tanaka have explored how this works in the agriculture of canola, a seed used to produce oil. "[G]rain grades link farmers to elevator operators. Seed quality tests link seed producers to farmers. Measures of oil content and composition link large sellers to buyers of canola. Measures of shelf life link processors to retailers.... Tests are measures of nature at the same time as they are measures of culture."[5] The role of measures they detect in the production and consumption of canola is found in nearly every other agricultural product.

Plato reminds us of two different ways of measuring. One involves numbers, units, a scale, a beginning point. It establishes that one property is greater than or less than another, or involves assigning a number to how much of a property something possesses. We can call this "ontic" measuring. Histories of measurement technology relate how ontic measurement developed from improvised body measures and disconnected artifacts into a single network that relates different kinds of measurements and soon will tie them all ultimately to physical constants.

[5] Busch and Keiko (1996), at 23.

Another kind of measuring does not involve placing oneself next to a stick or in a pan. This is what Plato said is guided by a standard of the "fitting" or the "right." This measuring is less an act than an experience; the experience that things that we've done, or we ourselves, are less than they could or should be. We cannot carry out this kind of measuring by following rules, and it does not lend itself to quantification. Is this only "metaphorical" measuring? It is comparison against a standard. Placed alongside the fitting or the right example, our actions—even our *selves*—do not have enough being; there is more to be. We feel we are not measuring up to our potential. We can call this "ontological" measuring.

Ontological measuring involves no specific property, in a literal respect, for it involves nothing quantitative. Calculate all we please, we will never produce this kind of measurement. No method can lead us to it. Ontological measurement connects us with something trans-human, something *in* which we participate, not something *over* which we command. While in ontic measurement we compare some object with another object exterior to it, in ontological measurement we compare ourselves, or something we have produced, with something in which our being is implicated, to which it is related—such as some concept of the good, the just, the beautiful. Ontological measurement is ontically measureless.

Scholars of the ancient world were mostly confident that standards for our potential existed, and that human beings could find such standards and use them as measures. Aristotle described the moral man as a "measure" in the *Ethics*. By this he meant, not that a moral man is something against which we can physically or even symbolically compare ourselves, but that our encounters with genuinely moral human beings "call us out of ourselves," making us want to be better humans.

Ontological measuring is the measuring that good examples invite. The history of literature and art is replete with great works and performances that each artist can experience as intangible yardsticks, so to speak, for measuring his or her own achievements. To be sure, traditions change, and with it ideas of what is good and what not. But tradition provides an authenticating horizon in which artists experience a measure of what is good and what not, what original and what an echo, what vibrant and full of life and what deficient. The "call of conscience" involves ontological measurement, a secular variation on the old spiritual idea of being "called back to yourself." Conscience, like other forms of ontological measurement, requires opening ourselves to being able to say, "I could be better," to being able to experience ourselves as ontologically deficient—a positive thing.[6] This is the foundation of ethics.

Human beings practice both kinds of measuring all the time. But the two kinds of measurements are often confused, with damaging results. Stephen J. Gould's book, *The Mismeasure of Man*, is about the fallacy that "worth can be assigned to individuals and groups by measuring intelligence as a single quantity." Shakespeare's play *Measure for Measure*—an allusion to Matthew 7:1–2—is about the need to temper literal application of legal measures with empathy and mercy in order to live up to what it means to be fully human. Moral thinking begins with the distinction between ontic and ontological measures.

---

[6] Crowell (2008).

Heidegger was fond of citing Holderlin's passage: "Is there on earth a measure? There is none." If so, it reflects an odd state of affairs: as the modern world has progressively improved and perfected its ontic measures, it has diminished its ability its ability to measure itself ontologically. How? The reason is that, in the modern metroscape, ontic measuring can distract from, and even have a corrosive effect on, ontological measurement.

The capacity and new tools for measurements in our lives seems continually on the increase, and can appear to be an unqualified good. A web site, "The Quantified Self," bills itself as providing "tools for knowing your own mind and body." These tools are means for collecting data about the times we spend in such activities as working eating, sleeping, having sex, worrying, cleaning up, having coffee, and every other aspect of everyday life. "Behind the allure of the quantified self," wrote the site's co-founder in the *New York Times*, "is a guess that many of our problems come from simply lacking the instruments to understand who we are."[7] How fortunate, therefore, that we are to be able to quantify every aspect of our lives in this high-speed, rapidly changing world! No ambiguity here. Measurement is an indispensable tool of self-knowledge. The better we do it, the more we know of our selves.

By contrast, "Vital Statistics of a Citizen, Simply Obtained," a 40-min video by the American artist Martha Rosler (1977), depicts measurement as simply dehumanizing. Most of the video consists of a 33-year old woman being measured by two white-coated men, one of whom makes the measurements, the other writes them down. At first they have her stand against the wall and draw a Vitruvian-man-like measured image of her with outstretched limbs. Then they ask her to take off more of her clothes as they measure more intimate parts of her body, culminating in her "vaginal depth." They have her lie down horizontally in front of her measured image. As she is being measured and the one male announces each of her measures to the other, he calls it "below standard" (whereupon the soundtrack has a razzing sound), "above standard" (a beep), or "standard" (pleasant chimes). Meanwhile, a feminine voiceover characterizes what's happening in apocalyptic terms, referring to rape, dehumanization, degradation, exploitation, eugenics, and tyranny; the voiceover says that the woman is being indoctrinated to manage her image, to view her body as parts, and to lose track of her self, and quotes Sartre to the effect that "Evil is the product of the ability of humans to make abstract that which is concrete." After the male measurers have their way with the woman, she puts clothes back on in two sequences: in one she dons a wedding dress and veil, in the other a hot little black dress. The wedding dress sequence has her returning to the wall to stand demurely and compliantly next to her measured image; in the black dress sequence she darts off in the opposite direction. The video ends by returning to the two male doctors addressing another woman: "Next!"[8]

---

[7] Wolf (2010).

[8] According to a web site about the film, "Rosler's distanced depiction of the systematic, institutionalized 'science' of measurement and classification is meant to recall the oppressive tactics of the armed forces or concentration camps, and to underscore the internalization of standards that determine the meaning of women's being."

No ambiguity here, either. Measuring is doing bad things to us. What is wrong is not simply misplaced precision; too much of a good thing. Measuring is far more sinister, a tool of oppression. It destroys our selves, or at least those of women; men evidently either had no selves to begin with, or shed theirs long ago. Best for those who still have selves to renounce measurement.

The metroscape means that the environment in which we measure is not neutral; this of course is Heidegger's point about the Gestell. In the modern atmosphere, measuring tends to dazzle and distract us. We tend to look away too much from what we are measuring, and why we are measuring, to the measuring itself. Measuring certainly works, and helps us to get around—but in the modern metroscape, it can lead us to think that it is all we need to get around.

The Vitruvian Man was an ideal image, something that connected human beings with beauty, perfection, and other trans-human goals, goals towards which measures could at best only serve as signposts. Joe and Josephine are something different; they are models, things that designers need as a means to achieve the efficient creation of interfaces between human beings and the world. Trans-human goals are absent; Joe and Josephine assist the aim of putting the world at the disposal of the wants and needs of human beings. The digital avatars that consumers can now create for themselves on 3D scanners in retail stores like Brooks Brothers and Victoria's Secret are still more remote from Vitruvian man and even Joe and Josephine. It is a means for us as individuals to purchase clothes whose measurements are perfect for our bodies.

How can we keep an eye on the difference between ontic and ontological measurement, and prevent the one from interfering with each other?

One way is to ask about what, if anything, is missing from the measurements delivered by the modern metroscape. Even in the modern metroscape, measuring does not thrust the rest of human life permanently in the background—the question, "Why do we measure?"—if we pay attention. We have to may more careful attention than ever to the goals we are trying to achieve with measurements, rather than simply to measurements. We have to focus on our dissatisfactions, on what measuring does not deliver. We have to address these dissatisfactions, not by discarding the measures we have and seeking to find newer and better ones, for these, too, will eventually turn out not to do what we want and eventually need to be renounced, nor by assuming that what we are after lies "beyond" measuring. Rather, the modern metroscape requires us to articulate more carefully what and where our measurements do not deliver.

Another way to keep an eye on the difference between ontic and ontological measurement in the modern metroscape is to reflect, not simply on how individual acts of measurement are carried out, but on the metroscape itself, and what it does to us. We can do this in part by retelling the story of measurement—reminding ourselves how the modern metroscape came to be, what the alternatives were, why we rejected them, and what we gained but also lost by rejecting them.

References

Busch, Lawrence, and Tanaka Keiko. 1996. Rites of passage: Constructing quality in a commodity subsector. *Science Technology & Human Values* 21:1(Winter): 3–27.

Crease, Robert P. 2011. *World in the balance: The historic quest for an absolute system of measurement*. New York: Norton.

Crowell, Steven. 2008. Measure-taking: Meaning and normativity in Heidegger's philosophy. *Continental Philosophy Review* 41: 261–276.

Kula, Witold. 1986. *Measures and Men* Trans. R. Szreter. Princeton: Princeton University Press.

Rosler, Martha. 1977. "Vital Statistics of a Citizen, Simply Obtained," video performance available at the Modern Museum of Art, New York City.

Vogel, Hans Ulrich. 1994. Aspects of metrosophy and metrology during the Han period. *Extrême-Orient Extrême-Occident* 16/16: 135–152.

Wolf, Gary. 2010. The data-driven life. *New York Times Magazine,* April 28.

Part II
Hermeneutic and Phenomenological Philosophy of Science and Technology

Consciousness, Quantum Physics, and Hermeneutical Phenomenology

Patrick Aidan Heelan

Abstract Two hundred years ago Friedrich Schleiermacher (See Wellmon 2006) modified Kant's notion of anthropology—'hermeneutically,' as he said—so as to make it inclusive of the tribes that Captain Cook found in the South Sea Islands. This paper honors the late Joseph J. Kockelmans for making a similar hermeneutic move to update Kant's notion of natural science so as to make it inclusive of the phenomenological lifeworld (For 'lifeworld,' see Husserl's *The Crisis of European Sciences and Transcendental Philosophy*, 1954, 121–148, and the 'lifeworld' theme throughout the *Crisis.*) syntheses of classical, relativity, and quantum physics. The new synthesis is in fact not alien to the views of some of the founders of quantum mechanics, notably Eugene Wigner, John von Neumann, Paul Dirac, Werner Heisenberg—possibly even Albert Einstein. In this hermeneutical move, the 'observer' is 'embodied consciousness,' and 'measure-numbers' represent 'observable presence.' The new theoretical synthesis of physics is a representation

NOTES ON SOURCES: Some of the sources used in this paper are listed in the references below. Most of the referenced Heelan texts can be found on the website, https://gushare.georgetown.edu/heelanp/ or http://fordham.bepress.com/phil_research/. In the field of mathematics and theoretical physics, I have learned from my physics mentors: from the lectures of Erwin Schrödinger and John Synge on classical non-Euclidean geometries, and from personal communications with and the publications of Nobelists Eugene Wigner (cf. 1963, 1967) and Werner Heisenberg (cf. 1950) on the role of subjectivity in assessing the rationality of the quantum theory. I have also profited from discussions on cognitive science with Karl Pribram and his writings (cf. 1971, 1991) on the building of a scientific model of human embodied consciousness. In linguistics, I have learnt much about language from my colleague in German Linguistics, Heidi Byrnes at Georgetown University. I owe a special debt to Babette Babich, at Fordham University, my former student, who has been a constant partner in scholarship for many years. These, among many others too numerous to mention, are the principal dialogical and dialectical sources of the rational heuristic I have used to explore the nature of the human consciousness and the Spirit that raises it up above pure Nature.

P.A. Heelan (✉)
William Gaston Professor of Philosophy
Georgetown University, Georgetown, USA
e-mail: Babich@fordham.edu

B. Babich and D. Ginev (eds.), *The Multidimensionality of Hermeneutic Phenomenology*, Contributions to Phenomenology 70, DOI 10.1007/978-3-319-01707-5_7,

of a physical system as a dynamic Hilbert Vector Space; empirical 'observables' are represented by projection operators, each of which maps a subspace of definite measurable values. Among these projection operators, some pairs are 'complementary' and share a common subspace of the Hilbert Space where they can be precisely measured together in a common laboratory setting. Some pairs, however, are 'non-complementary' and do not share a common subspace; these lead to Uncertainty Principles of the quantum mechanical kind. The quantum notion of an "observable" introduces into the discursive language of physics the common sense lifeworld notion of "contextuality." This analysis completes Husserl's analysis of science in the *Crisis*, so well articulated and developed by Kockelmans (See Kockelmans' contributions to the phenomenology of natural science in Kockelmans and Kisiel (1970)).

1 Phenomenology, Hermeneutics, and Quantum Theory

> It is only when we arrive at consciousness as the universal medium of access of whatever exists and has value, including the lifeworld itself, that our research for foundations reaches its final destination. In other words, our ontology of the lifeworld reaches its ultimate foundation only in the constitutive analyses of transcendental phenomenology.
>
> — Joseph Kockelmans[1]

Physicists and philosophers of science are persuaded that epistemologically quantum mechanics departs radically from classical physics. Few, however, take seriously the insight of three of the founders of quantum physics—Erwin Schrödinger, Werner Heisenberg, and Eugene Wigner[2]—that the strangeness of quantum physics involves the emergent epistemological and ontological role of embodied human consciousness in the process of measurement.

I am reminded of the following story told to me by Heisenberg in 1965. In April, 1926, before his paper on the Uncertainty Principles in Quantum Mechanics was published, Albert Einstein invited him to speak to the senior physicists in Berlin on this topic; Einstein presided at the meeting and, when Heisenberg had concluded his presentation, he spoke saying that all this uncertainty talk was nonsense. Much taken aback, Heisenberg responded that he was only applying the principle Einstein himself used in his 1905 relativity paper—that the measure-numbers of the mathematical theory described reality. To this, Einstein responded: "The

[1] Ibid, 67.

[2] I am a physicist who studied (1946–1948) relativistic cosmology with Erwin Schrödinger and John Synge at the Dublin Institute for Advanced Studies; later I studied as a post-doc (1960–1962) in high-energy quantum physics with Eugene Wigner at Princeton; and in 1962–1964, I visited frequently with Werner Heisenberg in Munich while writing a book on Heisenberg's philosophy of science (Heelan 1965). Out of my many discussions with them, I developed an interest in the way these three Nobel Prize physicists, interested in Husserl's philosophy, attributed a fundamental role to human consciousness in quantum physics.

measure-number is to that which is measured as the number on your cloakroom ticket is to your overcoat—it tells you nothing about your overcoat." Einstein then invited Heisenberg to walk with him while they discussed the problem of quantum uncertainty. They walked and talked for an hour, and when they returned to the conference room, those who awaited their return asked what the outcome was. Einstein replied that he and Heisenberg now understood one another and were in agreement. When pressed to state what their agreement was, Einstein refused to speak about it. At that moment I asked Heisenberg what was their agreement. He replied that a distinction has to be made between the 'presence' of the 'real' and any 'intuition' that may have accompanied the sense of presence. I will return to this topic below. The aim of this paper is to reflect on and attempt to articulate the content of the agreement between Heisenberg and Einstein on that occasion in 1926.

In the tradition of classical physics, the observer is a disembodied mind external to nature, and the objective essence of nature is revealed in the mathematical intuition of its structure. In quantum mechanics, however, the observer is not a disembodied mind, nor does the mathematical intuition represent an 'objective' presence but the probability of a presence.[3] In quantum mechanics, the observer is a human consciousness embodied in instruments that serve as a bodily extension of the observer's embodiment. The empirical observations of quantum physics consequently involve a bilateral relation between the observer's enhanced embodied consciousness (on the subject side) and what is observed (on the object side), each having its place and context within the lifeworld of Nature as culture.[4] The concepts and judgments of quantum physics consequently are contextualized subjectively and objectively by the 'natural world' as structured by science.

Human consciousness makes meanings from its sensory engagement with nature through practices that are learned and later function at an unconscious anticipatory intentional level.[5] For a deeper understanding of this process, I draw, not from Carnap's logical empiricism, but from the post-kantian German philosophies contemporary with the development of quantum mechanics. While quantum mechanics was taking shape in Göttingen, Leipzig, and Munich, where Heisenberg studied and worked, the phenomenology of Edmund Husserl was academically prominent. Later, Heisenberg was also one of a circle of professional scientific intellectuals who met regularly during the summer with Martin Heidegger in the Black Forest.[6] In Germany, phenomenology and hermeneutics were *wissenschaftlich* approaches to psychology, art, literature, music, and natural science, establishing them both as 'scientific' and 'philosophical,' on a par academically with the role of logic and analysis in the contemporary USA.[7] In addition to Husserl and Heidegger,

[3] See Kockelmans (1970a, c) in Kockelmans and Kisiel (1970).

[4] See ibid, Kockelmans (1970b).

[5] See ibid, Kockelmans and Kisiel (1970).

[6] Heisenberg contributed an essay on the Uncertainty Principle to a *Festschrift* to honor Heidegger on his 70th birthday.

[7] See Kisiel (1970a, b) in Kockelmans and Kisiel (1970).

Dilthey's work on history, Maurice Merleau-Ponty's work on perception,[8] and Hans-Georg Gadamer's work on literature and art also inspired a strong current in European academic culture before and after the war. These currents of thought also inspired Michael Polanyi, who was Wigner's scientific mentor. Looking back on his life as a physicist, Wigner told his chronicler, Andrew Szanton:

> My chief scientific interest in the last twenty years has been to somehow extend theoretical physics into the realm of consciousness.... Consciousness is beautifully complex. It has never been properly described, certainly not by physics and mathematics. It is shrouded in mysteries. And what I know of philosophy and psychology suggests that these disciplines have never defined consciousness either. (Szanton 1992, p. 309).

2 Human Consciousness as the 'Governor of Mental Life'

What is 'human consciousness'? Few cognitive scientists are willing to define it, perhaps, because a human subject trying to define it objectively leads to an infinite series of recurrent questions! Human consciousness certainly processes information signals—but so does Deep Blue, the IBM computer chess champion; but, in addition, it has sensory experiences, produces new insights, tests for relevant truth in the world, and makes free value-laden decisions on the basis of the information it gets—Deep Blue lacks all of these. I think the best functional account of human consciousness is given by the distinguished Canadian neuropsychologist, Merlin Donald. He calls it "The Governor of Mental Life" which functions as the meaning-maker and manager in science, culture, and religion. About this he wrote:

> What consciousness is really about, at least in the human species ... is much deeper than the sensory stream. It is about building and sustaining mental models of reality, constructing meaning, and asserting autonomous intermediate-term control over one's thought process, even without the extra clarity afforded by the explicit consensual system of language. The engine of the symbolic mind, the one that ultimately generates language to serve its own representational agenda, is much larger and more powerful than language, which is after all its own (generally inadequate) invention.[9]

Meaning-making—otherwise called "meaning-constitution" or "intentional activity"—is the making of concepts, predications, judgments against an appropriate a priori background of lifeworld, context, and practices. They all involve dialogically the specific subjective embodiment of the speaker as well as an intended environmental context for the discourse. For scientific discourse, a dialogical community lives in the context of a theoretical language and a scientific laboratory. In this analysis, I follow the way of hermeneutical and phenomenological thinking according to Edmund Husserl ([1952] 1989, 1966, [1901–1913] 1970a,

[8] The terms "perception" and "observation" are used in this article as synonymous.

[9] Donald (2001), 75.

[1954] 1970b),[10] Martin Heidegger (1962, 1967, 1982, 1995, 1999, 2002a, b), Maurice Merleau-Ponty (1962), Kockelmans and Kisiel (1970), and others.[11]

3 Hermeneutics as the Universal and Transcendental Process of Meaning-Making

The universal and transcendental process of meaning-making is a circular or cyclic process that is often called the 'hermeneutical circle/cycle.'[12] Each circle—or cycle—follows a sequence of four phases—a. *experiencing/observing*, b. *theory-making*, c. *theory-testing*, and d. *deciding*—each phase giving access to new insights; each cycle leading to a partially transformed beginning of a new cycle in which further development is made. Each cycle revises and improves the previous cycles of inquiry until the basic queries have been sufficiently explored dialogically.

To exemplify the process of the hermeneutical circle, I will tell the story of the distinguished psychologist James J. Gibson's discovery of the non-Euclidean geometry of human vision while training young pilots to fly during the war. So many of these young men killed themselves when landing their planes that he came to suspect that the problem was not an engineering problem, but a human one related to spatial vision. He suspected that 'natural' human vision systematically estimated vertical altitudes differently from the way they are estimated by scientific measurement. Thus he was led to the hypothesis that the visual space of humans had a different geometry from the Euclidean. Some time in the early seventies, I was invited by the MIT Psychology Department to speak on the occasion of the celebration of Gibson's 70th birthday. I spoke about the work I was doing on the curved Riemannian geometry of Van Gogh's paintings and I spoke of the experimental studies of von Helmholtz and others on the non-Euclidean geometries of visual space.[13] Gibson was pleased with my talk and responded by telling his story about why so many student pilots killed themselves when trying to land their planes because human vision without instruments is not adequate for flying. He said that based on this experience, he formulated a rule—now universally mandatory for all pilots—that, when landing a plane, they must rely exclusively on technological guidance, such as on-board instruments, instructions from the airport tower—or

[10] For an excellent guide to Husserl, see Welton (2000).

[11] All of these are linked with the ancient Greek and scholastic tradition through Bernard Lonergan's reflection on the transcendental process of meaning-making, and the importance of what he calls, 'interiority' Lonergan ([1957] 1992, [1972] 1990); 'interiority' is the awareness of oneself as being an embodied consciousness and as such, the Governor of one's Mental Life.

[12] See Kisiel (1970b); also Heidegger (1962). The scholastic tradition is a bridge that connects the classical tradition and phenomenology; for this reason, I find Bernard Lonergan helpful; see Lonergan ([1957] 1992).

[13] See Heelan (1983/1987).

lacking these—they must follow the now standard "Gibsonian" markings on the ground approaching the airport.[14]

The four phases of the hermeneutical circle can easily be discerned when applied to Gibson's story: the *experience* of pilots' failures[15]; the *theory/hypothesis* of non-Euclidean vision; the *theory-testing* in experimental studies of binocular visual geometry[16]; and the *decision* to apply the consequences of binocular visual geometry to piloting planes.[17]

In his reflections on visual space, Gibson also asked himself about binocular vision in the context of human evolutionary history—whether a 'natural' binocular space, which is curved and of finite size, would have served early human communities in their 'natural' environment better than an infinite flat Euclidean space to which modern culture is accustomed. He concluded that 'natural' binocular curved visual space would be more useful, first because it highlights a nearby quasi-Euclidean frontal zone for good eye-hand coordination, while more distant objects are projected without depth onto the visual dome, the one that rests on the horizon and rises to become the background for the clouds during the day and the stars during the night. From the point of view of cognitive science, however, the account of pure vision given above seems to be consistent with the dual visual neurological pathways that neuroscientists have found.[18] For our early ancestors, however, and for ourselves today—should we strip away what science teaches us—the 'natural' meaning of pure vision is neither Galilean nor Einsteinian, but what comes from Grimm's fairy tales.[19]

Perhaps of even greater critical importance is the hermeneutical criticism of classical scientific research on human vision—such as Galileo's—in overlooking the dual role played by light—for light is a physical *medium* subject to electromagnetic laws and it also carries a visual *message* about the environment. This dual function is often overlooked and—in the familiar words of Marshal McLuhan—"the medium is the message." The objects of visual experiences—what we see—are not just the photons/rays of light falling on the retina but the

---

[14] See Gibson (1979).

[15] Ibid.

[16] See Heelan (1983/1988), passim, and the Appendix in which the history of the geometry of curved visual spaces is presented.

[17] Gibson found the hypothesis was reasonable in the light of biological evolution; that many everyday phenomena seemed to support it, and that the laboratory scientific made by H. von Helmholtz (c. 1876) and others such as R. Luneburg, A. Blank, T. Indow, J. M. Foley and others provide positive evidence.

[18] Jacob and Jeannerod (2003), Jacob (1988), Pribram (1991).

[19] The *Visual Space* of our early human ancestors was constituted by a nearby virtually Euclidean zone that Arnheim (1974) called the 'Newtonian Oasis,' and a far zone that surrounds it where the depth of field diminishes rapidly to zero Heelan (1972, 1983, [1983] 1988), Part I and Appendix; Luneburg (1947, 1985). In theory, the non-Euclidean geometry of natural human visual space can be derived a priori from stereoscopy. The characteristics of this general structure have been confirmed by testing (Luneburg 1947, 1895 ; Heelan 1972, 1983, [1983] 1988).

information they carry about the environment; what the photons/rays show is not themselves but the presence of distant three-dimensional environmental bodies which are their source. As human visual organs receive the incoming stream of photonic messengers, they draw from them environmental information appropriate for action. Among such action are the coordination of hands, eyes, limb movements, and possibly instrumental controls.[20] As physical entities the photon-messengers move in Galilean/Euclidean physical space, but they invite interpretation by human embodied consciousness, who consequently sees an illuminated space of physical objects in the curved visual space inherited from our biological ancestors. In this curved space, there is a local privileged zone where hand–eye coordination is quasi-Euclidean. Distant objects, however, are given only in superficial profiles on the surrounding celestial dome. The geometric family of such visual spaces, as I have said, can be inferred a priori from the theoretical treatment of binocular stereoscopic vision. The curvature of such visual spaces plays an active—and often disconcerting and dangerous role—particularly, in engineered environments, such as modern highways, and—as Gibson found—in guiding planes to safe airport landings. The conclusion that Gibson came to was, that 'natural' human vision was shaped for terrestrial living and not for living in the air like birds.

The *hermeneutical circle,* as I have said, is the structure of the *transcendental rationality of dynamic human consciousness*.[21] This is not simply what is usually understood as *Enlightenment Reason* or *objective science*. Self-awareness of this transcendental dynamic embodied function constitutes a rare virtue that Bernard Lonergan calls '*interiority*'[22] which is discernable in the writings of ancient and modern authors, from Plato and Aristotle to Aquinas, and up to the present time. Heraclitus once said that *human consciousness loves to hide itself*—a sentiment shared with many psychologists, cognitive scientists, social scientists, physicists, and philosophers.[23] Such a sense of the *embodied-self-in-the-world* is reflected in a special way in the phenomenological writings of Husserl, particularly in his later works, also in Merleau-Ponty's *Phenomenology of Perception* (1962), and in Heidegger's *Being and Time* (1927/1962).

'*Interiority*' ts a virtue of human consciousness that is also exemplified in the views of at least the four physicists I mentioned at the head of this article, namely, Schrödinger, Wigner, Dirac, and Heisenberg. '*Interiority*' makes deep demands on philosophers and cognitive scientists, especially on those concerned with the rationality of contemporary physics, cognitive science, ethics, and religious faith.

---

20 See Berthoz and Petit (2008).

21 See Heelan (1994, 1998).

22 See Lonergan (1957/1992).

23 See Hadot (2006), Chap. 1.

4 The Governor of Mental Life and Meaning-Making

The Governor of Mental Life—human consciousness—makes meanings of different kinds. We turn next to *meaning-making in the natural sciences.* Many different kinds of meanings are made in the natural sciences, such as *concept and category formation, theory formation, theory testing in the laboratory, and theory affirming.*

About *concept/category*[24] *formation*: We ask first: What are 'concepts' ontologically? How are they constituted? Are they 'local/contextual' *invariants/likenesses/ symmetries*[25] of an *a posteriori* set of particular *empirical* instances/events held in the memory as alike in some categorical way and likely to be changeable over time? Or are they a priori 'unchangeable/transcendental' ideals, expressed, say, in *mathematics or pure logic*, with respect to which any *empirical* instance/event absolutely and necessarily conforms?

Whatever concepts are, and however constituted, they are represented by mathematical and linguistic media of communication: It is then necessary to distinguish the two uses of the representing medium: the *medium* as messenger, and the *information* carried by the messenger for delivery to appropriate interpreters—speakers and hearers. The nub of Heisenberg's and Einstein's problem referred to above was how to distinguish and relate the medium and the message in order to make sense of the quantum Uncertainty Principles.

About *theory formation*: In particular, what are the distinguishing linguistic roles of mathematics and logic in the formation and use of theory? In terms of "grammar" and "lexicon," the 'lexicon' of a science refers to what is 'observable' in the process of measurement, and the 'grammar' of the science refers to its mathematical theoretical structure where 'intuition' has its place.

About *theory testing*: Theory testing leaves a residue of *meaning uncertainty* due to the contingency of empirical evidence. *Contingency* is a function of the variety of possible *contextual*[26] *circumstances implicitly intended in the instantiation of an 'observable,'* such as *the evidentiary horizon of the laboratory*, *the social demand for cultural and institutional agreement, the historical dimension of languages, practices, cultures, institutions, etc.* These implicate social, cultural, and historical aspects of natural science, as well as, say, the philosophical and theological culture of the local environment. They demand of the speakers/hearers an attitude of continual prudent review and revision. The natural sciences are evidently not finished products. Ethical, aesthetic, and religious meaning-making, as well as other value-added aspects of decision-making, serve to condition the choices of the inquirer as well as the chosen circumstances of the inquiry.

---

[24] For the purposes of this paper, I do not distinguish between "concept" and "category."

[25] I use the terms "invariant," "likeness," and "symmetry" interchangeably; they define the same group-theoretical quality which remains constant despite merely perspectival changes—represented usually by group-theoretic transformation laws of space and time.

[26] See Heelan (1974, 2003) and Hasan (2010).

The complex canvas sketched out above is large, but I intend to cover just as much as is necessary to show how phenomenological and hermeneutical considerations force one to move beyond current science-speak to find the observable (ontological) reality that science noetically and epistemologically intends.[27]

Evolutionary Concept/Category Formation: Descriptive category formation is part of the general story of human evolution! What is it about category-making that makes the human sciences hide the emergence of human consciousness in the story of evolution? Human infants do not enter the world conscious of knowing anything about it, but they enter equipped with all that is necessary to learn from their environment. They learn from adults around them by 'reading their minds,' communicating by 'mimicry,' and later by 'language,' exploring their environment for observable content, and eventually expressing what they mean in the language of the family or caregivers. Finally, they learn to collaborate with their family and caregivers who by their natural authority introduce them to their local world, and to the means to share it, and to represent it through language as members of a human community.[28]

Concept/Category Formation of Observable Objects: The process of observation (perceptual recognition) supposes a descriptive category that is associated with a lexical name, an observational praxis, and a standard sensory medium of representation. How is the category that goes with that lexical name constructed? A Husserlean phenomenological analysis[29] would describe its intentional constitution as the creation or recognition of a *symmetry* (an invariant and repeatable pattern or likeness) present in a set of individuals 'given' to observation amid the flux of sensations by the human learned art of interpreting visual stimuli.[30] Learning of this kind is an interpretative/hermeneutical process structured both by nature and by culture.[31] It is more primitive than, say, the reading of a text, since the reading of semiotic textual signs already presupposes an acquired cultural resource from which to draw. One function, then, of the Governor of Mental Life is to reveal an *intuited meaningful symmetry* that is 'given' in observation because 'found' in a flux of local embodied sensation within a local enframing practical context of observer and observed. The mathematical structure in this account supposes an intuited group-theoretic symmetry, made present by a learned praxis of observing, by rendering meaningful by 'interpretation,' the sensory flux. Something—let us call it a "symmetry"—is found

[27] Kisiel and Kockelmans address these philosophical questions from within the language of Husserl and Heidegger; I approach them here from the scientific side, showing how scientists have failed to reach out hermeneutically beyond their models and their "data" in order to re-discover what is ontologically present but hidden in the measured "datum"; ref. Kisiel and Kockelmans (1970), especially Kisiel (1970c) and Kockelmans (1970b).

[28] See Tomasello (1999).

[29] See Jacob and Jeannerod (2003), Jacob (1988), Pribram (1991).

[30] Husserl makes an important distinction between (1).'experience' which is intentional in relation to ontological reality and the core of the pure phenomenology of experience, and (2). 'experience' which is 'inner consciousness/perception' and the content of the former, see Husserl (1970a), *Investigation V*, 542–545. See also Cassirer (1944).

[31] See Tomasello (1999).

and defined by a common likeness among the set of canonical exemplars, chosen to be held in memory. Each canonical exemplar held in memory is related to the others then as (Husserlian) 'profiles' of the same symmetry. The members of the canonical exemplifying set are updated periodically, with new exemplars replacing old ones, leading thereby to a shift in the meaning of the symmetry. The category is then defined by the symmetry exemplified by an appropriate set of canonical particulars ('profiles'). The category represents an invariant that involves the observer, a canonical set of observed exemplars in memory, and a standard enframing of physical and cultural context.

However, canonical exemplars which exemplify a particular symmetry, say, being a 'ball,' can nevertheless fail to exemplify other symmetries, such as "roundness"—for a 'football' (in the USA and in the case of rugby in the UK) is not round, though it is round in the rest of the world. The category of 'ball' then has an uncertainty relation to 'roundness.'

Concept/Category Formation by Measurement: Measurement gives a 'numbered datum.' Returning to the Heisenberg/Einstein problem referred to above, a numbered datum in quantum physics could be no more than a present messenger. The message it carries, however, has to be 'read' from the messenger-taken-as-code, the messenger as 'information.' To get the message from the information, the coded message has to be 'interpreted.' In the case of a measurement, the message is the datum; the datum is real, and present in the laboratory (together with the other theoretical observables functionally related to it). However, nothing more is communicated by the messenger-as-code than the ontological presence in the laboratory of the 'observable' —now as the 'observed.' But while 'observation' is generally accompanied by the intuition of place, shape, size, color, *etc.* these common lifeworld qualities are absent and seemingly irrelevant. An act of quantum measurement then is an 'observational' act, performed by an 'observer' conscious of being embodied in the laboratory, but blind to the common sensual intuitions of the lifeworld; in the paradigmatic way, the quantum observer 'embodied' in the laboratory has the "consciousness of a blind man" 'embodied' in his cane, inhabiting it with his bodily sensibility, and capable, for instance, of intuiting his local lifeworld space, but incapable of intuiting its colors.[32]

5 The Role of Theory and Laboratory Context in Meaning-Making

Pure Mathematics, Anschaulichkeit/intuition, Meaning Uncertainty: Puremathematics is the pure science of meaningful structure, it is a set of defined formal relationships among a lexicon of postulated mathematical entities—whether numbers, figures, or patterns—that inhabit the space of the mathematical (algebraic or geometrical)

[32] See Merleau-Ponty (1962), Heelan ([1983] 1988).

imagination. These mathematical entities exist only as *intuited*—in German, as *anschaulich*—in the esthetic space of the mathematical logic or imagination.

Theoretical Physics and Mathematical Models in Physics: The essence of modern science is historically the mathematizing of the measured world. Mathematical points, lines, and surfaces, however, are not empirical bodies in the world; they are pure—non-empirical—elements defined within *mathematical intuition* by algebraic or geometric functions. In relation to the real sensible world, an intuited representation in the imagination is no more than a *semiotic element* like a lexical word of text or like a syntactical structure of grammar.[33] It can be used, however, in a predication of experience the way a pure concept is used; such a predication instantiates the *mathematical representation of a physical exemplar.*

How is this done? A theoretical computation is a function in the field of mathematics. The classical mathematical field is the field of *Anschaulichkeit*, the field of logical structures and functions intuitable in Space and Time as imagined.[34] Mathematical formulas can be used to symbolize operations in the 'real' sensible world, usually through the instrumentality of measurement, thereby associating a network of measure-numbers with a network of related and named physical properties. Through such a mathematical model, real aspects of the empirical spatio-temporal world of human culture can be ordered and controlled.[35]

As a function related to human evolution, mathematical intuition is a cultural development of the primordial human ability to see, hear, touch, taste, and feel the world perceptually by recognizing recurrent patterns in the sensory flux, such as the numbers—measure-numbers—supplied by a laboratory measurement. These real patterns of measure-numbers, accessible by measurement, can then be used to represent an ontological entity—sometimes misleadingly called a 'theoretical entity.' The measure-number is not 'what is meant,' it is itself no more than a messenger that is a symbol of the empirical presence of something real in the space-time of the laboratory but which possibly is not imaginable or intuitable. Observation in this way is organized mathematically by measurement as a medium to explore 'what is' but what may not be imaginable/intuitable in any of the space-times with which we are familiar. Mathematics as coding introduces the essential evolutionary function of mathematics which has its own esthetic and practical value while it also has the capacity to point beyond itself to something ontological. Though mathematics has a transcendent esthetic beauty for professional mathematicians, it is not a divine language, as some distinguished physicists have piously speculated. It certainly is a human language as its history shows; it is one, however, that serves rather the function of a 'grammar' than that of a 'lexicon,' and is closely connected with the way we embodied humans organize our world by number codes, naming recurrent patterns among exemplars despite evident differences among them that produce uncertainties.

---

[33] See Hasan (2010). For a more phenomenological presentation, see Kockelmans (1970a) in Kockelmans and Kisiel (1970).

[34] For the *hermeneutic* foundations of mathematics, it is worth looking at Lakoff and Nunez (2000).

[35] See Ryckman (2005), Heelan (2003, 2004).

Among the basic organizational skills we have is the native ability to find patterns in the sensory flux to which we assign a meaning that is public, and shared through language (or a language substitute) with our cultural community.[36] Such shared and recurrent meanings are based on two kinds of recognized patterns in the sensory flux: *anschaulich* (intuitively meaningful) space-time patterns, and intensional,[37] (categorically meaningful). The former is the symmetry (invariant) that characterizes abstract mathematical intuitions that are universally valid in principle for all mathematically oriented communities; the latter is the symmetry (invariant) that characterizes observation and measurement, both of which are contextualized by local empirical circumstances, communities, needs, and goals.

Classical physics is the natural science which has faith in assuming that the observational world is simply the instantiation of culture-free a priori ideal mathematical objects. It is clear, however, that in certain situations elementary particles have to be treated differently from geometric points in space and time. Quantum physics seems to have good theoretical and experimental reasons for giving up faith in the identity of physics with mathematics.[38]

This should not have been a surprising discovery in the context of human evolution, since there is little likelihood that human visual and tactile perception would have been shaped by any other practices than those that coordinate the local actions of eyes, ears, hands, and legs which privilege a range of what turns out to be non-Euclidean finite visual spaces.[39] Cosmological matters of human interest, such as seasonal and weather changes, were treated by reading the signs in the heavens and in other ways; while matters of health and nourishment were managed by taste and smell, and by reading Nature's 'signatures' in plants and animals. There is then no a priori reason from evolutionary principles to justify universal scientific trust in the *Anschaulichkeit* criterion of the modern scientific human imagination. Such a trust was inherited mostly from the early modern period of European cultural history when the mathematization of the physical world—small, medium, and large—came to be incautiously accepted as fundamental. In recent times, it has gradually become evident that the very small and the very large need their own lexicon—linked possibly to a common overarching transcendental 'grammar.' An important contribution

[36] For the grammar of scientific discourse, see Rheinberger (1997), Berthoz and Petit (2008), and also below.

[37] The terms 'extension' and 'intension' belong to mathematics and classical logic; *extension* connotes *quantitative meanings* (numbered or spatio-temporal), *intension* connotes *cognitional* (conceptual, logical) meanings. However, contrast this with the term 'intention,' differing slightly in spelling, on which account it is regularly confused with 'intension.' 'Intention' connotes purpose or intent and is related to action and experience. A derivative term, 'intentionality,' is central to a kind of philosophy that deals with how the meanings we make involve human action and experience. This is the philosophical 'phenomenology' associated with Edmund Husserl, Maurice Merleau-Ponty, and Martin Heidegger.

[38] See Heelan (1965, 1974, 1975, 1979, 1987, 1988).

[39] See Heelan ([1983] 1988), Appendix.

to this end was made by Wigner and others[40] in introducing (what is now called) the 'standard' form of the quantum theory; this is a Hilbert (infinite-dimensional vector) representation of Space that functions like a 'grammar' in which both classical and quantum entities can be represented, the classical by universal symmetries, and the quantum by local contextual symmetries.[41]

Hilbert Vector Space as the Grammar of a Science: Operators on Hilbert Space vectors represent practical measurement procedures that link human consciousness observationally with the micro-systems represented by the vectors in a Hilbert Space.[42] The 'grammar' of those micro-systems represents not 'what is,' but how 'what is' is structured and structurally related. The scientific world we live in is then constituted existentially of universal 'absolute' symmetries, such as free classical entities, and 'local contextual symmetries'; this is the micro-structure of the laboratory world. Quantum physics has discovered a strange new property of *spin* that reaches across all Space and Time to function as a global link among micro-systems in cosmic nature. This global linkage exemplifies one of the kinds of global "entanglement"[43] in the scientific world. Such properties, while they stretch the powers of scientific intuition and observation beyond their natural (instrument free) human limits, serve to supply the intelligible foundation for the difference between the stable objects familiar in the everyday world we live in, and the *instability of its dynamic foundations*. Classical Space and Time can be seen in this perspective as the invariant (or symmetry) of a stable, but historically changing, human environment, rather than an invariant of the pre-existing unstable foundational world, the existence of which humans have come to recognize only lately. Other spaces, such as the variety of cultural visual and musical spaces in the course of historical time, belong to the domain of local contextual spaces and times. Quantum entities—as we know them today—seem to belong contextually to the unstable dynamic foundation of stable local historical cultural worlds.

How then are we to understand and represent to ourselves the 'quantum micro-realities' that appear fleetingly in laboratory experiments, or the anomalous 'cosmological macro-realities'—such as 'dark matter' and 'dark energy'—that appear in astronomical studies of the cosmos? Each makes its ontological 'presence' known in the laboratory where its measure-numbers appear. *According to Heisenberg and Einstein* (see above), however, the measure-number is no more than a *'code for presence' of what Heisenberg called "the observable," and not a 'description' of what exists ontologically.* The *ontology* then—if knowable—has to be expressed in a (more) fundamental grammatical/mathematical language—(let us call it) "F-space-time"—which is 'beyond' and 'deeper than' the 'mathematical space-time' of laboratory measurement. 'What is'—namely, the observable—is not describable in the space-time occupied by the messenger-lexicon, but (presumably) in some currently

[40] See Wigner (1962, 1963, 1967), also Wheeler and Zurek (1983), Dirac (1930), von Neumann (1955).

[41] See Heelan (1974, 1979), Bracken (2003).

[42] See Wheeler and Zurek (1983), Heelan (2004).

[43] See Aczel (2001), Shimony (1997), Gernert (2005).

unknown "F-space-time," which is the context of the lexicon's message; F-space-time would then be the (currently unknown) foundational 'grammar' for the lexicon's message-language.

In summary, in the present state of micro- and cosmo-physics, both affirm the ontological laboratory presence of fundamental physical systems, but fail to be able to describe them ontologically in any current intuitive version of laboratory space-times. Let us suppose that there is yet a more fundamental logical and epistemological space-time beyond that of any current laboratory—F-space-time—in which those entities can be described, and about which the measure-numbers 'speak.' Such a fundamental space-time would (presumably) continue to be characterized by (grammar-based) epistemological intuition regarding the context of that about which the measure-numbers 'speak.'

This conclusion agrees with the outcome reached by Heisenberg and Einstein in Berlin in 1926. They agreed that the laboratory measure-numbers indicate the ontological presence of a micro-system—the 'observable'—within a context of measurement that does not provide an ontological description of the 'observable' micro-system in terms of the space-time of the measuring laboratory. What does the latter part of this claim mean?

In a phenomenological analysis, it means first, that the laboratory with its measuring instruments belongs to 'the extended body' of the observer, and thus, that the *observer* is 'the embodied human consciousness so extended.' Consequently, the *observer* lives in and through the laboratory measuring instruments as oriented towards practice; this is the channel of his/her 'noetic' intentionality. Under such circumstances, the measure-number becomes just a coded messenger, and like the photons received by the eyes in seeing, the message delivered by the measure-numbers is interpreted within the context of the measurement. In vision, it is not the photons which are 'observed' by the culturally prepared viewer, but the illuminated objects; so also in measurement, the measure-numbers are not what are '*observed*' by the culturally prepared 'observer' but what they point to beyond themselves, namely, the presence of micro-systems in the laboratory situation, unaccompanied, however, by any intuitive description of them in the space-time of the laboratory.

Clearly then the ontology of quantum micro-entities in current quantum physics is not defined by human intuition [*Anschaulichkeit*]. To the extent that mathematical theory is the formal structural criterion of the 'language of physics,' its function is closer to that of *grammar* in linguistics that structures the *lexicon* of the lifeworld antecedent to observing and describing events in the lifeworld. The *lexicon,* however, names the categories of the things, actions, and values which exist for a local dialogical community. The criterion of *Anschaulichkeit* then is not, in this historical phase of human scientific culture, the basis of universal natural laws.

The Uncertainty of Meaning-making: 'Thin' versus 'Thick' Descriptions.[44]

1. *'Thin' description*: Laboratory science and other abstractive academic disciplines give *thin* descriptions. These are descriptions that are narrowly contextualized, and

[44] See Geertz (1973), Chap. 1, and Williams (1985), 129–152.

theoretically (abstractly) defined. The life of a modern scientific community, shielded as it is from having to take account of the diversity of surrounding cultures, horizons, agenda, styles, and goals, *etc.* is permeated with *thin* descriptions

2. *'Thick' description*: *Thick* descriptions are practical 'world-guided' descriptions of actions or events that take account of the local, historical, multi-contextual, and multicultural niches where dialogue takes place. *Thick* descriptions permeate practical common life. Cultural knowledge is thick because it involves inter-contextual discourse among speakers and hearers whose skill in such discourse is not narrowly disciplinary but culturally dialogical. Aristotle included it under the character of prudence, "*phronesis*." This kind of discourse requires respect for the complexity and diversity of issues and authorities that characterize human cultural activity. The structure of the communicative exchanges in this kind of discourse is more like that of quantum theory, that is, one based on the choice of some relevant localized contextualized symmetry shared by all parties to the discourse.

6 Dialogical Syntheses and the Grammar of Hilbert Space

'*Thick*' dialogical discourse is exemplified in the history of science, for example, in the work of Ludwik Fleck (1979) who discovered the nature and source of the venereal disease, syphilis. He narrates how this discovery came to him. It was by re-interpreting a selection of old 'facts' from dialogical sources as diverse as 'old wives tales' and odd pieces of popular medical lore, in addition to the outcomes of his experimental work. He found in his community's public memory many of the ingredients from which he retrieved the insights which led him to define the two new scientific facts that have made him famous: the disease now known as 'syphilis,' and the 'spirochete' that causes it.

Rational Dialogical Synthesis: Classical and Quantum Science. The initial conflict between classical and quantum physics is a version of the pre-Socratic question: If Nature hides behind classical physics, what of the Nature that hides behind quantum physics? Can they be the same Nature? The key terms in the dispute often seem to be 'objectivity' and 'subjectivity,' but in fact the key notions relevant to the transition from classical to quantum physics are 'intuition' [*Anschaulichkeit*] in the mathematics of quantum physics and 'measurement' in data acquisition and management. These terms, basic for understanding Niels Bohr's notion of 'complementarity,'[45] are today clamoring for re-examination.

Wigner and Dirac proposed a resolution of the dialogical conflict: The grammar and lexicon of both kinds of physics could be represented by a Hilbert Vector Space. In such a Space, a quantum system in pure (unobserved) motion is represented as developing under the influence of the appropriate Schrödinger equation; observed

[45] See Beller (1999).

data, however, are represented mathematically by the '*eigen*' (definite proper) values of the data operators acting on the state vector representation; these *eigen* values are codes for observable states of the system.

A dialogical conflict arose about the ontological and epistemological criteria of '*truth*' and '*reality*' in physics; it came to focus on the question of how mathematical *intuition* can be reconciled with *observations* to produce a coherent human understanding of the 'natural' world. Mathematical intuition is the a priori working space of the theoretical physicist; it structures—as it were grammatically—the a priori of the quantum narrative. Measurement defines the a posteriori observational space of the experimental physicist by providing the codes—measure-numbers—which enable the narrative of the observed data to be told. By introducing 'complementarity,' Bohr fudged the answer by attributing reality—of a classical kind—to the pure (ideal) objects of mathematical theory, while placing limits on observational access to these classical realities by measurement. In Kantian (and Neokantian) terms, mathematics describes the *noumenon*, while 'complementarity' restricts the observed *phenomenon.*

Deeply involved in all of this is the function of mathematics in any pure (non-empirical) discourse about the real world. In classical physics, its traditional function was to provide the intuition—*Anschaulichkeit*—of a generalized universal pure Space and Time that comprehensively 'represented' the physical world. This intuition was the sole reality guarantee of the 'representation' provided by physical theory. Such a view of mathematics goes back to Plato's 'likely story' in the *Timaeus*. Two millennia later, we find ourselves telling a different story!

Currently, there is a breakdown in dialogical common sense, because the traditional classical (Platonist) connection between a universal *anschaulich* space-time and its atomic contents fails in both quantum and relativity physics. The breakthrough was made by Eugene Wigner, Paul Dirac, and John von Neumann. Wigner[46] was the key figure in this proclamation.[47] He was trained as a physical chemist and crystallographer, familiar with the chemical laboratory. His mentor was Michael Polanyi[48] also a physical chemist and later a well-known philosopher of science. Wigner applied to quantum theory the kind of dynamic group theoretic mathematics that crystallographers use to describe the production of crystalline symmetries in a solution, and to explain mixed crystalline forms in ways that foreshadowed the uncertainties of quantum physics. Wigner represented the state of a quantum system by an infinite-dimensional vector in a Hilbert Space governed by a dynamic law of change and development inspired by Schrödinger.

Wigner was Hilbert's assistant at the University of Göttingen from 1927 till 1931. There he reflected deeply on the higher purpose of physics, it was "to elevate the material side of the world, to make daily life easier for all the world's people."[49]

[46] See Szanton (1992), 309.

[47] Wigner was also the brother-in-law of Dirac, and a schoolboy chum of von Neumann in his native Hungary.

[48] See Scott and Moleski (2005).

[49] Szanton (1992), 111; see also 308–309.

The academic environment at the university was steeped in Husserl's philosophy and the notion of "invariance/symmetry" that defined its core. Wigner introduced what is now the standard form of the quantum theory. In this representation, states of the quantum system are represented by an infinite-dimensional vector in mathematical Hilbert Space. In the new mathematical representation, *observables*—properties of the system empirically accessible—are represented by *projection operators* on the Hilbert Space. A projection operator generates a subspace of the Hilbert Space in which the relevant observable has a definite ('*eigen*') value. Other observables applied to this subspace are permitted to have definite values provided their projection operators are compatible with it (by algebraic commutation), but not otherwise. What is new about the Hilbert Space representation is its capacity to represent and encode *contextuality*: pairs of observables which share a common context can be measured together, while pairs which do not share a common context, cannot be measured together with certainty. Relative then to quantum systems, *observables are context-dependent*; some pairs can be measured accurately together, other pairs cannot be measured accurately together and suffer uncertainty. The great achievement of Wigner and Dirac was to generate a mathematical representation of the context-dependency of observables in quantum mechanics.[50]

In the standard [Hilbert Vector Space] view of quantum physics, the '*properties*' of physical objects are expressed as a combination of locally context-dependent 'observables' (represented by contextual 'invariants' or 'symmetries') within the universal symmetry group of space-time. Classical physics knows nothing about, and ignores local context-dependent properties. Hilbert Space quantum physics revolutionized the science of physics because, by allowing for different local and mutually incompatible contexts or horizons of empirical research, the physicist can unify her search by drawing on a broader heuristic question: whether the unknown X (that is sought) defines a universal symmetry of the Hilbert Space or a local context-dependent symmetry—representing the relevant subspace—of the Hilbert Space. If I am researching a local symmetry, then I must design the kind of laboratory bench that allows me to observe the context-dependent effects of the local symmetry. The real world model now unifies both universal and local symmetries in a common synthesis.

The synthesis takes note of the presence and role of the embodied consciousness of the local observer as it functions in the measurement process, and who by choice and agency enters into the definition of the local group theoretic symmetry. Quantum physics then has to recognize the dependence of observables on the context-dependent physical platform of the chosen laboratory bench, insofar as the laboratory bench is an extension of the embodied character of the consciousness of the scientific observer. In contrast with observations in classical physics, observations in quantum mechanics from different local physical platforms (represented by different subspaces of the Hilbert Space) do not simply add up to a coherent objective culture-free scientific account.

[50] For these insights, Dirac received the Nobel Prize in physics in 1933; Wigner, for his part, in 1963.

One consequence is that in quantum physics, it is not possible in principle to give a comprehensive empirical description of a unified and objective 'world' for all observers. There is no more than a unified and comprehensive *grammar* of the pure world-for-any-observer—namely, the Hilbert Space with its systems of vectors and operators; but for any individual observer, the observables of the world constitute context-dependent branches peculiar to that observer and that observer's choice of what to measure. All such knowledge is partial, relative, scientific, but culturally perspectival.

Is this, you ask, the little we are left with when natural science promised so much more? We should not forget that the little we know of the quantum world is accompanied by the bonus of knowing something about ourselves—namely, awareness of the *interiority of the work of human consciousness embodied in the process of scientific measurement, just as in every act of perception/observation.* This reflection warns us that we are an evolutionary product of Nature and cannot assume that we can study Nature 'objectively' from beyond the horizon of Nature. We rule Nature from within Nature—not as monarchs of Nature—but as its gardeners.

7 Is a Theory of Human Consciousness?

Phenomenology is concerned with human consciousness. Human consciousness, the Governor of Mental Life, is the agent that produces and recognizes the categories for a strict theoretical science. Can human consciousness produce the category to define itself?

Human consciousness produces categories and recognizes instances of the various categories by becoming aware of recursive patterns of form and function in the sensory flux. If the only source of categorial knowledge is the embodied sensory flux, then the question becomes: Has human consciousness access to the kind of embodied sensory flux that reveals the category to which human consciousness belongs? A category is a'symmetry' ('invariant' or 'property') among a core set of embodied exemplars; it is normative for all instances of its kind; a categorial symmetry is theoretical to the extent that elements of the set can substitute for one another as exemplars of the symmetry. Human consciousness is peculiar in that while we have no trouble in practice recognizing exemplars of human consciousness—we call them "persons"—we are nevertheless hesitant to say that they can substitute for one another under a 'category' of 'persons,' in the way we count instances of cups, saucers, birds, and beasts. Persons can substitute for one another in specific ways, as car buyers, as music lovers, as sports' fans, as Democrats, as Republicans, Yes! but … as 'human consciousnesses'? No! because people do not share their personal identity. Personal identity is expressed existentially in the individual's living '*interiority*' or '*being-in-the world*.' If others are 'persons' like me, they will be fundamentally different knowers and doers, though sharing many cultural contexts. Isn't this then what it means to understand oneself as a person, and as an exemplar of human consciousness? But this is not a categorial understanding of human consciousness, for there are few if any universally true and absolute deductive inferences that can be drawn from the descriptions of individual persons. Phenomenology, as the study of human consciousness, leads to a theoretical study of the dynamic, normative structures of intentionality, meaning-making, and decision-making, but not to the existential choices and practices that individually shape the human consciousness that is a Governor of Mental Life.

8 Recapitulation

To summarize what has been said in this paper: This paper seeks to exemplify how hermeneutical phenomenology can analyze the implicit meaning and context of the natural sciences, in particular the epistemological and ontological problems of quantum physics. In this light quantum physics is shown to be fundamentally a return from the transcendental world of classical physics to the lifeworld of human experience, in which the contextuality of scientific discourse is exemplified in observation, and the human subject becomes the human embodied consciousness of the observer which mediates the epistemic engagement with the ontology of the observed datum. New insight is given on the deeper meaning of the quantum uncertainty principles when examined philosophically from the viewpoint of hermeneutical phenomenology. The new emphasis is on the distinction between embodied subjectivity and observable objectivity, from which flow the epistemology and ontology of quantum physics. The ontology of quantum physics is clearly not culture-free and history-free, but whether progress can be beyond the present social and historical inventions of human consciousness within Nature still remains to be seen.

In this essay I claim that a *Hilbert Space, exemplified by quantum physics, represents an existential grammar of the physical properties of the quantum system that is capable of synthesizing hermeneutically both universal and existential context-dependent concepts/categories.* Quantum theory is then a phenomenological and hermeneutical theory of the phenomena of human consciousness, as claimed by Schrödinger, Wigner, and Heisenberg—and possibly Einstein. It is a theory of how embodied human consciousness—which is the Governor of Mental Life—acts as a new norm, extending the notion of human rationality beyond the classical norms of rationality to include hermeneutic norms and so to show the fundamental dependence of science on the lifeworld, as Husserl claimed and Kockelmans so brilliantly defended.

References

Aczel, Amir. 2001. *EN Tanglement: The Greatest Mystery of Physics*. New York: Four Walls Eight Windows.

Arnheim, Rudolf. 1974. *Art and Visual Perception*. Berkeley: University of California Press.

Babich, Babette. 1997. Against postmodernism and the 'new' philosophy of science: Nietzsche's image of science in the light of art. In *Issues and Images in the Philosophy of Science*, ed. Dimitri Ginev and R.S. Cohen, 27–46.

Babich, Babette (ed.). 2002. *Hermeneutic Philosophy and the Philosophy of Science, Van Gogh's Eyes and God. Essays in Honor of Patrick A. Heelan, S.J.* Dordrecht: Kluwer.
Beller, Mara. 1999. *Quantum dialogue: The story of a revolution*. Chicago: University of Chicago Press.
Berthoz, Alain, and Jean-Luc Petit. 2008. *The Physiology and Phenomenology of Action*. New York: Oxford University Press.
Bracken, Anthony J. 2003. Quantum mechanics as an approximation to classical mechanics in Hilbert space. *Journal of Physics A: Mathematics and Theoretical* 36 #23(13 June): L329–L335.
Byrnes, Heidi. 2002. The dialogism of meaning: The discursive embeddedness of knowledge, the colloquy of being. In *Hermeneutic Philosophy and the Philosophy of Science, Van Gogh's Eyes and God*, ed. B.E. Babich, 411–422. Dordrecht: Kluwer.
Cassirer, Ernst. 1944. The concept of group and the theory of perception. *Philosophy and Phenomenological Research* V/1: 1–35.
Dirac, Paul A.M. 1930. *The principles of quantum mechanics*. Oxford: Clarendon Press.
Donald, Merlin. 2001. *A Mind so Rare: The Evolution of Human Consciousness*. New York: Norton.
Eger, Martin. 1999. Language and the double hermeneutic in natural science. In *Hermeneutics and science*, ed. M. Feher, O. Kiss, and L. Ropolyi, 265–289. Dordrecht: Kluwer.
Einstein, Albert. 1950. *Out of My Later Years*. New York: Philosophical Library.
Farre, George. 1998. Characteristics and implications for the philosophy of nature. *Acta Polytechnica Scandinavica* 91: 3f.
Fleck, Ludwik. 1979. *Genesis and Development of a Scientific Fact*. Trans. T. Trenn. Chicago: University of Chicago Press.
Geertz, Clifford. 1973. *The Interpretation of Cultures*. New York: Basic Books.
Gernert, Dieter. 2005. Conditions of entanglement. *Frontier Perspectives* 14: 8–13.
Gibson, James J. 1979. *The Ecological Approach to Visual Perception*. Boston: Houghton Miflin.
Ginev, Dimitri, and Roert Cohen (eds.). 1997. *Issues and Images in the Philosophy of Science*. Dordrecht: Kluwer.
Hadot, Pierre. 2006. *The Veil of Isis*. Cambridge, MA: Harvard University Press.
Hasan, Ruqaiya. 2010. *Describing Language: Form and Function* (Collected works of Ruqaiya Hasan). London: Equinox Publishing.
Heelan, Patrick. 1965. *Quantum Mechanics and Objectivity: A Study of the Physical Philosophy of Werner Heisenberg*. The Hague: Nijhoff.
Heelan, Patrick. 1972. Towards a new analysis of the pictorial space of Vincent van Gogh. *Art Bulletin* 54: 478–492.
Heelan, Patrick. 1974. Quantum logic and classical logic: Their respective roles. In *Logical and Epistemological Studies in Contemporary Physics*, Boston studies in the philosophy of science series, vol. 13, ed. Robert S. Cohen and Wartofsky Marx, 318–349. The Hague: Reidel.
Heelan, Patrick. 1975. Heisenberg and radical theoretic change. *Zeitschrift Für Allgemeine Wissenschaftstheorie* 6: 113–138.
Heelan, Patrick. 1979. Complementarity, context-dependence and quantum logic. In *The Logico-Algebraic Approach to Quantum Mechanics*, University of Western Ontario series on the philosophy of science, ed. C. Hooker, 161–179. Dordrecht: Reidel.
Heelan, Patrick. 1983. Perception as a hermeneutical act. *Review of Metaphysics* 37: 61–75.
Heelan, Patrick. 1987. Husserl's later philosophy of natural science. *Philosophy of Science* 54: 368–390.
Heelan, Patrick. 1983/1988. *Space-Perception and the Philosophy of Science*. Berkeley: University of California Press.
Heelan, Patrick. 1988. Husserl, Hilbert and the critique of Galilean science. In *Edmund Husserl and the Phenomenological Tradition*, ed. Robert Sokolowski, 157–173. Washington, DC: The Catholic University of America Press.
Heelan, Patrick. 1994. Galileo, Luther, and the hermeneutics of natural science. In *The Question of Hermeneutics: Festschrift for Joseph Kockelmans*, ed. Timothy Stapleton, 363–375. Dordrecht: Kluwer.

Heelan, Patrick. 1998. Scope of hermeneutics in the philosophy of natural science. *Studies in the History and Philosophy of Science* 29: 273–298.

Heelan, Patrick. 2002. Faith and reason in philosophical perspective. In *La Responsibilité de la raison: Hommage à Jean Ladrière*, ed. J.-F. Malherbe, 149–175. Leuven: Peeters.

Heelan, Patrick. 2003. Phenomenology and the philosophy of the natural sciences. In *Phenomenology World-Wide*, ed. A.T. Tymieniecka, 631–640. Dordrecht: Kluwer.

Heelan, Patrick. 2004. The phenomenological role of consciousness in measurement. *Mind and Matter* 2: 61–84.

Heelan, Patrick. 2009. The role of consciousness as meaning-maker in science, culture, and religion. *Zygon* 44: 467–486.

Hefner, Philip. (ed.) 2006. On topics related to the quantum theory, reality, consciousness, spirit, and mind. *Zygon: Journal of Religion and Science* 41: #3

Heidegger, M. 1962. *Being and Time*. Trans. J. Macquarrie, and E. Robinson. London: SCM Press.

Heidegger, M. 1967. *What is a Thing*? Orig. title, *Die Frage nach dem Ding* [1954] (trans: W.B. Barton, Jr and V. Deutsch). Chicago: Regnery.

Heidegger, M. 1982. *On the Way to Language*. Trans. P.D. Hertz. New York: Harper and Row.

Heidegger, M. 1995. *Ontology (Hermeneutics of Facticity)*. Trans. J. van Buren. Bloomington: Indiana University Press.

Heidegger, M. 1999. *Contributions to Philosophy (From enowning)*. Orig. title, *Beiträge zur Philosophie* [1989] (trans: Parvis Emad, and Kenneth Maly). Bloomington: Indiana University Press.

Heidegger, M. 2002a. *Off the Beaten Track*. Orig. title: 1950. *Holzwege* (trans. J. Young, and K. Haynes). Cambridge: Cambridge University Press.

Heidegger, M. 2002b. *Supplement: From the Earliest Essays to Being and Time and Beyond*, ed. J. van Buren. Albany: Starte University of New York Press.

Heisenberg, Werner. 1950. *Physical Principles of the Quantum Theory*. Trans. C. Eckart and F. C. Hoyt. New York: Dover. Orig. title: Physikalische Prinzipien der Quantentheorie [1930].

Heisenberg, Werner. 1971. *Physics and Beyond: Encounters and Conversations*. New York: Harper and Row. Orig. title, *Der Teil und das Ganz*. [1970].

Husserl, Edmund. 1966. *Analyses Concerning Passive and Active Synthesis (Analysen zur passiven synthesis)*. Trans. A. Steinbock. Dordrecht: Kluwer.

Husserl, Edmund. [1901–1913] 1970a. *Logical Investigations*. Trans. J. N. Findley. New York: Humanities Press.

Husserl, Edmund. [1954] 1970b. *The Crisis of European Sciences and Transcendental Philosophy: An Introduction to Phenomenological Philosophy*. Trans. D. Carr. Evanston: Northwestern University Press.

Husserl, Edmund. [1952] 1989. *Ideas II*. Trans. R. Rojcewicz, and A. Schuwer Eng. Dordrecht: Kluwer.

Jacob, Francois. 1988. *The Statue Within: An Autobiography*. Trans. Franklin, Philip. New York: Basic Books.

Jacob, Pierre, and Marc Jeannerod. 2003. *Ways of Seeing: The Scope and Limits of Visual Cognition*. Oxford: Oxford University Press.

Kisiel, Theodore. 1970a. Phenomenology as the science of science. In *Phenomenology and the natural sciences*, ed. J. Kockelmans and T. Kisiel, 5–44. Evanston: Northwestern.

Kisiel, Theodore. 1970b. Husserl on the history of science. In *Phenomenology and the natural sciences*, ed. J. Kockelmans and T. Kisiel, 68–92. Evanston: Northwestern.

Kisiel, Theodore. 1970c. Merleua-Ponty on philosophy and science. In *Phenomenology and the natural sciences*, ed. J. Kockelmans and T. Kisiel, 251–273. Evanston: Northwestern.

Kockelmans, Joseph. 1970a. The mathematization of nature in Husserl's last publication, Krisis. In *Phenomenology and the natural sciences*, ed. J. Kockelmans and T. Kisiel, 45–67. Evanston: Northwestern.

Kockelmans, Joseph. 1970b. Merleau-Ponty on space perception and space. In *Phenomenology and the natural sciences*, ed. J. Kockelmans and T. Kisiel, 274–311. Evanston: Northwestern.

Kockelmans, Joseph, and Theodore Kisiel. 1970. *Phenomenology and the Natural Sciences*. Evanston: Northwestern University Press.

Lakoff, George, and Rafael Nuñez. 2000. *Where Mathematics Comes from: How the Embodied mind Brings Mathematics into Being*. New York: Basic Books.
Lonergan, Bernard. (1957) 1992. *Insight: A Study of Human Understanding*. Toronto: University of Toronto Press.
Lonergan, Bernard. [1972] 1990. *Method in Theology*. Toronto: University of Toronto Press.
Lonergan, Bernard. 1950. The metric of visual space. *Journal of the Optical Society of America* 40: 627–642.
Luneburg, Rudolf. 1947. *Mathematical Analysis of Binocular Vision*. Princeton: Princeton University Press.
Luneburg, Rudolf. 1895. *The metric of visual space. Journal of the Optical Society of America* 40: 627–642.
Merleau-Ponty, Maurice. 1962. *The Phenomenology of Perception*. New York: Routledge and Kegan Paul.
Penrose, Roger. 1994. *Shadows of the Mind: A Search for the Missing Science of Consciousness*. Oxford: Oxford University Press.
Pribram, Karl. 1971. *Languages of the Brain: Experimental Paradoxes and Principles in Neuropsychology*. Englewood Cliffs: Prentice-Hall, Inc.
Pribram, Karl. 1991. *Brain and Perception: Holonomy and Structure in Figural Processing*. Hillsdale: Erlbaum.
Rheinberger, H.-J. 1997. *Toward a history of epistemic things*. Stanford: Stanford University Press.
Ryckman, Thomas. 2005. *The Reign of Relativity: Philosophy in Physics 1915–1925*. Oxford: Oxford University Press.
Scott, William T., and Martin X. Moleski. 2005. *Michael Polanyi: Scientist and Philosopher*. Oxford: Oxford University Press.
Shimony, Abner. 1997. On mentality, quantum mechanics and the actualization of potentialities. In *The Large, the Small and the Human Mind*, ed. Penrose Roger, A. Shimony, N. Cartwright, and S. Hawking, 144–160. Cambridge, UK: Cambridge University Press.
Szanton, Andrew. 1992. *Recollections of Eugene Wigner*. New York: Plenum.
Tomasello, Michael. 1999. *The Cultural Origins of Human Cognition*. Cambridge, MA: Harvard University Press.
Von Neumann, John. 1955. *Mathematical Foundations of Quantum Mechanics*, ed. R.T. Beyer. Princeton: Princeton University Press. Chap.VI: 417–445.
Wellmon, Chad. 2006. Poesie as antropology: Schleiermacher, colonial history, and the ethic of ethnography. *The German Quarterly* 79: 423–442.
Welton, Donn. 2000. *The Other Husserl: The Horizons of Transcendental Phenomenology*. Bloomington: Indiana University Press.
Wheeler, John A., and Wojciech H. Zurek. 1983. *Quantum Theory and Measurement*. Princeton: Princeton University Press.
Wigner, Eugene. 1962. Remarks on the mind-body question. In *The Scientist Speculates*, ed. I.J. Good, 284–301. London: Heinemanm.
Wigner, Eugene. 1963. The problem of measurement. *American Journal of Physics* 31: 6.
Wigner, Eugene. 1967. *Symmetries and Reflections: Scientific Essays of Eugene P. Wigner*. Bloomington: Indiana University Press.
Williams, Bernard. 1985. *Ethics and the Limits of Philosophy*. Cambridge, MA: Harvard University Press.

Die ewige Wiederkunft wissenschaftlich betrachtet. Oskar Beckers Nietzscheinterpretation im Kontext

Michael Stöltzner

Abstract Within the literature on Nietzsche's thesis of eternal recurrence, the discussion of its ethical or anthropological aspects prevails. From the viewpoint of today's physical cosmology, this is hardly surprising. Yet during the first half of the 20th century a couple of eminent scholars have treated eternal recurrence as a serious if speculative scientific idea, either to justify its validity or to find it worthy of an elaborated criticism within the science of the day. My paper critically investigates the 1936 attempt of the logician and philosopher Oskar Becker to justify both a physical and a logical argument he spots in Nietzsche's writings. While Becker endorses Abel Rey's 1927 book *Le retour éternel e la philosophie de la physique,* he remains silent about his former colleague at the University of Bonn, the mathematician Felix Hausdorff and his 1898 book *Das Chaos in kosmischer Auslese* (published under the pseudonym Paul Mongré). The reasons for this silence were, to my mind, not only that Hausdorff had been forced out of his job under the Nuremberg laws while Becker showed a constantly growing sympathy for the Nazi regime, but also in Becker's intention to emphasize the pro-scientific side of the newly elevated state philosopher Nietzsche. I analyze Becker's arguments against the backdrop of Rey's and Hausdorff's considerations and argue that these severely limit the validity of Becker's conclusions. To discuss Nietzsche's physical proof first, Becker argues that assuming a finite space, a finite number of possible particle positions and energy states, a finite number of atoms, and Laplacian determinism indeed yields a periodic recurrence of all world states. Being aware of Rey's discussion of Nietzschean recurrence within the context of statistical mechanics, Becker concedes that by relaxing determinism one might only be able to prove that nature cannot avoid recurring to any given single state. And thus he passes the buck to the sciences. But as the discussions ensuing Zermelo's recurrence objections have shown, the statistical

M. Stöltzner (✉)
Department of Philosophy, University of South Carolina,
901 Sumter, Columbia, SC 29208, USA
e-mail: stoeltzn@mailbox.sc.edu

B. Babich and D. Ginev (eds.), *The Multidimensionality of Hermeneutic Phenomenology*, 113
Contributions to Phenomenology 70, DOI 10.1007/978-3-319-01707-5_8,

character of the second law requires a more radical departure from the established Laplacean conception of world state. On this basis, Rey argued that recurrence illustrates a more profound opposition between subject and object, in which the second law becomes the measure of the subjective and eternal recurrence the measure of the objective. Becker's mathematical argument takes its point of departure by distinguishing, on the one hand, *regressus in infinitum* and *progressus ex infinito* and, on the other hand, *progressus in infinitum* and *progressus ex infinito.* Thus the inconsistency of a *progressus in infinitum usque ad certam finem (resp. nunc)* does not fault the *progressus ex infinito usque ad certam finem (resp. nunc)* because the world states could be periodically arranged, as Nietzsche has claimed. Becker's argument, for one, rests upon a constructivist stand in the foundations of mathematics and the Heideggerian underpinning of it by the temporality of mathematical thought that he had already given in his 1927 *Mathematische Existenz.* But Becker also assumes that, for periodic motions, one can (in thought) reverse the order of time. A look at Hausdorff's book and a proto-set-theoretic argument presented there shows, however, that such a reversal does not work without invoking what Hausdorff calls transcendent reality.[1]

Nietzsches Verhältnis zu den Wissenschaften und seine Rolle für eine wissenschaftliche Philosophie sind bis heute umstritten. Da ist zum einen der 'positivistische' frühe Nietzsche, der die Kantische Metaphysikkritik radikalisierte, der eine überkommene Ethik zu zertrümmern sich anschickte und der die Geschichtsteleologie des 19. Jahrhunderts verspottete. Dieser Nietzsche erfreute sich sogar unter neopositivistischen Philosophen einer gewissen Wertschätzung,[2] und zwar gerade dort, wo deren eigene Metaphysikkritik noch rhetorisch angefacht werden sollte. Scheinbar auf der anderen Seite steht der Autor des *Zarathustra* und insbesondere jener Manuskripte und Fragmente, die als *Wille zur Macht* kompiliert für die Ideologie des Nationalsozialismus eingespannt werden sollten, und—man denke an die Rede von Züchtung und Übermensch—es wohl auch konnten.

Es gibt gute philosophiehistorische Gründe, einer solchen Zweiteilung von Nietzsches komplexem Denkweg zu widersprechen. Ich habe aber dennoch mit ihr begonnen, weil sie die Brisanz jenes Versuchs aufzeigt, den Oskar Becker 1936 unternommen hat, nämlich "Nietzsches Beweise für seine Lehre von der ewigen Wiederkunft"[3] aus der Perspektive von Mathematik und Naturwissenschaft ernst zu nehmen und sie mit den Mitteln einer wissenschaftlichen Philosophie zu untersuchen. Denn die ewige Wiederkunft war, Nietzsches eigenen Äußerungen

[1] A shorter English version of this paper has appeared as Stöltzner 2012. The current, original version was presented at the 2008 Beckertagung in Hagen (Germany). I am indebted to the participants for their manifold comments and suggestions, especially to V. Peckhaus, D. Piecha and A. Gethmann-Siefert.

[2] Vgl. insbes. von Mises (1939).

[3] Becker (1936), 368–387. Ich benutze den Wiederabdruck in Becker (1963), 41–66.

zufolge, die "Grundconception"[4] des *Zarathustra* und ganz allgemein der Angelpunkt seines Spätwerks, ausersehen, den Nihilismus zu überwinden und die Umwertung der Werte zu ermöglichen. Eine positivistische Grundstimmung ist dabei zunächst nicht zu bemerken. Glaubt man jedoch der Biographie von Lou Andreas-Salomé, so wollte Nietzsche nach der Auffindung des Gedankens 1881 in Sils-Maria sich zehn Jahre in Wien oder Paris in die zeitgenössische Naturwissenschaft vertiefen, um "eine wissenschaftlich unverrückbare Basis dafür zu gewinnen".[5] Doch sei Nietzsche sehr bald klar geworden, "daß die wissenschaftliche Fundamenentierung der Wiederkunftslehre auf Grund der atomistischen Theorie nicht durchführbar sei."[6] Daher habe er in der Folge ihre ethischen und religiösen Konsequenzen betont und sie zunehmend als eine mystische Offenbarung begriffen. Auch Becker zitiert eingangs seines Aufsatzes "die bekannte biographische Tatsache"[7] von Nietzsches Studienabsichten, ohne allerdings zu erwähnen, dass Andrea-Salomés Lesart heftig umstritten war, und die ewige Wiederkunft noch zu Nietzsches Lebzeiten ins Zentrum der Konflikte um die Herausgabe seines Nachlasses geraten war. So bestand der (später geschasste) erste Herausgeber Heinrich Köselitz auf der mechanistischen Interpretation. Ebenso Rudolf Steiner, der wie Becker die Verbindung der ewigen Wiederkunft zu Thesen aus Eugen Dührings *Cursus der Philosophie*[8] betonte, jedoch zusammen mit Andrea-Salomé annahm, dass sich Nietzsche selbst recht schnell von der wissenschaftlichen Unhaltbarkeit des Wiederkunftgedankens überzeugt hatte.

Becker möchte dem Leser durch eine detaillierte, auf den Schriften und dem seinerzeit publizierten Nachlass basierende Philologie vorführen, dass sich in Nietzsches Werk zwei gültige Beweise für die ewige Wiederkunft des Gleichen finden, ein physikalisch-kosmologischer und ein mathematisch-logischer. Diese sollen im Zentrum meiner Ausführungen stehen und auf dem Hintergrund des damaligen Wissensstandes kritisch gewürdigt werden. Auf die Frage, inwieweit Beckers Rekonstruktion von Nietzsches Gedankengang auch tatsächlich den Intentionen des Autors gerecht wird und ob alle verwendeten Quellen im Lichte der heutigen Editionsmaßstäbe überhaupt zulässig sind, kann ich hingegen nicht eingehen.[9]

Becker war sich durchaus bewusst, dass sein Unterfangen nicht zum Mainstream der Nietzscheforschung gehörte. Zu Beginn des Aufsatzes benennt er auch die Gründe für das geringe Interesse an den wissenschaftlichen Grundlagen von Nietzsches Wiederkunftsgedanken, ja die Ablehnung, die dieser erfahren habe, nämlich die in Deutschland besonders ausgeprägte Trennung zwischen Natur- und Geisteswissenschaften. Unter Nietzsches französischen Interpreten, insbesondere bei

[4] Vgl. Friedrich Nietzsche: *Ecce homo*. Nietzsche (1980), Band 6, 335. Für den historischen Hintergrund des Wiederkunftsgedankens bei Nietzsche und seiner Rezeption stütze ich mich auf die Darstellung von Werner Stegmaier (2004), insbes. 37–49.

[5] Lou Andreas-Salomé, zitiert nach Stegmaier (2004), 44.

[6] Ebd.

[7] Becker (1936), 41.

[8] Dühring (1875).

[9] Ich habe daher nur die im vorliegenden Aufsatz verwendeten Nietzschezitate weiter nachgeprüft.

Charles Andler[10] und Abel Rey, bestehe hierfür weitaus mehr Interesse. Reys umfängliche Studie *Le retour éternel e la philosophie de la physique* aus dem Jahre 1927 ist dabei für mich die wesentlichere.[11] Nicht nur weil sie aus der Feder des wichtigsten Protagonisten der zweiten Garde der französischen Konventionalisten stammt. Sondern vor allem auch, weil Rey über das philosophiehistorische Anliegen hinaus mit Nietzsches Wiederkunftgedanken in wissenschaftsphilosophischer Hinsicht Ernst machen wollte, ebenso wie Becker. Ich werde *Le retour éternel* daher als systematischen Kontrapunkt zum von Becker akzeptierten physikalisch-kosmologischen Beweis heranziehen. Meine Schlussfolgerung wird sein, dass Becker im Gegensatz zu Rey kaum die Konsequenzen aus der atomistischen bzw. statistischen Interpretation des zweiten Hauptsatzes der Thermodynamik rezipiert hat, jene Fragen des Atomismus, die Nietzsche, Andrea-Salomé zufolge, zehn Jahre lang hatte studieren wollen. Im Vergleich zu Beckers detaillierter Textarbeit und der überzeugenden Einordnung der Nietzscheschen Gedanken in die Philosophiegeschichte fällt Reys Buch allerdings deutlich ab. Dem heutigen Leser werden die kosmologischen Überlegungen, die sich bei Becker und Rey finden, als hochspekulativ und geradezu unwissenschaftlich vorkommen. Doch ist hier Vorsicht geboten. Noch bis in die 1950er Jahre wurde von verschiedenen Seiten bestritten, ob es eine physikalische Kosmologie im streng wissenschaftlichen Sinne überhaupt geben könne bzw. ob die relativistische Feldphysik dafür bereits eine geeignete Basis biete, selbst wenn seit den 1920er Jahren explizite Vorschläge kosmologischer Modelle diskutiert worden waren.[12]

Als Kontrapunkt des zweiten von Becker akzeptierten Beweises der Wiederkunftsthese—bzw. des dritten, den er bei Nietzsche findet—dient mir ein anderer vorgängiger Interpretationsversuch, der ebenfalls mit Begriffen und Methoden arbeitet, die den Ende des 19. und zu Beginn des 20. Jahrhunderts aufgekommenen Untersuchungen über die Grundlagen der Mathematik, insbesondere der Mengenlehre, entstammen. Diesen Versuch erwähnt Becker allerdings nicht, obwohl er ihn meines Erachtens hätte kennen müssen. Schließlich war dessen Autor sein Kollege an der Bonner Universität und der bedeutendste Mathematiker daselbst. Die Rede ist von Felix Hausdorff und seinen Jugendwerken *Sant' Ilario. Gedanken aus der Landschaft Zarathustras* und *Das Chaos in kosmischer Auslese*,[13] die dieser nach seiner Habilitation über ein Problem der mathematischen Astronomie unter dem damals weitgehend bekannten Pseudonym Paul Mongré (1897, 1898) publiziert hatte. Hausdorff/Mongré stand auch in enger Verbindung zum Nietzsche-Archiv und kommentierte in mehreren an prominenter Stelle erschienenen Zeitschriftenartikeln kritisch den Editionsfortschritt. Während *Sant'Ilario* schon rein äußerlich als eine von vielen zeitgenössischen Nachahmungen des *Zarathustra* 1936, mithin fast vierzig Jahre später, wenig Interesse erheischen mochte, war *Das*

[10] Andler (1920–1931).

[11] Rey (1927).

[12] Vgl. Kragh (1999), insbes. Kap. 5.2.

[13] Paul Mongré (1897); ders.: (1898). Beide wurden im bereits zitierten VII. Band der *Gesammelten Werke* von Felix Hausdorff wieder abgedruckt und werden im Folgenden nach dieser Ausgabe zitiert.

Chaos eine umfangreiche philosophische Analyse über die Begriffe von Raum und Zeit sowie allgemeiner eine Verteidigung eines 'besonnenen Empirismus'—so nennt Hausdorff seinen eigenen Standpunkt—gegen einen metaphysischen Realismus, die z.B. auch bei Moritz Schlick positive Erwähnung fand, gerade wegen ihrer Vorwegnahme der konventionalistischen Analyse des Raumproblems.[14]

Oskar Becker kannte Hausdorffs mathematische Arbeiten bereits zu Zeiten von *Mathematische Existenz,* wo er dessen *Grundzüge der Mengenlehre* mehrmals zitierte.[15] Und dass in Bonner Professorenkreisen des stillen Kollegen Hausdorffs frühes literarisches und philosophisches Schaffen gänzlich unerwähnt geblieben sein sollte, erscheint mir unwahrscheinlich—wenn es auch nicht das erste Beispiel eines derartigen akademischen Nebeneinanderherlebens wäre. Falls Becker *Das Chaos* kannte, so liegt nahe, dass er Hausdorff nach dessen Emeritierung 1935 in Folge der nationalsozialistischen Rassenpolitik nicht unbedingt zitieren wollte. Hausdorff emigrierte übrigens nicht aus Deutschland, sondern lebte bis 1942 unter zunehmenden Schwierigkeiten weiter in Bonn. Als man ihn zusammen mit seiner Frau als einen der letzten noch in Bonn verbliebenen Juden nach Theresienstadt deportieren wollte, begingen beide Selbstmord.[16]

Doch mir geht es hier nicht um ausgleichende Gerechtigkeit in Sachen Nietzsche. Vielmehr zeigt der nicht geführte Bonner Dialog, dass Beckers Rekonstruktion von Nietzsches mathematischem Beweis zu starke implizite Voraussetzungen enthält und es andere mathematische Formulierungen des Wiederkunftgedankens gibt, die diesen als unhaltbar erweisen, jedenfalls im von Becker intendierten immanenten (d.i. nicht-transzendenten) Sinn. Eines von Hausdorffs Argumenten basiert zudem auf informellen mengentheoretischen Überlegungen und ist der Beckerschen Begründung von Nietzsches Beweis nicht unähnlich.

Indem meine Analyse für beide Beweise negativ ausfällt, und zwar auch auf Basis des damaligen wissenschaftlichen Kenntnisstandes, wird deutlich, dass Beckers Aufsatz im Kontext der zentralen Rolle zu verstehen ist, die er der ewigen Wiederkunft an anderen Orten seines Werkes gibt, und der großen Bedeutung, die er Nietzsche ganz allgemein zuweist. Und hier wird sehr wohl auch die sogenannte ethische Lesart des Nietzscheschen Gedankens schlagend, seine Rolle für das Verständnis des Historischen und der Zeitlichkeit, wie sie bereits 1927 in *Mathematische Existenz* hervortritt. Aber auch die ideologische Ausbeutbarkeit von Nietzsches Gedankengut wird in Beckers Kriegsvortrag von 1942 *Gedanken Friedrich Nietzsches über Rangordnung, Zucht und Züchtung* deutlich. Auf der Grundlage einer Kritik an Nietzsches lamarckistischem Verständnis des Selektionsprinzips versucht Becker dort, den Gedanken der Züchtung im Sinne des

[14] Vgl. den Briefwechsel zwischen Hausdorff und Schlick aus den Jahren 1919/1920 sowie Epple (2006), 263–289.

[15] Becker (1973) (erste Auflage 1927 in *Jahrbuch für phänomenologische Forschung*, Bd. VIII) zitiert Hausdorff (1914), auf 175, 356 und 359. Auf 162 findet sich auch ein Verweis auf eine frühere Arbeit von Hausdorff (1907), 84–159. Die Seitenangabe bei Becker (217) ist falsch, so dass der Verweis hier nicht ohne weiteres präzisiert werden kann.

[16] Zur Biographie findet man Näheres in Brieskorn (1996).

Nationalsozialismus weiterzuentwickeln, wobei sich gerade dasjenige Volk als das überlegene erweist, das den Wiederkunftsgedanken mit all seinen Konsequenzen erträgt.[17] Man kann daher retrospektiv wohl mit einem gewissen Recht Beckers beide Nietzscheaufsätze auch als eine publizistische Auseinandersetzung mit Alfred Baeumlers Nietzschebild betrachten, in der Becker einen wissenschaftsfreundlicheren Nietzsche innerhalb des nationalsozialistischen Deutschlands hoffähig machen wollte.[18] Wenn dem tatsächlich so war, dann verbot sich jeder Verweis auf Hausdorff, insbesondere in den *Blättern für Deutsche Philosophie*.

1 Die ewige Wiederkunft in *Mathematische Existenz*

Hauptthese von Beckers *Mathematische Existenz* war, dass die phänomenologische Analyse den Streit um die Grundlagen der Mathematik zugunsten des Intuitionismus entschied. Doch ging Beckers Grundlegungsprogramm über die reine Rechtfertigung von Brouwers Konstruktivismus insoweit hinaus, als es die allgemeine Rolle von Zeitlichkeit und Historizität in der Mathematik betonte. Im Gegensatz zum Cantorschen transfiniten Progressus, den Becker unbegründet und als in seiner Ganzheit, dem Kontinuum, für den Verstand nicht fasslich fand, entsprach die Brouwersche Wahlfolge, z.B. die Augen fortgesetzter Würfe mit einem Würfel, einem genuin zeitlichen Prozess, der wie ein historischer Prozess zukunftsoffen war. Folgen, die durch ein explizites Bildungsgesetz konstruktiv darstellbar waren, lebten zwar nicht außerhalb der Zeit, doch waren Aussagen über sie unabhängig von jeweiligen Zeitpunkt im konkreten Fortschreiten der Folge gültig. Indem Becker den Vollzug in den Mittelpunkt der mathematischen Phänomene rückt,

> ist das eigene (historische) menschliche Dasein als ausschlaggebend hingestellt. Die Mathematik erhält damit eine '*anthropologische*' Fundierung. Nicht ein ordnungsmäßig gegliedertes, 'objektives', im traditionellen Sinn 'an sich' seiendes Universum ..., sondern das faktische Leben des Menschen, das jeweils eigene Leben des Einzelnen ... ist das ontische Fundament, auch für das Mathematische.[19]

War die mathematische Folge somit auf die historische Zeit bezogen und in ihrem Fortschreiten unwiederholbar, so erwies sich die Naturzeit als die Domäne der ewigen Wiederkunft, auch wenn diese nicht außerhalb der Zeitlichkeit stand, sondern nur einem anderen Prinzip der Zeitigung folgte.

> Das Kennzeichnende der Naturzeit gegenüber der historischen Zeit ist das Bestehen der Möglichkeit der *Wiederkehr des Gleichen*, des *Sich-Wiederholens* des gleichen Ereignisses. Dagegen kennt die historische Zeit die Wiederkehr nicht; man könnte zugespitzt sagen: (echte) Zukunft schließt (echte) Wiederkunft aus. ... Damit ist (in der historischen Zeit)

[17] Becker (1942).

[18] Für diesen Hinweis danke ich Detlev Piecha. Zu Baeumlers Nietzschebild, vgl.: Piecha (1998), 132–193.

[19] Becker (1973), 196.

> genau genommen auch die Kategorie des *Gleichen* ausgeschlossen, es gibt dort nur *Nämliches* (Identisches im strengen Sinne). Die Zeit ist beide Male in ganz verschiedener Weise *principium individuationis*. Naturzeit ermöglicht das Wiederauftreten des genau Gleichen ... "zu verschiedenen Zeiten." ... Die genaue Gleichheit ist empirisch ... zwar niemals verwirklicht, aber sie ist (ideal) *möglich.*[20]

Ich stelle den Unterschied zwischen historischer Zeit und Naturzeit an dieser Stelle bewusst heraus, weil Becker neun Jahre später in seinem mathematischen Beweis für Nietzsches Wiederkunftsgedanken das Problem aus dem anthropologischen Bereich in die Natur selbst verschiebt, indem er Nietzsches These wissenschaftlich ernst nimmt und sie nicht nur als theoretische Möglichkeit in ihrem ('ethischen' oder phänomenologischen) Bezug auf den Menschen belässt. Diese Objektivierung der Wiederkunft gilt natürlich erst recht für den physikalisch-kosmologischen Beweis. In *Mathematische Existenz* scheint Becker nicht so weit gehen zu wollen und unterstreicht: "'Wiederkehr' ist nicht gleichbedeutend mit echter 'Wiederholung', wie schon das Phänomen der musikalischen 'Reprise' zeigt."[21] Beckers Versuch, mit Nietzsche wissenschaftlich ernst zu machen, scheint mithin aus dem Bereich der Phänomenologie hinauszuführen. Und so schreibt Antonello Giugliano in seiner kenntnisreichen Studie über "Zahl und Zeit: Becker zwischen Nietzsche und Heidegger" schlicht: "Die Annäherung Beckers an Nietzsche war sehr problematisch, wie gerade sein—gegenüber Löwiths radikaler Thematisierung der ewigen Wiederkehr des Gleichen—'unphilosophischer' Aufsatz von 1936 ... erweist."[22] Auch wenn ich im Folgenden Beckers konkrete Lösungsansätze kritisch beurteilen werde, so möchte ich doch dieser Einordnung widersprechen. Es ist gerade das Verdienst von Beckers Nietzscheaufsatz, dass er die ewige Wiederkunft im Stile derjenigen Verbindung aus Wissenschaftsphilosophie, Phänomenologie und Philosophiegeschichte angegangen ist, die auch *Mathematische Existenz* charakterisiert hatte.

2 Ein unvollständiger und ein gültiger physikalisch-kosmologischer Beweis

Becker identifiziert 1936 drei Beweise für die ewige Wiederkunft in Nietzsches Werk. Doch erscheint ihm der erste, in der Nietzscheliteratur gelegentlich zitierte Beweisgang zurecht als "zu unbestimmt, um sachlich gewürdigt zu werden."[23] Denn er beruht auf der Alternative zwischen einer endlichen und konstanten Kraftmenge des Universums, die bereits die ewige Wiederkunft impliziere, und dem willkürlichen Eingreifen Gottes zur Veränderung der Kraftmenge, d.h. des Energieinhalts des Universums.[24]

[20] Becker (1973), 224f.

[21] Becker (1973), 318.

[22] Antonello Giugliano (2005), 47–58, Zitat auf 53. Gemeint ist Karl Löwith (1935).

[23] Becker (1936), 42.

[24] Mit 'Kraft' ist dabei im Sinne der auf Leibniz zurückgehenden und noch bei Helmholtz gängigen Terminologie das gemeint, was zu Beckers Zeit ebenso wie heute als 'Energie' bezeichnet wird.

Aber, so wendet Becker zurecht ein, schon die Dezimalbruchen-twicklungen für irrationale Zahlen gäben ein Beispiel für "ins Unendliche gehende nicht-periodische Entwicklungen ..., die nach einer bestimmten gesetzlichen Regel ablaufen, ohne göttlich schöpferische Willkür."[25]

Allerdings könnte man Nietzsches Aphorismus "Wer nicht an einen Kreisprozeß des Alls glaubt, muß an einen willkürlichen Gott glauben"[26] auch physikalischer und im Sinne der in den 1880er Jahren verbreiteten, vom Chemiker und Monisten Wilhelm Ostwald propagierten Energetik so interpretieren, dass alle Prozesse in der Welt Umwandlungen von Energiearten sind und das Universum als Ganzes ein abgeschlossenes System darstellt. Nietzsche hätte dann schlicht nicht zwischen den beiden Carnotschen Kreisprozessen—und den beiden Hauptsätzen der Thermodynamik—unterschieden und vergessen, dass Arbeit irreversibel in Wärme umgewandelt werden kann. In dieser Sichtweise würde der Beweisgang schlicht die der Wiederkunft entgegengesetzte Alternative eines Wärmetodes des Universums zudecken, was ihn zweifellos nicht besser macht.

Der zweite Beweis lautet in Beckers Rekonstruktion wie folgt:

> Vorausgesetzt, der Weltraum sei [(i)] *endlich,* gestatte auch [(ii)] nur eine *endliche* Zahl von Lagen (d.h. die möglichen Orte im Raum bildeten eine diskrete Mannigfaltigkeit), entsprechend gebe es nur [(iii)] eine *endliche* Zahl von "materiellen" Atomen, die als Kraftzentren zu denken sind (nach Boscovich), und endlich vorausgesetzt, nur *endlich* viele Abstufungen der Kraft (Energie) seien möglich [diese mithin (iv) von endlichem Betrage und (v) diskret] – so ergibt sich eine bloß *endliche Anzahl von möglichen Kombinationen* dieser einen bestimmten Weltzustand festlegenden Elemente, also nur eine endliche Zahl möglicher Weltzustände. Nimmt man nun weiterhin [(vi)] an, ein bestimmter Weltzustand lege eindeutig den nächsten fest (d.h. es bestehe eine exakte, nicht bloß eine statistische Kausalverknüpfung), so ist ein *streng periodischer* Gang des Weltgeschehens unbedingt notwendig. Denn nach Erschöpfung aller möglichen Weltzustände muß notwendig einer von ihnen einmal wiederkehren. Da dieser nun einen eindeutig bestimmten Nachfolger hat und dieser wiederum ebenso, so laufen die Weltzustände auch in der festgelegten Reihenfolge so lange ab, bis sich der zuerst wiedergekehrte zum zweitenmal wiederholt usw., in alle Ewigkeit.[27]

Becker bezeichnet diesen Beweis als tadelsfrei, sofern man noch zur Präzisierung des Begriffs 'Nachfolger' die Zeit selbst als unstetig auffasst. Während Becker die anderen Schritte durch Zitate aus Nietzsche belegen kann, findet sich diese These nirgends auch nur angedeutet. Sie sei aber auch nicht zwingend notwendig, "denn bei einer diskreten Raumbeschaffenheit sind ohnehin nur unstetige Bewegungen denkbar."[28] Da Beckers Bemerkung ohnehin nur im Rahmen der klassischen Mechanik Sinn macht—in der Relativitätstheorie wird die Zeit ja dem Raum im Wesentlichen gleichgestellt—ist allerdings zu bedenken, dass ohne unabhängige

---

[25] Becker (1936), 42.

[26] Zitiert nach Becker (1936), 42. In Nietzsche (1980), Band 9, 561, lautet Aphorismus 11 [312] jedoch: "Wer nicht an einen *Kreisprozeß des Alls* glaubt, *muß* an den [sic] *willkürlichen* Gott glauben." Und es findet sich auch ein Verweis auf die Behandlung des Themas bei Vogt (1878). Auch Becker erwähnt an anderer Stelle den Einfluss Vogts auf Nietzsche.

[27] Becker (1936), 43.

[28] Becker (1936), 45.

Diskretisierung der Zeit, die Bewegungen beliebig schnell ablaufen können. In diesem Falle griffe Beckers Argumentation zu kurz.

Mit Bezug auf die anderen Bedingungen bleibt Becker betont vorsichtig. “Es fragt sich aber, ob jene Annahmen in der wirklichen Welt zutreffen. Dies zu entscheiden, ist Sache umfangreicher physikalischer Forschungen. Nietzsches zweiter Beweisgang ist also ein bloßes Programm, der kühne Entwurf eines *möglichen* Beweises für seine Lehre.” In der Fußnote betont er jedoch die “direkte phänomenologische (bzw. psychologische) Bedeutung”[29] seiner Annahmen. Trotz der formalen Defizite hält Becker Nietzsches Argumentation für hinreichend plausibel, so dass er einige Erläuterungen gibt, die sich auf die Physik seiner Zeit beziehen—jedoch nicht alle, die er hätte geben können und sollen.

Die diskrete Natur von Raum und Kraft entnahm Nietzsche, Becker zufolge, der Atomenlehre von Boscovich und Dührings These, dass die vorhandene Zahl der Weltkörper und ihrer Selbsttätigkeiten in jedem Zeitpunkt endlich sein müsse.[30] Hinter Dührings These, so sei hinzugefügt, steckt jedoch auch eine radikaler mathematischer Konstruktivismus, den dieser schon in seiner *Kritischen Geschichte der allgemeinen Principien der Mechanik* propagiert hatte.[31] So umstritten der Atomismus in den 1880er und 1890er Jahren gewesen war, so war er doch 1936 vollständig akzeptiert. Mit der weitergehenden These über die “Unanwendbarkeit der Geometrie im Kleinen,”[32] d.h. der Diskretheit des physikalischen Raumes, war der Physiker Erwin Schrödinger zwei Jahre zuvor allerdings weitgehend alleine geblieben. Er hatte sie als eine mögliche Konsequenz aus den Interpretationsproblemen der Quantenmechanik erwogen.

Als Nietzsches Quelle für die Endlichkeit des Raumes zitiert Becker den Leipziger Astrophysiker Johann Carl Friedrich Zöller, in welchem er auch den ersten erblickt, “der von Riemanns sphärischer Geometrie eine ernsthafte physikalische Anwendung machte, vielleicht die einzige bis zum ‘kosmologischen Glied’ der Einsteinschen Gravitationsgleichung!”[33] Dabei geht es um die Möglichkeit eines unbegrenzten aber nicht unendlichen Raumes konstanter positiver Krümmung, in dem sich nach endlichen Zeitintervallen die endlich vielen Körper letztlich einander wieder annähern. Allerdings bezieht sich die (pseudo-)Riemannsche Struktur in der Einsteinschen Theorie auf die $3+1$-dimensionale Raumzeit, so dass die ewige Wiederkunft nicht in diesem Sinne begründet werden kann.[34]

[29] Ebd.

[30] Vgl. Becker (1936), 46.

[31] Dühring (1873/1877/1887). Während die erste Auflage betont sachlich war und ihrem Autor den Benekepreis einbrachte, ergingen sich die zweite und dritte in wüsten Polemiken gegen die Berliner Mathematiker, wodurch Dühring letztlich auch seine venia legendi verlor.

[32] Schrödinger (1934), 518–520.

[33] Becker (1936), 47. Vgl. Zöller (1872).

[34] Diese tritt letztlich nur in solchen Welten auf, die Reisen in die eigene Vergangenheit ermöglichen, oder wenn ein Universum nach einem “big crunch” wieder über einen “big bang” sozusagen neu entsteht.

Zöllners eher bildhaft zu nehmendes Argument für die Endlichkeit von Raum und Materie am Beispiel eines Gasballes von rein gravitativ wirkenden Massekörpern bringt Becker zwar nicht zu Unrecht mit dem kosmologischen Glied Einsteins in Verbindung. Doch in beiden Fällen handelt es sich zuallererst um ein Stabilitätsproblem. Denn Einstein hatte das kosmologische Glied sozusagen von Hand zur Stabilisierung seines stationären kosmologischen Modells in die Feldgleichungen eingeführt und später wieder verabschiedet. Zu Beckers Zeit waren bereits spekulative Modelle diskutiert worden, wie Materie etwa infolge radioaktiver Prozesse immer weiter entstehen könnte, so dass ein unendliches Universum stabil bleibt und der gefürchtete Wärmetod vermieden werden kann.[35]

3 Eine probabilistische Modifikation des Beweises und die Brücke zu Abel Rey

In Beckers zuvor skizzierter Rekonstruktion von Nietzsches Beweisgang spielt die eindeutige Kausalverknüpfung zwischen den Weltzuständen eine zentrale Rolle. In einer Modifikation des Beweises schwächt Becker diese Bedingung dahingehend ab, dass die Natur lediglich die Wiederkehr eines herausgegriffenen Weltzustandes *A* nicht 'vermeiden' dürfe, indem sie etwa durch ständige Vermehrung der Kraft die Mittel gewänne, sich 'davor zu hüten'. Der Weg der Wiederkunft von *A* nach *A* müsse daher nicht immer über dieselben Zwischenzustände *B,C,D*, ... gehen, sondern könne über ganz beliebige Wege erfolgen.

> Dann herrscht lediglich das *Gesetz des Zufalls*, d.h. dann findet das statt, was die *Wahrscheinlichkeitsrechnung* voraussagt. Und diese sagt in der Tat, bei endlich vielen möglichen Zuständen *A,B,C* ... der Welt, das schließliche Wiederauftreten *jedes* von ihnen voraus—wenn auch keineswegs den geregelten Kreislauf der Zustände, der sich immer in identischen Zyklen wiederholt.[36]

In dieser Weise schließe man auch

> an die sonst gelegentlich vorgetragene Lehre Nietzsches [an], daß nämlich die Wahrheit in der Physik der Wahrscheinlichkeit Platz zu machen habe und das Unterworfensein unter Natur-'Gesetzen' mit dem Willen zur Macht streite. ... Aber diese Wahrscheinlichkeitsphysik und Kraftmechanik widerspricht keineswegs der ewigen Wiederkunft, sondern erfordert sie gerade![37]

Im modifizierten Beweisgang finden wir die Vorstellungen der Boltzmannschen statistischen Mechanik angedeutet: eine zufällige Bewegung einer großen Zahl einzelner Atome, für deren jeweilige Bewegungen die Gesetze der Mechanik weiterhin gelten. Der Raum als solcher ist in der statistischen Mechanik nicht diskretisiert, wiewohl der späte Boltzmann die Diskretisierung der Zeit

---

[35] Vgl. Nernst (1921).

[36] Becker (1936), 50.

[37] Becker (1936), 50–51.

erwogen hatte.[38] Auf atomistischer Basis konnte Boltzmann eine statistische Herleitung des zweiten Hauptsatzes der Thermodynamik geben. Diese war jedoch nicht ganz frei von Zusatzannahmen und Interpretationsproblemen, und Boltzmann hat seine Argumentation auch nach und nach im Lichte zweier klassischer Einwände präzisiert. Beide sind im Kontext der Wiederkunftsthese relevant. Boltzmanns Wiener Kollege Josef Loschmidt hatte bestritten, dass die gegen Zeitumkehr invariante Newtonsche Mechanik einen Satz abzuleiten gestatte, der die Gerichtetheit allen Naturgeschehens behaupte. Man könne jedoch schlicht die Bewegungen aller einzelnen Atome umkehren. Boltzmann entgegnete, dass die zur umgekehrten Bewegungen gehörenden Anfangszustände zwar aufträten, jedoch viel seltener seien als diejenigen, die eine im Sinne des zweiten Hauptsatzes ablaufende makroskopische Entwicklung ergeben. Die Gläser in unserer Welt zerspringen; sie bilden sich nicht spontan aus Scherben. Dennoch betonte Boltzmann stets, das derartige Ereignisse nicht unmöglich, sondern nur extrem unwahrscheinlich seien.

Es ist jedoch vor allem der zweite Einwand, auf den Becker anspielt, das von Ernst Zermelo gegen Boltzmann verwendete Poincarésche Wiederkehrtheorem. In einem geschlossenen System von endlich vielen wechselwirkenden Massepunkten kehrt ein bestimmter Zustand (d.i. ein Punkt im Phasenraum) nach endlicher Zeit zwangsläufig wieder, oder präziser gesagt, er kommt dem ursprünglichen Zustand eine vorgegebene beliebig kleine Distanz nahe. Die auch heute noch gängige Antwort Boltzmanns bestand darin, auf die unglaublich langen Wiederkehrzeiten zu verweisen. Schon für Systeme weniger Massenpunkte und nicht allzu hohe Genauigkeiten übersteigen diese das uns heute bekannte Alter des Universums. Wenn man wie Becker annimmt, und er war zu seiner Zeit hier nicht allein, dass das Universum ewig besteht, wandelt sich Poincarés Wiederkehrtheorem scheinbar zu einem Beweis des Nietzscheschen Wiederkunftgedankens. Doch die vom zweiten Hauptsatz der Thermodynamik behauptete Einstellung des thermischen Gleichgewichts (im Mittel) liegt auf einer ganz anderen, weitaus kürzeren Zeitskala als die Wiederkehr. Die Wiederkehr bewahrt mithin nicht vor dem Wärmetod. Allerdings geschieht die Annäherung an das Gleichgewicht nicht ohne Ausnahmen; Fluktuationen des Zustandes bleiben stets bestehen.

Derartige Fluktuationen spielen auch eine wichtige Rolle in Boltzmanns hochspekulativer kosmologischer Lösung des Wärmetodproblems. Wir leben in einem Teil des Universums, wo Ordnung, Galaxien und biologisches Leben, deswegen entstehen konnte, weil diese lokale Entropieabnahme durch andere Regionen des Universums kompensiert wurde. Da Boltzmann allerdings den makroskopischen Zeitpfeil auf dem zweiten Hauptsatz gründete, folgt daraus die etwas merkwürdige Tatsache, dass in diesen Regionen des Universums die Zeit umgekehrt verläuft als in der unsrigen.[39] Wenn diese Charakterisierung zuträfe, würde der Wiederkunftsthese gleichsam der kosmologische Boden entzogen. Sie wäre dann entweder eine lokale Eigenschaft gewisser Punktgebilde und käme kosmologisch

---

[38] Vgl. dazu Stöltzner (1999), 85–111.

[39] Vgl. Fasol-Boltzmann (1990), p. 292.

gesehen lokalen Fluktuationen gleich, oder aber sie wäre eine Eigenschaft des gesamten Universums, die sich auf einer Zeitskala abspielt, innerhalb derer die lokale Eigenzeit ihre Kapriolen schlägt, so dass die Verbindung zwischen der anthropologischen und der kosmologischen Ebene zerbräche.

Weitaus präsenter als bei Becker ist die eben skizzierte probabilistische Perspektive bei Abel Rey, auf dessen "ausgezeichnete Erörterungen"[40] Becker verweist. Nimmt man den Reyschen Gedankengang konsequent auf, so verweist er Becker jedoch meines Erachtens wieder auf den in *Mathematische Existenz* eingenommenen Standpunkt zurück, insofern die angestrebte Objektivierung bzw. 'Naturalisierung' der ewigen Wiederkunft nicht wirklich gelingt.

Rey seinerseits betrachtet die These von der ewigen Wiederkunft als die notwendige Konsequenz aus der Krise der Physik in der 2. Hälfte des 19. Jahrhunderts. Denn die Newton-Laplacesche Vorstellung, das Universum sei eine große mechanische Maschine, wurde durch die Arbeiten von Carnot und Clausius radikal gebrochen. Die These von der ewigen Wiederkunft bedeute daher gerade eine Ausdehnung der Carnotschen Kreisprozesse, insbesondere des Wärme umsetzenden zweiten, auf das gesamte Universum.[41] Sie sei wie die Energieerhaltung eine gigantische Extrapolation aus unserem empirischen Wissen.[42] Rey nimmt die ewige Wiederkunft nicht als einen Einwand gegen die statistische Ableitung des zweiten Hauptsatzes der Thermodynamik, sondern als notwendiges Postulat der kinetischen Theorie. "La théorie cinétique des gaz, la mécanique statistique, cette application pure et simple de la théorie des probabilités, postulent nécssairement la retour éternel."[43] Daran ändert auch nichts, dass sich im Prinzip auch das Unwahrscheinlichste beliebig oft wiederholen kann. Innerhalb eines hinreichend langen Zeitraums finden wir einen jeden Wiederkehrzyklus vor, habe dieser nun eine hohe oder eine sehr geringe Wahrscheinlichkeit. Selbst eine für uns unvorstellbare Zeit von $10^{10^{10}}$ Jahrhunderten sei nichts im Vergleich zur ewigen Dauer des Universums. Andererseits geschehe ein Ereignis, das sich jedes Jahrhundert wiederholt, also bestenfalls einmal pro Menschenleben, in derselben Zeit $10^{10^{10}}$ mal, mithin unvorstellbar oft.[44]

Dies illustriert, was Rey als die fundamentale philosophische Konsequenz des Wiederkunftsgedankens betrachtet: die Präzisierung des Verhältnisses zwischen Subjekt und Objekt. Dieses wird Rey zufolge zunächst von Gegensatzpaaren beherrscht:

> Inconscience-conscience; déterminisme-liberté; inertie-activité; réductible à la quantité-irréducible à la quantité ou qualité pure; ... hasard-finalité; déductible-intuitif: causalité fonctionelle-causalité efficiente.[45]

[40] Becker (1936), 50, Fußnote.

[41] Vgl. Rey (1927), 272 bzw. 267, 278.

[42] Vgl. Rey (1927), 301.

[43] Rey (1927), 304.

[44] Vgl. Rey (1927), 306.

[45] Rey (1927), 285f.

Zu diesen komme nun das Paar "retour éternel-vieillissement; réversibilité-irréversibilité"[46] hinzu. Fasse man die ewige Wiederkehr jedoch im Sinne der statistischen Mechanik auf, so verschwinde der radikale Gegensatz zwischen rein mechanischen und (scheinbar) reversiblen Zyklen wie den Planetenbahnen und dem fortschreitenden Bewusstseinsfluss im Sinne Bergsons. Infolge der unterschiedlichen Zeitskalen werde einerseits die ewige Wiederkunft zum Maß des Objektiven. Hier gibt es keine Evolution, sondern nur Revolutionen im astronomischen Sinne; nichts entsteht und nichts geht verloren.[47] Die Unmöglichkeit der ewigen Wiederkehr auf der menschlichen Skala, die erfahrungsmäßige Geltung des zweiten Hauptsatzes begründe andererseits das Maß des Subjektiven. Die ewige Wiederkehr liege daher an der Schnittstelle zweier differenter, jedoch einander nicht widersprechender Welten: dem Subjektiven und dem Objektiven. Durch die Einsicht in die ewige Wiederkunft werde auch Raum für die Spontaneität des Subjekts geschaffen, sie erlaube eine optimistische Weltsicht, weil der Geist sich das höchste Gesetz vorstellen könne. Auch die objektive physikalische Zeit werde so mit der subjektiven Zeit im Sinne von Bergsons Bewusstseinsfluss vereinbar.[48]

Reys Betonung der Differenz zwischen Subjekt und Objekt weist zurück auf die Differenz zwischen historischer Zeit und Naturzeit, die Becker in *Mathematische Existenz* noch wichtiger gewesen war als in seiner auf die objektive Wissenschaft aufbauenden, fast ein Jahrzehnt später publizierten Rekonstruktion von Nietzsches Beweisgang.

4 Beckers mathematischer Beweis

Becker ist sichtlich froh, noch einen weiteren Beweisgang zu haben, der "von den mannigfachen schwer zu beweisenden *Voraussetzungen* des zweiten Arguments unabhängig" ist. Er sei daher auch weniger ein bloßes Programm, sondern "in allen entscheidenden Zügen vollkommen ausgebildet" und trage damit das "sachliche Schwergewicht"[49] der Begründung der Wiederkunftsthese. Unter den zitierten Nietzsche-Aphorismen finden sich:

> Man gehe einmal rückwärts. Hätte die Welt ein Ziel so müsste es erreicht sein: gäbe es für sie einen (unbeabsichtigten) Endzustand, so müßte er ebenfalls erreicht sein. ...
>
> Wäre ein Gleichgewicht der Kraft irgendeinmal erreicht worden, so dauerte es noch. Der augenblickliche Zustand *widerspricht* der Annahme ... denn bis jetzt ist schon eine Unendlichkeit verflossen. Wenn das Gleichgewicht möglich wäre, so müßte es eingetreten sein.[50]

[46] Rey (1927), 286.

[47] Rey (1927), 307f.

[48] Vgl. Rey (1927), 310–312.

[49] Alle Becker (1927), 52.

[50] Zitiert nach Becker (1927), 51f. Vgl. Nietzsche (1980), Band 9, 553 für den ersten (dort 'Ziel' und 'Endzustand' gesperrt) und 534 für den zweiten Aphorismus.

Auch in diesem Fall war Becker zufolge wieder Dührings *Cursus der Philosophie* Nietzsches Hauptquelle. Beckers Rekonstruktion des Nietzscheschen Arguments beruht auf zwei Unterscheidungen, die sowohl bei Dühring als auch bei Kant und Schopenhauer nicht ausreichend gewürdigt worden seien.

1. Ein *regressus in infinitum* und ein *progressus ex infinito* sind nicht dasselbe. Deshalb läßt sich nicht von der Möglichkeit des ersten auf die des zweiten schließen.
2. *Progressus in infinitum* und *progressus ex infinito* sind verschieden; deshalb läßt sich nicht aus der Widersinnigkeit eines *progressus in infinitum usque ad certam finem (resp. nunc)* auf die Unmöglichkeit eines *progressus ex infinito usque ad certam finem (resp. nunc)* schließen.[51]

In Bedingung 1. drückt sich die Gerichtetheit der Zeit aus, die für Becker eine fundamentale Eigenschaft des Zeitbegriffs darstellt. Mit einer reinen Parameterzeit, einer B-Zeit im Sinne von McTaggarts berühmter Klassifikation,[52] wäre phänomenologisch auch wenig anzufangen, da die fundamentale Auszeichnung der Gegenwärtigkeit wegfällt. Bedingung 2. unterscheidet Anfangs- und Endlosigkeit. Becker stimmt dabei Schopenhauers Kritik an Kants Beweis der Thesis der Ersten Antinomie, die Welt habe einen Anfang in der Zeit, zu. Denn das Argument beruhe kurz gefasst darauf, dass eine Reihe ohne Anfang sich niemals vollenden lasse. Doch es lasse sich sowohl das Ende einer anfangslosen Reihe als auch der Anfang einer endlosen Reihe widerspruchslos denken. Auch Nietzsche verfehle wie sein Vorbild Dühring die zweite Unterscheidung. Doch der Beweis lässt sich Becker zufolge reparieren.

Es scheint zunächst so, als dürfe Becker Nietzsches Beweis eigentlich gar nicht akzeptieren. Denn dass ein Gleichgewicht oder anderes Ziel schon erreicht sein müsse, weil alles mögliche schon wirklich geworden, setzt ja voraus, dass die bis jetzt verflossene Ewigkeit eine aktuale Unendlichkeit darstellt. Doch es gibt selbst für den Konstruktivisten eine Möglichkeit, den eigentlich paradoxen, weil unvorstellbaren Begriff eines *progressus ex infinito usque ad certam finem* anzuerkennen, "und zwar einzig und allein durch die Ansetzung des unendlichen Weltverlaufs als eines *streng periodischen.* Denn ein streng periodischer Prozeß ist *vorstellungsmäßig umkehrbar*. Er besteht ja aus *endlichen* Perioden (z.B. a b c), die, jede für sich, umkehrbar sind (in c b a)."[53] Dann und nur dann können wir den *regressus in infinitum* 'cba, cba, cba, ...,' anschaulich dargestellt durch '$\leftarrow$... abc, abc, abc.' zu einem *progressus ex infinito usque ad finem* '$\rightarrow$... abc, abc, abc.' umkehren, indem wir einfach die Ablaufrichtung in den einzelnen endlichen Wiederkehrperioden umkehren, aus denen der unendliche Weltlauf zusammengesetzt ist. Die 1. Unterscheidung zwischen *regressus in infinitum* und *progressus ex infinito* ist daher überbrückbar.[54] Indem nun der *progressus ex infinito usque ad nunc* denkbar sei,

[51] Becker (1927), 55.

[52] Vgl. McTaggart (1908), 457–474.

[53] Becker (1927), 57.

[54] Vgl. Becker (1927), 57.

falle der Beweis von Kants Thesis und es treffe die Antithesis zu: das Universum hat keinen Anfang.

> Im periodischen Vorgang ist überhaupt kein Punkt ausgezeichnet, weder unter den homologen Punkten der verschiedenen Perioden, ... noch unter den Zeitpunkten innerhalb derselben Periode. ... Die leere Zeit selbst kann anschaulich gemacht werden als ein Vorgang mit 'unendlich kleiner' oder minimaler Periode '...*aaaaaaa*...', man denke etwa an des Ticken einer Taschenuhr.[55]

Mir scheint es, dass Becker hier eigentlich zwei Dinge beweist. Nimmt man die Unendlichkeit der Welt an, so folgt erstens die Nietzschesche Wiederkunft. Weil in dieser Weise ein unendlicher Weltverlauf in für den Konstruktivisten akzeptabler Weise gedacht werden kann, wird zweitens die Kantische Antinomie entschieden. Für Kosmologien wie die heutige, in denen das Universum ein endliches Alter besitzt, folgt in beiden Fällen nichts.[56]

Welchen Status hat nun Beckers Beweis der Wiederkunft? "Wollte hier jemand einwenden, es sei nur von 'gnoseologischen' oder bestenfalls 'phänomenologischen', aber nicht von 'ontologischen' Sachverhalten die Rede, so sei ihm erwidert, daß man die 'ontische' Struktur des periodischen Weltverlaufs nicht unter dem Bild einer Schwingung (Sinuskurve), sondern eines Kreises zu fassen hat. 'Die Zeit selber ist ein Kreis', wie Nietzsche sagt."[57]

Meines Erachtens scheint Becker die ontische Struktur hier durchaus mittels eines transzendentalen Arguments zu begründen. Das Vergangenheitskontinuum ist überhaupt nur vorstellbar, wenn es aus lauter Perioden besteht. Doch werden hierfür eine ganze Reihe von Voraussetzungen gemacht. Zunächst ist hier natürlich die Ablehnung des Cantorschen Kontinuums zu nennen, aber es finden sich meines Erachtens auch zwei der nach Beckers Worten schwer zu beweisenden physikalischen Voraussetzungendes zweiten Beweisganges wieder. Da ist zum einen die Diskretheit der Weltzustände; denn ansonsten wäre ja auch durch die Periodisierung das Kontinuum nicht entschärft. Zum anderen aber fällt Becker hinter die Modifikation des zweiten Beweisgangs im Sinne der statistischen Mechanik zurück und nimmt wieder eine eindeutig bestimmte Kausalverknüpfung und Zustandsabfolge '*abc* usw.' an. Würde lediglich der Zustand *a* nach einer gewissen Zeit wieder erreicht, jedoch auf einem anderen Wege, gelänge die Umkehrung nicht und es wäre sehr wohl der Zustand *a* ausgezeichnet, was Becker ja bestreitet. Würde *a* im Sinne des Poincaréschen Wiederkehrtheorems nur bis auf einen kleinen vorgegebenen Rest wieder erreicht, so könnte Becker das Argument nicht ins Unendliche iterieren und die Differenz im Sinne der 1.

[55] Becker (1927), 59.

[56] Allerdings könnte Becker auf Schopenhauers Einsicht verweisen, dass in diesem Falle "die Zeit selbst angefangen haben muß, was widersinnig ist" (zitiert nach Becker (1927), 55). Im Rahmen der heutigen Urknallkosmologie ist dies zwar eine formale Konsequenz der Verschmelzung von Raum und Zeit in der allgemeinen Relativitätstheorie. Doch betrachten viele diese Konsequenz als nicht annehmbar und verweisen darauf, dass die Kosmologie des frühen Universums eine Quantenkosmologie sein muss. Zu den heutigen Debatten, vgl. Stadler, Stöltzner (Hg.) (2006).

[57] Becker (1927), 58; Die Passage aus dem *Zarathustra* findet sich in Nietzsche (1980), Band 4, 200.

Unterscheidung (zwischen *regressus in infinitum* und *progressus ex infinito*) bliebe bestehen. Denkt man an Boltzmanns Argument über die Auszeichnung von Anfangsbedingungen, wäre dies auch nicht überraschend. Würde man diesen Weg gehen, so denke ich, ergäbe sich die Einsicht, dass es zwar Wiederkehrbahnen gibt, diese jedoch sehr selten, mathematisch gesprochen ein verschwindendes Maß besitzen. In diese Richtung weist nun gerade die Analyse von Felix Hausdorff alias Paul Mongré.

5 Ein nicht geführter Dialog unter Bonner Kollegen

Bereits 1893 fiel dem gerade frisch promovierten Mathematiker Felix Hausdorff auf, dass Nietzsches Argument, die endliche Zahl von möglichen Weltzuständen könne nur dann durch die unendliche Zeit erfüllt werden, wenn das Weltgeschehen periodisch sei, unhaltbar war. Der Literat und Nietzscheverehrer Paul Mongré teilte dies auch brieflich dem Herausgeber Köselitz mit, nicht ohne zu verschweigen, dass er Nietzsches Bemühungen um die Wissenschaft ebenso ernst nehme wie Andreas-Salomé.[58]

> Rein mechanistisch betrachtet ist der Gedanke einer Erschöpfbarkeit der Weltzustände sogar falsch und beruht auf einer Amphibolie des Unendlichkeitsbegriffs. ... Die Zeit ist eine eindimensionale Mannigfaltigkeit, ein ∞^1, die Gesamtheit aller Weltzustände [der Phasenraum] eine unendlichdimensionale, ein ∞^∞; jene kann also diese nur partiell realisiren, keinesfalls erschöpfen. Bestünde die Welt aus der Bewegung eines einzigen Punktdreiecks, so hätten wir ∞^3 mögliche Weltzustände ... und zu ihrer Unterbringung nur ∞^1 Zeitaugenblicke; es würden also ∞^2 Weltzustände unrealisirt bleiben, geschweige dass Zeit zu Wiederholungen übrig wäre.[59]

Hausdorff-Mongré hat diesen Gedanken auch im *Sant'Ilario*[60] breiter ausgeführt. Und der entsprechende Aphorismus endet ganz im Sinne von Andreas-Salomé: "Die ewige Wiederkunft ist eine gewaltige Conception, ein Mysterium, das schon als Möglichkeit aufregt, erschüttert, ungeheure Folgerungen zulässt. Wir treten diesem 'abgründlichen Gedanken' nicht zu nahe, wenn wir seinen oberflächlichen Beweis verwerfen."[61]

Auf den ersten Blick scheint Mongrés Dimensionsbetrachtung auch Beckers physikalisch-kosmologischen Beweisgang zu Fall zu bringen, doch sie ist nicht allgemein genug. Durch sein Studium der Cantorschen Mengenlehre erkannte Hausdorff kurz darauf, dass Punktmengen unterschiedlicher Dimension dieselbe Stufe der Unendlichkeit haben können, weil zwischen ihnen bijektive Abbildungen bestehen.

[58] Zu Hausdorff-Mongrés engen Beziehungen zum Nietzsche Archiv einschließlich der Tatsache, dass er vorübergehend als Herausgeber von Nietzsches naturwissenschaftlichen Schriften im Gespräch war, vgl. Stegmaier (2004), 66–70.

[59] Zitiert nach Stegmaier (2004), 46.

[60] Mongré: *Sant'Ilario*. In: Hausdorff (2004), 443–448. Aphorismus 406.

[61] Mongré (1897), 448.

In *Das Chaos* finden sich daher mathematische Argumente anderen Typs, die jedoch zum Teil auf eine ähnliche Intuition hinauslaufen. Bevor ich diese vorstelle, möchte ich jedoch noch Hausdorff-Mongrés Überlegungen zum Verhältnis zwischen Nietzsches Wiederkunftgedanken und dem Poincaréschen Wiederkehrtheorem vortragen und zu Beckers physikalisch-kosmologischem Beweis in Beziehung setzen.

Mongrés Aufsatz "Nietzsches Lehre von der Wiederkunft des Gleichen," publiziert in der *Zeit* im Jahre 1900, beginnt wie folgt: "Warum schreibt nicht einer unserer Philosophen die Geschichte der '*sphäro-cyklischen* Weltanschauung'? Auf diesen neuen Namen sei es erlaubt eine uralte Sache zu taufen, nämlich alle jene Versuche, mit dem Schlüssel irgend einer Kreis-Kugel-Ring-Symbolik das Weltgeheimnis zu entziffern."[62] Dem Riemannschen sphärischen Raum mit konstantem Krümmungsmaß stellt Mongré dabei die zyklische Zeit an die Seite.

> Wohlgemerkt, ich glaube nicht, daß die ewige Wiederkunft in der üblichen Form sich halten läßt; aber ich finde nicht, daß die moderne Philosophie darüber hinaus sei, solche Speculationen nach Für und Wider gründlich durchzudenken. ... Viel eher wäre die Philosophie berechtigt, sich in diesem Falle unzuständig zu erklären und die Acten über die ewige Wiederkunft an die *Naturwissenschaft* weiterzugeben.[63]

Die Naturwissenschaft entscheidet gegen die ewige Wiederkunft: "in einem sich selbst überlassenen System sind vollkommene Kreisprocesse unmöglich. Der Kosmos 'altert', vorausgesetzt, daß er ein sich selbst überlassenes System ist."[64] Während die Ostwaldsche Energetik überhaupt nur nichtumkehrbare Vorgänge zulasse und die umkehrbaren lediglich als ideale Grenzfälle betrachte, könne sich die immer noch dominante mechanische Weltanschauung mit der Wiederkehr noch am besten befreunden.

> Hier ist die ... Wiederkehr eines gegebenen Zustandes principiell möglich, nur äußerst unwahrscheinlich; ... gewissermaßen ein statistisches und kein deductiv nothwendiges Ergebnis. ... Periodicität des Weltlaufs ist hier also immerhin möglich, *angenäherte* Periodicität, nach einem Satze der neueren Mechanik, sogar gewiß, aber nur in einem System von endlichen Dimensionen und Geschwindigkeiten, also Voraussetzungen, die für das [mechanische] Universum nicht zutreffen dürften.[65]

Hier sprach jemand, der sich fünf Jahre zuvor über mathematische Probleme der Astronomie habilitiert hatte und Poincarés Arbeiten zur Himmelsmechanik genau kannte.

Im eben zitierten Aufsatz wird auch auf eine bereits anderenorts veröffentlichte mathematische Analyse verwiesen. Diese findet sich im zwei Jahre zuvor publizierten Buch *Das Chaos in kosmischer Auslese* (1898). Der Mathematik hatte Mongré bereits in *Sant'Ilario* die Bestimmung einer "Selbstkritik der Wissenschaft"[66] zugewiesen.

---

[62] Mongré: "Nietzsches Lehre von der Wiederkunft des Gleichen". In: Hausdorff (2004), Band VII, 897.

[63] Mongré Ebd., 897–898.

[64] Mongré Ebd., 900.

[65] Mongré Ebd., 900f.

[66] Mongré (1897), 436 (Aphorismus 401).

In *Das Chaos* dient sie nun vor allem zur Kritik einer realistischen Metaphysik, die eine Welt der 'Dinge an sich' unabhängig von unserem Bewusstsein postuliert. Hier der vernichtende Schluss des Buches:

> Die ganze wunderbare und reichgegliederte Structur unseres Kosmos [so wie wir ihn erfahren] zerflatterte beim Übergang zum Transcendenten in lauter chaotische Unbestimmtheit; beim Rückweg zum Empirischen versagt dementsprechend der Versuch, die allereinfachsten Bewusstseinsformen als notwendige Incarnationen der Erscheinung aufzustellen.[67]

Mithilfe mathematischer Kritik möchte Hausdorf den metaphysischen Realismus ad absurdum führen. "Zu diesem Zwecke greifen wir irgend eine unserer Bewusstseinswelt anhaftende Eigenschaft heraus, übertragen sie unverändert auf das absolut Reale und suchen sie dann, bei vorgeschriebener und festgehaltener empirischer Wirkung, möglichst stark umzuformen."[68] Ziel ist es insbesondere, die maximale Vieldeutigkeit jener absolut realen bzw. transzendenten Welt bei unverändertem empirischem Gehalt bzw. Bewusstseinsinhalt zu erweisen, in manchen Fällen sogar im mathematisch verheerend präzisen Sinne einer unendlichen Vieldeutigkeit.[69] Dieses Chaos in den Dingen an sich wird durch die Existenz unseres Bewusstseins auf den in ihm vorgefundenen Kosmos reduziert, was Mongré im Sinne einer epistemischen Auslese durch eben dieses Bewusstsein versteht.

Auch der Wiederkunftsgedanke erweist sich als Möglichkeit in der transzendenten Welt. Mongrés für das vorzustellende Argument entscheidender Zeitbegriff scheint auf den ersten Blick gar nicht so weit von Beckers phänomenologischem erntfernt. Er unterscheidet den 'Zeitinhalt', die kontinuierliche Reihe von Weltzuständen oder das materiale Substrat der Zeit, vom 'Zeitablauf' als desjenigen "räthselhaften formalen Prozesses, durch den jener Weltzustand die Verwandlungsfolge Zukunft, Gegenwart, Vergangenheit erfährt."[70] "*Weltzustand* ist eine *erfüllte* Zeitstrecke von der Länge Null, sowie *Augenblick* eine *leere* Zeitstrecke von der Länge Null ist."[71] Der Punkt entspricht einer Linie der Länge Null. "Was eine erfüllte Zeitstrecke ist, darüber ist unmittelbar unser zeitlich erlebendes Bewusstein zu befragen, das receptiv und productiv mit nichts anderem als mit der Erfüllung der Zeitform beschäftigt ist."[72] Führt man wie in der Mechanik die realistische Hilfsvorstellung einer absoluten Zeit ein, so ruht der Zeitinhalt und der Zeitablauf spielt sich bewegungsartig ab. Der Gegenwartspunkt bewegt sich auf einer Zeitlinie, deren Punkte die Weltzustände beschreiben. Doch aus transzendenter Sicht gesehen muss diese Bewegung keinesfalls konstant, ja nicht einmal stetig sein. "Die transcendente Succession der Weltzustände ist willkürlich und fällt

---

[67] Mongré (1898), 803.

[68] Mongré (1898), 602.

[69] So etwa im Falle der Skaleninvarianz des gesamten Universums, vgl. Mongré (1898), 605.

[70] Mongré (1898), 605.

[71] Mongré (1898), 605f.

[72] Mongré Ebd., 606.

nicht in unser Bewusstsein."[73] Der Zeitinhalt ist also indifferent gegenüber dem Zeitablauf im transzendenten Sinne. Der mit unserem Bewusstseinsinhalt übereinstimmende transzendente Weltverlauf ist singulär. "Weder der beharrende einzelne Weltzustand noch eine beharrende Vielheit von Weltzuständen ist Bewusstseinserscheinung, sondern allein jenes strömende oder gleitende Continuum von Weltzuständen",[74] das Werden.

Mongré untersucht nun den Fall, dass wir eine Strecke AB der Zeitlinie in umgekehrter Richtung BA durchschreiten.

> Kann die Zeitstrecke BA aus AB durch irgendwelche Reversion der Bewusstseinsvorgänge, also etwa durch physiologische, chemische und zuletzt mechanische Untersuchungen hergeleitet werden…? Das wäre das erreichte Ziel des Materialismus, die vollendete Übersetzungskunst, die Bewusstseinsvorgänge als Bewegungsgleichungen und diese wieder als jene verdolmetscht.[75]

Durch eine mathematische Kritik zeigt Mongré, dass dies nicht möglich ist. Wenn wir AB in sehr kleine Teilstrecken zerlegen (1,2,3,…,n-1,n), so erhalten wir durch einfache Umkehrung die Strecke (n,n-1,…,3,2,1). Im Limes $n \rightarrow \infty$ erhalten wir daraus in der Tat die Strecke BA, doch müssen wir dazu durch den Weltzustand hindurch gehen, was eine Fülle kompliziertester Vorgänge involvierte. "Fragen wir nun gar, wie es bei jener Reversion um die Erhaltung oder Veränderung von Relationen höherer Gattung steht, ob beispielsweise der Complex räumlich und materiell zusammengehöriger Einzelerscheinungen, der in positiver Richtung ein organisirtes bewusstes Lebewesen bildet, nach der Umkehrung sich zu einer ähnlichen räthselhaften Function wieder vereinigt, so sinken wir tief unter die Schwelle des Denkbaren und stossen auf Probleme,"[76] die nur der ideale Materialismus im Besitz einer Weltformel lösen könnte.

> Aus demselben Material von Weltzuständen sind demnach *zwei* Welten herstellbar, die als rein mechanische Erscheinungen aufgefasst sich nur im Vorzeichen ± unterscheiden würden, die von aussen betrachtet durchaus identisch sind, und die trotzdem den in ihrem inneren Connex eingesponnenen Wesen keine Communication über die Grenze herüber ermöglichen. … Wir sind als subjektive Träger der Zeit, unserer innersten Structur nach in die Structur der positiv gerichteten Zeitlinie eingeflochten und von der negativen Richtung ausgeschlossen.[77]

Hausdorff-Mongré beharrt mithin auf der ersten von Becker gemachten Unterscheidung (zwischen *regressus in infinitum* und *progressus ex infinito*), und die Umkehrbarkeit wird auch dadurch nicht erleichtert, dass es sich hier ja nur um eine endliche Strecke AB und nicht um einen unendlichen progressus/regressus in der Zeit handelt. Mir scheint insgesamt gesehen Mongrés Argumentationsgang eine stichhaltige Kritik an Beckers mathematischem Beweis, die nicht davon abhängt,

---

[73] Mongré Ebd., 610, Hervorhebung im Original.

[74] Mongré Ebd., 616.

[75] Mongré Ebd., 620.

[76] Mongré Ebd., 621.

[77] Mongré: Ebd., 622.

dass Mongré in *Das Chaos* eine Konzeption des Bewusstseins vertritt, die Beckers Ablehnung des Cantoscher Kontinuums zuwiderläuft. So heißt es einige Seiten zuvor: “Was in einem Augenblicke erhalten ist, ist niemals Vorstellung [im Sinne Kants]; die einfachste Bewusstseinsaction setzt immer unendlich viele, stetig aufeinander folgende Augenblicke, kürzer ein Continuum von Weltzuständen voraus.”[78]

Aus der Sicht unseres Bewusstseins ist daher die Umkehrung unmöglich, aus tranzendenter Sicht ist sie trivial. Der Weltbeobachter in der absoluten Zeit sieht jenes “beharrende Substrat des Wandels und Wechsels, das die Möglichkeit einer *ewigen Wiederkunft des Gleichen* in sich trägt, … das Reservoir des Daseins, dessen Realitätsgehalt durch den Process des zeitlichen Ablaufs nicht vermindert und durch millionenfache Wiederholung dieses Processes nicht erschöpft wird.”[79] Die ewige Wiederkunft verbildlicht die Zeitlosigkeit des materiellen Zeitsubstrats. Mongré unterstreicht den Unterschied seines Arguments von demjenigen Nietzsches.

> Mit der von Nietzsche neuerdings aufgestellten Formel der ewigen Wiederkunft, die eine inhaltliche Periodicität innerhalb des Weltgeschehens behauptet, ist die unsrige nicht identisch; Nietzsches Aussage bezieht sich auf die innere Structur der Zeitlinie (die er sich geschlossen, in sich zurücklaufend vorstellt), unsere gilt, ohne Rücksicht auf den Inhalt, von jeder Einzelstrecke. Nietzsches Hypothese … unterliegt schliesslich dem Richterspruch der Erfahrung; unser Satz redet von der transcendenten Möglichkeit.[80]

Indem Mongrés mathematische Analyse des Wiederkunftgedankens, im Gegensatz zu derjenigen Beckers, negativ ausfällt, wird die Frage wieder an die Physik zurückverwiesen—wo sie weniger leicht zu widerlegen ist, als sich das Mongré in seinem Brief an Köselitz vorgestellt hatte.

Mongré nutzt nun die Nietzschesche Hypothese zu einer Klassifikation metaphysischer Entwürfe in ontologische und genealogische, ein Gedanke, der sich mit anderer Terminologie auch in den historischen Betrachtungen am Ende von Beckers Aufsatz findet. Am Ende des Kapitels “Gegen die Metaphysik” kritisiert Mongré auch jegliche ‘Ethik der ewigen Widerkunft’ etwa in dem Sinne, “dass der dionysische Mensch nicht nur das Leben, sondern die unendliche Wiederkehr des Lebens bejahe.”[81] Aber die ewige Wiederkunft geht uns praktisch gar nichts an, da sich die identische Wiederholung unserer Wahrnehmung entzöge. “Wenn von dieser Wiederholung die leiseste Spur einer Empfindung oder Ahnung in die Zeitlinie dränge, … so wäre das eben keine identische Wiederkehr.”[82]

Mongré zufolge erweist sich die ewige Wiederkunft als metaphysischer Begriff, dessen epistemologischer Sinn darin besteht, dass er abgestreift werden kann, nicht als üsinnloser Begriff im Sinne des Neopositivismus, sondern nach erfolgter Selbstvergewisserung des Subjekts in seinem zeitlichen Bewusstsein.

[78] Mongré: Ebd., 607.

[79] Mongré: Ebd., 626.

[80] Mongré: Ebd., 632–633.

[81] Mongré (1898), 647.

[82] Ebd.

Wir finden mithin wie bei Rey am Wiederkehrgedanken wieder eine Trennung in objektive und subjektive Ebene, die dem Phänomenologen Becker nicht behagen konnte.

Das Thema der ewigen Wiederkunft hat auch den Mathematiker Hausdorff später nicht losgelassen. Moritz Epple[83] zitiert ein Manuskript aus dem Jahre 1908 in dem Hausdorff den Gedanken in das folgende Problem übersetzt. Gibt es Abbildungen aus den reellen Zahlen $\mathbb{R}$, in die Menge $\{1,2\}$, die vollständig wiederkehrfrei sind in dem Sinne, dass für keine Intervalle $I,J \subset \mathbb{R}$, für die f beide Werte unendlich oft annimmt, f sich in I und J 'gleich' verhält, d.h., dass eine ordnungserhaltende Abbildung existiert. Eine Lösung ist nicht überliefert.

6 Epilog

Paul Mongré hat in der *Neuen Rundschau* auch die Herausgabe des *Willens zur Macht* kommentiert. Es sei beklagenswert, dass Nietzsche gerade in seinen schwachen Punkten derzeit so populär sei, und nicht als der gütige, verstehende Freigeist, der er auch gewesen.[84]

> In Nietzsche glüht ein Fanatiker. Seine Moral der Züchtung, auf unserem heutigen Fundamente biologischen und physiologischem Wissens errichtet: das könnte ein weltgeschichtlicher Skandal werden, gegen den Inquisition und Hexenprozeß zu harmlosen Verirrungen verblassen. "Das Leben selbst erkennt ... kein gleiches Recht zwischen gesunden und entarteten Theilen eines Organismus an: letztere muß man *ausschneiden* ...". Ungefähr sagen das die päpstlichen Ketzerrichter auch ... Sollen wir wieder einmal erleben, wie man mit dem Hexenhammer philosophiert? ... [D]erselbe Philosoph, der den Moral-Castratismus der Guten und Gerechten verabscheut, befürwortet den eigentlichen chirurgischen Castratismus im Dienste der Auslese. Um das zu würdigen, müssen wir einmal, was schneller gethan als gesagt ist, unser heutiges Wissen von der Vererbung zusammenzählen. Daß Trunkenbolde häufig idiotische Kinder zeugen, daß Geisteskrankheiten, Tuberkulose, Neurasthenie sich der Disposition nach manchmal vererben, manchmal auch nicht. Auf diesen verblüffenden Reichthum an Kenntnissen hin soll man der Biologie das gesegnete Messerchen in die Hand geben?[85]

Als Becker im Jahr 1942 seine *Gedanken Friedrich Nietzsches über Rangordnung, Zucht und Züchtung* vortrug, hatte sich diese Befürchtung nur allzu sehr bewahrheitet. Die Biologie war auch damals, nach Aussage der meisten nicht-deutschen Genetiker und Eugeniker, keinesfalls ausreichend vorangekommen.[86] Der wissenschaftlichen Vorsicht wollte sich Becker nun aber offenbar nicht mehr anschließen.

[83] Vgl. Epple (2006).

[84] Mongré (1902), in Hausdorff (2004), 909.

[85] Mongré (1902), 907.

[86] Vgl. vor allem die Argumentation des Wieners Julius Bauer in Hofer (2007), 31–65.

Literaturverzeichnis

Andler, Charles. 1920–1931. *Nietzsche, sa vie et sa pensée*. 6 Bände. Paris: Éditions Bossard.

Andreas-Salomé, Lou. 1894. *Friedrich Nietzsche in seinen Werken*. Wien: zitiert nach Stegmaier, Einleitung, 44.

Becker, Oskar. 1936. Nietzsches Beweise für seine Lehre von der ewigen Wiederkunft. In *Blätter für Deutsche Philosophie* 9: 368–387; wieder abgedruckt in In Dasein und Dawesen. Gesammelte philosophische Aufsätze, 41–66. Pfullingen: Neske, 1963.

Becker, Oskar. 1942. *Gedanken Friedrich Nietzsches über Rangordnung, Zucht und Züchtung*. Bonn: Gebr. Scheur. Kriegsvorträge der Rheinischen Friedrich-Wilhems-Universität Bonn, Heft 97, Vortragsreihe: Führungsformen der Völker.

Becker, Oskar. 1973 [1927]. *Mathematische Existenz. Untersuchungen zur Logik und Ontologie mathematischer Phänomene*. Tübingen: Niemeyer. Erste Auflage [1927] in *Jahrbuch für phänomenologische Forschung*, Bd. VIII.

Brieskorn, Egbert (Hg.) 1996. *Felix Hausdorff zum Gedächtnis. Aspekte seines Werkes*. Braunschweig/Wiesbaden: Vieweg.

Dühring, Eugen. 1873/1877/1887. *Kritische Geschichte der allgemeinen Principien der Mechanik*. Berlin: Grieben/Leipzig: Fues.

Dühring, Eugen. 1875. *Cursus der Philosophie*. Leipzig: E. Koschny.

Epple, Moritz. 2006. Felix Hausdorff's Considered Empiricism. In *The architecture of modern mathematics. Essays in history and philosophy*, ed. Ferreiros, J. and J.J. Gray, 263–289. Oxford: Oxford University Press.

Fasol-Boltzmann, Ilse M. (Hg.) 1990. *Ludwig Boltzmann. Principien der Naturfilosofi. Lectures on Natural Philosophy 1903–1906*. Berlin/Heidelberg/New York: Springer.

Giugliano, Antonello. 2005. Becker zwischen Nietzsche und Heidegger. In *Oskar Becker und die Philosophie der Mathematik*, ed. Volker Peckhaus, 47–58. München: Fink.

Hausdorff, Felix. 1907. Untersuchungen über Ordnungstypen IV, V. *Verhandlungen der Königlich Sächsischen Gesellschaft der Wissenschaften zu Leipzig. Mathematisch-physikalische Klasse* 59: 84–159.

Hausdorff, Felix. 1914. *Grundzüge der Mengenlehre*. Leipzig: Veit.

Hausdorff, Felix. 2004. *Gesammelte Werke*, vol. VII. Berlin/Heidelberg/New York: Springer.

Hofer, Veronika. 2007. Positionen und Posen eines Experten. Der Konstitutionsforscher Julius Bauer (1887–1973) und die Eugenik in der Wiener Zwischenkriegszeit. In *Eugenik in Österreich. Biopolitische Strukturen von 1900–1945*, ed. Gerhard Baader, Veronika Hofer und Thomas Mayer, 31–65. Wien: Czernin Verlag.

Kragh, Helge. 1999. *Cosmology and controversy: The historical development of two theories of the universe*. Princeton: Princeton University Press.

Löwith, Karl. 1935. *Nietzsches Philosophie der ewigen Wiederkunft des Gleichen*. Berlin: Verlag Die Runde.

McTaggart, John M.E. 1908. The unreality of time. *Mind* 17: 457–474.

Mongré, Paul [Felix Hausdorff]. 1897. *Sant'Ilario. Gedanken aus der Landschaft Zarathustras*. Leipzig: C. G. Naumann; wieder abgedruckt in Hausdorff 2004, S. 85–473.

Mongré, Paul [Felix Hausdorff]. 1898. *Das Chaos in kosmischer Auslese. Ein erkenntniskritischer Versuch*. Leipzig: C. G. Naumann; wieder abgedruckt in Hausdorff 2004, S. 589–807.

Mongré, Paul [Felix Hausdorff]. 1902. Der Wille zur Macht. *Neue Deutsche Rundschau* 13 (12): 1334–1338; wieder abgedruckt in Hausdorff 2004, 905–909.

Nernst, Walter. 1921. *Das Weltgebäude im Lichte der neueren Forschung*. Berlin: J. Springer.

Nietzsche, Friedrich. 1980. *Sämtliche Werke. Kritische Studienausgabe,* Hg. v. Giorgio Colli und Mazzino Montinari. Berlin: de Gruyter.

Piecha, Detlev. 1998. Nietzsche und der Nationalsozialismus'. Zu Alfred Baeumlers Nietzsche-Rezeption. In *Nietzsche in der Pädagogik? Beiträge zur Rezeption und Interpretation, Beiträge zur Theorie und Geschichte der Erziehungswissenschaft*, Band 19, Hg. C. Niemeyer, H. Drerup, J. Oelkers, and L.v. Pogrell. Weinheim: Deutscher Studien Verlag.

Rey, Abel. 1927. *Le retour éternel e la philosophie de la physique*. Paris: Flammarion.

Schrödinger, Erwin. 1934. Über die Unanwendbarkeit der Geometrie im Kleinen. *Die Naturwissenschaften* 22: 518–520.

Stadler, Friedrich und Michael Stöltzner (Hg.). 2006. *Zeit und Geschichte/Time and History*. Frankfurt am Main: Ontos.

Stegmaier, Werner. 2004. Einleitung. In Felix Hausdorff, *Gesammelte Werke*, vol. VII, 1–83. Berlin/Heidelberg/New York: Springer.

Stöltzner, Michael. 1999. Vienna Indeterminism: Mach, Boltzmann, Exner. *Synthese* 1–2, 119: 85–111.

Stöltzner, Michael. 2012. Taking eternal recurrence scientific: A comparative study of Oskar Becker, Felix Hausdorff, and Abel Rey. In *Nietzsches Wissenschaftsphilosophie*, ed. H. Heit, G. Abel, and Marco Brusotti, 353–366. Berlin: de Gruyter.

Vogt, J.G. 1878. *Die Kraft. Eine real-monistische Weltanschauung*. Leipzig: 90.

von Mises, Richard. 1939. *Kleines Lehrbuch des Positivismus. Einführung in die empiristische Wissenschaftsauffassung*. Den Haag: W. P. van Stockum & Son.

Zöller, Johann Carl Friedrich. 1872. *Über die Natur der Cometen. Beiträge zur Geschichte und Theorie der Erkenntnis*. Leipzig: W. Engelmann.

Heidegger and Our Twenty-first Century Experience of Ge-Stell

Theodore Kisiel

Abstract I propose an etymological translation of *Ge-Stell*, Heidegger's word for the essence of modern technology, from its Greek and Latin roots as "syn-thetic com-posit[ion]ing," which presciently portends our twenty-first century experience of the internetted WorldWideWeb with its virtual infinity of websites in cyberspace, Global Positioning Systems, interlocking air traffic control grids, world-embracing weather maps, the 24-7 world news coverage of cable TV-networks like CNN, etc., etc.—all of which are structured by the complex programming based on the computerized and ultimately simple Leibnizian binary-digital logic generating an infinite number of combinations of the posit (1) and non-posit (0). The sharp contrast between the global time-space technologically foreshortened into instantaneity and simultaneity and the radically local time-space of our situated historical existence—in short, the temporal-spatial tension between *Ge-Stell* and *Da-Sein*—is examined for ways and means of bringing them together in contemporaneous compatibility.

Martin Heidegger got as far as the atomic-space-cybernetic age in his meditations on technicity and modern technology. We ourselves have been able to experience the marvels of the twenty-first century advance into the internet revolution and its instantaneous global reaches, such that, for example, we and the entire world with us were virtual witnesses of the recent events that transpired in Abbotabad, Pakistan, almost immediately after they happened.[1] We twenty-first century citizens of the world take for granted the convenience of stratospheric transportation networks and

[1] This talk, delivered on May 25, 2011 to the Heidegger Forschungsgruppe meeting in Messkirch, Germany, took as its example of virtually instantaneous global communication the raid on the compound of Osama bin Laden that took place in the early hours of May 2 East Asian time.

T. Kisiel (✉)
Northern Illinois University, 27W761 Washington Av,
Winfield, IL 60190, USA
e-mail: tkisiel@wpo.cso.niu.edu

B. Babich and D. Ginev (eds.), *The Multidimensionality of Hermeneutic Phenomenology*, Contributions to Phenomenology 70, DOI 10.1007/978-3-319-01707-5_9,

the satellitic transmission of instantaneous media events that enwrap the "global village" at every hour of every day by CNN. But modern technology had advanced sufficiently in Heidegger's day for him to be struck by the same drastic foreshortening of time and space and its global reach brought on by the radio technology of his time. Accordingly, what he had to say to us about the essence of modern technology in the twentieth century appears to apply as well, with some minor adjustments in terminology, to the more enhanced and advanced technological realities of the twenty-first century.

Such adjustments can easily be made in the single hyphenated word by which he defines the essence of modern technology, almost as ingenious as the single hyphenated word that defines his entire way of thought, namely, *Da-Sein*. For modern technicity, his one word is of course *Ge-Stell*. In the last three decades of his life, Heidegger repeatedly tells us what *Ge-Stell* is, and repeatedly notes that it is to be sharply distinguished from the ordinary everyday senses of *Gestell*, as in *Büchergestell* (bookcase) and *Brillengestell* (frame for eyeglasses). It must therefore be emphatically stated that *Ge-Stell* is simply NOT "frame, framework or enframing," the current English translations drawn directly from German-English dictionaries. What then is *Ge-Stell* in its global essentiality? It is, in Heidegger's breakdown of this single word, "*die versammelnde Einheit aller Weisen des Stellens*/the collective unity of all modes of setting in place, positioning, positing."[2] "*Im Ge- spricht die Versammlung, Vereinigung, das Zusammenbringen aller Weisen des Stellens*/The prefix *Ge-* speaks to the gathering, unification, bringing-together of all kinds of placing and positioning."[3] "*Das Ge-Stell ist die Versammlung, die Gesamtheit aller Weisen des Stellens, die sich dem Menschenwesen in dem Maße auferlegen, in dem es gegenwärtig ek-sistiert*/ Ge-Stell is the gathering, the integration of all the modes of placing, positioning, and positing that im**pose** themselves upon the human being in the manner in which the human being presently ex-sists."[4]

Against the current English favorite of "enframing," I therefore propose an etymological translation of *Ge-Stell* from its Greek and Latin roots as "syn-thetic com-posit[ion]ing," where the Greek-rooted adjective 'synthetic' adds the note of artifactuality and even artificiality to the system of positions and posits. For me, *Ge-Stell* as "syn-thetic com-posit[ion]ing" presciently portends the twenty-first century globalizations of the internetted WorldWideWeb with its virtual infinity of websites in cyberspace, Global Positioning Systems (GPS), interlocking air traffic control grids, world-embracing weather maps, the 24-7 world news coverage of cable TV-networks like CNN, etc., etc., all of which are structured by the complex programming based on the computerized and ultimately simple Leibnizian

[2] Heidegger (1977b), 104. The citation is taken from the seminar at Le Thor in 1969.

[3] Ibid., 129. Citation taken from the seminar at Zähringen, 1973.

[4] Ibid., 126, Zähringen, 1973. The same point was already made in a rich note circa 1955, whose first sentence reads: "Im Wort 'Gestell' spricht die Versammlung des Stellens, in der 'Versammlung' spricht das Echo zum *Logos*, im 'Stellen' spricht das Echo der *Thesis* (*Poiesis*)." Heidegger (2009), 320; see also 327 and 365. Hereafter cited as GA76.

binary-digital logic generating an infinite number of combinations of the posit (1) and non-posit (0). The synthetic compositing of computer logic thus maps out the grand artifact of the technological infrastructure that networks the entire globe of our planet Earth.

The phenomenon of technological globalization was already apparent by the time of the so-called "Great War" of 1914–1918, which was accordingly renamed the World War. One of the heroes of this highly mechanized war, Ernst Jünger, in his accounts of "totale Mobilmachung," the total mobilization that occurred in the last year of the war, began to attribute this phenomenon to *planetarisches Technik* and its use in the struggle for *planetarische Herrschaft*. This becomes Heidegger's word for globalization in this period to phenomenologically describe the human experience that results from the network of grids constructed by modern technology to guide and control the so-called "air waves" which harness the natural electromagnetic radiation occurring across the surface of our planet Earth for human use and consumption. Globalization is essentially a time-space term, a dynamic term that spells out a quasi-infinite velocity in nanoseconds through its virtual abolition of space into bi-locative simultaneity and its instantaneous reduction of all time differences. By the early twentieth century, radio technology had advanced sufficiently for Heidegger to be struck by the drastic foreshortening of time and space and its global reach. In the famous 'pincers' passage of SS 1935, Heidegger dramatically describes the global geopolitical as well as philosophical situation of a postwar Germany being squeezed by two international movements, both of them technological juggernauts, on the western front by American capitalism and on the eastern front by Bolshevistic communism, in the following words:

> Russia and America, when viewed metaphysically, are both the same: the same hopeless frenzy of unchained technology and of the groundless organization of the average man. When the farthest corner of the globe [*der Erdball*, the *terrestial globe* versus Heidegger's beloved *terra firma*] has been technically conquered and can be economically exploited; when any incident you like, in any place you like, at any time you like, becomes accessible as fast as you like; when you [by way of radio] can simultaneously "experience" an assassination attempt against a [Yugoslavian] king in France and a symphony concert in Tokyo; *when time is nothing but speed, instantaneity, and simultaneity, and time as history has vanished from the Dasein of all peoples*; when a boxer counts as the great man of a people; when the tallies of millions at mass meetings are a triumph; then, yes then, there still looms like a specter over all this uproar the question: what for?—where to?—and what then? [...in short, the question of be-ing in the twentieth century...][5]

Clearly, Heidegger was suspicious of this instantaneity and simultaneity of the time technologized by global communication primarily because it abolishes the time of situated history, the time of *Da-sein*. In 1935, this time-space abolition results from the medium of the radio along with the wire services of newspapers, but it just as readily reflects with uncanny foresight the more advanced digital-media systems of the twenty-first century. As Heidegger observes in 1949, by plane

[5] Heidegger (1953), 28f. English translation by Gregory Fried and Richard Polt (2000a), 40. Emphasis added.

and by radio and soon by TV, "all distances in time and space are shrinking."[6] He calls this the phenomenon of the distanceless [*das Abstandslose*]. Distant locales and exotic places are shown on TV or film so realistically that you may even feel that YOU ARE THERE [as we were, most recently, in Abbottabad, Pakistan] there and there and everywhere in a technologically induced bi-locative simultaneity. Heidegger asks: "What is happening here when, as a result of the abolition of great distances, everything is equally far and equally near? What is this uniformity in which everything is neither far nor near, is, as it were, without distance? Everything gets lumped together into uniform distancelessness [*Abstandslosigkeit*]. How? Is not this merging of all into the distanceless more unearthly than everything blowing up [by way of the atomic bomb]?"[7] What Heidegger misses in this all-too-familiar modern experience is a genuine experience of nearness, the proximity of be-ing. Because the experience of nearness fails to materialize with this abolition of all distances, the phenomenon of the distanceless has come to dominate our lives in the twenty-first century.[8]

Heidegger's own examples of *Ge-Stell* begin in a farmer's field about to be exploited for its mineral deposits, be it for coal or uranium ore. Instead of being cultivated, the land is now being challenged [*sich gestellt*] to yield energy, where we **set upon**[9] the land in order to extract coal or ore from it, then store this energy resource in order to have it ready for use. The hydroelectric plant is **set into** the river Rhine, thereby damming it up to build up water pressure which then **sets** the turbines turning whose thrust in turn generates and **sets** the electric current going into the network of long-distance cables, where the systematic transforming, storing, distributing and switching of electrical energy takes place.[10] Be it coal or hydroelectric power or atomic energy, in each case "Nature is positioned for its energy," nature is forced to yield its energy. Nature, thus held up to yield energy, emerges henceforth as the "storage-place of energy," like a global fuel depot or gigantic gas station.

Storage of resources, be it energy or information, becomes a central feature of the Ge-Stell, which Heidegger calls its fundamental unconcealment. "Everywhere, everything is ordered to stand by [*es wird bestellt, auf der Stelle zu stehen*], to be immediately in position for use, in fact to stand there to be on call for a further ordering [*Bestellen*]. [...] Whatever is ordered about in this way has its own standing. We call it the standing-reserve [*Bestand*]." And now comes the perhaps surprising denouement of *Ge-Stell* from the philosophical perspective: "Whatever stands

[6] Heidegger (1994), 3, citing from the preface to the lecture, "Das Ding." English translation by Albert Hofstadter in Heidegger (1971b), 165.

[7] Ibid., 4/166.

[8] Ibid., 20/181.

[9] Here, *stellen* is translated in various idioms of "to set." The typical translations of *stellen* are "put, place, set, stand," with strong overlaps with the verbs *setzen* and *legen*.

[10] Heidegger (1954b), 23–24, citing from the 1953 version of "Die Frage nach der Technik." English translation by William Lovitt in (1977a), 16.

by in the sense of standing-reserve no longer stands over against us as object."[11] "Thus when man, in investigating and observing, ensnares nature as an area of his own conceiving, he has already been claimed by a way of revealing that challenges him to approach nature as an object of research, until even the object disappears into the **objectlessness of standing-reserve**."[12] Heidegger in a parallel essay also notes that the most recent cyclotron experiments in nuclear physics likewise encounter this phenomenon of the complete disappearance of the object, which hitherto had been the very hallmark of modern science. But "that does not mean that the subject-object relation vanishes, but rather the opposite: it now attains its most extreme dominance, predetermined from out of *Ge-Stell*, syn-thetic com-positioning. It becomes a standing-reserve [*Bestand*] to be commanded and placed on order."[13] The subject-object relation now reaches, for the first time, its purely 'relational' character, that is, its character of orderability [*Bestellungscharakter*], in which both the subject and the object are claimed as standing-reserves [*Bestände*].

The more modern technology unfolds and develops, the more objectivity transforms itself into *disposability* (into a making-itself-available). *Gegenständlichkeit* is transformed into *Beständlichkeit*. Now there are no more objects (no more beings standing over against a subject that takes them into view)—there are only *Bestände*, to wit, reserve resources positioned for orderability: stock on hand, stored inventory, warehoused supplies and provisions, capital holdings, assets, funds held in reserve **(in short, beings held ready for plan-directed use)**. Political economists in fact no longer deal with objects but instead systematically order the space with an overall plan toward maximizing the utility of resources. Beings as a whole are aligned and ordered within a horizon of usefulness, domination or, better still, the disposability [*Beständlichkeit*] of all that needs to be placed under control. The planners themselves are no longer scientifically oriented toward a field of objects but now emerge in *their true gestalt as technicians and even technocrats*, i.e., humans who see beings a priori in the horizon of making-them-useful. It can no longer appear in the objective neutrality of an over-against. There is nothing other than reserve resources: warehoused stock, inventories of goods, stores of supplies, stockpiles of uranium, reserves of provisions, energy reserves, capital reserves, federal reserve funds,[14] not to speak of the quasi-infinite store of information in the so-called memory banks of the internetted WorldWideWeb. "The ontological definition of reserve stock is not the persistence of durable goods but their character of disposability, the constant possibility of being offered and ordered, i.e., of enduring availability. Its constancy is not that of objectness but that of the standing reserve, a constancy defined in terms of syn-thetic com-positioning. In disposability, the being is *posited* as being exclusively available from the ground up, available for use in the planning

[11] Ibid., 24/17.

[12] Ibid., 27/19.

[13] Ibid., 61/173, citing from the essay "Wissenschaft und Besinnung."

[14] Heidegger (1977b), 105–6.

of the whole."[15] There are no longer any objects but only 'production resources' and 'consumer goods' at the disposal of everyone, who themselves are put into service in the business of production and consumption. In universities (now called "knowledge industries") as well as in corporations, personnel departments are now called departments of human resources. And since all resources are disposable, they are at once replaceable. This is clearly manifest in the industry of consumer goods with its abundance of substitutes and, in an era of mass production, leads to the tendency to replace rather than repair used goods.[16] But extending the same attitudes to human resources is fraught with all manners of abuse, the extremes of which we have witnessed under the totalitarian regimes of the twentieth century.

The recent disruption in the global flow of standing reserves caused by the Japanese earthquake illustrates another phenomenon unique to modern technicity, namely, that Heidegger's broken hammer experience has gone global. The widely adopted Toyota strategy of just-in-time inventories for its production lines led, as a result of the earthquake, to drastic disruptions in the supply lines of numerous automobile production lines around the world. Massive power outages and recent identity thefts of mega-lists pirated on the internet are further examples of the broken hammer experience gone global. Recall the fears of massive attacks on the Internet and WorldWideWeb by cyber-terrorists in the millennial year of Y2K. Among other things, it conjures the image of the lightning-speed electronic circulation of vast sums of currency whipping around the world's financial markets in a global cash flow whose reverberations sometimes verge on a cascading collapse. Such a globally impelled crash, whether by impersonal market forces or computer hackers, would make the worldwide depression of 1929, at least in its velocity of impact, pale in insignificance.

To be sure, all of these examples of global disruption occur in the high-velocity time-space of modern technicity, which is not at all comparable with the more vitally measured time-space of the broken hammer experience. Recall that the broken hammer experience retrospectively reminds us of the referential context and its vital connections that the broken hammer interrupts, say, in the work world of the carpenter. At one point, Heidegger asks what exactly is the "basic referential context" (GA76, 302: *Grundverweisungszusammenhang*) of a "world" of machination and notes its radical difference from the referential world of handwork and hand tools by pointing to the regulated and uninterrupted repeatability "in exactly the same way" of the "mechanical" motions of the machine and the more calculative referential relations necessary for its manufacture.[17] The "machine is not an 'imitation' of handwork and natural processes but rather a self-standing organization of all the processes of beings."[18] And this "organization of all the processes of beings" in its deliberately calculated mechanical design is not even a world.

[15] Ibid., 106.

[16] Ibid., 107.

[17] GA76, 307.

[18] GA76, 308.

Heidegger thus speaks of an "unworlding and unearthing of beings" in the machinations of Ge-Stell,[19] where beings stand in a state of total abandonment by be-ing [*Seinsverlassenheit*].[20]

We are accordingly moving from the epoch of objectivity [***Gegenständlichkeit***] to the *epoch of disposability* [*Beständlichkeit*], the most extreme gestalt of the history of the metaphysics of constant presence since the Greeks. "Because we no longer encounter what is called *Ge-Stell* within the horizon of representation, the view that allows us to think of the be-ing of beings as presence, *Ge-Stell* no longer approaches us as something present and thus seems at first alien and strange."[21] As the most extreme gestalt of the history of the metaphysics of constant presence, and so the completion and fulfillment of this metaphysics, the *Ge-Stell* assumes a strange constant absence which in effect serves to point it in another direction, to serve as a passage from metaphysics to another thinking governed by the properizing event, *das Er-eignis*. The *Ge-Stell* is "Janus-faced, it is essentially double-sided [...] it is so to speak the photographic negative of the event, *das Ereignis*."[22] Accordingly, "an outstanding way to draw near to *das Er-eignis*, the properizing event, would be to look deeply into the essence of *Ge-Stell*."[23] The *Ge-Stell* thus prompts *Be-sinnung*, a meditation on its meaning. It is therefore not a matter of regarding the emergence of technology as a negative event (and certainly even less as a positive event, as if it were a paradise on earth). "That in and from which man and be-ing approach and challenge each other in the technological world claims us in the manner of *Ge-Stell*, syn-thetic com-positioning. In the reciprocal self-positing [*Sichstellen*] of man and be-ing we discern the claim that defines the constellation of our age."[24] With the *Ge-Stell*, it appears that we are on the verge of overcoming the subject-object relation and entering into the mutual ownership of man and being that the properizing event is.

> The intimate *be-longing* together of man and be-ing in the manner of a mutual escalating challenge brings us in startling fashion nearer to that and how man is delivered over to the ownership of be-ing and be-ing is appropriated to the essence of man. Within *Ge-Stell* there prevails a rare and exceptional ownership and appropriation. We must simply experience this owning in which man and be-ing are proper for one another, i.e., we must enter into what we call the *event of enownment and properizing, das Ereignis* ... a *singulare tantum* ... unique ... What we experience in *Ge-Stell* as the constellation of be-ing and man through the modern world of technology is a *prelude* to what is called *Er-eignis*. For in the event there resides the possibility that it may turn the sheer prevalence of *Ge-Stell* into a more inceptive appropriating. Such a transformation of *Ge-Stell* into *das Er-eignis* would by virtue of this event bring the appropriate recovery—appropriate, thus never to be made by

---

[19] GA76, 307.

[20] GA76, 297.

[21] Heidegger (1957), 28. English translation by Joan Stambaugh (1969), 35f.

[22] Heidegger (1977b), 104.

[23] Ibid.

[24] Heidegger (1957), 27f./35.

man alone—of the world of technology out of its domination to servitude in the realm by which man reaches more properly into the properizing event.[25]

Presuming that we could wait in anticipation for the possibility that *Ge-Stell*, the reciprocal challenge of man and be-ing in the calculation of the calculable, would address itself to us as the appropriating event that first expropriates man and be-ing into their proper [character]; then **a** path would be freed for man to experience beings in a more inceptive way—the totality of the modern technological world, nature, and history, and above all their be-ing.[26]

In Heidegger's depiction, therefore, at the most extreme extremity of the history of the metaphysics of constant presence, we find ourselves poised at the very threshold of crossing over into an authentic experience of be-ing in the propriating event, *das Er-eignis*. But despite the apparent and so tantalizing proximity of this ex-perience, we are not given to expect a smooth gradual crossing over to it simply because of the extremities at which we are poised: the machinations of technology have resulted in the complete abandonment of beings by be-ing [*Seinsverlassenheit*] and the human being is in peril of not only forgetting his essential be-ing but even of having forgotten this forgetting of be-ing. "But in this extreme extremity of destining peril the most intimate relationship [of man and be-ing] shows itself, but shows itself only as a completely veiled hint."[27] It is necessary to push the ex-p*eri*-ence of the *peril* of technology to the extreme to glimpse the e-vent emerging in the *Ge-Stell*. Accordingly, Heidegger recommends not attempting to arrest or to master technology but to drive it to its extreme in order to ex-peri-ence it in its full peril to the human being, and at the same time to meditate on the meaning of its destining essence.[28] To put this extreme experience in another way, technology in its essence is the "most extreme neglect [*Ver-wahr-losung*] of the under-cut of difference [*Unter-schied*, of be-ing and beings]…. Technology—the neglect of (nearness), yet accordingly in this neglect [we find] the nearing of the turn of the forgottenness of the under-cut of difference."[29] Finally, Heidegger, following Hölderlin, prompts the "sons of the Alps" to make the perilous crossing "over the abyss on lightly built bridges" by invoking these encouraging lines from Hölderlin's *Patmos*: *"Wo aber Gefahr ist, wächst/Das Rettende auch//*But where peril is also grows the saving." How the extreme peril of technology might allow us to glimpse "the growing light of a saving [power]" is suggested by the hint that the Greek word *techne* is the common root of both technology and art, even the fine arts.[30] By way of this hint, *Ge-Stell* at its extreme of unworlding [*Entweltung*] and unearthing [*Enterdung*] may well be transformable into the world and earth of *das Geviert*.

This crossing over from *Ge-Stell* to *Geviert* once again operates between extremes that, in their very contrast, provide clues for the crossing. How? Consider,

[25] Heidegger (1957), 28f./36f.

[26] Ibid., 32f./40.

[27] GA76, 327.

[28] GA76, 255.

[29] GA76, 370.

[30] Heidegger (1954b)/(1977a), 36–43/28–35.

for example, the abolition of time and space that comes with modern technology, where everything is equally far and equally near, inducing a uniformity in which everything is neither far nor near, is, as it were, without distance, such that everything gets lumped together into a uniform distancelessness [*Abstandslosigkeit*]. What is missing in this all-too-familiar modern experience of time and space is a genuine experience of nearness, the proximity of be-ing. But that very experience of missing the near opposed to the far in their authentic presential sense is the beginning of meditative thinking—for which nearness can become conspicuous by its very absence—and of the turn toward moving beyond the essence of modern technology as *Ge-Stell*, which in its essence does not admit of any qualitative nearness or farness.[31] *Ge-Stell* in its essence disallows nearness. And what nearness [*Nähe*] truly nears is the intimacy of a world as a neighborhood [*Nähe*] in which we can dwell meaningfully.[32] "*Ge-Stell* as the completed destining of the forgottenness of the essence of be-ing inconspicuously radiates a ray of the distant arrival of world. The fact that world withholds its worlding here does not mean that nothing happens with world: the withholding itself radiates the lofty nearness of the most distant farness of world."[33]

A crucial opposition is clearly emerging in our consideration of modern technicity, namely, the contradistinction between the technical time-space of the distanceless **versus** the time-space of historical Dasein. In SS 1928 Heidegger characterized the historical world as a temporal playing field [*Zeit-Spiel-Raum*] that grants Da-sein the freedom of movement within a finite world of distinct historical possibilities. One is tempted nowadays to compare this basic contradistinction with that between the cyberspace of virtual reality and the concrete space of historical reality, by way of the many recent crossovers from virtual to historical reality in organizing protest movements on line, be it environmental, economic, and most recently, that of the "Arab spring." The most recent twenty-first century technologies like the internet have by and large had a liberating effect as compared to the twentieth century, which often employed technology as totalitarian tools of domination like the propaganda propagated by newspapers/radio/film and the leveling of *das Man* to uniformity and conformity. Have '1984' and 'Big Brother' become figments of the past now overcome, at least on the global scale in which they were fictionally portrayed?

On other occasions, Heidegger describes this contradistinction in terms of technical-functional relations versus vitally lived relations, or, a bit more deeply, as the contradistinction between a technical world of functionality and a lived world of meaningfulness, which are the topics of two radically different kinds of thinking, calculative thinking and meditative thinking [*be-sinnendes Denken*], which accordingly meditates on the meaning [*Sinn*] of be-ing. In the Spiegel Interview of 1966, for example, where Heidegger admits to being frightened [*erschrocken*] when he first saw the pictures of the earth taken from the moon, he remarks: "We do not

[31] Heidegger (1994), 45.

[32] Ibid., 46.

[33] Ibid., 53.

need atomic bombs at all [to uproot us]—the uprooting of man is already here. All our relationships have become merely technical ones. It is no longer upon an earth that man lives today."[34] He finds it uncanny to be living in a world in which everything is pure function, and this functioning simply leads to more and more functioning, and this technicity increasingly dislodges man and uproots him from the earth and native roots. This takes us to another formulation of our contradistinction, that of the global versus the local, which came into currency with the generation that lived through the PC (personal computer) revolution but is quite apt to the old Heidegger's concerns, as he meditates on the impact of technological giganticism on local traditions and on the rhythms and ways of life of the "good old days."

1 Autochthony in the Atomic Age

Heidegger assumes a less terrified and more meditative and placid [*gelassene*] tone toward Ge-Stell in his 1955 talk in Messkirch memorializing the hometown composer Conradin Kreuzer, published under the title *Gelassenheit* but whose original title for the hometown crowd that first heard it was "*Bodenständigkeit im Atomzeitalter*," "Autochthony in the Atomic Age."[35] He notes here that it is not only *schwäbischer Boden—der Geniewinkel*—that has produced great poets and thinkers, but also the Boden of Middle Germany, East Prussia, Silesia as well as Bohemia has inspired its great poets and thinkers.[36] What is this ground that produces great poets and thinkers? Nothing less than the native language in which one finds oneself rooted, the earth of language in its dialects in their tonality, rhythms, and song, in short, the down-to-earth language of original experience.[37]

To come to terms with the inexorable onslaught of modern technology on his hometown and environs, Heidegger recommends that his *Landsmenschen* should

[34] Heidegger (2000b), 669–670; translated by William Richardson as "'Only a God Can Save Us': The Spiegel Interview (1966)," Heidegger (1981), 56.

[35] The adjective *bodenständig* is typically translated as "indigenous, native" so that the more abstract *Bodenständigkeit* etymologically suggests being native to a land or a nation and, even more starkly (and mythologically), having one's roots in native soil. Whence the clear possibility of using this term for nationalistic and even for racist ends, as was the case in Nazi Blubo (=*Blut und Boden*) propaganda. And Heidegger here is speaking directly to a post-war native German audience. But it should be noted that Heidegger first used the word often enough in the twenties in a phenomenological and so non-nationalistic context to connote the re-duction "back to the origins, roots, native ground" of original experience. This is important to note when we try to redirect his suggestions toward our own unique situation of being caught up in our twenty-first century *Ge-Stell*.

[36] Heidegger (1959), 16; translated by John M. Anderson and E. Hans Freund as *Discourse on Thinking* Heidegger (1966), 47.

[37] It might be noted here that Colonel Claus von Stauffenberg, who was born and raised not too far from Messkirch, also developed his poetic sense of the Germany for which he was willing to fight and die directly from *schwäbischem Boden*, inspired especially by the poetry of Hölderlin and Stefan George.

strive to cultivate two basic comportments to meditatively confront the flood of technical devices that were already working their way into the life and fabric of the town and gradually making themselves more or less indispensable. The first comportment involves affirming the unavoidable use of technical devices but denying them the right to dominate our lives, i.e., of letting technical things be what they are but then of willing to let them go to avoid becoming slavishly dependent on them. Heidegger identifies this yes-no comportment toward technical devices as the releasement toward things [*Gelassenheit zu den Dingen*]. "Having this comportment we no longer view things merely in a technical way. … We notice that while the production and use of machines *demands of us another relation to things*, it is not a meaning-less [*sinn-los*] relation. Farming and agriculture, e.g., have now become a motorized food industry. Thus here, evidently, as elsewhere, **a** *profound change is taking place in man's relation to nature and to the world*. But the meaning [*Sinn*] that reigns in this change remains obscure."[38] The issue here, accordingly, is to make sense of all this high tech infiltrating into our lives by way of meditative [*be-sinnendes*] thinking. For example, what are we to make of the fact that "Nature is becoming a gigantic gas station, an energy source for modern technology and industry,"[39] a storage-place for energy, thus a "natural resource" subject to the calculations of those wishing to exploit it for profit or conquest?

> There is then in all technical processes a meaning, not invented or made by us, which lays claim to what we do and leave undone. We do not know the significance of the uncanny increasing dominance of atomic technology. *The meaning pervading technology hides itself.* But if we explicitly and continuously heed the fact that such *hidden meaning touches us everywhere in the world of technology* we stand at once within the realm of that which hides itself from us, and hides itself just in approaching us. That which shows itself and at the same time withdraws is the essential trait of what we call the mystery. I call the comportment that enables us to remain open to the meaning hidden in technology, *openness for the mystery* [*Offenheit für das Geheimnis*].[40]

Releasement to and from technical things and openness for the mystery of the meaning of modern technicity: These two comportments combined serve to promote meditative thinking and so to counter the threat of becoming so bedazzled by the marvels of modern technology that calculative thinking comes to be accepted as the only way of thinking. Humans would thereby deny and throw away their essential nature of being meditative beings and no longer nurture their capacity for meditative thinking.[41] In our present situation, we are called upon to be open to the mystery of the global domination of technology and to meditatively ponder the profound changes that it is exacting upon our relations with nature and the world in order that we might find meaningful ways for us to live in this new world. For these two comportments "grant us the possibility of dwelling in the world in a totally different way. They promise us a new ground and foundation [*Boden*] upon

[38] Heidegger (1959), 25; (1966), 54f.

[39] Ibid., 20/50.

[40] Ibid., 25f/55.

[41] Ibid., 27/56.

which we can stand and endure in the world of technology without being imperiled by it. … They give us a vision of a new autochthony [*Bodenständigkeit*] that someday might even be fit to bring back the old and now rapidly disappearing autochthony in a transformed gestalt."[42] "If releasement toward things and openness toward the mystery awaken within us, we might arrive at a path that will lead to a new ground and foundation [*Boden*]. In that *Boden* the creativity that produces lasting works could strike new roots."[43]

What would such "lasting works" created out of the new autochthony look like? Would they involve some sort of fusion of technology and art, some sort of "tech art," as suggested by the Greek *techne*, which means both art and technology? At one point, Heidegger does hint broadly that an autobahn bridge might be a candidate for gathering the fourfold.[44] But can a Boeing-787 taking off ever gather the fourfold? We know that Heidegger developed an appreciation for Paul Klee and modern art later on in life. Or would it involve an Eastern approach to art, like the Taoism that comes into play in the jug that jugs? Then there is the *feng shui* approach to architecture, which Heidegger spontaneously applied in his account of how a Schwarzwald Bauernhof gathers the fourfold.[45] Since the resolution to modern technicity is bound to pass to some extent through art, it is worth concluding by examining Heidegger's sense of the artwork for clues to the possible transition from *Ge-Stell* to *das Er-eignis*.

2 How the Artwork Works in a Historically Local Context

Heidegger's early use of the hyphenated word *Ge-stell* in 1935 as it operates in the gestalt of an artwork evokes a 1956 cautionary note from him to distance this more focused "local" sense from the modern meaning of *Ge-Stell* operative on a global scale in modern technicity. But it also opens the opportunity for us to examine the different sort of gathering of modes of *stellen*, the different kinds of settings and positioning that are operative in an artwork.

First of all, "To be a work means to set up [***aufstellen***] a world."[46] In setting up the world, the work sets forth [***her-stellt***] the earth, accordingly with *herstellen* being taken in the strict etymological sense of the word. The work sets itself back [*sich* ***zurückstellt***] and thereby puts the earth into the openness of a world.

[42] Ibid., 26/55.

[43] Ibid., 28/56f.

[44] Heidegger (1954a), 153; translated by Alfred Hofstadter as "Building Dwelling Thinking" in Heidegger (1971b), 152.

[45] Ibid., 161/160.

[46] Heidegger (1950), 33; translated by Alfred Hofstadter as "The Origin of the Work of Art" in Heidegger (1971b), 44.

> That into which the work sets itself back [*zurückstellt*] and which it lets come forth in this setting back of itself we called the earth. … In setting up a world, the work sets forth the earth. … To set forth the earth means to bring it into the open as the self-closing.[47]

"The setting up of a world and the setting forth of earth are two essential traits of the work-being of the work. They belong together in the unity of work-being."[48]

> The world is the self-opening openness of the broad courses of the simple and essential decisions in the destiny of a historical people. The earth is the spontaneous coming forth of the continually self-closing and accordingly covering and sheltering. World and earth are essentially different from one another and yet are never separated. The world grounds itself upon the earth and the earth towers through the world.[49]

"The opposition of world and earth is a strife."[50] "Inasmuch as the work sets up a world and sets forth the earth, it is an institution of this strife." "The work-being of the work consists in the strifing of the strife between world and earth."[51] The strife here is between the self-opening openness of the world and the self-closing closedness and so covering sheltering of the earth, in short, the strife between unconcealing and concealing, the happening of truth. "Truth happens only by establishing itself in [both] the strife and the playing space [*Spielraum*] that it itself opens up."[52] "Truth establishes itself in the work. Truth comes to presence [*west*] only as the strife of clearing and concealing in the opposition between world and earth."[53]

One final setting [*Stellen*] must be made for the work to do its work as a happening of truth. Having set itself up [*aufstellt*] as world and set itself forth (her-stellt) as earth by setting itself back [*zurückstellen*] into the earth, the work must now set and fix in place [***feststellen***] the strife of truth in the ***gestalt***. Put another way, the truth must establish itself by being fixed in place in the gestalt of an artwork. "*Art is the setting and fixing in place of self-establishing truth in the gestalt.*"[54] The Greek sense of *morphe* as gestalt or form is made clear by Ge-stell, understood as the gathering together of the various settings of truth in the rift-design of the bounding outline (*peras*) of the gestalt.

In the creating of the work, the strife as rift must be set back [*zurückgestellt*] into the earth, and the earth itself must be set forth [*hervorgestellt*] and used as the self-closing. Such use, however, does not use up or misuse the earth as matter, mere stuff, but rather frees the earth to be just itself. This use of the earth is a working with it that indeed looks like the employment of matter in handicraft. Hence the appearance that artistic creation is also craft activity. It simply is NOT. But it is

[47] Ibid., 35/46f.

[48] Ibid., 36/48.

[49] Ibid., 37/48f.

[50] Ibid., 37/49.

[51] Ibid., 38/49.

[52] Ibid., 49/61.

[53] Ibid., 51/62.

[54] Ibid., 59/71. Emphasis added.

always a use of the earth in the *setting and fixing in place of truth in the gestalt*. In contrast, the making of tools and equipment is never immediately the effecting of the happening of truth. The production of equipment is finished when a material has been sufficiently formed to have it ready for use. The equipment's readiness for use means that it is released beyond itself to disappear into usefulness.[55]

In the artwork, by contrast, its matter is not used up and does not disappear but is rather set forth as earth into the openness of the world. Rather than using up words in the manner of everyday discourse, the poet uses the word "such that the word truly becomes a word and remains a word" in all its glory and brilliance. This is the *Bodenständigkeit* or earth-rootedness of language so cherished by Heidegger.

"The poetizing project of truth, which sets itself (*sich stellt*) into the work as a gestalt, is never enacted in an indeterminate void. Rather, the truth in the work is projected to the coming preservers, i.e. to a historical humanity [and not a Volk!]."[56] The preservers in their Dasein now take their place in the in-between and in the middle of the strife of world and earth, unconcealment and concealment. With the artwork we are in a historical world of a historical people in search of its destiny, not in the uniform technological time-space of the distanceless, but rather in the time-space of historical Dasein. It is the temporal playing field [*Zeit-Spiel-Raum*] of history that grants us freedom of movement in and through a historical world of distinct finite possibilities. And the artwork itself is just one of the forms of the historical happening of truth, along with philosophical questioning, state-founding deeds, and essential sacrifice, like the "people-saving death" of Albert Leo Schlageter. "The world is the self-opening openness of the broad courses of the simple and essential decisions in the destiny of a historical people."[57] Such a historical world with its tradition of deeds and sacrifices and concepts offers a people an appointed task [*Aufgegebenes*] which points them to their future world of possibilities. This appointed task unique to a people at once discloses to them a native endowment [*Mitgegebenes*] already given to them on the basis of what they have been. Clearly, the appointed task of today's historical humanity is to ponder the profound change that is taking place by way of the essence of modern technology, Ge-Stell, and to ready itself to cope with these changes in a way that remains true to our own unique proper situation of be-ing, in which "das Leben selbst legt sich aus," life itself lays itself out, interprets itself, explicates itself. This domain of original meaningfulness which precedes the subject-object relation is what must be repeatedly retrieved and retained so that we may once again learn to live poetically on the earth in a post-modern world of technology.

[55] Ibid., 52/64. Emphasis added.

[56] Ibid., 63/75.

[57] Ibid., 37/49.

References

Heidegger, Martin. 1950. Der Ursprung des Kunstwerkes. In *Holzwege*, ed. Martin Heidegger. Frankfurt am Main: Klostermann.

Heidegger, Martin. 1953. *Einführung in die Metaphysik*. Tübingen: Max Niemeyer.

Heidegger, Martin. 1954a. Bauen Wohnen Denken. In *Vorträge und Aufsätze*. Pfullingen: Neske.

Heidegger, Martin. 1954b. *Vorträge und Aufsätze*. Pfullingen: Neske.

Heidegger, Martin. 1957. *Identität und Differenz*. Pfullingen: Neske.

Heidegger, Martin. 1959. *Gelassenheit*. Pfullingen: Neske.

Heidegger, Martin. 1966. *Discourse on Thinking*. Trans. John M. Anderson and E. Hans Freund. New York: Harper & Row.

Heidegger, Martin. 1969. *Identity and Difference*. Trans. Joan Stambaugh. New York: Harper & Row.

Heidegger, Martin. 1971a. Building dwelling thinking. In *Poetry, Language, Thought*, ed. Heidegger. Trans. Alfred Hofstadter. San Francisco: Harper & Row.

Heidegger, Martin. 1971b. *Poetry, Language, Thought*. Trans. Albert Hofstadter. San Francisco: Harper & Row.

Heidegger, Martin. 1971c. The origin of the work of art. In *Poetry, Language, Thought*, ed. Heidegger. Trans. Alfred Hofstadter. San Francisco: Harper & Row.

Heidegger, Martin. 1977a. *The Question Concerning Technology and Other Essays*. Trans. William Lovitt. San Francisco: Harper & Row.

Heidegger, Martin. 1977b. *Vier Seminare*. Frankfurt am Main: Vittorio Klostermann.

Heidegger, Martin. 1981. 'Only a god can save us': The Spiegel interview (1966). In *Heidegger: The Man and the Thinker*, ed. Thomas Sheehan, 56f. Trans. William Richardson. Chicago: Precedent.

Heidegger, Martin. 1994. Das Ding; Das Ge-Stell; Die Gefahr. In *Bremer und Freiburger Vorträge*, Gesamtausgabe, vol. 79, ed. Jaeger Petra. Frankfurt am Main: Klostermann, pp. 5–21, 24–45, 46–67.

Heidegger, Martin. 2000a. *Introduction to Metaphysics*. Trans. Gregory Fried and Richard Polt. New Haven: Yale University Press.

Heidegger, Martin. 2000b. Spiegel-Gespräch mit Martin Heidegger. In *GA 16: Reden und andere Zeugnisse eines Lebensweges*, 669–670. Frankfurt am Main: Klostermann.

Heidegger, Martin. 2009. *Leitgedanken zur Entstehung der Metaphysik, der neuzeitlichen Wissenschaft und der modernen Technik*, Gesamtausgabe, vol. 76, ed. Strube Claudius. Frankfurt am Main: Klostermann.

References

Constellating Technology: Heidegger's *Die Gefahr/The Danger*

Babette Babich

Abstract Heidegger's question concerning technology was originally posed in lectures to the Club of Bremen. This essay considers the totalizing role of technology in Heidegger's day and our own, including a discussion of radio and calling for a greater integration of Heidegger's thinking and critical theory. Today's media context and the increasing ecological pressures of our time may provide a way to think, once again, the related notions of event [*Ereignis*] and ownedness [*Eigentlichkeit*].

1 Constellating Technology

»Die Konstellation des Seyns spreche uns an.«

— Heidegger, *Die Kehre*

On December 1, 1949, Heidegger addressed the Club of Bremen under the title: *Insight Into That Which Is*, featuring four sub-lectures, each one lengthy enough to count as a lecture in its own right.[1] A few months later, Heidegger reprised the colloquium in Baden-Baden on two successive days on the 25th and 26th of March, 1950. A popular account of the Baden-Baden lectures in *Der Spiegel* invokes Heidegger's influence on Sartre and the French Existentialist movement,[2] but reflects that if it is the image of the philosopher in his Black Forest cabin that "makes

[1] Martin Heidegger (1994). Cf. Heidegger (2012) and see for translations of "The Question Concerning Technology" and "The Turn" as well as the additional essays, "The Age of World Picture" and "Science and Reflection," Heidegger (1977b).

[2] "Heidegger. Rückfall ins Gestell," *Der Spiegel*, 14: April 6, 1950.

B. Babich (✉)
Department of Philosophy, Fordham University,
113 West 60th Street, New York, NY 10023, USA
e-mail: babich@fordham.edu

B. Babich and D. Ginev (eds.), *The Multidimensionality of Hermeneutic Phenomenology*, Contributions to Phenomenology 70, DOI 10.1007/978-3-319-01707-5_10,

headlines,"[3] the most newsworthy event would be the two day lecture series: described as an "absolute exception," and emphasizing that Heidegger was technically still banned from teaching. There is an obvious dispute about the dates of the official ban[4] yet what is not disputable is that Heidegger would not resume university teaching at Freiburg until 1951.[5] What is also not in dispute is that under the Nazis, Heidegger was deemed insufficiently important ("scientifically" or as a scholar) and he was relieved from service in university and re-assigned to service in the *Volksturm* following the heavy bombing attack on Freiburg.

Towards the end of the war itself Heidegger managed to get permission to relocate his papers to Messkirch and he also offered a conflict-laden reading of Hölderlin in a lecture held in a castle above Beuron to which he and other university faculty retreated, speaking there not on needfulness [*Die Not*] or desperate times [*dürftiger Zeit*] but (and much rather), *Die Armut*, poverty.[6] Still or in any case, the *Spiegel's* assertion of an 'absolute exception' seems less than accurate for two days of lectures reprising the one day Bremen lectures held three months earlier.[7] Indeed Heidegger tells us that he would repeat the Bremen lectures on other occasions, the most well-known of which being a presentation of these lectures in Munich at the Bavarian Academy of Fine Arts in Munich, on June 6, 1950, where he presented the first, second, and last lecture of the series of four lectures presented in Bremen and repeated in Baden-Baden.

The first lecture was titled *Das Ding* [The Thing], the second *Das Ge-Stell*—which may be variously translated, most popularly, as "The Enframing" or, more recently, as "The Positionality" or even, with a Brooklyn (and I hope suitably gangster accent) "The Set-Up,"—the third, *Die Gefahr* [The Danger], and the fourth, *Die Kehre* [The Turn]. Five years later, in 1954, Heidegger featured the central themes from these lectures in his *Vorträge und Aufsätze*, published in 1954, in which *Die Frage nach der Technik*, "The Question Concerning Technology," has pride of place as the first chapter, followed by "Science and Reflection" and so on.[8] Indeed, had Heidegger scholarship been differently, hermeneutically minded, rather as Joe Kockelmans has been able to

---

[3] Ibid.

[4] The suspension of Heidegger's right to teach was imposed 1945–1949 but Heidegger would not resume teaching until 1951, as Heidegger's own comment on Richardson's "Appendix" to his *Heidegger: Through Phenomenology to Thought* indicates, Richardson (1993), 678–679. The recommendation of a period of 5 years appears in Jaspers' *Gutachten* but as Günter Figal has noted, among others like Rüdiger Safranski, the prohibition was indeed lifted as of 1949, although Heidegger would not officially "resume teaching until after assuming emeritus status in 1951." In: Figal (2006), 38. See for an overview of relevant primary sources, Martin (1989).

[5] There is some ambiguity as to what might be meant by a *Berufsverbot* or *Lehrverbot* and the Spiegel article suggests that this refers to university as well as general or public lectures, such that Heidegger's commemorative lecture *Wozu Dichter?*, presented in 1946 in honor of Rilke would/should also be counted as ‚lecturing.'

[6] Heidegger's June 27, 1945 Beuron lecture "Die Armut," is apotheosized by Lacoue-Labarthe in his introduction to Heidegger (2004).

[7] Here too, if we are counting the ways Heidegger might be considered as 'teaching,' one may also count a radio broadcast in 1951. Heidegger (1951); courtesy of Klett-Cotta und WDR.

[8] See Heidegger's (1978 [1954]a, b).

read Heidegger, along with a few others like Ted Kisiel, like Hans Seigfried and Patrick Heelan, all of whom read and foregrounded Heidegger's thinking in the mid-1960s through to the early 1980s on the topic of technology and modern science, Heidegger's collection of his *Lectures and Essays* (as yet untranslated as such) might well have set the tone for the post-war Heidegger reception.

But as it happens the history of the reception of a thinker's ideas is often the history of the reception of the translation of those ideas. Thus Ralph Manheim's translation of Heidegger's *Introduction to Metaphysics*, first translated in 1959 and thus in advance of Macquarrie and Robinson's translation of *Being and Time* in 1962 along with the 1971 translations of the studies of poetry, language, and above all the essay on the origin of the work of art, would entail for Anglophone readers that Heidegger's reflections on science and technology were relegated to second tier in Heidegger scholarship.[9] Yet things are not all that different in France, though one may note Dominique Janicaud as exception and Rainer Bast, Ewald Richter, and Carl-Friedrich Gethman in Germany.[10]

Today, in English language studies we may have the preconditions for a change in English language Heidegger scholarship with Andrew Mitchell's new translation of the Bremen and Freiburg lectures.[11] But the comparison of French and German studies tells us that we should expect to take some time to add the question of Heidegger and science to the issue of technology, a compound concern that and along with his thinking on art Heidegger always saw in terms of what I am here seeking to articulate as a constellation.

It was this same constellation that was in view for Kockelmans himself who, along with the already mentioned Hans Seigfried and Patrick Heelan, authored important early studies of Heidegger and the sciences.[12] Kockelmans also went on, together with Ted Kisiel, to dedicate an important collection to framing this thought constellation within continental philosophy of science, with the alas relatively utterly unreceived but indispensable collection, *Phenomenology and the Natural Sciences*,[13] together with Kockelmans' own single authored *Heidegger and Science*,[14] which Kockelmans was able to explore as a central theme of his own

[9] Cf. Heidegger (1959, 1962) as well as Heidegger (1971).

[10] See, in particular, Janicaud (1985), as well as (patently: in addition to others, both earlier and since): Bast (1986), Richter (1992) and see too Gethman's (1991) as well as Seigfried's (1991), respectively.

[11] Heidegger, *Bremen and Freiburg Lectures*, as cited above.

[12] Instructively, the American tendency to fail to mention German and French scholarship on the topic of Heidegger's philosophy of science and above all to exclude mention of work done by Kockelmans or Heelan, see for a recent instance, Heelan (2012) or Richardson as well as Seigfried in favor of voices supposed to be received (at the time the names mentioned in passing were Hesse, Lakatos, and Feyerabend, although the article's actual citations were limited to Kuhn) characterizes Jack Caputo's essay (1986). To be sure, Heidegger's philosophy of science cannot be discussed apart from Heidegger's engagement with Husserl and Kant and above all perhaps with Nietzsche. See for this context, Babich (2010a).

[13] Kockelmans and Kisiel (1970).

[14] Kockelmans (1985a).

research while also publishing in the same year a wide ranging study in Nijhoff's influential *Phaenomenologica* series on *Heidegger on Art and Art Works*.[15]

The story of continental philosophy of science and Heidegger is a complicated one, not able to be related here but at the same time unable to be dispensed with as it very directly affects the reception of Heidegger in philosophy of science in particular but also in philosophy in general.[16] Thus the fortunes of continental philosophy as such and in contest with analytic philosophy and the overarching *ressentiment* of things French and especially in the post-war years of things German make a difference as well. In addition, analytic philosophy (as I argue elsewhere)[17] has tended to be especially suspicious of Heidegger's focus on questioning or critique. To this it should also be acknowledged that critique per se had been associated ever since Immanuel Kant himself with the encroaching danger of nihilism, thus Heidegger's 1939 lecture courses on Nietzsche's epistemology (entitled "The Will to Power as Knowledge") and 1940 course on "Nihilism" hardly helped matters in this regard.[18] But as with many things, there is much more than a single influence or factor.[19] That these factors continue to interweave and play in current understanding is also something I hope to foreground in what follows.

The Bremen lectures for their own part draw on formulations unpublished (the *Beiträge*) as well as published as we recall Heidegger's 1946 "Letter on Humanism,"[20] a letter composed in reply to the Jean Beaufret's question to him in the wake of the devastation of World War II, prompted in part in response to Jean-Paul Sartre's Paris lecture in the same post-war year: *Existentialism is a Humanism*.[21]

---

[15] Kockelmans (1985b).

[16] See, again, in general, Babich, "Towards a Critical Philosophy of Science" and with specific reference to Heidegger, see Babich (2012, 159–192 and 2013b). In addition to Trish Glazebrook's introductory overview: "Why Read Heidegger on Science?" in: Glazebrook, ed., *Heidegger on Science*, 13–26, see too in the same collection Richter, "Heidegger's Theses Concerning the Question of the Foundations of the Sciences" (67–90) as well as important contributions by Heelan, "Carnap and Heidegger: Parting Ways in the Philosophy of Science" (113–130) as well as Ute Guzzoni "*Gelassenheit*: Beyond Technoscientific Thinking" (193–204) and Kiesel's "A Supratheoretical PreScientific Hermenutics of Scientific Discovery" (239–260).

On Heidegger and the disciplinary profession of philosophy as such, especially but not only in Anglophone culture, see Babich (2003), 63–103.

[17] See, for one example, a recent interview, Babich (2011), 37–71.

[18] See for these courses: Heidegger (1991).

[19] Kleinberg's (2005) is, I think, a useful addition here, especially in the postwar context, but see too for the pre-war context the now-standard reference on Heidegger-Carnap, Friedman (2000)—cf. Heelan's essay "Carnap and Heidegger" cited above—enhanced in depth by Gordon's (2012).

[20] Heidegger (1954). Additional elements, were we tracing the history of the lectures themselves can also be found in Heidegger, *Beiträge zur Philosophie* and so on. Heidegger's (1977c) and the same translation is also included in the English edition of Heidegger's *Wegmarken* by MacNeill (1998).

[21] Jean-Paul Sartre (1946, 2007).

The Thing (the first of the lectures later reprised in Munich), is also included in *Vorträge und Aufsätze*, together with Heidegger's prefatory "Hinweis" or contextualizing reference to the 'shrinking' of time and space through the same well-known technological means that continue to shrink distances to this day. And as already noted, eight years later, a little contribution based on the Bavarian lectures also appears as the first in the Neske series *Opuscula aus wissenschaft und dichtung*, entitled *Die Technik und die Kehre* and duly citing the context of the original lectures.[22]

As a consequence, by 1962 all but one of the original four lectures had been published, in one variant or reprise or another. My theme here focusses on that otherwise unpublished essay, "The Danger," *Die Gefahr*, although and of course parts of the text are assimilated into *The Question Concerning Technology*. As this point of assimilation also makes clear, a discussion cannot but include reference to all four, especially *Das Ge-Stell*.

The thoughts Heidegger gathers together in these lectures, given as we are told, and let it be noted again, over the course of a single day, and hence in a single breath (the German celebrates just this capacity, doubtless due to the length of their sentences: *der lange Atem* being a term of approbation), go back to the *Beiträge*, Heidegger's supposed second major work, but a work scholars now largely disregard (after the initial flurry of interest).[23] These days and already for some time we have tended to focus on what we take to be the early Heidegger—roughly the pre-*Being and Time* Heidegger, this being the bailiwick of either the very pious, literally so, Heideggerians, or else those who follow and trace the origins of Heidegger's original thinking in the spirit of Ted Kisiel's genealogical, phenomeno-philological brand of Heidegger-hermeneutics. Then there is the later Heidegger, corresponding largely to the Heidegger discussed here, but many people, especially in literature departments also take this to be the Heidegger of *Poetry Language Thought* and *On the Way to Language*, and so on all the way to *Time and Being* and the *Discourse on Thinking* as well as the later seminars.

And yet the division into early Heidegger and late Heidegger, corresponding to Heidegger I and Heidegger II, is problematic. Heidegger himself politely points this out by foregrounding entanglement, rather in the guise of his *Being and Time* discussion of future temporalization (out of the past) in his "Letter to Father Richardson," telling us (not really very helpfully) that

> only by way of what [Heidegger] I has thought does one gain access to what is to-be-thought by [Heidegger] II. But the thought of [Heidegger] I becomes possible only if it is contained in [Heidegger] II.[24]

[22] Heidegger (1962).

[23] Heidegger's originally unpublished *Beiträge: vom Ereignis* was published in his collected works in advance of the schedule Heidegger had envisaged. It is also available in English in different editions, under two species of translation.

[24] Heidegger (2003), 8.

To take up Heidegger's third lecture, *Die Gefahr*, it will be necessary to refer to the lecture just preceding it on the technological frame or setup, *Das Ge-Stell*. Here, I'd like to speak of the language thematic of both lectures (*Ge-fahr, Ge-Stell*) in a way that is not made easier by the limitation of addressing the question in English, as I am inevitably doing and just to the extent that the English translations cannot but efface the prefixes in either case. The patent point is that these two words, as different as they are, share the same prefix Ge- and that this is relevant as a word form and substantively. Although it is not often done, it's important to take note of this because Heidegger's mode of thinking through what he calls his *Insight Into That Which Is* tacks a path through related notions (i.e., that which is). In this respect he includes as the core of his lectures, two themes formed with a prefix, the "Ge-," a prefix, as we will all remember from Heidegger's *The Question Concerning Technology*, that he considers so very important that he talks about it there just as he does in *Das Ge-Stell*, focusing on the painfully ungainly *Ge-Stell*, taking it apart, literally by hyphenation and at what can appear to be surprising length. This gives (or should give) a translator pause and William Lovitt, to his credit, thought about the challenge it presented in his translation of *The Question Concerning Technology*[25] and Andrew Mitchell, who has just published his translation of the original four lectures with Indiana Press also gives his reasons for his rendering (though some may have wished for more detail than the few lines he offers).[26]

The rendering of *Gestell/Ge-Stell* as "framework"/"positionality" may be due to little more than the politics of re-translations, for and after all, a translator has to change enough in order to justify the effort, and it can seem that where Lovitt has "enframing," Mitchell simply inserts, it can appear to have been a kind of cut and paste, "positionality." Thus Mitchell's translation, which is a fluid one, has a dangerous side of its own as it tends to favor a one-to-one style of translation of the sort that today's Cambridge University Press translations have made into a kind of analytic gold or plastic standard, perhaps this begins with the Fichte and Hegel translations, but it is also (with some considerable and disastrous consequences) in evidence in the Cambridge Nietzsche editions. According to this standardizing standard, one finds an equivalent and settles for it, and to this extent the glossary in the Mitchell translation is more literal than say the listing to be found in Macquarrie and Robinson, for example.[27]

Gestell, a kind of physical array or constellation, means framework or structural outline or scaffold. The word is significant because *Gestellung* also means muster, and one can be ordered to such a mustering, commandeered or called up to service. As is familiar to those of us who know his concern with the fortunes of technology, traditional and modern, what Heidegger wishes to do here, after he has set up his initial tracing reflections on modern technology per se, is to tease out the

[25] In addition to his note on the transforms affected by such prefixes in his introduction (p. xx), see William Lovitt's footnote 17 in his translation of "The Question Concerning Technology," in: Heidegger (1977a), 3–35, here p. 19. Cf. note 14, p. 13, as well as notes pp. 15–16, pp. 16–17.

[26] Andrew Mitchell (2012), xi.

[27] Mitchell, "English German Glossary," in: Heidegger, *Bremen and Freiburg Lectures*, 173–198.

determining or destining *set up* that is part and parcel of modern technology as this intricately ordered and dependent set up is opposed to the cognate fitted-togetherness but individually separable configuring of old-fashioned equipmentality, as such. Tools qua tools have always involved referentiality. This is what Heidegger calls *Bewandtnis* and it is the subject of his memorable analysis of handiness—*handhabbarkeit*—in Heidegger's in *Being and Time* discussion of *Zuhandenheit* (BT 98/69), namely readiness-to-hand and in turn and presuming such a readiness in its modality as "unreadiness-to-hand," the revelation of "being-just-present-at hand-and-no-more" (BT 103/73) as these fit together precisely in such a work context. Using a hammer for a given project, whether it involves the kind of complexity that would have engaged Heidegger's own father as a cooper or joiner (these are related carpenterly professions, but the unions to this day keep them well distinct), or just hanging a picture on the wall, one is referred to a nail or, if this is a metal-free project, think of a trip to IKEA or more romantically, think of Eric Sloane's America where nails were expensive and using wood's properties part of Yankee or New England ingenuity (read thrift or cheapness), with the hammer will go the pegs or cleats.[28] The difference however is that the same claw hammer that nails nails, removes nails (note that this does not apply to German hammers, they do not come with a claw as one is meant to remove one's nail with the proper tool) and the same hammer, German or Sears Craftsman style can be used to break through a wall if one wishes to remodel a kitchen or for other purposes of the sort and in my classes on Heidegger's *Question Concerning Technology*, I sometimes like to imagine circus acts, *cirque du soleil* meets gas station mechanic, juggling with three and then four hammers and so on—these have to be claw hammers for the sake of showmanship and counterpoise. Modern technology quite specifically does not work like that. If you misplace the charger for a new cellphone, you will find that using one of the chargers in your collected array of chargers from cellphones gone by will be an exercise in futility. Connectivity is the point. Modern technology, Heidegger argues, goes beyond the traditional in-order-to of particular kinds of equipmentalities, the kind of practical ordering or for-the-sake-of-which that Aristotle lists for us with reference to the bridler's art in the very first section of the *Nichomachean Ethics*. In *Being and Time*, Heidegger refers to the aforementioned workshop array of tools but he also lists the items on his own desk as tools of a kind: paper, desk blotter, fountain pen, ink and so on. So today we might add to all those desk items, a computer, printer, internet connection, surely all this is the same—just update. Heidegger thinks not and his four Bremen lectures, "Insight Into That Which Is," try to explore what is different in modern technology and that is to say to raise the question regarding technology as a question, just as we might remember that he has been at pains to point out just how hard it is to ask after anything at all beginning with *Being and Time*, section two of which unpacks what it means to question.

Thus and just to offer a contrasting illustration of the romantic sort that we can use to document modern progress, Heidegger used a handsaw of the kind that requires two workers to cut wood for use as fuel in the cabin his wife had arranged

[28] See for example, Sloane's (2004 [1965]).

Fig. 1 Hans-Georg Gadamer and Martin Heidegger, Todtnauberg, 1923. Bildagentur dpa

to have built for him in 1922 and for which wood-cutting task he required the efforts of one of his students, my own teacher Hans-Georg Gadamer.

Heidegger later sent a picture commemorating this moment to Gadamer as a gift on his 75th birthday in 1975 and Gadamer thus includes it, with Heidegger's note, in his *Philosophische Lehrjahre*.[29] The picture dates from 1923, that is: pre-*Being and Time*, which would thus make this image, for those who like these terms, a picture of Heidegger I (Fig. 1).

Let me note just because it matters in the current context that pants of the kind worn by both Heidegger and Gadamer in this picture did not in fact testify to some kind of back-to-the-land fascist movement but were standard for the time and there are photos of my own father, who was born in New York City in 1935, wearing short pants (i.e., not shorts) of a similar fashion, in pre-war NYC, circa 1940 or '41. Details like these, ontic as they are, do not deter folk who have assumed that this picture must date from at least a decade later, say circa 1933, or must even be a postwar image, those who might claim that it provides iconic evidence for Heidegger's nostalgia for the past. For my part, I take the irony to be the labor itself as, like Tom Sawyer, Heidegger commandeers Gadamer's assistance to help him cut some wood, ironic because of Gadamer's later recollection that when he first met Heidegger he took him for a manual-laborer—a Hausmeister—in NYC that would be a super.

The thing about a two handed bow saw is that the 'Gestell' involved to support the wood being sawed has as such no particular connection to the saw or the piece of lumber. It is called a saw horse, technically, just as other Gestell types count as clothes horses or racks, umbrella stands and the like, and you can buy these too at

[29] Hans-Georg Gadamer (1995), 33.

IKEA and a pair of them will help you cut plywood but can also serve to hold a dinner table perfect for a fashionable loft kitchen. The components can be used together or not, they are severable with respect to use but also distance and thus they are more rather than less self-contained. Heidegger was therefore using the support of such frames to position wood to cut with a bowsaw, given Gadamer's help, given his wife's gift to him of a house in the high hills of the black forest (they are not really mountains), rural land that was then, as a lot of land still is, without convenient access to electricity,[30] for example, although there was, and that would be a sine qua non, water afforded by the famous spring to which Celan would dedicate his poem *Arnica, Eyebright, Arnika, Augentrost.*[31]

By contrast and this is the point Heidegger seeks to make throughout, modern technology, modern tools, power tools are different and everything turns on power and its dependencies: thus nature in the purview of modern technoscience becomes on Heidegger's analysis something that it never was until modernity: a giant gas station, a source for the development of natural resources, meaning energy, meaning electricity. In the case of a power tool you are tied to that referentiality by the cord, even if you have a cordless drill, because as Hurricane Sandy reminded us in New York City, you really need to charge cordless tools, including laptops and iPads and cellphones. So whether it is an outlet (this becomes a kind of holy grail for students looking to plug in their laptops or travelers looking to do the same), or extension cord, they all point to the need for electricity, and all the stuff you will have to think about if one gets a job at NYU (at NYU pay) and wishes to build a cabin of one's own upstate in New York's Putnam county, say, you'll need water, cable, the works, and all that will be a pre-requisite before you can get to reflect upon Heidegger's observation that a mechanical tool "is nothing that separately presences for itself."[32] In other words, that is to say that even in its components, i.e., qua taken apart, as he also speaks about automobiles broken down for shipment, modern technology requires far more than just completeness unto itself to be able to be set in motion. Thus contrasting the modern technological apparatus with a self-propelled wheel assembly, like the spinning wheel or else like the "bucket-wheel in the rice fields of China" as he invokes these still in use in rural china,

---

[30] By the time the cabin was built it likely had electricity. Germany had electric lighting since the 1880s and by 1913 a good many households as well as the university in Freiburg itself used electricity. See for instance, Chickering (2007).

[31] Celan's poem was written after his 1957 visit to Heidegger's hut in the Black Forest and was included in a collection of Celan's poetry entitled *Lichtzwang* published shortly after the poet's death in 1970. The title of the poem, *Todtnauberg,* is a metonymic allusion to place and the rest of the poem seems to do the same: Arnika, Augentrost, der/Trunk aus dem Brunnen mit dem/Sternwurfel drauf,//in der/Hütte,/die in das Buch/—wessen Namen nahms auf/vor dem meinen?—,/die in dies Buch/geschriebene Zeile von/einer Hoffnung, heute,/auf eines Denkenden/kommendes/Wort/im Herzen,//Waldwasen, uneingeebnet,/Orchis and Orchis, einzeln,/Krudes, später, im Fahren,/deutlich,/der uns fährt, der Mensch,/der's mit anhört,//die halb-/beschrittenen Knüppel-/pfade im Hochmoor,//Feuchtes,/viel."Paul Celan (1980), 240–241 and (2000), Vol. 2, 255–256. See for one discussion, Lyon (2006). See too Herman Rapaport's chapter "Forces of Gravity" in his *Is There Truth in Art?* (1997), 110–143.

[32] Heidegger (1994), 34.

modern technological machinery only "stands" or works as such "when it goes."[33] If the machine is out of order, if requisite parts are missing, it is worse than nothing and now we are back to the sheerly present at hand (or the irremediably present at hand in the case of those old power cords that connect to appliances or tools one no longer has). Here Heidegger is concerned to attend to the ordering of both the machine and the mechanical network into which it is set just in order that it might be a mechanism of this or that kind. Thus as noted, he also gives the example of the automobile, pointing out that the automobile is more than a tool made of separable parts into which it can be broken down and out of which it can be assembled but exemplifies modern technology to the extent that its use, and intriguingly this has been the subject of several politically theoretic studies of technology, requires an entire schema, a constellation or network, all of it sine qua non. This is not merely a matter of fuel and and a network of fuel stations, of building a network of roads for automotive use and redesigning entire downtown urban areas to include parking garages and highways that pass over or pass through a city and so on. Thus Langdon Winner and others talk about the concerted efforts in the early decades of the last century to demolish street cars and established forms or networks of public transport to shift consumers of public transportation, which cost whatever it cost for a ticket, to consumers of private transportation which required a whole lot more in the way of direct and indirect costs.[34]

Private vs. public transportation underlines Heidegger's point. Hitler built the Autobahn and his system of roads (still a fetish factor in Germany—*Freie Fahrt für freie Burger*, where the emphasis is on free, meaning no speed-limit) was as beneficial for the nation in peacetime as in wartime. Thus Heidegger can remind his Bremen businessmen that unlike the jug that he uses to illustrate the thing in his first lecture, the automobile does not "just" stand there even when it is parked. Instead it is "at the ready," precisely available for use in every potential or possible sense. Hence the automobile, and by extension, the truck for industrial transportation "is able to be challenged forth precisely for a further transport, which itself sets in place the promotion"—and in good, Rotary Club, English we might prefer to say that this potential to be challenged forth drives the wheels—"of commerce"[35]

Here Heidegger goes on to clarify the way in which we are today set up, as it were, to be consumers of precisely the technological schema or framework or, to use Jacques Ellul's term for the very same thing, the technological system, because the point concerning technology is that there is no having of it by halves. You cannot opt out, you cannot take it or leave it—the later Heidegger—Heidegger III we could say—suggests in his *Discourse on Thinking* that we might do a kind of zen thing with technology, a kind of mindfulness he called *Gelassenheit*, but like zen and like

---

[33] Ibid.

[34] Langdon Winner offers a discussion of this point along with a number of references to classical political studies of the shift from public to private transport on the eastern and western seaboards in Winner (1986).

[35] Heidegger (1994), 35.

mindfulness (Heidegger called this thinking), *Gelassenheit* turns out to be more elusive or harder than it sounds.

For Heidegger, "the forester who surveys the wood to be felled"—the line here is reproduced in its entirety in "The Question Concerning Technology"—traces and does not trace the path followed by his grandfather just to the extent that the wood he cuts is ordered, set up for and into the lumber industry which is ordered or fit into producing "cellulose stock" for the paper industry which in turn is set up for delivery "to the newspapers and tabloids that impose themselves on the public sphere in ordered to be devoured by it."[36] If the Frankfurt School were not disposed to reject everything Heidegger notes (after all Horkheimer would still have all the priority one might wish) there is a useful critical analysis in the next paragraph, which does not indeed appear to the same extent in the later essay *The Question Concerning Technology.* Thus Heidegger here touches upon themes echoing those of Horkheimer and Adorno in their own 1944 *Dialectic of Enlightenment*—elements of which grew out of Adorno's work, begun in 1941 on Lazarsfeld's Princeton radio project[37]—along with Friedrich Georg Jünger as well as Herbert Marcuse, in addition to Günther Anders (the stepchild of the Frankfurt School) as indeed Rudolf Arnheim, points also approached from a different point of view by Edward Bernays and Vance Packard.[38]

The point is media, and Heidegger goes on to talk about radio and film in order to explain the very way that the human being him- or herself is disposed of, imposed upon, precisely with respect to his or her disposition as such:

> Radio and film belong to the standing reserve of this commandeering [of the human being] through which the public sphere [*Offentlichkeit*] is set up, challenged forth, and thereby installed in the first place.[39]

For Heidegger, this is not merely the work of the "radio broadcast advisory council" but is already at hand in "the standing reserve called the radio, i.e., challenged forth to the ordering of the broadcast industry."[40]

My point is to call attention to a remark that Heidegger offers in a phrase uncannily similar to Adorno's physiognomic observation regarding the twirling of the

[36] Heidegger (1994), 37.

[37] See Adorno (1945). See for the results of the Princeton Radio Project, Adorno (2006) and see too Thomas Y. Levin's contextual discussion, which to be sure does not connect Adorno with either his contemporary Anders much less, given the same contemporaneity, Heidegger: Thomas Y. Levin with Michael von der Linn (1994), 316–324. See for further discussion and further references Babich (2013a), Chap. 6.

[38] Vance Packard's (1957) is a popularized discussion of the then-well-established effects of Edward Bernays' (1928). Bernays' work is better known under the rubric of Public Opinion Research or Motivation Research, and is of course all about advertising or marketing but which was originally developed (and is still used) for the political purpose of shaping public opinion—as its original name indicates. For a discussion with respect to television, see Günter Anders cited below as well as independently of Anders, the Canadian political theorist, Dallas Walker Smythe (1954), 143–156.

[39] Heidegger (1994), 37.

[40] Ibid.

radio dial but also in a context akin to the "homeworker" analysis that would be offered by Günther Anders in 1956, which piecework manufacturing in turn produces or generates the media consumer qua media consumer, a point to be taken up by the Canadian media-political theorist Dallas Smythe, arguing and in the process explaining why commercial broadcast access is of value to manufacturers, that, in Heidegger's words here

> every radio listener who turns its dial is insulated as part of the component character of the parts of the standing reserve, locked in as a piece of the standing reserve, in which he remains confined even if he still thinks he is utterly free to turn the device on and off.[41]

Paralleling his trademark tool example, Heidegger observes that even if one were to turn off the radio, one would remain connected or bound to it. Indeed as I have argued to be typical for Heidegger's style of intensification, he emphasizes the point with an iconically philosophical thought example: were a cosmic miracle suddenly to silence all radio broadcasts, so Heidegger argues, the very same connection would still persist.[42] On this extreme supposition, even if:

> suddenly everywhere on earth in everyplace, radio receivers were to disappear—who could comprehend the cluelessness, the boredom, the emptiness that would at a blow assault the human being and thoroughly unhinge their routine affairs.[43]

This is also, though that is a paper of its own, the reason for Heidegger's extended reflection on what is involved when a particular tract of land is challenged forth to produce coal, which is in turn demanded by the electrical industry which itself deploys a massive set up just to be able to convert coal into steam, into power for industrial and private use. Heidegger uses this example because such industries and their interconnections (especially all the details we tend not to think about) were transparent to him as they were to every German, every Frenchman, etc., etc., after the war. Thus the competing desire to use land for mining (raw materials) clashed with the need to use land for agriculture (foodstuffs), but the technization of both handcrafts, only meant that the one application namely mining or as we call it today: land use development, demanded vastly more land than ever before, and the second application, farming, also took more land in its mechanized variety than had been traditionally needed.

But the economics of competing land applications and how they might be parceled out and to which interest groups concerned Heidegger less than the very complicated array or constellation of modern scientific, technologized industry as such.

---

[41] Ibid. Anders himself offers a sustained discussion of this counter-example in "Die Welt als Phantom und Matrize. Philosophische Betrachtungen über Rundfunk und Fernsehen"in his 1956 book, Anders (1980), 97–214.

[42] Heidegger's thought example has been 'real' (or Baudrillardian 'integratedly real') for some time and as newspaper reports of New York residents reported (and my own students attested) during Hurricane Sandy, when they couldn't charge cell-phones and usual avenues of internet access were down—today that would be the wireless equivalent of what radio was in 1949—there was great anxiety.

[43] Heidegger (1994), 39.

Thus in addition to his coal example, or airplane example (in the original lecture as we have just cited it, he talks about automotive components packed for export as items of so much standing reserve—present at hand we could say—and parking lots and highways, as components of the automotive industry, all very patently ready to hand). Likewise as also noted, Heidegger focusses on forestry, the woodsman today as compared to his forebears and with that he is off with a discussion of forest management practices, which means harvesting, i.e., cutting down the trees for the sake of and exactly as cued to the needs of other industries as we have just detailed these: like the enormous need for paper after the war, be it for planning or for journalism, which industry also catches Heidegger's attention as it is this same industry (this is the point he makes about radio) includes human beings who are themselves parts of this same industry, ordered into it, set up into it, to the extent that both paper journalism and radio are so many culture industries to use the language that Horkheimer and Adorno and Anders also employ to speak of these media enterprises, as such public industries, as Heidegger explains, are used to direct or set up the "public sphere" so that it may be challenged forth and ordered, i.e., so that public or political planning can proceed according to political design. Indeed as Heidegger certainly knew—the political fate of Germany depended upon it—such public sphere planning was quite explicitly at issue. The question at hand was at the time: what kind of government would rebuild the country? What direction would it take?[44] If it can be argued that in West Germany, excluding socialism would have to be politically overdetermined, Marx himself had offered serious critiques of the kind of advantage capitalism takes in the time of crisis and had already analysed that the only efficiency served was that of profit. Heidegger makes this point in his own lectures, an emphasis repeated in his "Building, Dwelling, Thinking,"[45] which was of course all about the urgent misery of the housing crisis, which was also at the time a food crisis and his "Letter on Humanism" culminates with references, among other things, to Marx, as we note by considering his contrast between thinking and doing, contending that "thinking is a deed" and continuing by emphasizing such a deed "also surpasses all praxis."[46] For Heidegger, however the thought in question, the 'understanding' of the world that Marx had famously attributed to all philosophy heretofore in his *Theses on Feuerbach*, would not be marked by anything like "the grandeur of its achievements" or indeed efficacy as such "but through the humbleness of its inconsequential accomplishment."[47] Here in the *Letter on Humanism*, and presumably Heidegger would have known exactly what he was saying by writing this, the conclusion points to the same constellation of philosophy as a project of understanding the world or changing it, and Heidegger suggests that theory itself

---

[44] The beautiful German coin, a 50 pfennig piece issued the same year and featuring a young woman planting a small bush, offers an iconic illustration of this very concern.

[45] Cf. Heidegger, "Building, Dwelling, Thinking" in Heidegger (1971), 143–161.

[46] Heidegger (1977c), 274. I note that Heidegger already is in dialogue with communism, and its anticipated threat in his lecture *Die Armut*.

[47] Ibid.

can use a bit of reflection on itself and what it is capable of: "It is time to break the habit of overestimating philosophy."[48]

The problem here is already one I have been framing out: that is the problem of the Ge-Stell as this parallels the frames set up to re-build houses or indeed cathedrals in Freiburg as the cathedral there was damaged during the war[49]—if you visit and climb to the top you can see that the Freiburg residents set a plaque to thank the stones, as it were, for not falling. And to be sure, as those of us who live in the city know all too well, once a scaffolding goes up around a building to repair it or what have you, its durability seems guaranteed.[50] The scaffolding, the framework, the set up, is not only indispensable but all-pervasive.

Thus when we read the essay *Das Ge-stell*, the set up or the setting up, you can also say the enframing (I have already noted that my concerns about 'the positionality,' just to the extent that it can sound like a Kama Sutra move or some Deepak Chopra trademarked approach to heated or Bikram yoga), or indeed when we read *Die Gefahr*, we are confronted with Heidegger's most notorious comments on the technized transforms of industry and its consequences. Heidegger looks at what the mechanization of anything and everything does, and points out that it does not fail to affect us in the most basic way.

Thus Heidegger writes about the requisitioning and planning that characterized a wartime and a postwar *Nachkriegszeit* Germany, and he would certainly know about both as he himself (qua dispensable) had been set into, conscripted into service at the end of the war. For in wartime everything was placed at the disposal of this kind of ordering and everything came to be regarded, this is the effect of the transformation of this kind of ordering, as so much standing reserve. We even may remember, it's a postmodern meme, and certainly my grandparents would have remembered, various wartime advertisements encouraging the average American to do his or her "part" during the second world war. Now we already know from reading Marx's *Capital* if we did not know it from Adam Smith or others that just such a transformation of nature and human relations is the heart of economic ordering. All the war shows, as if it had needed to be shown, is the calculation of the same order and the details of dependencies. The things one tends not to notice (that the amount of wood that will be needed to be managed in the Black Forest will be directly dependent upon the proliferation of journalism and propaganda and information tracts—pick your euphemism—so that, once again, rather than serving the lumber industry the woodsman is more accurately or actually serving the pamphlet or leaflet industry) and such superficially counter-intuitive relations were made more transparent in the years during and especially after the war. This way of commandeering the resources

[48] Ibid., 176. See on this Graeme Nicholson (1987), 171–187.

[49] I adverted to this at the start and Philippe Lacoue-Labarthe underscores this as well when he recounts Heidegger's dispensation from university responsibilities in order to relocate his manuscripts to a safe place (in Messkirch), following "the (heavy) bombardment of Freiburg by English and American aerial forces." Lacoue-Labarthe (2004), 9.

[50] Many of us will have known, as I have known, urban scaffolds of the supposedly 'temporary' kind that have managed to endure for decades and decades…

of the world for such further purposes "endures" and Jacques Ellul will take a leaf from Heidegger (and Friedrich Georg Jünger) to insist on what he calls the "autonomy" of technology and technique, noting that once it is set in motion, today's modern technology cannot be arrested. It might perhaps, and however unlikely, be diverted, but never simply stopped.

For Heidegger this setting up endures insofar as the set up is in turn imposed for the sake of other purposes, to which it is ordered (raw materials are raw materials for something, although and of course they can be stockpiled more generically within that same framework). Deployment or utilization "sets everything up in advance such that what is set up conduces to success ....But the resultant is arranged as success beforehand."[51] And for Heidegger the resultant schema cannot but be self-reinforcing, and what is defined as "success," as he goes on to elaborate this, "is that kind of resultant that is itself allied to the production of further results. We call it ordering/requisitioning/com-portment [*das Be-Stellen*]."[52]

For Heidegger, and if this were another paper, I might go in another direction, there is a difference between the kind of productivity of the village carpenter (we began by noting that Heidegger's father was one such) who might make or produce a table or who might for another purpose, make a coffin, a *Todtenbaum*, which itself would be destined, fitted not into the productive time and cares of the carpenter's industry but into another schema of another kind of temporality and care—here Heidegger uses the language of the cares or concerns of *Being and Time*—and that means into the constellation and intimate engagements of another world that is not the world of the manufacturer's workshop but the different world directionalities and setting of the "peasant's farm, the house and the land, the ones who dwell there, their kin, and the neighborhood."[53]

There is no connection with any of that today, and intriguingly, we can cross the distance in time between Heidegger's 1950 lecture and 2013 without needing to change a thing. The "mechanized burial industry of the metropolis"[54] as Heidegger goes on to say by contrast does not lend itself to peasant rituals, themes or terminology. And if you want to see a French take on some of that, I recommend the climax of the wonderfully existentialist (not existential) 1986 film by Claude Berri, *Jean de Florette* when Jean (Gérard Depardieu in perhaps his most sympathetic role) is destroyed by his own *techne* (his dynamite) and his lack of *techne* (peasant experience) and above all by the failure of *techne* as what Aristotle named *phronesis* which would be knowing the difference between the two (that said, the technological critique of *Jean de Florette* is more Jacques Ellul than Martin Heidegger).

Comparing in a swift analogy the peasant's placing of his ox, positioning the animal in his traces just so, in order to advance the work he needs to get out of the ox, Heidegger writes that "Men and women must report themselves to a work service [*Arbeitsdienst*]. They are conscripted. They are met by a constellation [*Stellen*] that

[51] Heidegger (1994), 26.

[52] Ibid.

[53] Ibid.

[54] Ibid.

places them, i.e., commandeers them."[55] Heidegger thus goes into, as is his wont as we recognize this strategy from *Being and Time* to the later work on language, the meaning of the word, asking what das Ge-Stell means and and answering "to place, to position, to set'" so as to experience what comes to pass in that requisitioning and accountability through which a given stock arises and is thus a standing reserve.

Heidegger's analysis concerns civilian conscription, during and after a war on the most human level, whereby what is deployed are human beings as troops contra human beings as troops and of course and most lamentably contra those civilians who happen to be the enemy, and as part of that the requisitioning of whatever is at hand for the purposes of war. Like those summoned to do their part during war, Heidegger's point is that the approach is a total one, and there are parallels with Friedrich Georg Jünger, not unlike the parallels Walter Benjamin draws out in his reflection on the world of art in the age of technological reproduction with regard to the consequences of the first world war, when Benjamin cites the Futurist Manifesto of the Italian artisti, Marinetti in his own reflections.[56] As Benjamin then goes on to explain the object contradiction that is the work of art as such:

> the aesthetic of modern warfare appears as follows: if the natural use of productive forces is impeded by the property system, then the increase in technological means, in speed, in sources of energy will press toward an unnatural use. This is found in war, and the destruction caused by war furnishes proof that society was not mature enough to make technology its organ, that technology was not sufficiently developed to master the elemental forces of society.[57]

Benjamin continues by invoking what appears to be the fascist aesthetic, the aesthetics of pure politics: "*Fiat ars—pereat mundus*" and he explains this is as a direct consequence of technology and points out, too flatly for the nuanced sensibilities of a Horkheimer: "This is evidently the consummation of *l'art pour l'art*."[58] Invoking the cliché sublime converted here into the art-spectacular of a humanity converted from divine object to a subject absorbed with "its own annihilation as a supreme aesthetic pleasure," we are still far from thinking through the caesura, the space between the themes of his conclusion: "Such is the aestheticizing of politics, as practiced by fascism. Communism replies by politicizing art."[59]

To bring the point from a period after the first world war to Heidegger's time after the second world war (and still to this day, however we wish to understand 9/11 and the war on Iraq, Afghanistan, Syria and so on, and however we wish to

---

[55] Ibid.

[56] See the conclusion of Benjamin's "The Work of Art in the Technical Age of Reproducibility." I recommend the version (the second) of Benjamin's essay that appears in Benjamin (2008) despite the great advantages of Arendt's (1968) contextualization of the version that appears in the Shocken edition, because of the specific and useful secondary apparatus provided for this essay. Benjamin's discussion of photography including an allusion to war and to the origins of the technique, is varied, albeit without reference to Benjamin, in Friedrich Georg Jünger (1946).

[57] Benjamin (2008), 42.

[58] Ibid.

[59] Ibid.

understand the Keystone Pipeline to the US coast or the relentlessly stupid use of fracking), Heidegger points to these everyday circumstances and these everyday ontic consequences when he observes that "a tract of land is coopted, namely for the coal and ore that subsists in it."[60]

This notion of cooption and it should be clear here that Heidegger is talking about newly requisitioned tracts, newly requisitioned by the Nazis and then again in the postwar era, rather than offering some merely nostalgic musings in praise of the farmer's traditional field. For us today and to be sure, all this is a matter of 'development,' one thereby sets up a coal or another mining industry (we can add, if we like, that just such cooption sets up a fracking industry for extracting natural gas, requiring the use of vast quantities of fresh water, yet further evidence of the 'perfection' of technique in Jünger's sense as the engineering science of fracking requires pure rather than 'recycled' water, which is then an industry, paralleling Heidegger's awful agricultural example, that is/becomes an industry for the production of contaminated aquifers along with the production of contaminated soil and of course—because we are talking about gas—the production of polluted and poisoned air). In this way or "through such requisitioning [*Bestellen*] the land becomes a coal reserve, the soil a mineral deposit." Immediately contrasting this with the farmer's practice with respect to the land and to nature, as a kind of allowing, this is the meaning of *Gelassenheit*, "the crops to grow as nature itself allows,"[61] Heidegger thus seeks to raise the question concerning the difference made by modern technology in this contrasting opposition, and here we need the entire quote

> In the meantime, however, even the tending of the fields [*die Feldbestellung*] has gone over to the same re-quisitioning [*Be-Stellen*] that imposes upon the air for nitrogen, the soil for coal and ore, the ore for uranium, the uranium for atomic energy, and the latter for destruction on command.[62]

The lineage traced is that of modern technology and the efficiency of a technological world order. Everything is regarded, and we know this, we take this for granted, for the purposes of development, by which we mean if we are doing development studies: technological orderability or usability in the same schema or setup. Everything fits into this frame and there is no outside. If Marx saw the dynamic of the machine as reducing the needed labor of the worker to no more than an appendage, a fitted extra, and thus the stupidification of the human as a necessary part of capital and its mechanized deployment, as part of the complex relation of the human being to nature within the sane very material dialectic, Nietzsche himself points to a similarly coordinate structure when he argued that we humanize nature and everything else by cutting it to our measure (these are the "bounds" of sense in Nietzsche's articulation of the critique of reason in the third book of *The Gay*

---

[60] Heidegger (1994), 26–27.

[61] Ibid., 27.

[62] Ibid.

Science) but also as he goes on in *On the Genealogy of Morals* to highlight the numbing of the mind that, as he teased, is called "the blessing of work."[63]

Still it is one thing again to guess at the brutalizing direction of technology, and Nietzsche's language of the "God of machines and smelting pots,"[64] seems to capture the high regard we have for the priests of the same god, the engineers and technicians and indeed the scientists and theorists of all kinds. By contrast, and this is where the practical level, the ontic matters of the ordinary come into play, it is quite another thing again to live through the pains of such a transformation of the world in the image of technology, as Heidegger lived through this transfiguration, through two world wars, even if one could argue that we are today still, as we are, and not that we give it a thought, living through wars all the same. For the work of this transfiguring force is now largely consummate and we ignore, we do not live through, the wars we have consistently been fighting. If Heidegger could ask if the victims relegated to annihilation camps 'died' or if (and let us not forget that for Heidegger the word and the meaning of the word in each case makes all the difference) they did not much rather and simply 'perish'? The word he uses, the technical term as historians also use it, is liquidation. And whatever fate that is, what it is not, what does not have a chance to touch it (and those so condemned are bereft of exactly this on Heidegger's account) is death: a death, and above all, not one's own death: a death that one might take up, or and this is pure luxury, as we see, refuse to appropriate, refuse to live. The inauthentic death is also what one does not die in such camps.

For our part, we also ignore, as Baudrillard argued that we should not but that we cannot but fail to see, the political realm, which "political" we take to be all about what the journalism cum culture industry serves up to us.[65] Baudrillard's term was 'integrated reality,' which we ought today rename embedded reality, all the while unaware of what really happened to close down the OWS movement (New York City, after all, is where it began) and it is worth noting that I offered an earlier talk scheduled during the events of the original Occupy Wall Street[66] to the same group at the New School that initially invited me to give the talk on which the current essay is based. Here what matters with this detail and allusion to "real life" is that we scholars and citizens, journalists and consumers barely notice today that Wall Street is no longer "occupied," and we do not bother to attend to such routine and

[63] Nietzsche (1980), Vol. 5, 382. The full citation is useful: "Viel häufiger als eine solche hypnotistische Gesammtdämpfung der Sensibilität, der Schmerzfähigkeit, welche schon seltnere Kräfte, vor Allem Muth, Verachtung der Meinung, »intellektuellen Stoicismus« voraussetzt, wird gegen Depressions-Zustände ein anderes training versucht, welches jedenfalls leichter ist: *die machinale Thätigkeit*. Dass mit ihr ein leidendes Dasein in einem nicht unbeträchtlichen Grade erleichtert wird, steht ausser allem Zweifel: man nennt heute diese Thatsache, etwas unehrlich, »den Segen der Arbeit«.

[64] Nietzsche (1980), Vol. 1, 114f. Nietzsche is here, in his first book, coordinating the allure of a metaphysical comfort with the ideal of an "earthly consonance."

[65] See Baudrillard's (2005a).

[66] The earlier talk in question combined a lecture originally given in Dublin and a lecture entitled "Requiem" given at Boston College. The first lecture is forthcoming: as Babich (2013a).

such ontic details unless a Facebook post is sufficiently annoying to compel us to do so, likewise we are oblivious to our torture of our prisoners as we still detain them in Guantanamo, all that after electing a president on the explicit mandate that such detention centers follow the rule of law (hasn't happened and we elected that same president again, anyway), and we certainly think nothing of the overkill (tanks in the street, martial law, the complete shutdown of the town) required to catch two college students in Boston (called terrorists), killing one and leaving another at least initially unable to speak (Aristotle's *Nicomachean Ethics* goes on at length about the fortune of that circumstantiality): we as a media populace followed the manhunt in Boston with the avidity usually reserved for a contest on *American Idol*. Politics for us, as Baudrillard wrote again and again, alas with Gallic impenetrability, is all about the issues that are presented to us as news.[67] History may well tell a different story, but this is doubtful, and this too was also Baudrillard's point, Kittler's too when he could turn his attention from his Greeks and other dissipations.[68] After everything is digitized which means to be sure, after every record can be infinitely revised or changed at will—according to whoever's whim, whatever, the point to be remembered (no one will be able to make it) is that no one will be able to demonstrate/prove/notice the effects or consequences of such limitless alterability (this is the real meaning of the Leibnizian difference that makes no difference).

Yet one should be skeptical: we remain in need of a critical theory for our times and the current practitioners of the same, be they in Frankfurt or New York or Chicago, have fallen silent on anything that resembles critique. And these titulary practitioners control all the journals (*Critical Theory*, *Constellations*, etc.) and they control all the fellowships and they control all the books that are published in supposedly respectable presses. And did I say professorial posts too? No, because I did not have to: this goes without saying.

Repeated twice in these two core lectures, *Das Ge-Stell* and *Die Gefahr*, which may now be taken as the locus of Heidegger's abyssal politics, is (again) his un-speakable, claim about death and technology and we have heard about this and about its untenability all our intellectual lives. The most incendiary locus for this twice-repeated provocation might be as expressed in *Das Ge-Stell*. This is the locus that one scholar quoted out of context after gaining access to the then-not-yet published text (this is the fun of plundering archives, not that there are all that many chances for those doing archival work to do comparable things), after promising not to quote it out of context. But by breaking a promise (and one makes such promises in order to break them, as Kant tells us, namely as we seek to gain an advantage and because we know or tell ourselves that without just that false promise, breaking in our unsovereign mouths, as Nietzsche says calling us windbags, even as we utter it [this is the point of the aphorism on the Nietzsche's 'sovereign individual at the start of the second part of *On the Genealogy of*

[67] See Baudrillard (2005b) but see too one of his final essays available in English, Baudrillard (2009).

[68] This is not a matter of being for (or against) the media as it is also not a matter of being for or against technology.

Morals]), that same advantage is denied us. The advantage won by Wolfgang Schirmacher yielded the quote that generated a small book industry, large if you count Wolin, huge if you count Tom Rockmore's books, which is of course the Heidegger scandal, beginning with Levinas, Lacoue-Labarthe, Derrida, Habermas too.[69] In fact Heidegger makes two similar declarations, but the first one is the most notorious and it runs as follows

> Agriculture is now a mechanized food industry in essence the same as the production of corpses in the gas chambers and extermination camps, the same as the blockading and starving of countries, the same as the production of hydrogen bombs.[70]

All of these things, for Heidegger, hence our horror, in essence: *the same*. For Heidegger this sameness is so because it cannot but be so: everything is drawn into the gyre, the "centre cannot hold" indeed we need the whole array of Yeats' rebuke of historicity and modern fatefulness or futurity because the essence of modern technology in our world happens to remain as that which Heidegger saw it as being, and to which insight into that which is, he sought to call our attention.

The setting upon of modern technology is critical, crucial, indispensable for Heidegger and that is how he can utter such an offensive comparison: for him modern technology is all about such equations, such calculations, such reductions. Thus we noted with respect to a different kind of land-use, switching agrarian land over, opening it up, literally so, to the coal industry, that Heidegger writes that with that the coal itself (he has the Rilkean poem to the wealth of the kings slumbering in the mountains in his mind), is ordered, set upon: "challenged forth for heat, as the ground is challenged forth for coal." Here the constellating point in question will be that heat itself, today we would say energy,

> is already set to set up steam, the pressure of which drives the turbines, which keep a factory productive, which is itself ordered to set in place machines that produce tools by means of which again, machines are set to work and maintained.[71]

The subsequent and for environmental studies indispensable reflective array to which Heidegger then turns only offers an elaboration of this point:

> The hydroelectric plant is placed in the river. It imposes upon it for water pressure, which sets the turbines turning, the turning of which drives the machines, the gearing of which imposes upon the electrical current through which the long-distance power centers and their electrical grid are positioned for the conducting of electricity. The power station in the Rhine river, the dam, the turbines, the generators, the switchboard, the electrical grid—all this and more is there only insofar as it stands in place and at the ready, not in order to be there (presence), but to be positioned, and indeed solely to impose upon still others.[72]

---

[69] There is no shortage of discussions of the same: I list this literature myself in several essays, as do many, many, many others, but see, for a start, Babich (2009), 227–243 as well an earlier essay, on Babich (1992), 83–106.

[70] Heidegger (1994), 27

[71] Ibid.

[72] Ibid.

Heidegger could not understand the engineering array or constellation any better, maybe this what our culture industry means when it praises German engineering to this day, usually in a Volkswagen ad or just an advertisement for a coffee machine.

Heidegger goes on to notice that this includes human being in deep ways and he speaks of the machination, "mechanization of the human,"[73] "the human being is ordered by and for the requisitioning."[74]

All this can seem to be taking us rather far afield, and as *Das Ge-Stell* serves as prelude to Heidegger's lecture on *Die Gefahr*, we turn to consider, as promised, Heidegger's reflection on the *Ge*.

> We name the collection of mountains [*die Versammlung der Berge*] that are already gathered together, united of themselves and never in retrospect, the mountain range [*das Gebirge*]. We name the collection of ways according to which we are disposed to such and such, and can feel ourselves so disposed, our frame of mind [*das Gemut*]. We now named the self-gathered collection of placing, setting [*das Stellens*], wherein everything orderable essences in the standing reserve, das Ge-Stell.[75]

Here for Heidegger everything is harrowed, harvested, arranged, disposed to standing reserve and industry, and in this sense he can claim that "*das Ge-Stell* is the essence of technology."[76]

1.1 Die Gefahr/The Danger

As is typical for Heidegger, as we already know if we have learned to follow the rhetorical didacticism that characterizes the strategic articulations of *Being and Time,* Heidegger repeats the moves he introduces in *Das Ge-Stell* in the following lecture *Die Gefahr*, and he does so in a thoroughly scholastic fashion. To be sure, the reason that Jack Caputo and others can undertake to read Heidegger and Aquinas together is because of Heidegger's scholastic formation, not unlike Kant's own formation and indeed and to be sure as Heidegger admires Kant throughout his life.[77] Here Heidegger closes his fourth lecture on the turn by invoking Kant on the ultimate practical question, the ground of being qua being and as such: that would be God even for the godless, as (the believing) Kant himself is usually blamed for being the instigator of nihilism, at least according to Fichte and Jacobi.[78] For his part, Kant was already writing in a godless time, after Newton, after Laplace's

[73] Ibid., 28.

[74] Ibid., 29.

[75] Ibid., 32,

[76] Ibid., 33.

[77] See on Heidegger and Aquinas Jack Caputo's often cited study (1982). See on Heidegger and Kant, as an overview, Daniel Dahlstrom (2010). Willi Goetschl (1994) offers a useful background for the (very differently) hermeneutically contextualizing framework to which I am adverting here.

[78] This is complicated even beyond the constellations Freerick Beiser has tracked in his work. I discuss this, citing Beiser and others, in some of my footnotes to Babich (2010b), 231–256.

Mécanique Céleste (finished in 1725, Kant would draw upon this for his own nebular hypothesis in 1755), and Heidegger's schoolman's (and hence classically didicatic strategy) is simply to tell us what he is doing and then to do so and then to reprise what it is that he has done.[79] In this sense "The Turn" inevitably has nothing to do with the way typical Heideggerians seeking to divide their bit of Heidegger into something manageable tend to speak of it, as if there might be a change in Heidegger's thinking (Heidegger as we know is famous for saying that a thinker thinks only one thought), and we have already noted that where we might need to locate such a change or turn we do not need to wait for these lectures for it is already noted in Heidegger's *Letter on Humanism* (and it is of course albeit in a secret, esoteric, or unpublished way already present in the *Beiträge*).[80] In the *Letter on Humanism* Heidegger declares that "everything is reversed,"[81] or turned around, but scholars will find such a translation or reflexive turning in his *Introduction to Metaphysics*, or indeed in the 1935 lectures on *The Origin of the Work of Art*, which are themselves, as they have to do with nothing other than the Greek notion or meaning of *techne,* likewise indispensable for the four lectures on technology.

The focus on calculation with which Heidegger ends his lecture on *Das Ge-Stell* is replaced with a reflection on worlding in terms that we recognize as the terms of the fourfold, and which if we keep Heidegger's reflections on the happening or event of truth in his lecture on the artwork highlights "worlding" coming to presence:

> World is the fourfold of earth and sky, divinities and mortals. In the uniting whole of its presence, the mirrorplay of the fourfold guards everything that thingingly presences and absences between the four.[82]

As we also recognize from *Being and Time*, Heidegger gives nothing—he is not a Hegelian, as it happens, for nothing—without simultaneously also taking it away. Thus after indicating the importance of the safeguard, of sheltering (and we recall that this is at the heart of his reflections on *physis*), Heidegger observes that "The world still refuses itself as world. World still withdraws into the concealment proper to it."[83] The difficulty for any discussion here as we recognize this immediately from our familiarity with *Being and Time* but also from our rather persistent unfamiliarity with Heidegger's 1930 *Essence of Truth*, is that we are confounded by lighting and concealing, showing hiding, *aletheia/lethe*.

The problem as Heidegger writes here, nicely concisely, is that "*aletheia* does not properly guard itself in its own essence it lapses into concealment, *lethe*,

[79] See on this: Babich (1993), 239–260.

[80] See for an important and subtle discussion of this complex theme, Richard Polt (2006) and see too in this context Babich (2010c), 397–415.

[81] Heidegger (1977c). See for a discussion of this politicized political context along with further references, see Babich (2013c).

[82] Heidegger (1994), 48.

[83] Ibid., 49.

aletheia falls into forgetfulness."[84] By this means to be sure, Heidegger both introduces the danger qua danger as well as recuperating his own reflection on the sheer forgottenness of being which he has in the interim (as we know from the *Beiträge*) begun to write as *Seyn*.

What Heidegger here calls the "refusal" of world, which he expresses as the *Ereignis*, happening or event, also sometimes rendered as "appropriation" " Diese Ereignis besteht darin das Welt als die Wahrnis des Wesens des Seins sich verweigert."[85] World thus refuses itself as the preserver, guardian, harborer of the essence of Being. Heidegger now offers us two references to temporality, one to the then-current dispensation of world-affairs, as the "unfolding of planetary totality," observing as the defeated party to the previous contest for world-domination (i.e., the Germans as the losers in the second world war) could not but be, however awkwardly, perfectly placed to observe that "the modern battle for mastery of the earth is concentrated upon the position of the two contemporary 'world' 'powers." (51) This is complicated to the extent that Heidegger coordinates the refusal of world as manifest as eventuated via or through the defenseless of the thing noting that in this relation one to another they are "the same if to be sure not the identical." (Ibid.) But the distinction is not idle for Heidegger: "the same [*das Selbe*]," he will go on to emphasize "is never the identical [*das Gleiche*]." (52) At this point what is at issue for Heidegger is the refusal of world and the vulnerability of the thing in the prevailing turn of the set up he has analysed as modern technology. Everything but everything is presented as the ordered 'items at hand' or standing currency of standing reserve. "Ge-Stell" he writes adumbrated in this play on standing reserve "is" this disposition and is accordingly "the essence of modern technology." (51) But this conjunction is one of the moment, the present time, the insight is into that which is, in its immediacy, thus Heidegger goes on to observe that this holds not 'as such' or 'from all time' but very literally 'here,' just to the extent that it is here and now that we find that the "oblivion of the essence of being is consummate." In the same way, and now we see why so many commentators inevitably turn here to a reflection on The Origin of the Work of Art," Heidegger also writes that "World and Ge-Stell are the same." (52)

Calculation, a concern for Heidegger from the start, both in his reflections on truth in *Being and Time* as in the *Essence of Truth*, as in his reflections on "Science and World Picture," all originating from his original and enduring interest in science and his interest, inevitable for anyone who works on Dilthey's account of history but also anyone in philosophy who is both a contemporary as Heidegger was, roughly speaking here, as you are whether you like it or not as students a contemporary of my ancient self just as I was when I was a 23 year old student when I met first met William Richardson as well as being the contemporary of my even more ancient teacher, the same Gadamer at 80, so similarly was Heidegger a contemporary of Max Weber as well as from its outset to its flourishing with the same Rudolf Carnap we already began by noticing, and beyond to its current

[84] Ibid.

[85] Ibid., 50–51.

world-dominion within philosophy proper in so-called analytic philosophy, logical positivism, the issue of values was for Heidegger a matter of weighting and weighing the same. We cannot count the time of life with a clock, we cannot calculate it at all. Thus Heidegger writes, playing on the banality of banality, the indifference of the diffident—*Alles gilt gleich*—same old, same old, we might say. (52)

If a further discussion of calculation cannot here be considered, what is important to note is that the same preface, the *Ge-* that remains at issue, is also to be considered in the danger, *die Ge-fahr*. Two coordinate and even nested claims make this clear: "The essence of technology is the Ge-Stell. The essence of the Ge-Stell is the danger." (54) In effect, it is the *Ge-Stell* as such, the enframing, the set up that "sets after the truth of the essence of being with forgetfulness." (53) This harrying, harrowing is Heidegger's "pur-suit'—here the word is not *Gestellen*, but *Nachstellen*. For Heidegger, as he writes, in "Old High German, to pursue is called *fara*." (53) The *Ge-Stell*, the set-up, or the en-framing "gathered in itself as pursuit is the danger [*Das in sich gesammelte Stellen als Nachstellen ist die Gefahr*]." (53) What is key here just as in the folded, referentiality or integral orderedness of the Ge- in *Ge-Stell*, is the gathered in itself of the pursuit in question, as the danger. The Hegelian move here brings us him to reflect "that Beyng (Being or *Sein* spelled with a y, in an ancient mode, as *Seyn*) is the danger. Beyng is unqualifiedly in itself, from itself, for itself" (can't get more Hegelian than this) "the danger. As this pursuit, which pursues its own essence with the forgetting of this essence"—here, again, we recognize aletheia—"beyng as beyng is the danger." (53)

Here Heidegger's definition of the danger summarizes the lectures to this point:

> The danger is the collected pursuit [*sich in sich versammelte Stellen als Nachstellen*] as which en-framing/set up [*als welches das Ge-Stell*] in the guise of unguardedness of the thing, pursues the self-refusal of world with the forgetting of its truth." (54)

For Heidegger, and note that our reading through an English language lens challenges us, we are left to reflect that we do not experience [*Erfahren*] the danger as danger." (55) It is in this context that Heidegger presents the currency of need and desperation, that is: he lists a litany of death, as indeed of pain that is to say suffering, and also of poverty, all and each as what confronts us and at the same time manages not to touch us, leaving us unmoved, unchanged, in a terrifying sense. The phenomenon to which Heidegger refers here continues to this day as we well know, all you have to do is read the paper, check Facebook and note how many awful things and then note how little any of those things affect you really or at all: talk about the oblivion of being as much as you like.

For Heidegger in the midst of extraordinary need and desperation, and from 1945 onward, certainly unabated by 1949 in Germany, that is then pretty much everywhere in that defeated land, precisely to the extent that the businessmen and city fathers to whom Heidegger spoke in Bremen, just to the degree that they did indeed address this need and that need, as people organized to respond to devastations in this way and that, remedying problems in this way and that, that precisely in the midst of "ameliorating pain and tending to neediness" (55–56) what remains critical for Heidegger is that precisely while so engaged "one does not attend to *the* need." (65) Heidegger has a name for this—which he explores already in *Being and Time*, errancy, *die Irre*.[86] It is our amazing ability not to be where we are, which (remember that we are for Heidegger Dasein), only means that we are not who we are. In this sense, Heidegger observes here, „Das Wesen der Irre beruht im Wesen des Seyns als der Gefahr." [The essence of errancy subsists in the essence of Seyn as the danger.] (56)

This same errancy plagues us when in the same paradoxical sense in which the paradox of neediness prevails such that we all have needs, we all have our desperations, but we do not in midst of our worries actually because we cannot begin to attend to needfulness as such. In the context of this reflection on death, suffering or pain and neediness or needfulness and all the heedlessness of the same in the midst of an abundance of the same, we encounter the second version, or variant upon Heidegger's seeming insensitivity (which we now see to be an insensitivity in his words on insensitivity as we hear him). This locus, situated in postwar needfulness, is the most grim, and it is perhaps because of its time, harsher in tone than Heidegger's more popular (it was a radio) lecture "Building, Dwelling, Thinking," broadcast in 1951 with its own respective (and indeed more hopeful) reflections on need and needfulness. Here in *The Danger* we read:

> Hundreds of thousands die in masses. Do they die? They succumb. They are put down. Do they die? They become inventory items of a standing reserve for the fabrication of corpses. Do they die? They are unremarkably liquidated in annihilation camps. And even without such—millions now in China, end pathetically in starvation. (56)

In this context, it can be argued that the Heidegger of *Being and Time* thereby reclaims his own reflection on what he had offered for reflection on death, that is being-towards-death, as this specifically characterizes human beings in their mortality, as beings concerned with their being and aware of their vulnerability in being in the mode (this is high Heidegger, esoteric Heidegger) of disattending, flight, forgetfulness of being.

> Death shelters [*birgt*] the essence of being. Death is the highest re-fuge [*Gebirg*] of the truth of being itself, the refuge that in itself shelters [*birgt*] the concealment [*Verborgenheit*] of the essence of being and gathers together the sheltering [*bergung*] of its essence.... To be capable of death in its essence means to be able to die. (Ibid.)

But for Heidegger: "The human is not yet the mortal." (56) Since much of Heidegger's project in *Being and Time* was all about explaining life in terms of living and in terms of the vanity of mortal beings who take themselves to be immortal, as we do, proximally and for the most part, what fascism took from its others was what made them human, even in its constant, as it is pretty much always, default.

Here I want to emphasize as this essay moves toward its conclusion that the same technique, the same modus, asks us to attend to Heidegger's very overtly hermeneutic phenomenology (he is not—despite the *Spiegel*'s sensationalist insistence on the same, an insistence shared by numerous junior college professors—an 'existentialist') with respect to our obliviousness, thoughtlessness. "Immeasurable suffering shifts

[86] *Die Irre*, or errancy has been a lasting concern for William Richardson as one can read beginning with his (1993). And see too the contributions to Babich (1995).

and surges across the earth. But the essence of pain conceals itself. …Everywhere we are besieged with countless and boundless suffering. We however are not pained [*schmerzlos*], we are not appropriated to the essence of pain."(57)

We are not pained and today there is more of this un-moved, painlessness than ever. Who bothers to watch animal rights videos, if one ever did, who is really concerned about the plight or fate (pick any word you like) of the Palestinians, the Syrians, the Nigerians, etc. and etc. and etc.?

> Death, the mountains of *Seyn*, pain, the schema of *Seyn*, poverty, the liberation into the ownership of *Seyn*, are features allowing the danger to be remarked, that needfulness is excluded in the midst of the greatest neediness, that the danger is not as the danger allowing the danger to be noted. (Ibid.)

We are unpained, we do not sense what is all around us, as Heidegger who will turn in his last essay on the turn to language by which as he explains he means our need to lay claim to it. And today, I would argue, we are no further advanced: we still need to recall Hölderlin's warning to us, whether as scholars of being or of language as all those who have lost their tongues in foreign (and native) lands.

Here Heidegger seeks to differentiate his reading from those who contend that "technology is the catastrophe of the modern world" (58) and so on. For Heidegger it is already problematic to offer a critique of technology in a technological age, no matter in what voice one seeks to do so. The point here is that whether one praises or damns it, "at the same time one greedily scurries after the latest technological advance, perhaps one cannot but run after it in this way." (58) Yet to this same extent "judgment and inclination with respect to technology contradict themselves and the same contradiction is taken as objection." (Ibid.)

Heidegger's perhaps best known claim that the "essence of technology is itself nothing technological" (60) remains, as he reprises it in the final lecture, *The Turn*, almost in the same words, arguing that everything that is "merely technical" can "never attain to the essence of technology" (*Die Kehre*, 76). We do not grasp and hence cannot begin to articulate what he calls the "insight into that which is," to the extent that we do not even ask after the import of the times as they unfold around us.

"But," for Heidegger, "we do not yet hear, we, under the dominion of technology, whose hearing and seeing decay through radio and film." (*Die Kehre*, 77) Here we can and should add the internet (why on earth not?), but for Heidegger what we do not yet hearken to or see is occluded not simply by way of our thuggishness or inattention: "The constellation of beyng is the refusal of world as world."

If earlier, Heidegger had responded to a question on humanism by recalling a related request for a contribution to ethics by distinguishing between the modern notion and the ancient Greek sense of the same, he also took care to be blunt about the circumstances of such thought, as we have already referred to his earlier lecture on poverty. For Heidegger as he goes on to note in his letter to a former enemy in 1946, philosophizing or thinking "about being shattered is separated by a chasm from a thinking that is shattered."[87]

[87] Heidegger (1954/1977c), 223/340.

Maybe, and in the spirit of the small, the slight recommendation, we might begin, after all this time, to take up Heidegger's more complex question. That is his question concerning the "world, worlding," as this would be "the nearest of everything that nears," now heard as we perhaps should always have heard it as the question of *Ereignis*, that is in terms of what Heidegger called *Eigentlichkeit*: appropriation appropriated as it were, qua the "ownership of appropriation." (Ibid.) For as we also know, from the start, what Heidegger meant by *Eigentlichkeit* was never 'authenticity' (and it is easy to remember that German has a term for authenticity, *die Authenticität*) but owned ownedness, appropriated appropriation.[88]

Ereignis.

References

Adorno, Theodore. 1945. A social critique of radio music. *The Kenyon Review* 8(2): 208–217.

Adorno, Theodore. 2006. In Adorno, *Current of music*, ed. Robert Hullot-Kentor. Frankfurt am Main: Suhrkamp.

Anders, Günther. 1956. Die Welt als Phantom und Matrize. Philosophische Betrachtungen über Rundfunk und Fernsehen. In Anders, *Die Antiquirtheit des Menschen, Erster Band. Über die Seele im Zeitalter der zweiten industriellen Revolution*, 97–214. München: Beck.

Arendt, Hannah. 1968. Introduction. In Walter Benjamin, *Illuminations: Essays and reflections*, 1–58. New York: Schocken Books.

Babich, Babette. 1992. Heidegger's Silence. In *Ethics and danger: Currents in continental thought*, ed. Charles Scott and Arleen Dallery, 83–106. Albany: State University of New York Press.

Babich, Babette. 1993. A musical retrieve of Heidegger, Nietzsche, and technology: Cadence, concinnity, and playing brass. *Man and World* 26: 239–260.

Babich, Babette (ed.). 1995. *From phenomenology to thought, errancy, and desire : Essays in honor of William J. Richardson, S.J.* Dordrecht: Kluwer.

Babich, Babette. 2003. On the analytic-continental divide in philosophy: Nietzsche's lying truth, Heidegger's speaking language, and philosophy. In *A house divided: Comparing analytic and continental philosophy*, ed. C.G. Prado, 63–103. Amherst: Humanity Books.

Babich, Babette. 2009. From the ethical alpha to the linguistic omega: Heidegger's anti-semitism and the question of the affinity between ancient Greek and German. In *Words in blood, like flowers: Philosophy and poetry, music and eros*, 227–243. Albany: State University of New York Press.

Babich, Babette. 2010a. Towards a critical philosophy of science: Continental beginnings and bugbears, whigs and waterbears. *International Journal of the Philosophy of Science* 24(4): 343–391.

Babich, Babette. 2010b. Ex aliquo nihil: Nietzsche on science and modern nihilism. *American Catholic Philosophical Quarterly* 84(2): 231–256.

Babich, Babette. 2010c. Heideggers Widerstand. Nietzsche lesen als eine »Konfrontation« mit dem Nationalsozialismus. In *Heidegger und der Nationalsozialismus II. Heidegger Jahrbuch 5*, ed. Holger Zaborowski and Albert Denker, 397–415. Freiburg im Briesgau: Alber.

Babich, Babette. 2011. An Impoverishment of Philosophy. *Purlieu: Philosophy and the University* (1): 37–71.

[88] This essay was originally presented at the New School at the invitation of Christopher Merwin to the New York Phenomenology Group on 25 April 2013. I am grateful to Chris Merwin, Tracy Strong, and to the participants at that meeting for an especially engaging discussion.

Babich, Babette. 2012. Heidegger's philosophy of science and the critique of calculation: Reflective questioning, *Gelassenheit*, life. In *Heidegger on science*, ed. Trish Glazebrook, 159–192. Albany: State University of New York Press.

Babich, Babette. 2013a. *The Hallelujah effect: Philosophical reflections on music, performance practice and technology*. Surrey: Ashgate.

Babich, Babette. 2013b. The 'New' Heidegger. In *Heidegger in the 21st century*, ed. Tziovannis Georgakis and Paul Ennis. Frankfurt am Main: Springer. Forthcoming.

Babich, Babette 2013c. Shattering the political or the question of war in Heidegger's *Letter on Humanism. Proceedings of the Heidegger circle 2013*. Convenors: David Pettigrew and Rex Gilliland. New Haven: University of Southern Connecticut, 25–49.

Bast, Rainer A. 1986. *Der Wissenschaftsbegriff Martin Heideggers im Zusammenhang seiner Philosophie*. Stuttgart: Frommann-Holzboog.

Baudrillard, Jean. 2005a. *The Intelligence of Evil or the Lucidity Pact*. Trans. Chris Turner. London: Bloomsbury.

Baudrillard, Jean. 2005b. *Simulacra and Simulation*. Trans. Sheila Faria Glaser. Ann Arbor: University of Michigan Press.

Baudrillard, Jean. 2009. *Why Hasn't Everything Already Disappeared?* Trans. Chris Turner. Chicago: Seagull.

Benjamin, Walter. 2008. *The Work of Art in the Age of Its Technological Reproducibility, and Other Writings on Media*. Trans. Edmund Jephcott and Harry Zohn. Cambridge: Belknap Press of Harvard University Press.

Bernays, Edward. 1928. *Propaganda*. New York: IG Publishing.

Caputo, John. 1982. *Heidegger and Aquinas: An essay on overcoming metaphysics*. New York: Fordham University Press.

Caputo, John. 1986. Heidegger's philosophy of science: The two essences. In *Rationality, relativism, and the human sciences*, ed. Joseph Margolis et al., 43–60. The Hague: Nijhoff.

Celan, Paul. 1980. *Poems*. Trans. Michael Hamburger. Persea: New York.

Celan, Paul. 2000. *Gesammelte Werke*. Frankfurt am Main: Suhrkamp.

Chickering, Roger. 2007. *The great war and urban life in Germany: Freiburg, 1914–1918*. Cambridge: Cambridge University Press.

Dahlstrom, Daniel. 2010. The critique of pure reason and continental philosophy: Heidegger's interpretation of transcendental imagination. In *The Cambridge companion to Kant's critique of pure reason*, ed. Paul Guyer, 380–400. Cambridge: Cambridge University Press.

Der Spiegel. 1950. Heidegger. Rückfall ins Gestell. 14: April 6.

Figal, Günther. 2006. *Einführung. Heidegger Lesebuch*. Frankfurt am Main: Klosterman.

Friedman, Michael. 2000. *A parting of the ways: Carnap, Cassirer, and Heidegger*. Chicago: Open Court.

Gadamer, Hans-Georg. 1995. *Philosophische Lehrjahre. Eine Rückschau*. Frankfurt am Main: Klostermann.

Gethman, Carl Friedrich. 1991. Der existentialen Begriff der Wissenschaft. Zu Sein und Zeit §69b. In *Lebenswelt und Wissenschaft: Studien zum Verhältnis von Phänomenologie und Wissenschaftstheorie*, ed. Carl Friedrich Gethman, 181–209. Bonn: Bouvier.

Glazebrook, Trish. 2012. Why read Heidegger on science? In *Heidegger on science*, ed. Trish Glazebrook, 159–192. Albany: State University of New York Press.

Goetschl, Willi. 1994. *Constituting Critique: Kant's Writing as Critical Praxis*. Trans. Eric Schwab. Durham: Duke University Press.

Gordon, Peter. 2012. *Continental divide: Heidegger, Cassirer, Davos*. Cambridge: Harvard University Press.

Guzzoni, Ute. 2012. *Gelassenheit*: Beyond technoscientific thinking. In *Heidegger on science*, ed. Trish Glazebrook, 193–204. Albany: State University of New York Press.

Heelan, Patrick Aidan. 2012. Carnap and Heidegger: Parting ways in the philosophy of science. In *Heidegger on science*, ed. Trish Glazebrook, 113–130. Albany: State University of New York Press.

Heidegger, Martin. 1951. *Bauen Wohnen Denken*, Westdeutsche Rundfunks, Cologne. CD edited by Eduard Führ and the Waxman Verlag.
Heidegger, Martin. 1954. *Über den Humanismus. Brief an J. Beaufret*. Bern: Francke.
Heidegger, Martin. 1959. *Introduction to Metaphysics*. Trans. Ralph Manheim. New Haven: Yale University Press.
Heidegger, Martin. 1962. *Die Technik und die Kehre*. Pfullingen: Neske.
Heidegger, Martin. 1962 [1927]. *Being and Time*. Trans. John Macquarrie and Edward Robinson. New York: Harper & Row.
Heidegger, Martin. 1971. *Poetry, Language and Thought*. Trans. Albert Hofstadter. New York: Harper.
Heidegger, Martin. 1977a. The question concerning technology. Trans. William Lovitt. In: Heidegger, *The Question Concerning Technology and Other Essays*, 3–35. New York: Harper.
Heidegger, Martin. 1977b. *The Question Concerning Technology and Other Essays*. Trans. William Lovitt. New York: Harper & Row.
Heidegger, Martin. 1977c. The letter on humanism. Trans. Frank A. Capuzzi. In: Heidegger, *Basic Writings*. New York: Harper & Row.
Heidegger, Martin. 1978a [1954]. « Die Frage nach der Technik ». In Heidegger, *Vorträge und Aufsätze*, 9–40. Pfullingen: Neske.
Heidegger, Martin. 1978b [1954]. « Wissenschaft und Besinnung » In Heidegger, *Vorträge und Aufsätze*. 41–66. Pfullingen: Neske.
Heidegger, Martin. 1991. Heidegger, *Nietzsche. Volumes Three and Four*. Trans. Joan Stambaugh and Frank A. Capuzzi. San Francisco: Harper.
Heidegger, Martin. 1994. Einblick in das was ist. In Heidegger, *Bremer und Freiburger Vorträge*, 1–77. GA 79. Frankfurt am Main: Klostermann.
Heidegger, Martin. 1998. The letter on humanism. Trans. Frank A. Capuzzi. In Heidegger, *Pathmarks*, ed. Will MacNeill. Cambridge: Cambridge University Press.
Heidegger, Martin. 2003. Letter to Richardson. In William J. Richardson, *Heidegger: Through phenomenology to thought*. New York: Fordham University Press.
Heidegger, Martin. 2004 [1945]. *la pauvréte (die Armut)*. Trans. Philippe Lacoue-Labarthe and Ana Samardzija. Strasbourg: Presses Universitaires de Strasbourg.
Heidegger, Martin. 2012. *Bremen and Freiburg Lectures. Insight into that Which is and Basic Principles of Thinking*. Trans. Andrew J. Mitchell. Indiana: Indiana University Press.
Heidegger, Martin. 2003 [1989]. *Beiträge zur Philosophie* (Vom Ereignis). Frankfurt am Main: Vittorio Klostermann.
Janicaud, Dominique. 1985. *La puissance du rationel*. Paris: Gallimard.
Jünger, Friedrich Georg. 1946. *Perfektion der Technik*. Frankfurt am Main: Vittorio Klostermann.
Kiesel, Theodore. 2012 [1962]. A supratheoretical prescientific hermenutics of scientific discovery. In *Heidegger on science*, ed. Trish Glazebrook, 239–260. Albany: State University of New York Press.
Kleinberg, Ethan. 2005. *Generation existential: Heidegger's philosophy in France, 1927–1961*. Ithaca: Cornell University Press.
Kockelmans, Joseph J. 1985a. *Heidegger and science*. Washington, DC: Center for Advanced Research in Phenomenology and the University Press of America.
Kockelmans, Joseph J. 1985b. *Heidegger on art and art works*. Dordrecht: Martinus Nijhoff.
Kockelmans, Joseph J., and Theodor Kisiel (eds.). 1970. *Phenomenology and the natural sciences*. Evanston: Northwestern University Press.
Lacoue-Labarthe, Philippe. 2004. Prélude. In: Heidegger, *la pauvréte (die Armut)*, 7–65. Strasbourg: Presses Universitaires de Strasbourg.
Levin, Thomas Y., and Michael von der Linn. 1994. Elements of a radio theory: Adorno and the Princeton radio research project. *The Musical Quarterly* 78(2): 316–324.
Lyon, James K. 2006. *Paul Celan and Martin Heidegger: An unresolved conversation, 1951–1970*. Baltimore: Johns Hopkins University Press.
Martin, Bernd (ed.). 1989. *Martin Heideggerund das "Dritte Reich": ein Kompendium*. Darmstadt: Wissenschaftlich Buchgesellschaft.

Mitchell, Andrew. 2012. Translator's Foreword. In Heidegger, *Bremen and Freiburg Lectures*, Vii–xvi. Bloomington: Indiana University Press.
Mitchell, Andrew. English German glossary. In Heidegger, *Bremen and Freiburg Lectures*. 173–198. Indiana: Indiana University Press.
Nicholson, Graeme. 1987. The politics of Heidegger's rectoral address. *Man and World* 20: 171–187.
Nietzsche, Friedrich. 1980. *Kritische Studienausgabe*. ed. Giorgio Colli and Mazzino Montinari. Berlin. de Gruyter.
Packard, Vance. 1957. *The hidden persuaders*. New York: David McKay Co.
Polt, Richard. 2006. *The emergency of being: On Heidegger's contributions to philosophy*. Ithaca: Cornell University Press.
Rapaport, Herman. 1997. Forces of gravity. In Rapaport, *Is there truth in art?* 110–143. Ithaca: Cornell University Press.
Richardson, William J. 1993. [1963]. *Heidegger: Through phenomenology to thought*. 4th edition. New York: Fordham University Press.
Richter, Ewald. 1992. *Heideggers Frage nach dem Gewährenden und die exakten Wissenschaften*. Berlin: Duncker & Humblot.
Richter, Ewald. 2012. Heidegger's theses concerning the question of the foundations of the sciences. In *Heidegger on science*, ed. Trish Glazebrook, 67–90. Albany: State University of New York Press.
Safranski, Rüdiger. 1999. *Martin Heidegger: Between Good and Evil*. Trans. Ewald Osers. Cambridge: Harvard University Press.
Sartre, Jean-Paul. 1946. *L'existentialisme est un Humanisme*. Paris: Les Éditions Nagel.
Sartre, Jean-Paul. 2007. *Existentialism is a Humanism*. Trans. Carol Macomber. New Haven: Yale University Press.
Seigfried, Hans. 1991. Heideggers Technikkritik. In *Lebenswelt und Wissenschaft: Studien zum Verhältnis von Phänomenologie und Wissenschaftstheorie*, ed. Carl Friedrich Gethman, 209–244. Bonn: Bouvier.
Sloane, Eric. 2004. [1965]. *A reverence for wood*. New York: Dover Publications.
Smythe, Dallas Walker. 1954. Reality as presented by television. *The Public Opinion Quarterly* 18(2): 143–156.
Winner, Langdon. 1986. *The whale and the reactor. A search for limits in an age of high technology*. Chicago: Chicago University Press.

Heidegger and the Reversed Order of Science and Technology

Lin Ma and Jaap van Brakel

Abstract Contrary to the common view that modern technology derives from modern science, Heidegger presents a reverse picture in which science originated in the essence of technology, wherein Being speaks. We argue that it is in this sense that Heidegger speaks of the Same [*das Selbe*] of science and technology, both being ultimately grounded in the history of Being. In the long span from 1938 to 1976, Heidegger has continuously delved into the relation of science and technology. In our research we also engage ourselves with various claims made by philosophers of technoscience. We show that Heidegger has always kept himself well-informed of traditional as well as new types of technology and science, including quantum physics, atomic technology (as used in nuclear reactors), and biophysics (including speculations about genetic manipulation in the 1950s). Nevertheless, one cannot ascribe to Heidegger the view that these new developments originate a new Epoch of Being.

1 Introduction

Two weeks before Heidegger's death on May 26, 1976, the tenth annual meeting of the "Heidegger Conference" of the North American Heidegger Society was held at DePaul University in Chicago. Heidegger's letter of greetings to this conference has been reported as the last philosophical text by his hand. In this letter, he requested that the participants take up the following question as "stimulation" [*Anregung*] for their discussion:

> Is modern natural science the foundation [*Grundlage*] of modern technology — as assumed — or is it, for its part, already the basic form [*Grundform*] of, the determining fore-conception

L. Ma (✉) • J. van Brakel
Hoger Instituut voor Wijsbegeerte, Renmin University of China, Kardinaal Mercierplein 2, 3000 Leuven, Belgium
e-mail: lin.ma.2007@gmail.com; jaap.vanbrakel@hiw.kuleuven.be

B. Babich and D. Ginev (eds.), *The Multidimensionality of Hermeneutic Phenomenology*, Contributions to Phenomenology 70, DOI 10.1007/978-3-319-01707-5_11,

> [*Vorgriff*] and incessant incursion [*Eingriff*] of technological representation into the realized and arranged [*ausführende und einrichtende*] machinations of modern technology? ([1976], p. 747/3)

Here, carefully formulating in terms of a question, Heidegger made the crucial point that, instead of the common-sense idea that science lays the foundation for technology, the technological essence may well be that from which science is derived and on behalf of which science functions.

What is perhaps most perplexing is the fact that Heidegger singled out this question, from among many other questions, as an indispensable inquiry relative to "the asking of the question of Being" (748/4) Toward the end of his 1976 letter, Heidegger suggests that fruitful reflection upon the relation between science and technology could help prepare a transformation of man's dwelling in this world, which he claims to be what the question of Being *is* "in truth" (748/4).

At the 11th "Heidegger Conference" in 1977 as well as the 35th "Heidegger Conference" in 2001, the question of the being of science and technology was made a central theme of the meeting. However, the contributions to these conferences have *hardly* addressed Heidegger's question *directly*.[1] In this paper, we attempt to shed light upon this enigmatic aspect of Heidegger's thinking and its essential connection with "the mystery of what *is* today in truth in the technologically determined world" (ÜTS 138). First, we trace the genesis of this question in Heidegger's thinking since about 1940 and expose relevant materials that disclose his major concerns in ruminating upon this question (Sect. 2). In this section, quite a number of substantial citations come from *GA* 76, *Zur Metaphysik—Neuzeitlichen Wissenschaft—Technik* (henceforth *MWT*) published in 2009,[2] where one finds abundant preparatory notes for public lectures on the essence of science and/or technology. We argue that it is in the year 1938 that Heidegger developed some *Wegmarken* for a response to this question, and his pondering on it did not come to an end until his death.

Second, using a text from the Winter of 1945/1946 as an initiating clue, we lay out a three-fold structure of the way in which Heidegger sets forward his questioning (Sect. 3), which leads to the crucial thesis that modern science and technology are the "Same" [*das Selbe*].

When it comes to Heidegger's thinking on modern science and technology, the prejudice that Heidegger "deeply contested" science and technology remains prevalent (Ihde 2010, p. 110). We emphasize, in the very beginning of the present paper that this is only one facet of the profile. It is true that Heidegger often express

[1] [Most of the participants at the 2001 conference, organized 25 years after Heidegger's initial 1976 question, did address Heidegger's question, including Heidegger's student, Ute Guzzoni in her presentation. A German version of her essay appears in the proceedings and an English translation of her presentation is included in Glazebrook 2012. Holger Schmid also takes up Heidegger's question. See Schmid's chapter to follow in the present volume.—BB].

[2] For other abbreviations of Heidegger's works see the list of references included at the end of this chapter. Page numbers of English translations if available follow the page numbers of the original.

worries about such things as "the possible self-destruction of the human being," but immediately adds: "it is not a matter of hostility toward science as such" (*ZS* 124/94). In many other places, Heidegger reiterates that "the sciences are in themselves something positively essential" (*WhD* 16/14), and stresses that our comportment toward technology should be "yes" and "no" at the same time: "We can use technical devices, and yet with proper use also keep ourselves free of them, that we may let go of them any time" (G 526f/46).

In a discussion in 1955 about one of his favourite poets, J.P. Hebel, Heidegger argues that science and technology can well co-exist with "*simple naturalness*" (Hebel145/97). Referring to Hebel as a "friend to the house which the world is" (139/93), Heidegger suggests that his poetry exemplifies the ideal of co-existence. After citing Goethe's review of Hebel's *Allemanic Poems* in which he says that Hebel "thoroughly countrifies the universe," Heidegger adds that Hebel *also* shows nature in its scientific calculability. The current problem, as Heidegger points out, is precisely that "calculable nature" and "natural nature" have been separated into "two alien realms," the latter being degraded and the former being "offered as the sole key to the mystery of the world" (146/98).

In recent literature, some scholars ascribe to Heidegger a prescience of what is now called "technoscience,"[3] while some others suggest that Heidegger's work provides a clue leading to the thesis that the "non-locality" of quantum mechanics points toward a non-totalizing science and an other Epoch in the history of being. As to the first claim, Heidegger was prescient of what is now called technoscience, and in this respect his writings are still relevant to current debates in Science & Technology Studies, but this is a rather unimportant issue in the context of his thinking on the relation of science and technology (Sect. 3). We discuss the specific issue of quantum physics and nuclear technology in Sect. 5.

Yet another piece of criticism coming from philosophers of technoscience, is that Heidegger did not engage himself in concrete, empirical study of actual technologies (Ihde 2010, pp. 2–5).[4] Therefore, his more conclusive statements about science and technology lack a firm ground. This criticism seems to be far-fetched. It is true that Heidegger has never carried out any prolonged "case study," which is currently a shared practice among philosophers of technoscience. However, one can see from

[3] The term 'technoscience' is a term that is now widely used in "Science & Technology Studies." Most contemporary philosophers of technology (such as Feenberg, Ihde, Latour) subscribe to the thesis that science and technology cannot (anymore) be separated and should be studied via detailed case studies (i.e. philosophy of science and technology "after the empirical turn"). They are critical of Heidegger because he is considered an essentialist, determinist, and pessimist. Ihde's *Heidegger's Technologies: Postphenomenological Perspectives* is a systematic discussion of important differences between Heidegger's approach and his own "pragmato-phenomenological account" (which "leaves in shambles the metaphysical Heideggerian tale") Ihde (2010), 113.

[4] Latour paints Heidegger as a thorougly pessimistic technological determinist. For discussion see Kochan (2010). According to Feenberg (1999), vii, 15–17, 29, Heidegger is an essentialist on three counts: a-historical, substantivist, and one-dimensional. For discussion see Thomson (2000).

Heidegger's working notes, conversations, and lectures that he has always kept himself well-informed of traditional as well as new types of technology.[5]

After a discussion of Heidegger's use of the words 'epoch' and 'Epoch' (Sects. 4 and 5) and the (alleged) essential difference between pre-modern and modern technology, we propose in Sect. 7 a slightly revised version of Heidegger's view of the relation of science and technology, concluding that his slogan "Science is applied Technology" (*FG* 10/15) is better replaced by "Science and Technology are the Same."

In other writings, Heidegger has employed the expression "the Same" [*das Selbe*] sometimes as a unique notion, sometimes in its more ordinary sense, and sometimes as a pun that denotes both senses. As a unique notion, the Same speaks of the "*belonging*-together" of being and thinking (*ID* 92/29). It is that which lets Being and thinking relate and concern one another mutually. To think of the belonging-together is not to assimilate the components into a unity, as a simplistic understanding of the principle of identity [*Gleichheit*] would presume. What is at issue is to *experience* the togetherness in terms of belonging. The Same is the matter of thinking, the task of thinking, and the way of thinking. As he writes in "Moira," "What is silently concealed in the enigma-word τὸ αὐτό is the revealing bestowal [*entbergende Gewähren*] of the belonging-together of the duality [of Being and beings] and the thinking that comes forward into view within it" (*M* 251/95).

Heidegger also uses the notion of the Same [*das Selbe*] to explicate relations in various areas, for example, the relation between poetic and philosophical thinking: "Only poetry is of the same order [*in derselben Ordnung*] as philosophical thinking, although thinking and poetry are not identical [*gleich*]" (*IM* 28/28). The bond between poetry and thinking is a gathering, that is, the Same, that shelters and grounds both of them. That here is such a bond is because both poetry and thinking arise as a correspondence to the claim of being. As Heidegger states, "Essential thinking must always say only the same, the old, the oldest, the beginning, and must say it primordially."[6]

In this light, Heidegger's saying that science and technology are the Same (ÜTS 16/136), presumably, would entail that, in exploring their relation, one needs to reflect upon their essence, and in reflecting upon their essence, one needs to think back upon their belonging-together as a gathering from out of which both science and technology come into being and sustains themselves. Heidegger defines the primordial Same as the belonging-together of being and thinking; therefore, speaking of the Same of science and technology inevitably directs our thinking back to their essential connection with the history of Being.

---

[5] See for example two pages of insightful notes on the steam engine (*MWT* 367–8), including "the politics of artifact." How the steam engine moved the working place of women and children from home to the factory, from country side to town. Examples regarding more recent technologies are given in Sects. 4 and 5.

[6] *GA* 54, 114/77. On Heidegger's use of the Same in regard to intellectual relations between philosophers, see Ma (2008), 202.

2 Reversing the Assumed Order of Modern Science and Technology

On the basis of relevant materials, we suggest that it is in the year 1940 when the question concerning the relation of science and technology emerged as one of the essential questions that Heidegger continued to be seriously preoccupied with until 1976.[7]

In a note taken after having delivered the lecture "Besinnung auf die Wissenschaft" in Freiburg in June 1938, Heidegger writes,[8] "Modern science as 'technology'—This step in the lecture of 1938 not yet completed although everything ready."[9] This "step," not yet taken in that lecture, was soon taken resolutely.

In his notes of 1940 with the heading *‹Philosophie› und ‹Wissenschaft›*,[10] Heidegger directly identified modern science with technology, and technology with the completion of metaphysics: "What modern science is: 'technology'. What 'technology' is—completion of metaphysics" (*MWT* 126).[11] By way of clarification, he adds such statements as: "Inserting modern science into the essence of modern technology. The latter appears later, but from early on already rules in the essence [*im Wesen*]." "Pure natural science is an essential completion [*Wesensvollzug*] of technology." "The *unity* of modern science as technology."[12]

---

[7] According to Ihde (2010), 93, Heidegger's interest in science and technology is shown only in a few brief periods, that is: the period of *Being and Time*, the mid 1930s, the mid 1950s, and, after "a gap," his last statement of 1976. This picture is unconvincing in view of Heidegger's own writings, as the citations in this section show.

[8] At *GA* 16, 349, one finds a summary of this lecture, the last sentence of which reads: "neuzeitliche Wissenschaft [ist] eine Weise der Technik." This is the same lecture, elsewhere referred to as "Die Begründung des neuzeitlichen Weltbildes durch die Metaphysik" of 9 June 1938 (*GA* 16, 802), which led to the published text of ZWB.

[9] *MWT* 126: "Die neuzeitliche Wissenschaft als "Technik"—Dieser Schritt im Vortrag 1938 noch nicht vollzogen, obzwar alles bereit." The "step" was also prepared in *Beitr.* (1936/1938), but also there not fully taken. For example, in § 76 Heidegger provides a list of propositions on science. In proposition number 19 (155/107), he speaks of the "growing consolidation of the machinational-technical essence of all science", which can be considered as support for the claim that Heidegger was prescient concerning technoscience. In the revised list of propositions on science in his notes of 1940 he is more explicit (proposition 22, *MWT* 124–5): "Modern science is research because it has its essential foundation (*Wesensgrund*) in technology."

[10] The editors of the *Gesamtausgabe* don't give a date for this bundle of notes, except for indicating that it is from the period of ÜM (1936/1946). The years given for various text on *MWT* 126f suggest 1940 as the best guess.

[11] "Was die neuzeitliche Wissenschaft ist: »Technik«. Was »Technik« ist—Vollendung der Metaphysik." For a discussion of the relation of technology and metaphysics see §1 in Ma and van Brakel (2014).

[12] "die Einfügung der neuzeitlichen Wissenschaft in das Wesen der neuzeitlichen Technik, das später als sie erscheint, aber früher im Wesen waltet" (*MWT* 127); "reine Naturwissenschaft [ist] ein Wesensvollzug der Technik" (125). "Die *Einheit* der neuzeitlichen Wissenschaft als »Technik«" (128).

In various writings after 1940, Heidegger often mentions the technological essence of science. For instance:

> "The more plainly the sciences are carried along by their predetermined technological essence. . . ." This comes from «Nietzsche's Wort Gott ist tot» of 1943 (*HW* 211/159).
>
> "[M]odern science stems from the essence of technology" (*FG* 179/116). This is cited from "Der Lehrer trifft den Türmer an der Tür zum Turmaufgang" (Winter 1944/45).
>
> "Modern science is application of the essence of technology."[13] This comes from the *Bremer Vorträge* of 1949.
>
> "Modern science is grounded in the nature of technology" (*WhD* 140/135); and "we still seem to be afraid of facing the exciting fact that today's sciences belong in the realm of the essence of modern technology and nowhere else" (*WhD* 16/14). This claim is made in *Was heißt Denken* of 1951.
>
> "Today, that which modern science moves in its innermost essence, ... we can only incompletely characterize ... by giving it the name 'technology'."[14] This comes from a marginal note to "Wissenschaft und Besinnung" of 1953.
>
> "Modern technology is the supporting grounding feature [*tragende Grundzug*] of modern natural science" (ÜTS 18/137). This is cited from the lecture „Technische und Überlieferte Sprache" of 1962.
>
> "The fundamental character of [the] scientific attitude is its cybernetic, that is technological character."[15] This comes from „Das Ende der Philosophie und die Aufgabe des Denkens" of 1966.
>
> "The essence (*das Eigene*) of modern technology and in it the already grounding sciences: *die Gestellnis*."[16] This is found among his numerous notes made in the early 1970s, published in *Gedachtes* (*GA* 81, 349).

We can see that the issue has been on his mind during the whole period of 1938–1976. What is most prominent among these remarks is a reversal of the order of science and technology. Since modernity, it has been an accepted view that technology owes its birth to science. Modern technology emerged only when science let itself avail in a certain area. Against this assumption, Heidegger suggests that we consider science from its technological essence. In doing this, we would reverse their order in seeing modern science as the application of the essence of technology. In its essence, technology is more primordial, in that the Greek word τέχνε means precisely "a bringing forth of beings . . . *out of* concealment specifically *into* the unconcealment of its appearance" (ÜM 46/35; em. or.) The reason why technology is more primordial is none other than that it denotes unconcealment of beings.

[13] "Das Wesen der modernen Technik, das Ge-stell, begann mit dem wesensmäßigen Grundakt des Bestellens, insofern es zuerst die Natur als den Grund-Bestand im vorhinein sicher stellte. Die moderne Technik ist nicht angewandte Naturwissenschaft, vielmehr ist die neuzeitliche Wissenschaft Anwendung des Wesens der Technik, ..." (*BV* 43).

[14] Marginal note to W&B only in *GA* 7, 62. "Das, was die moderne Wissenschaft in ihrem innersten Wesen bewegt, das, wodurch sich der gezeigte unscheinbare Sachverhalt ereignet, können wir heute nur erst ganz unzureichend und überdies leicht mißdeutbar kennzeichnen, wenn wir dafür den Namen »Technik« nennen."

[15] EP 72/58. Cf. ZWB, 85/64: "From an inner compulsion, the researcher presses forward into the sphere occupied by the figure of, in the essential sense, the technologist."

[16] "das Eigene der neuzeitlichen Technologie und der in ihm schon gründenden Wissenschaften: die Gestellnis." On *Gestellnis* see Ma and van Brakel (2014).

Thus, we need to consider the relation between science and technology from the vantage point of the history of Being.

Among sciences, Heidegger pays particular attention to (mathematical, theoretical) physics, because it is assumed to be the foundation and origin of all (natural) sciences. In a letter to Takehiko Kojima of 18 August 1963, Heidegger states: "The grounding feature of modern mathematical science is the technological, which appears first in its new and essential Gestalt through modern physics."[17] That the essence of science is the technological, Heidegger presumes, finds a first reflection in physics. In "Die Frage nach der Technik" of 1953, Heidegger says: "Modern physics is the herald of Enframement [*Gestell*], a herald whose origin is still unknown."[18] In "A Triadic Conversation" (Winter 1944/45), Heidegger even claims: "Physics must be technology, because theoretical physics is *the* proper, pure technology" [*FG* 8/5; em. or.]. In the next section, we will see how Heidegger comes to this point.

3 Laying Out a Structure of Argument

Heidegger's earliest and perhaps the most important presentation of a detailed argument regarding the relation between science and technology can be found in "A Triadic Conversation between a Guide, a Scientist, and a Scholar" from the *Country Path Conversations* composed in the Winter of 1944/45, just before the end of the Second World War. The argument can be summarized in terms of the following steps.

First step: Experimental physics makes use of technology. The example Heidegger gives is "the machine that splits the atom" (*FG* 6/3). Hence, experimental physics in itself might be considered to be technology being applied, instead of technology coming into being through the application of experimental physics. On the basis of this observation (which is repeated in 1962—see below), one may intend to describe Heidegger as prescient of technoscience.[19] However, this is only a preliminary remark, and Heidegger has yet much more to say.

[17] "Die Natur wird daraufhin herausgefordert, d. h. gestellt, sich in einer berechenbaren Gegenständlichkeit zu zeigen. … Dieses Her Stellen, d. h. das Eigentümliche der Technik, vollzieht sich auf eine einzigartige Weise innerhalb der Geschichte des europäischen Abendlandes durch die Entfaltung der neuzeitlichen mathematischen Naturwissenschaft. Deren Grundzug ist das Technische, das zuerst durch die moderne Physik in seiner neuen und eigentlichen Gestalt zum Vorschein kommt" Heidegger (1963), 156.

[18] "Die neuzeitliche Physik ist der in seiner Herkunft noch unbekannte Vorbote des Ge-stells" (FT 23/303).

[19] Heidegger being prescient concerning "technoscience" was already apparent in *Beitr.* "Natural sciences will become a part of machine technology and its operations." To this could be added Heidegger's early awareness of the "knowledge economy": what counts is not anymore which country has the richest natural resources (minerals etc.), but the country which is most successful in technological innovation (Bedr 9).

Second step: Theoretical physics, which is "the foundation of 'fundamental research' in all the natural sciences" (*FG* 8f/4), does not employ instruments or machines. Hence, it is of course *not* applied technology in the sense in which experimental physics is applied technology. However, theoretical physics *is* (applied) technology in a different sense. Heidegger claims: "Physics must be technology, because theoretical physics is *the* proper, pure technology" [8/5], "the technological essence of physics lies precisely in that it is theoretical physics" (11/6). Why?

Third step: The technological and the theoretical are the selfsame (*das Selbe*) (*FG* 11/6). Why?

> Thinking [in theoretical physics and elsewhere in the sciences] sets nature toward itself as the spatiotemporally-ordered manifold of moving points of mass. … natural processes are re-presented [*vor-gestellt*]. In this fashion, nature is what is pro-duced [*Hergestellt*]… [thus] nature is as that which stands over against the human. … [this] producing is the basic trait of the objectification of nature … something objective for mathematical representation (11/7).

Heidegger's discussion of the essence of science in lectures and writings from the period 1935–1938 can help illuminate the idea embodied in the third step,[20] in particular his discussion of the notion *ta mathemata* and the impact of Descartes' metaphysics,[21] which reverses the meaning of *obiectum* and *subiectum* (*FD* 106/280). According to Heidegger, in such statements as modern science is factual, experimental, and measuring, we miss its fundamental characteristics, which consists in two things. One is the way of working with the things, that is, setting things on their proper foundation *in advance* by using axioms and by "precalculability."[22] The other is the metaphysical projection of the thingness of the things. This way of projection, in opening up a domain in which only things of a certain kind can show themselves, "sketches out in advance a blueprint of the structure of every natural body as well as of its relation, to every other body" (Kockelmans 1985, p. 150). Thus the task of science is to specify what can be accepted as scientific facts.[23]

As a consequence, those most commonly assumed features of science (experiments & instruments, exactness & law, calculation & mathematics, and more

---

[20] Kockelmans has given a still useful and insightful overview of Heidegger's view of science drawing on *FD* (1935/1936), ZWB (1938), WhD (1951/1952), W&B (1953), FT (1953), G (1955). See in particular ch. 5 "Toward an Ontology of the Modern Natural Sciences." Of course at the time of writing Kockelmans did not have access to sources which have become available from the many volumes of the *Gesamtausgabe* published up to now.

[21] For Heidegger's broad notion of "the mathematical" see *FD* 69-77/249-255, ZWB 78/58, Kockelmans (1985), 142f, 150–1, and Dea (2009). *Ta mathemata* means for the Greeks that which man knows in advance in observing entities and dealing with things. Carson (2010) characterizes it as a view of mathematization as prescription that things make their appearance as objects predictable, calculable, and governable in a technological sense.

[22] In "Wesen der Sprache" (WS 178/74) Heidegger ascribes to Nietzsche the insight that "method" is more essential than "result." Scientific method is not a mere instrument, "it has pressed science into its own service." To be contrasted with "thinking", where there is "neither method nor theme."

[23] Galileo poses conditions in advance to which nature must answer in one way or another. With Newton nature becomes the closed totality of the motions of the spatio-temporally related point-masses. See for discussion Kockelmans (1985) and Dea (2009).

recently: research enterprise, specialization, institionalisation) take on a broader meaning.[24] For example, calculation now means: taking something into account by setting it up as an object of expectation.

Heidegger argues that modern science is completely different from medieval and ancient (Western) science (*FD*), and that the experience of nature entailed therein is also completely different. However, it makes no sense to say that modern science is more exact or that Newton's theory is true and Aristotle's view false. What is essential of modern science is its method of precalculability and the metaphysical projection of the thingness of the things.

During one of the Zollikon seminars in 1965, again the essence of science is described in similar terms as in the late 1930s: it is the "connection between *measurability* and *method*" (*ZS* 134/103), where *measurability* is in fact the same as calculability, which "means *precalculability*." This is the method of science: securing the calculability of nature. The mind is reduced to a "technician of calculations" (139/107).[25]

Coming back now to the *Country Path Conversations*, the Scientist summarizes the argument thus far:

> Then the name 'technology' strictly speaking, refers to a kind of representing, that is, a kind of cognition, and hence to a kind of theoretical comportment. The essence and the dominance of technology consists in the fact that, through it, nature has become an object. Nature is set up by the human, halted by him, so that it may be accountable to him and to his plans for it. Technology is the objectification of nature. (*FG* 12/7)

Since "the word itself … harbor[s] the significance [*Deutung*] of the matter named by it" (*FG* 12/7), Heidegger, in the role of the Guide, explains how the word 'technology' in the modern sense is a particular kind of τέχνε: "Modern technology is that letting-see and setting-toward in which nature comes to appear as a mathematical object" (13/8f).[26] Hence, Heidegger's idiosyncratic statement that "physics is applied technology" (15/10) does not necessarily contradict the common assumption that "technology is applied physics." This is because: "In each of the two statements the words 'physics,' 'technology,' and 'application' signify something different. … Because physics is applied technology in the sense of τέχνε, 'technology' in the familiar sense can and must be applied physics (15/10).

---

[24] Heidegger may also be said to be prescient concerning the current "research" phase of modern science: research in groups, distinction of *Geisteswissenschaften* and *Naturwissenschaften* disappearing, institutionalisation, intertwined with industry, becoming an enterprise. See ZWB and Kockelmans (1985), 152–162.

[25] And this is further interpreted in *Ge-stell* terms (135/105): "the point is *control* and *domination* of the processes of nature" (em. or.). This is followed by a citation from the last part of Descartes' *Discourse on Method;* "we render ourselves the master and possessors of nature" (136/105). [Descartes, *Philosophical Writings*, 1: 119. "Nous render comme maîtres et possesseurs de la nature."].

[26] Heidegger is well aware that, as expressed by the Scholar in the conversation (14/9): "I just can't rid myself of the suspicion that you are interpreting the Greek word τέχνε in terms of your own dogmatically asserted definition of the essence of modern 'technology'."

In FT and ÜTS, Heidegger elaborated on the connection between his earlier statements concerning the essence of science and the later claims regarding the essence of technology in terms of the essence of the *Ge-stell*. They are of one and the same essence. What unites the theoretical (that is, science) and the technological is the metaphysical projection of things in setting them up as usable objects.

In 1953 Heidegger repeats twice, that according to chronological reckoning, modern technology is later;[27] nevertheless, from the point of view of the essence holding sway within it, it is earlier historically speaking.[28] At the same time, modern physical theory of nature prepares the way for the essence of modern technology in that the challenging gathering-together into ordering revealing holds sway in modern physics, although it does not at once come expressly to appearance. Similarly, man's ordering attitude and behavior have already displayed themselves in the rise of modern physics as an exact science.[29]

Heidegger further explains, as pure theory, modern science's way of representing already sets nature up (pursues and entraps nature) to exhibit itself as a coherence of forces calculable in advance. It orders its experiments precisely for the purpose of asking whether and how nature reports itself when set up in this way.[30] Physics sets up (orders) nature as what is pre-calculable.

In 1962 (ÜTS), Heidegger remarks that, among scientists and technologists, one now hears the view that science and technology are in a relation of "mutual support." For example, in nuclear physics, the technical instruments co-determine which phenomena will be observed, thus co-determining the process of knowing. However, he points out, the further question concerning their common origin is not raised. Such a mutual relation is only possible provided science and technology are co-ordinated [*gleichgeordnet*]. What is the thing in which S&T correspond to one another? Both share the comportment of challenging posing [*herausfordernde Stellen*] toward things. The essence of science and technology does not consist, as one might think, in a means-to-an-end structure, but in the fact that a demand [*Anspruch*] is made (a non-human power), which orders humans to challenge nature forth (19/137).[31]

[27] FT 303-4/21-2. Also *BV* 43.

[28] Kockelmans (1985), 177 interprets these passages by saying "although science is first 'in execution', technicity was first in a perhaps still unconscious intention."

[29] Given the typical Heideggerian view that "That which is primally early shows itself only ultimately to men" (FT 23/327).

[30] That revealing concerns nature, above all, as the chief storehouse of the standing energy reserve. "For physics, nature is the standing reserve [*Bestand*] of energy and matter" (*BV* 42).

[31] "Der Mensch selbst ist gestellt, ist daraufhin angesprochen, dem genannten Anspruch zu entsprechen" (ÜTS 20/138). Cf. FT 21/302: "It remains true, nonetheless, that man in the technological age is, in a particularly striking way, challenged forth into revealing."

4 Epoch and Epochs

In the secondary literature Heidegger's strict separation between premodern and modern technology has been disputed (see Sect. 6). It has also been argued that presently we don't live anymore in Heidegger's Epoch of the *Ge-stell*, because since his death there have been revolutionary changes both in science and technology. For example, Ihde claims that "today's technologies evidence a quite different flavor from what was prominent during Heidegger's lifetime" (5) and suggests that we have entered a new Epoch of Being (of technoscience and quantum physics). Therefore, he argues, Heidegger's position regarding technology is out-dated. This judgment does not seem to be fair to Heidegger. Certainly, since the 1970s, new types of technologies appear one after another, of which it had not been possible for Heidegger to have any knowledge. However, Ihde has neglected some of Heidegger's writings that testify to a broad vision concerning the future of technology. Heidegger has stressed many times that the "modern age" is in no way at an end (*WhD* 57/54). Given the "predetermined essence of science," we have to reckon with "a gigantic progress of science in the future" (*Beitr* 156/108; G 524/45).

In addition, several of Ihde's examples of technologies, about which he claims that Heidegger knew nothing, or which he failed to present in the right way, are not correct. In particular this is the case with respect to quantum physics (see next section) and what Heidegger calls "biophysics." In the seminar at Le Thor in 1969, Heidegger makes such a remark (*Sem* 358/55): "This means that the human being can be produced according to a definite plan just like any other technological object."[32] Ihde (2010, p. 111) notices this statement and suggests that it is at most reflective of the Nazi legacy of eugenetics. However, this abjucation is poor-grounded. Heidegger draws on contemporary statements from scientists. In 1955 Heidegger cites Stanley, the American chemist who was the Noble Laureate in Chemistry in 1946, who made the following prediction: "The hour is near when life will be placed in the hands of the chemist who will be able to synthesize, split and change living substance at will" (G 525/44). Stanley, as Heidegger notes, made this statement at a meeting taking place a few months before Heidegger wrote the relevant text. A decade later, in one of the Zollikon seminars held in March 1966, Heidegger refers to "present research on the technique of genetic mutation in humans" (*ZS* 177/135), referring to a book published 2 years before.[33] It is obvious that until late in his lifetime, Heidegger always kept himself well-informed about cutting-edge advances in science and technology.

Misunderstandings may arise because Heidegger sometimes employs the word 'epoch' for quite different purposes. This has caused confusions, such that some scholars assume that there might be different epochs of modern science

[32] Cf. *WM* 257/197: "Sometimes it seems as if modern humanity is rushing headlong toward this goal of *producing itself technologically*" (em. or.).

[33] The reference is to "[Friedrich] Wagner, *Die Wissenschaften und die gefährdete Welt.*[*Eine Wissenschaftssoziologie der Atomphysik*, München, 1964] pp. 225 ff., 462 ff."

(on Heidegger's terms). According to our reading, we need to draw a line between the presumably "ontological" Epoch and the "ontic"epochs. Modernity is a unique Epoch in the history of being, whereas there are a plurality of "epochs" *within* the Epoch of Modernity. Occasionally Heidegger also uses such words as 'era', or 'revolution' as alternative phrasing of epochs in the ontic sense. These refer to particular periods or phases *within* modern science or technology that is supposed to have a unilateral history. One has to guard against the tendency of ascribing ontological weight to such epochs. The Epoch of modernity is a much more encompassing epoch that embraces possible breaks between classical and quantum physics, pre-industrial, industrial, and postindustrial technology, early modern science and institutionalised research enterprises, and so on.

For example, within the history of modern technology, Heidegger distinguishes such developmental stages as: power machinery, electrical technology, and atomic technology (FT 23/303). In regard to atomic technology, Heidegger states: "If the taming of atomic energy is successful, and [I know] it will be successful, then a totally new era of technical development will begin" (Ibid.). Such a statement should not be understood as atomic energy signifying the beginning of another Epoch of Being. Similarly, in an undated note, probably from the early 1950s, he speaks of the "second industrial revolution," which amounts to "entering decision-making into the machine," and which will constitute "the era of automatisation after World War III."[34] Again this should not be taken as a new *ontological* phase in the history of technology/technoscience.

With respect to developments in science, Heidegger offers the epithet of "a revolution that belongs to the greatest in human thought" to Newton's first law of motion (*FD* 89/257). This "revolution" is one of the contributions to the beginning of the Epoch of Modernity.[35] Within this Epoch, he further identifies several "lower-level" revolutions and changes. In the Le Thor Seminar of 1969, he remarks that today, there are no objects as there were for eighteenth/nineteenth century scientists: "the further that modern technology unfolds, the more does objectivity [*Gegenständlichkeit*] transform in standing-reservedness [*Beständlichkeit*]" (*Sem* 367/61). In the 1973 seminar in Zähringen he refers to it in these words: "man has gone from an *epoch* of objectivity to an *epoch* of orderability [*Bestellbarkeit*]" (388/74, em. ad.). This transition parallels the transformation of science into "research." According to Heidegger, since the 1930s, science has been under threat (*Bedrohung*). This threat stems not only from science proceeding according to a certain pre-given method, but from science becoming "research," that is, becoming an externally financed institution and an enterprise, like the industries with whom it

[34] "Das automatische Zeitalter nach dem III. Weltkrieg" (*MWT* 368); "Die zweite industrielle Revolution. Die Eingabe des Entscheidens in die Maschine" (*MWT* 376).

[35] Heidegger has provided detailed accounts of the work of Galileo and Newton (FD 77-95/255-271). Dea (2009), 54 interprets these accounts in *Being and Time* terms: "Before Newton, the fore-understanding the scientist brought to his understanding of nature included an interest in individual entities and, hence, a hermeneutical opennes to Being; after Newton, the scientists' hermeneutical horizon is restricted by the fore-understanding that individual entities are the indifferent manifestations of universal laws."

collaborates (Bedr, ZWB).[36] We suggest the genesis and strengthening of the institutional character of academic research since the second half of the twentieth century and today has corroborated Heidegger's insight. But his distinction of an "epoch of objectivity" and an "epoch of orderability" should not be understood as two successive Epochs in the history of Being.

5 Quantum Physics

It has often been suggested that the transition of classical to modern physics is so fundamental that it would herald a new Epoch of Being, but this is not Heidegger's view.[37] Kockelmans has rightly stressed this point. With quantum physics, "nature must still set itself in place in advance for the objectifying and securing processes which science, as the theory of what is actually real, accomplishes" (Kockelmans 1985, p. 169). Quantum physics still aims at writing the "one single fundamental equation from which all the properties of all elementary particles, and therewith the behavior of all material things, follow."[38] The comportment toward nature embraced by quantum physics remains to be the same with that embraced by classical physics. This is so, even though Heidegger speaks of "epochs in modern physics" (W&B 54/172). In a passage in square brackets in W&B Heidegger adds (55/173):

> The subject-object-relation thus reaches, for the first time, its pure "relational"—, i.e., ordering, character, in which both the subject and the object are sucked up [*aufgezogen*] as standing-reserves. That does not mean: the subject-object-relation vanishes, but the opposite: it now attains to its most extreme dominance [*Herrschaft*], which is predetermined from out of Enframing [*das Gestell*].

Werner Heisenberg, whom Heidegger first met in 1935 in Todtnauberg, provided him with first-hand information about quantum physics. According to Carson (2010), in Heidegger's lectures in the fall of 1935, quantum physics was given a partial exemption from his apparently derogatory view of science. But from 1936 onward, Heidegger presented all forms of *Wissenschaften*, including quantum physics and the humanities, as an extension of the technological will. Carson further documents that, in the years 1949–1953, Heidegger had been looking for a public confrontation with Heisenberg, which was finally taking place at the week-long lecture series of 1953 on *Die Künste im technischen Zeitalter*. Heisenberg presented

[36] He asks, without implying that "an other beginning" is occurring: "What understanding of beings and what concept of truth is it that underlies the transformation of science into research?" (ZWB 86/65).

[37] According to Ihde (2010), 109, it is by the mid-fifties that Heidegger came to recognize that "quantum physics totally resituates the early modern subject-object distinction." This is incorrect, as we will see.

[38] Heidegger citing Heisenberg in W&B 54/172.

"Das Naturbild der modernen Physik."[39] Heidegger, delivered, "Die Frage nach der Technik."[40]

Already in 1937, soon after the Einstein-Bohr debate concerning the interpretation of quantum mechanics, Heidegger writes several pages concerning what he calls "statistical physics" and the uncertainty relation (*MWT* 175–181). "Can we say with N. Bohr, 'that here [referring to the uncertainty relation] the separation between observed object and observing subject starts to disappear'? No!" This statement is followed by a couple of reasons why Bohr is wrong (and Einstein is right?). For Heidegger, it is completely mistaken to base a new epistemology on the uncertainty relation (179). According to Heidegger "cause" and "rule" are presupposed in every experiment, including quantum physics.

Again he thought about the issue at length when preparing for the "meeting with Heisenberg." In a letter to Medard Boss (October 28, 1953), Heidegger writes: "I am kept very busy by the lecture in Münich ["Die Frage nach der Technik"] and with an interrelated correspondence with Heisenberg. At the hut I wrote a wide-reaching sketch and got deep into the question of causality."[41] The result of Heidegger's ponderings and exchange of ideas found its way in the printed text as follows (FT 24/304).[42]

> If modern physics must resign itself ever increasingly to the fact that its realm of representation remains inscrutable and incapable of being visualized, [it is still being] challenged forth by the rule of Enframing, which demands that nature be orderable as standing-reserve. Hence physics, in all its retreating from the representation turned only toward objects that has alone been standard till recently, will never be able to renounce this one thing: that nature reports itself in some way or other that is identifiable through calculation and that it remains orderable as a system of information.

In addition, there is the technological side of atomic physics: atom bombs, nuclear power stations, which he mentions in several places. In *Gelassenheit* of 1955 there is a critical discussion of several pages on nuclear science, from which we already cited: "If the taming of atomic energy is successful, and it will be successful, then a totally new era of technical development will begin."

In a marginal note to "Wissenschaft und Besinnung" Heidegger comes close to acknowledging the need of a "new epistemology" when he remarks that "through" Heisenberg's uncertainty relation the human is explicitly included in the artificiality

---

[39] Cf. Heisenberg (1955).

[40] Some time before the meeting, Heidegger had distributed a draft of the text that was later to be published with the title "Wissenschaft und Besinnung" (W&B).

[41] *ZS* 246f/310. From the letter a month earlier (dated September 30), it is apparent that Heidegger had been working on the preparation of his lecture and public discussion with Heisenberg for a considerable period of time. It would seem that his notes and drafts in this summer have not (all?/yet?) been published (cf. *GA* 76).

[42] Cf. *BV* 43: "Although atomic physics is differently disposed [*geartet*]—only statistical instead of determinate [*eindeutig*] calculability–, it is still the same physics." W&B (54/172): "atomic physics admits only of the guaranteeing of an objective coherence that has a statistical character."

of the instruments, becoming a part of it.[43] However, even in the case of quantum physics and nuclear technology, science and technology are still governed by "the Same" (that is, calculability, orderability, predictability). During one of the Zollikon Seminars in 1966 (*ZS* 177/134-5) Heidegger repeats his view that predictability is not invalidated by the indeterminacy principle.[44] He mentions the atomic bomb and the notion of upper and lower limits: without predictability, technical construction would be impossible. There is no a-causal world view (as Heisenberg advocated).[45]

Since Heidegger pondered about these matters, numerous publications on quantum mechanics have appeared, leading to a range of quite different interpretations.[46] Nevertheless, the foundational questions remain to be the same as those raised in the Bohr-Einstein debate in the 1930s: how to understand "measurement" of parameters in the quantum mechanical formalism. This is what is called "the measurement problem." However, until now there is no agreement about where to begin in order to have a provisional formulation of the nature of the problem. Hence, the question whether quantum mechanics would require a "new epistemology" is still open and Heidegger's rather "instrumentalistic" view (what counts is predictability) is still an option.

6 Premodern and Modern Technology

One important aspect of Heidegger's reflection upon technology (and hence upon the essence of science and technology) is a strict distinction between pre-modern and modern technology. Relevant scholars such as Zimmerman (1990) have embraced such a distinction. For Heidegger, pre-modern means of production is a "natural bringing-forth" that corresponds to *poiesis*. The work of a craftsman is not so much different from the work of an artist in that both attempt to let things come into unconcealment and reveal themselves, which is a matter of ἀλήθεια. With modern means of production, in contrast, things are challenged, controlled to reveal themselves as standing-reserves; nature is a resource to be exploited.

In a recent article, Riis (2011) disputes Heidegger's view. According to him, there simply does not exist an unbridgeable gap between pre-modern technology

[43] "Among all the objects he can only meet himself—but what is 'himself' in this case: the instrumentation!" *GA* 7, 57.

[44] Comments on Heisenberg etc. were occasioned after the participants of the seminar had objected to Heidegger using classical physics as a "general" characterization of science.

[45] Heidegger acknowledges that quantum physics has changed the notion of causality "once again .... It seems as though causality is shrinking into a reporting, 'a reporting challenged forth' of standing-reserves that must be guaranteed either simultaneously or in sequence" (FT 24/304).

[46] Including an empirically equivalent alternative to "standard" quantum mechanics, viz. Bohm mechanics, which presupposes a deterministic world and, by present standards, leads to the same predictions as "standard" quantum mechanics.

and modern technology.[47] For sure, Heidegger should be aware that the Greeks employed slaves to build temples and to work in silver mines, and the mode of controlling embodied in these forms of producing does not have essential qualitative difference from that embodied in modern technology.[48] Hence Riis concludes that, as Dasein, humans are essentially modern.[49]

We agree with Riis that a sharp distinction between pre-modern and modern technology is far from self-evident.[50] It is certainly not obvious that ancient ways of production are completely exempted from forms of extracting resources such as silver, as if resources are simply out there in nature ready for use. Similar ways of subjecting nature to human force and control already exist in ancient times. Consider the following passage from Heidegger's 1962 lecture (ÜTS 18/137), where we substitute the word 'silver' for the word 'energy':

> The [silver] that is locked in nature is unlocked [when silver ore is mined], what is disclosed is transformed [that is, with chemical means silver is extracted from the silver salts in the ore], what is transformed is reinforced [i.e. concentrated, purified, refined], what is reinforced is stored, what is stored is distributed [to the silver smiths]. These ways, according to which nature is secured, are controlled [e.g. (dying) slaves in the silvermines]. This controlling, in its turn, must secure itself further.

The main idea of this passage remains unchanged with the substitution of the word 'silver' for 'energy.' There is no silver as such in nature, but only silver ore. Controlling nature as well as human beings exists in all periods of time and in all regions of the globe.[51] One can as well be referred to the construction of pyramids and other "gigantic" architecture around the world, and the phenomenon of deforestation that persistently occurs throughout the history of humanity and across the globe.

For sure, certain ancient tools of production, such as a windmill to which Heidegger refers, appear to be far less challenging [*herausforderend*] than other tools, such as the power station "in" the Rhine to which Heidegger also refers. However, we would argue that the contrast between controlling and non-controlling modes of obtaining resources or of making products does not exactly correspond to a certain timeline of pre-modern age and modern age. This contrast also exists in

---

[47] Heidegger seems to be pointing to a similar direction when he writes (*Beitr* 132/92): "It is very difficult to grasp historically the emergence of what is machinationally ownmost to beings, because basically it has been effective in operation since the first beginning of Western thinking."

[48] Other "falsifying" examples (discussed by Ihde) could be clocks determining daily life in Europe long before the rise of modern science or technology. Also: technology used in navigation on the Atlantic for the past millennium (to be contrasted with "lifeform" navigation on the Pacific—which in some sense was equally successful). Ihde (2010), 68: "through the use of technologies, experience had already become prepared for the scientific experience of the world."

[49] "The rule of *das Gestell* has challenged humans as long as they have existed" Riis (2010), 116.

[50] In a somewhat different context Claude Lefort has argued that there is a definite connection between political discourse and modernity and that therefore the societies of ancient Rome and Athens are modern societies.

[51] What might be exceptions? Inuit, Aboriginals, Khoisan, Polynesians?

ancient times. The following parable from the *Zhuangzi*, a famous Chinese Daoist scripture, very well illustrates this point.

Zigong, a disciple of Kongzi (that is, Confucius), came across an old gardener, who was carrying water to the fields. He used up a great deal of energy but produced very little result. Zigong asked the gardener whether he would like to use a well sweep.[52] The gardener flushed at his words with anger and said (Watson 1968, p. 134):

> I have heard my teacher say, where there are machines, there are bound to be machine worries; where there are machine worries, there are bound to be machine hearts. With a machine heart in your breast, you have spoiled what was pure and simple; and without the pure and simple, the life of the spirit knows no rest. Where the life of the spirit knows no rest, the *dao* will cease to buoy you up. It's not that I don't know about your machine—I would be ashamed to use it!

A possible Confucian retort to the gardener might be that, all the same, the old gardener has to use a sort of gadget such as a bucket in order to get water, and most probably he was actually using one. However, a Daoist would reply that the use of the latter forms of tool conform to the *dao* of nature insofar as it does not exert itself upon water/nature but instead tries to integrate itself with nature in a calm and peaceful way; whereas the use of a well sweep to get water violates the *dao* of nature in exerting unnecessary force and attempting to subject water under its control. This parable from over 2,000 years ago vividly illustrates the contrast between controlling and non-controlling modes of obtaining resources from nature.

Now if the distinction between premodern ("natural") and modern ("calculative") technology is undermined, would this bear any influence upon Heidegger's claim that modern science and modern technology are the Same (ÜTS 18/137)?

7 A Different Slogan: Science and Technology Are the "Same"

Some scholars have presented Heidegger's strict distinction between pre-modern and modern technology as the ground on which his idea of a unique "Epoch of Modernity" is established. However, Heidegger's idea of a unique "Epoch of Modernity" is set in a much broader vision of the history of Being, rather than grounded on a more or less simplistic division between pre-modern and modern technology. In this section we propose a slight revision of Heidegger's story starting with the "ontic" observation that the current globalised world (Heidegger would say "planetarized") is unique and is to be fundamentally distinguished from earlier Epochs of Being, leaving open the possibility that there is more than one beginning of the history of Being.[53]

[52] "It's a contraption made by shaping a piece of wood. The back end is heavy and the front end light and it raises the water as though it were pouring it out, so fast that it seems to boil right over!"

[53] See on the latter issue Ma (2008), 92–99.

The most essential feature of the globalised world lies in the spread and domination of science and technology.[54] Heidegger calls the seventeenth and eighteenth century the "getting ready" of modernity, the nineteenth as getting breath, and the twentieth as the more obvious beginning of an Epoch of being that is there to stay for a long time (*MWT* 80). With the advantage of hindsight we can see that the essence of science and technology which comes to full fruition in the twentieth century (*Ge-stell*, "research enterprise") was already "getting ready" in the time of Galileo, Newton, and Descartes; while from the perspective (of Epochs) of the history of being, Modernity has already been underway since the time of Plato: "modern [science and] technology is completed metaphysics."

What is the Same in which the essence of science and technology is retained and united? That is, the application of a particular method of representation and precalculability of nature, and the challenging/controlling revealing of nature. The crucial word is "re-presenting."[55] Heidegger discussed the notion of "representing" in numerous places. Perhaps the following passage in ZWB (108/82) gives a condensed but rather complete characterization:

> To represent [*Vorstellen*] means here: of oneself, to set something before one and to make what has been set in place [*das Gestellte*] secure as thus set in place. This placing-in-securedness must be a calculating, since only calculation guarantees being certain, in advance and always, of that which is to be presented. … The being is no longer that which presences. Rather it is that which, in representation, is first set over and against [*entgegen Gestellte*], with the character of an object [*Gegen-ständige*]. Representation, setting-before, is a making everything stand over and against as object [*Ver-gegen-ständlichung*] which masters and proceeds against. In this way, representation drives everything into the unity of the thus-objectified.

This notion of representation "identifies" the Epoch of Modernity. Ontically it is first present in the rise of modern science as Heidegger's detailed discussions of the work of Galileo, Newton and Descartes show. Its consequences only became fully apparent in the rise of *das Ge-stell*, science as "research" and "planetarization."

> In the planetary imperialism of technologically organized man, the subjectivism of man attains its acme, from which point it will descend to the level of organized uniformity and there firmly establish itself. This uniformity becomes the surest instrument of total, i.e., technological, rule over the earth (ZWB 11/84).

Of the consequences of this unique form of re-presentation, which is shared by modern science and technology, perhaps the three most important ones can be highlighted as follows (starting with a collation of passages already cited in Sect. 3):

[54] Heidegger mentions five essential features of modernity, including science and machine technology (ZWB 75f/57f). According to him the fundamental event of modernity is the conquest of the world as picture (94/71).

[55] It has been said that the burden of science and technology lies not in their calculative style but rather in their insistent and aggressive spirit. Alderman (1978), 43. However, we agree with Rojcewicz (2006), 114 that "science attacks nature with experiments is not what is impositional, but the prime imposition is the representation of nature."

First, as to nature: natural processes are re-presented [*vor-gestellt*]. "Nature is what is pro-duced [*Hergestellt*]... [thus] nature is as that which stands over against the human. ... [this] producing is the basic trait of the objectification of nature. Nature is set up by the human, halted by him, so that it may be accountable to him and to his plans for it" (*FG* 12/7).

Second, the metaphysical projection of the thingness of the things which opens up a domain in which only things of a certain kind can henceforth show themselves; things are set on their proper foundation *in advance* and are set up as usable objects.

Third, the control and domination of the processes of nature; man's ordering attitude and behavior, his comportment of challenging posing (*herausfordernde Stellen*) toward things, the challenging gathering-together into ordering revealing, adding later that it is man himself who is *demanded* to do this. This consequence includes the "rule of *das Gestell*," which demands that nature be orderable as standing-reserve.

Hence, the "originary" bringing forth, in the context of the forgottenness of being, develops into a pro-ducing and re-presenting of a standing-reserve to be ordered and challenged. Instead of "Science is applied technology," this line of thinking may suggest that the answer to the question concerning the relation of science and technology should be: "The essence of science and technology is the Same."

As a matter of fact, on a few occasions, Heidegger explicitly speaks of the Same [*das Selbe*] of science and technology. We have already seen that, in the *Feldweg Gespräche* of 1944/45, he suggests: "The technological and the theoretical are the selfsame [*das Selbe*]" (11/6). In 1959, Heidegger begins "Aufzeichnungen aus der Werkstatt" by saying that science is 'identical' [*identisch*] with the ruling [*walten*] of modern technology.[56] And finally, in his technology lecture of 1962: "Now, what is that thing in which natural science and technology agree and thus is the Same? ... A reciprocal relation [*Wechselverhältniss*] between natural science and technology can only subsist if both are co-ordinated [*gleichgeordnet*] ...." (ÜTS 16/136).

8 Questioning Is the Piety of Thought

Contrary to the common view that modern technology derives from modern science, Heidegger presents a reverse picture in which science originated in the essence of technology, wherein Being speaks. It is in this sense that Heidegger talks about the Same of science and technology. As we have seen, on many occasions, it seems that Heidegger has already provided a more or less explicit answer to the question

[56] "Manche scheinen heute mit der Not zu ringen, für das Walten der modernen Technik und der mit ihr identischen Wissenschaft eine Vorstellung von der Geschichte zu finden, in die sich der durch jenes Walten bestimmte Weltzustand einordnen" *GA* 13, 151.

regarding the relationship of modern natural science to modern technology. However, he keeps on raising this question again and again, to such an extent that he singles it out for special attention in his letter of 1976. Why? Probably, it is because the very act of questioning itself is more important than the answer, whereas "we moderns" (*WhD* 140/135)[57] can at best *prepare* an answer. As is reported in the protocol of the Zähringen seminar of September 7, 1973, read and approved by Heidegger (*Sem* 390/75):

> The entry into this domain [i.e. "the ultimate decisions that this reality compels us to take up"] is not produced by the thought undertaken by Heidegger. To believe thinking capable of changing the place of man would still conceive of it on the model of production. Therefore? Therefore, let us say cautiously that thinking begins to prepare the conditions of such an entry.

In 1940 when first answering the question (see citations in Sect. 2), he expressed the answer in a questioning way, as if merely suggesting the possibility.[58]

> Technology. Is it merely application of natural science to …—to what then?—Or is modern natural science the consequence of "technology"? But the latter (powered machines technology) is chronologically later than modern natural science.—That does not exclude that the metaphysical essence of this technology, is as a matter of fact [*sachlich*] essentially earlier.

Some reasons for Heidegger's vacillation are intimated in "The Teacher Meets the Tower Warden" from the *Country Path Conversations*:

> Tower Warden: … technology possesses this power to alter actuality only because scientific representations, whose actualization technology is said to achieve, already arise out of the peculiar essence of technology.
>
> Teacher: I find it difficult to follow your thoughts every time you present the relation of science and technology in this manner.
>
> TW: This is not only the case for you. It will still take a long time before the human enters into an engagement with the insight that modern science stems from the essence of modern technology.
>
> T: Why do you place such weight on this insight?
>
> TW: Because only it allows the experiences through which the human could achieve a befitting relation to the technological world.
>
> T: If this is the case — which, to be frank, I do not entirely perceive — then there is no time to lose in the project of awakening the essential insight.
>
> TW: Certainly not — but it is also the case that we cannot force this insight through mere instruction and decree. (*FG* 179/116)

The essential origin of modern science and the essence of modern technology need be traced out through questioning and this can never be a facile matter. This is because, first of all: "A fog still surrounds the essence of modern science [and this fog arises from the fact] that we are still not thinking" (*WhD* 16/14). A similar remark concerning the essence of technology discloses Heidegger's profound

---

[57] We are citing the translator of *Was heißt Denken?* Heidegger himself wrote: "man heute."

[58] *MWT* 144: "'*Technik*' Ist sie nur die Anwendung der Naturwissenschaften auf …—worauf denn? Oder ist die neuzeitlichte Naturwissenschaft die Folge der 'Technik'? Aber diese (Kraftmaschinentechnik) zeitlich später als neuzeitliche Naturwissenschaft. Das schließt nicht aus, daß das metaphysische Wesen dieser Technik das sachlich wesensmäßig frühere."

evasion: "Meanwhile the essence of modern technology is even darker than the essence of science—so dark that probably we have not even once succeeded to question modern technology properly [*sachgerecht*]."[59] As late as 1969, in a letter to Roger Meunier (*Sem* 416/88),[60] Heidegger claims that "a sufficiently grounded insight into the relation of the two [that is, the interlocking of modern technology and modern science] has not yet been gained." And presumably his letter of 1976 expresses the same idea.

Only through the insight that modern science stems from the essence of modern technology could humans possibly obtain a befitting relation to the technological world. However, there could be no rush in asserting this point.

References

Martin Heidegger

1935. *Gesamtausgabe* 40: *Einführung in der Metaphysik*. 1983. *An introduction to metaphysics*. Trans. Gregory Fried, and Richard Polt. New Haven (Connecticut): Yale University Press (2000). *IM*.

1935/1936. Die Frage nach dem Ding, 66–108. Translated as "Modern science, metaphysics, and mathematics." In: *Basic Writings*, ed. David F. Krell 247–282. *FD*.

1936/1938. *Gesamtausgabe* 65: *Beiträge zur Philosophie (Vom Ereignis). Contributions to Philosophy (From Enowning)*. Trans. Parvis Emad, and Kenneth Maly. Bloomington: Indiana University Press (1999). *Beitr*.

1936/1946. Überwindung der Metaphysik. In *VA* [2000], 67–98. Overcoming metaphysics. In *The Heidegger controversy*, ed. Richard Wolin. 67–90. Cambridge MA: MIT Press (1993). ÜM.

1937/1938. Die Bedrohung der Wissenschaft. In *Zur philosophischen Aktualität Heideggers*, ed. Dieter Papenfuss, and Otto Pöggeler. Band 1. *Philosophie und Politik*. Frankfurt am Main: Klostermann (1991). Bedr.

1938. Die Zeit des Weltbildes. In *HW* [1969] 75-114/57-85. ZWB.

1942/1943. *Gesamtausgabe* 54: *Parmenides*. *Parmenides*. Trans. Andre Schuwer, and Richard Rojcewicz. Bloomington: Indiana University Press (1992). *GA* 54.

1944/1945. *Gesamtausgabe* 77: *Feldweg Gespräche* (1982). *Country Path Conversations*. Trans. B. W. Davis. Indiana University Press (2010). *FG*

1949. *Einblick in das was ist. Bremer Vorträge*. In *GA* 79: 3–78. *BV*

1951/1952. Was heisst Denken? Gesamtausgabe 8 (2002). *What is Called Thinking?* Trans. Fred D. Wieck, and John Glenn Gray. New York: Harper & Row (1968). *WhD*.

1952. Moira (Parmenides, Fragment VIII: 34–41). In *VA* [2000], 235–262. Moira (Parmenides, VIII, 34–41). In *Early Greek Thinking*. Trans. David F. Farrell, and Frank A. Capuzzi, 79–101. New York: Harper & Row (1975). M.

1953. Die Frage nach der Technik. In *VA*, 5–36. "The question concerning technology." In: *Basic Writings*. Trans. William Lovitt, 311–341. London: Routledge (1993). FT.

[59] W&B, marginal note only in *GA* 7, 62: "Das Wesen der modernen Technik ist indessen noch dunkler als dasjenige der Wissenschaft—so dunkel, daß wir vermutlich noch nicht einmal dahin gelangt sind, nach der modernen Technik sachgerecht zu fragen."

[60] Meunier is the translator of Heidegger's inaugural lecture "Was ist Metaphysik?" into French.

1953. Wissenschaft und Besinnung. In *VA* [2000], 37–66. Science and reflection. In *The Question Concerning Technology and Other Essays*. Trans. William Lovitt 155–182. New York: Harper & Row (1977). W&B.

1955. Gelassenheit. In *GA* 16 [2006]: 517–529. Memorial address. In *Discourse on Thinking*. Trans. John M. Anderson, and E. Hans Freund, 43–57. New York: Harper & Row (1966). G.

1955. *Hebel—der* Hausfreund. In *GA* 13 [2002]: 133–150. In English in *Contemporary German philosophy*, Vol. 3, 89–101. State College: Pennsylvania State University Press (1983).

1955/1957. *Identität und Differenz. Identity and Difference*. Bilingual edition. Trans. Joan Stambaugh. New York: Harper & Row (1969). *ID*.

1957/1958. Das Wesen der Sprache. In *Unterwegs zur Sprache*, 157–216. Stuttgart: Günther Neske (1959). "The nature of language" In *On the Way to Language*. Trans. Peter D. Hertz, 57–110. San Francisco: Harper and Row (1971). WS.

1962. *Überlieferte Sprache und Technische Sprache*, St. Gallen: Erker (1989). Traditional language and technological language. Trans. Wanda Torres Gregory. *Journal of Philosophical Research* 23 (1998): 129–145. ÜTS.

1963. Brief an Takehiko Kojima. In *GA* 11 [2006]: 155–161.

1964. Das Ende der Philosophie und die Aufgabe des Denkens. *Zur Sache des Denkens*, GA 14 [1951/52]: 67–90. "The end of philosophy and the task of thinking." In *On time and being*, 55–73. Chicago University Press. EP.

1969. *Gesamtausgabe* 5: *Holzwege*. *Off the Beaten Track*. Edited and Trans. Julian Young, and Kenneth Haynes. Cambridge: Cambridge University Press (2002). *HW*.

1976. Neuzeitliche Naturwissenschaft und moderne Technik. In *GA* 16 [2006]: 747–748. "Modern natural science and technology." *Research in Phenomenology* 7 (1977): 1–4.

1993. *Basic writings*, ed. D. F. Krell. London: Routledge.

2000. *Gesamtausgabe* 7: *Vorträge und Aufsätze*. *VA*.

2002. *Gesamtausgabe* 13: *Aus der Erfahrung des Denkens 1910–1976*. Frankfurt am Main: Klostermann. *GA* 13.

2004. *Gesamtausgabe* 9: *Wegmarken*. *Pathmarks*. Trans. William McNeill. Cambridge: Cambridge University Press (1998). *WM*.

2005. *Gesamtausgabe* 79: *Bremer und Freiburger Vorträge*. Frankfurt am Main: Klostermann. *GA* 79.

2005. *Gesamtausgabe* 15: *Seminare in Le Thor*. *Four Seminars. Le Thor 1966, 1968, 1969, Zähringen 1973*. Trans. Andrew Mitchell, and François Raffoul. Bloomington & Indianapolis: Indiana University Press (2003). *Sem*.

2006. *Gesamtausgabe* 11: *Identität und Differenz*. Frankfurt am Main: Klostermann. *GA* 11.

2006. *Gesamtausgabe* 16: *Reden und andere Zeugnisse eines Lebensweges 1910–1976*. Frankfurt am Main: Klostermann. *GA* 16.

2006. *Gesamtausgabe* 89: *Zollikoner Seminare, Protokolle—Gespräche—Briefe*. Herausgegeben von Medard Boss. *Zollikon Seminars. Protocolls—Conversations—Letters*. Trans. Franz Mayr, and Richard Askay. Evanston: Northwestern University Press (2001). *ZS*.

2007. *Gesamtausgabe* 81: *Gedachtes*. Frankfurt am Main: Klostermann. *GA* 81.

Others

Alderman, Harold. 1978. Heidegger's critique of science and technology. In *Heidegger and modern philosophy*, ed. Michael Murray. New Haven: Yale University Press.

Carson, Cathryn. 2010. Science as instrumental reason: Heidegger, Habermas, Heisenberg. *Continental Philosophical Review* 42: 483–509.

Dea, Shannon. 2009. Heidegger and Galileo's slippery slope. *Dialogue* 48: 59–76.

Feenberg, Andrew. 1999. *Questioning technology*. London: Routledge.

Guzzoni, Ute. 2012. *Gelassenheit*: Beyond technoscientific thinking. In *Heidegger on science*, ed. Trish Glazebrook, 193–204. Albany: State University of New York Press.
Heisenberg, Werner. 1955. *Das Naturbild der heutigen Physik*. Hamburg: Rowohlt.
Ihde, Don. 2010. *Heidegger's technologies. Postphenomenological perspectives*. New York: Fordham University Press.
Kochan, Jeff. 2010. Latour's Heidegger. *Social Studies of Science* 40: 579–598.
Kockelmans, Joseph J. 1985. *Heidegger and science*. Washington, DC: Center for Advanced Research in Phenomenology & University Press of America.
Ma, Lin. 2008. *Heidegger on East–west dialogue: Anticipating the event*. New York: Routledge.
Ma, Lin, and Jaap van Brakel. 2014. Out of the *Gestell*? The role of the East in Heidegger's *das andere Denken*. *Philosophy East & West* 64. Forthcoming.
Riis, Søren. 2011. Towards the origin of modern technology: Reconfiguring Martin Heidegger's thinking. *Continental Philosophical Review* 44: 103–117.
Rojcewicz, Richard. 2006. *The Gods and technology. A reading of Heidegger*. Albany: State University of New York Press.
Thomson, Ian. 2000. What's wrong with being a technological essentialist? A response to Feenberg. *Inquiry* 43: 429–444.
Watson, Burton. 1968. *The complete works of Chuang Tzu*. New York: Columbia University Press.
Zimmerman, Michael E. 1990. *Heidegger's confrontation with modernity. Technology, politics, art*. Bloomington/Indianapolis: Indiana University Press.

Logos and the Essence of Technology

Holger Schmid

Abstract The present study takes up Martin Heidegger's claim that today's technoscientific reality cannot be properly understood unless seen as the issue of a 2,300 year "incubation." Against long-lived clichés of romanticizing archaism—the "nostalgia for Greece" for example—this claim here appears in light of a consistently Pauline-Johannine futurism.

Accordingly, modern technology, that is "metaphysics" itself, is to be envisioned from a vantage point where, above all, world and language are known to arise from one and the same constitution, as implied in the key terms of *logos* and *poiesis*. Hence there must once again be talk of "the Greeks": respecting Heidegger's *Sache* as well as meditating upon his methods.

As technology today comes to be ever more identical with reality in general, we face a condition of "reality" which is patently indebted to the world-constitutive function of scientific knowledge (with its emphasis on the species of natural—or "physical"—science, to the extent that this mode of scientific thought has consequently all but absorbed its former antagonist, the "moral," i.e., the human sciences). The self-dissolution or melting into one, as it were, of these traditional antitheses seems inescapably to mark the ultimate peak of modernity. What used to appears to be nature turns out now to be a social construction of simulacra—or technologically generated "fictions." Nature, we are assured, does not exist. All of this, we believe, could not have been foreseen during the first half of the twentieth century, when, during and subsequent to World War I, Heidegger and many others encountered technology as a "planetary" problem. It is this coincidence of nature and technology that surely constitutes the most revolutionary aspect of the world-change we are currently undergoing. And yet, we are at the same time reminded of a passage in Plato,

H. Schmid (✉)
Université de Lille 3, 24, rue de Thionville, F-59000 Lille, France
e-mail: holger.schmid@univ-lille3.fr

B. Babich and D. Ginev (eds.), *The Multidimensionality of Hermeneutic Phenomenology*, Contributions to Phenomenology 70, DOI 10.1007/978-3-319-01707-5_12,

one which is precisely adduced by Heidegger in the course of his own enquiring after technology: "Everything that is responsible [*aitia*] for creating something out of nothing is a kind of poetry." (Symp. 205b)[1] Faced with such a coincidence between *physis* and *poiesis*, (which Heidegger will praise as "Greek"), we might well surmise that our most recent technological revolutions are but the perfection of Platonist metaphysics: nature finally recognized as illusion, as simulacrum. To that extent, Heidegger's question as a task of thinking the nexus between science *qua* "modern natural science" (reason-giving: *logos*) and technology does not involve us in the old cliché of Romanticism: of immediacy lost and regained. Much rather could one claim, perhaps with a touch of exaggeration, that Heidegger remains ever indifferent to the past as such, a Pauline futurist or eschatologist throughout—a disposition that seems in paradoxical contrast to his insistent recourse to the Ancients, to "the Greeks." To explore this apparent paradox with respect to the "essence" of technology is the goal of what follows.

1 Modes of Incubation

It is surely not difficult to concede that in the last 250 years of Western history (i.e., since the First Industrial Revolution) there has been a crucial link between modern science and technology: since the beginning of the world of machines, that is, the beginning of "modern technology."[2] But Heidegger's more specific point is that the principle of sufficient reason, as embodied in Leibniz' thought, is to be recognized as that which transpires today (having only today become visible in its unfolding), constituting as it were the metaphysical ground of our still (and especially) "metaphysically" informed present. Heidegger's anti-historicist question thus intends to be an anamnesis of the present and its aetiology. It is precisely in the reason-giving principle that Leibniz refers us back to Plato's Socrates and to his paradigm of the only life worth living: one that is perpetually examined and controlled by *logon didonai* (*Apol.* 38A). Thus the relevant time lag would actually intensify—and it will increase further. And yet the whole idea of a chronological sequence might seem misconceived, for the very point of Heidegger's anamnesis is that technology is not at all an independent entity to be set over against metaphysics or theory. Technology, such is the thesis, is to be found *qua* "practice" precisely at the core, and as the core, of metaphysics itself. Hence, part of Heidegger's anamnesis will be to ascribe to the principle of sufficient reason-giving what he calls an "incubation period" of no less than 2,300 years.[3] What breaks out like a disease (or like the brooding of an egg) has thus been prepared over the course of a very long era. "Older" than technology, in any event, is the "essence" of technics, which holds sway not only in modern science, but in European science as such. Thus we may

[1] See Heidegger (1977a), 10.

[2] See, e.g., Kockelmans (1985), 173.

[3] Heidegger (1996), first section.

say that this essence as it reigns today—to the extent precisely that it is itself nothing technological—is what began to rise 2,300 years ago: that is, in Heidegger's sense, *metaphysics*.

Is the principle of reason then to be qualified as a disease? Once again, we are reminded of Socrates who famously professes, a moment before his death, to owe a rooster to the healing god Asclepius. Is then Socrates' ever-examined life of *logon didonai* that very disease? What has this to do with Socrates' kind of open-ended questioning that might well be called the piety of his thought? Is *logon didonai* a condition, a factor, or the heart of the disease? Is the history of thinking or metaphysics the history of nihilism? In such anamnesis, would there not be, once again, a normative implication of nostalgia for a painless, pre-nihilistic state of "health"? But we should in any case be careful with our metaphors—and with the Pavlovian reflexes they are liable to provoke. A few things seem initially plausible—even before we begin to reflect on those 2,300 years:

1. What does it profit a physician nostalgically to wish away a "disease" that awaits diagnosis? 2. In Heidegger's incubation time the Pre-Socratics are conspicuously included: 2,300 years counted backwards from Leibniz' 1700 AD necessarily lead to 600 B.C.; hence not even the Seven Sages would be able to escape the verdict. 3. The notion of "disease" itself is historically conditioned, depending upon how an age or culture defines health. To Heidegger's mind, this is perfectly clear, bringing him closer to Ludwik Fleck than to Sigmund Freud: perhaps not a useless remark here, as mention will presently be made of King Oedipus. On the other hand, it is Husserl who envisages the genealogy of modernity as pathogenesis, culminating, as we know, in a "crisis" not of science alone but of modern life (or "European humanity") in general.[4] What distinguishes Heidegger, then, is his mode of recourse to antiquity. The Greek questioning experience and the essence of technology—how then are they to be conceived to hang together? And how are we to understand the "plague," the disease that haunts Thebes at the beginning of Sophocles' most famous tragedy?

Heidegger's questioning is guided by an observation which he shares, incidentally, with a number of other thinkers on metaphysics. The concept of philosophy—traditionally metaphysics—is itself preconditioned, at least since Plato and Aristotle, by an idea of "knowing" (or *scientia*) which is in turn shaped by a model of production, as typically embodied in Socrates' frequent references to artisans while inquiring after *techne*: thus production as "manufacture" (*Handwerk* is Heidegger's German term) and further, in Roman and Christian metamorphosis, as "creation" (with reference here to a creator).[5] The Greek term for this is *poiesis*. Its correlate is *techne* as the knowledge which is liable to become synonymous with all *episteme* in Plato: cognition or knowing in general, i.e., the noetic

[4] On this, see Müller (1976), 22f.,with further references. The concept of "lifeworld," in its therapeutic intention emphasized there, is shown in its provenance from Heidegger's early lecture courses by Schmitz (1996), 19f.

[5] Cf., e.g., Heidegger (1977b), 48.

relationship to entities as such.[6] But as "knowing," it is the very sphere constituted by traditional rationality, "reason" stemming from the Latin *ratio* or, in Greek, *logos*. In its sphere, then, the traditional plurality of those knowing modes (the *technai*) comes to be subsumed under the one heading of *techne*. It further ensues, according to Heidegger, that above all it is in what seems to be its very opposite, namely contemplation or *theoria*, that the model of knowing as fabrication (*poiesis*) achieves its sovereignty.

What thus emerges is the question specific to the later Heidegger, intertwined with the problems of *die Kehre*, the "turning" in the 1930s, with reference to poetry, to the critical battle around and against Nietzsche[7] (very much including Jüngerian recrudescences, the Will to Power having transformed itself, through the Gestalt of the worker, into the Will to Will). And arising in the midst of all this is the idea of "planetary technology" where Heidegger reinforces and/or abandons the philosophical project of restituting or restoring the sciences—that is, the university—to a lost or obscured "essential ground," and thereby first completing metaphysics. As this is also the context of Heidegger's political disaster, it is clear that there is room for a number of serious questions, which would center on the issue of Heidegger's insight concerning National Socialism as opponent or embodiment of planetary technology, before and after 1938. This would be the insight, philosophically speaking, regarding "metaphysics" which represents itself as the problem, not the solution. And that is how technology comes to appear as the basic trait or structure of metaphysics itself. It is therefore a strikingly contemporary interest that inspires Heidegger to turn to a renewed anamnesis "of the Greeks" in order to lay bare the core of technology.

This of course involves the further issue of the form and fashion of our own Heidegger exegesis. It is clear that were we pledged, consciously or not, to a research model of the philosophy of technology (*qua* assembling expert knowledge and information data concerning technology), a Heidegger would have little to teach us: and least of all by distinguishing technology from its "essence." Not only has he no ethics, even worse, he has no logic—and no physics either. Thus, in particular, our perspective would not be disturbed by self-critical afflictions and suspicions, suggesting, e.g., that in so thinking we might simply re-iterate what Heidegger, it is to be hoped with an eye to his own *Machtrausch*, calls "busy-ness" [*Betrieb*]. In that case, we would not reflect upon but merely exemplify the expert's *hysteron proteron* of confusing the problem with the solution, just as it happened to the problem-solving hero Oedipus, who had to mistake the Theban plague as an outside "thing" or research object to be investigated. Yet what remains most interesting therein may be the fact that Heidegger, while still endeavoring to teach "metaphysics" in 1935, describes this very Oedipus, crushed between the assault of appearance and the advent of truth in his furious search for identity, as the exemplary "Greek" Dasein: "we must see him as the embodiment of Greek being-there, who most radically and wildly asserts its fundamental passion, the passion for disclosure of being, i.e.

[6] See, for example, Heidegger (1984), 179.

[7] On this see Babich (1993), 239–260.

the struggle for Being itself."[8] Here indeed the connection between the "Greek" and the catastrophic procedure of Oedipus' expert questioning *techne* seems immediately to point to the non-apparent "challenge" which Heidegger will attempt to think as the essence of technology.

2 Poiesis and the Un-poetic

The crucial aspect of the Heideggerian inquiry will turn out to be that the state of the *world* and the state of *language* are one and the same; and this is precisely what is expressed by the problem of *logos*. In principle, therefore, the question as to the essence of technology becomes ever more identical with the problem of an originary creation or production, as a constitutively "Greek" *poiesis* in contrast with, and obstructed by, the traditional metaphysical model of production (the same constellation likewise explains Heidegger's ongoing preoccupation with the poets, further extending well into the 1980s in the shape of Hans-Georg Gadamer's belated self-critical musings on poetry). The essence of technology according to Heidegger thus expresses a lack of, or a retreat from, or a refusal of a world. In that sense, if the task of thinking be to conceive this refusal "as such" (in terms of the knowing, *techne*, as essentially related to *poiesis*), the state of the world then appears by definition as "unpoetic." Now the term Heidegger introduces for this world-state is, of course, the notoriously provocative term: *Ge-stell*, usually rendered as "enframing" or else as "setting-upon" with connotations of trapping or entrapping. It may accordingly be assumed that as the name for technology's essence, *Ge-stell* must also be the formula for the question of how to distinguish technology from this essence, to the extent that, as a definition of this essence, it is contrasted with *poiesis* in the originary or Greek sense. In Heideggerian terms, then, there is implied a reciprocity or coincidence of an experience of language and of Being, proximally corresponding to the Greek versus the modern era (as *explanans* vs. *explanandum*). Less obvious and in the background, as it were, there is also in *Stellen* a crucial reference to the ancient Greek *thesis*, the counterpart of *physis*, and thus to the nature-culture dyad, famous since the Sophists and Aristotle, and recalling, via the "thetic" activity of *techne* in bringing things to stand (i.e., to be), the distinction between the "positive" and the "natural" in the Western tradition.[9] To capture this proximity, which will presently be recognized to imply the *Ge-stell* as the self-desisting Fourfold, it would seem tempting to render *Ge-stell* by the term "Sistence."[10] More remarkably, however, *Ge-stell* would seem to possess a polemic edge against Ernst Jünger's

[8] Heidegger (1959), 107.

[9] See here Heidegger (1994), 62ff.

[10] *Stellen* corresponds to the verb "to sist," taken in its old, broad sense: "to cause to stand, to order one before a court, to place or posit, etc." OED. One may add that versions such as 'positionality' are utterly misleading, all the more so because the term pertains to Helmuth Plessner's anthropological definition of human specificity as eccentric positionality.

Gestalt of the worker,[11] signalizing the same epochal signature or state of affairs—namely, the "total mobilization" or Will to Will in a contrasting light.[12]

This should serve as a rough characterization of Heidegger's point of departure for his inquiry concerning technology. Here, one might still have the impression that much of this talk about "world" and world-refusal looks frighteningly familiar: the good, poetic, ancients versus the bad, world-deprived, technological moderns, exactly as romantic cliché would paint its nostalgia for Greece. It is all the more striking, however, if, as documented by one of Heidegger's seminars held at Le Thor as late as 1969, we should then find the thinker still emphasizing what he terms the "fundamentally *un-poetic* nature of the interpretation of language by the Greeks"—an assertion that appears to border on the paradoxical (the Greeks being, of course, the poetic nation *par excellence*) as long as we do not raise the question: what—or better: whom—does Heidegger mean by the Greeks? What does "poetic" mean here? (What does the Greek interpretation of language have to do with the Greek experience of Being?) What would a more "poetic" interpretation look like? All these aspects will turn out to have to do with Heidegger's treatment of *logos*.

3 Logos, the Constitution of World and Language

A first step towards characterizing the paradox would seem to reside in the assumption that what is meant is the Greeks' philosophy, *qua* philosophy, that carries and embodies this "un-poetic" interpretation (whereupon Aristotle, for example, could only be seen as the one who rehabilitated the poets banished by Plato): philosophy as such would be at stake—to the extent that it adopted in its entirety the epistemic model of production or manufacture as described earlier. Entities as *physis*, composed of form and matter, are thereby reduced to *thesis* (that is, to a product of work), negating the genuinely natural, physical character of standing and growing in itself. Aristotle, to be sure, does distinguish between the two kinds of "movement," the natural and the cultural. But as the ontological conception of the thing as ensemble of matter and form (or possibility and actuality) is retained, this continues to serve to reaffirm the demiurgic or poietic model of thinking and knowing

---

[11] On the Heidegger-Jünger relationship in general, see Franco Volpi (1990), 9–45. Here, 32 for Jünger's reaction to *Gestell.*

[12] Such use of the term *Gestell* would then be datable as subsequent to "The Origin of the Work of Art" (1936), where it had simply designated the "thetic" stance of the artwork in the strife of world and earth. Thus 1938, as the the time of Heidegger's renewed (and by then decidedly critical) reflection upon Jünger's "worker" seems to suggest itself. Precision of insight into Heidegger's inner history during and after the Hitler empire seems occasionally hampered by negligence of Friedrich Georg Jünger's pivotal role therein, especially with regard to the book *Die Perfektion der Technik* (2010 [1939]) and its significance for Heidegger's changing view of technology: F. G. Jünger's name is absent from, e.g., Zimmerman (1990); Milchman and Rosenberg (1996); Rockmore (1992); Rockmore and Margolis (1992); Macann (1996); Pöggeler (1994); Jamme and Harries (1992); Seubold (2000), 119–132.

(and it is against Aristotle's dichotomy of entities that Heidegger had evoked the unitary Platonic thought of *physis* as itself the highest *poiesis*!).

Hence the paradox persists: Greek philosophy would thus be unpoetic precisely owing to its manufacturing or *poietic* paradigm. At any rate, this "manufacturing of knowledge" is both technological and ancient, i.e., "Greek." The consequence becomes obvious in the problem of language and its Greek interpretation, which is the problem of *logos*. As Heidegger explains in 1969, it is the reduction of *aletheia* to the field of *legein* (in the sense of speech, as *verbum dicendi*) which characterizes the Greek inception from its beginning, "always already, in advance," i.e., ever since Homer's epic language: this is what constitutes the "unpoetic" interpretation of language, in view precisely of the fact that there is, according to Heidegger's conviction, no higher-ranking poetic practice than that of the Greeks.[13] Thus it is indeed in the *legein* itself that the unpoetic comes to be founded. At Le Thor, for once directly criticizing Aristotle's *Poetics*, Heidegger still adds (or perhaps has a participant add) a quotation of an apophthegm once uttered in conversation by Stéphane Mallarmé: "poetry has entirely lost its course since the great Homeric aberration." A gloss he leaves unexplained, advising the reader to meditate upon its implications. But it is clear that, at this point in Heidegger's reasoning, *logos* itself, in order to be freed from its metaphysical reduction to the apophantic and semantic, must be envisaged in terms of a more originary, "more Greek" and hence more "poetic" meaning of *legein*, and that the obvious locus of such an attempt must be the exemplary thinking of *logos*, in Heraclitus.

Thus in the Western tradition, "reason" and "language" are brought to hang together in *logos*, and that is why *logos* must be at the core of Heidegger's sustained reflection on the essence of technology: that is: *Ge-Stell* or "Sistence," as this essence, determined by its contrast with the "world" that it refuses or of which it constitutes the self-desisting event. Its counterpart will then be Heidegger's vision of that world in describing which he regresses, according to some commentators, into archaicizing mythology: the famous *Geviert*, the Fourfold, as the structure of the world formed by the interdependent, inseparable, resonating tetrad of "regions": divinities and mortals, sky and earth. The essence of the entity or the thing, as obliterated and left unthought by Plato as well as Aristotle (both spell-bound by the pattern of production) and thereby *a priori* annihilated by science, is now conceived as that which hosts or assembles the Fourfold, reminiscent of "thing" in Old High German, meaning "assembly" (around a 'cause' or 'matter' of dispute, in 'council').

In such apparent mythologizing, the suspicion of escapism and irrationalism is naturally bound to arise; and we seem to be back precisely to that romantic and nostalgic picture of a lost unity of the world. Are we then dealing with a new philosophy of *Ur-Gemütlichkeit*, as a sharp tongue commented regarding one of Heidegger's lectures? Yet it is also true that such a perfectly sober mind as that of the Prussian statesman, designer of the very notion of the liberal arts and theorist of language, Wilhelm von Humboldt, will find, a century earlier, surprisingly Heideggerian terms for describing the "assembling" bent of the Greek mind: "when

[13] Heidegger (1977a), 73f.; see also Heidegger (1967), 271.

choosing an object," he writes, "they always take together [compare *legei*], as much as possible, the terminal points of all spiritual existence, heaven and earth, gods and humans, vaulting them in the idea of fate [*Schicksal*] as keystone." One could surely surmise a common, probably Platonic, source for this coincidence between von Humboldt and Heidegger, which may in fact come somewhat unexpectedly.[14] Hence in all of this there may be rather less irrational mysticism than much more structural thinking. But how are we to go about expounding and clarifying the problem of the refused "world" in what looks like a welter of paradox and contradictions, where Heidegger in addition attempts to think much more rigorously and radically than Humboldt, the enlightened humanist? Together, Enframing and the Fourfold signify the unity of language and world—the "assembling" which is the more originary meaning of *logos* (to be dis-covered). The relevant and problematic aspect thereof (which is precisely that of production or *poiesis* as the "un-poetic") would now seem to contain the problem of Enframing as the essence of technology, accessible by means of elucidating "the Greeks," that is, the Greek experience of language alone. More precisely still, the "unpoetic" (derivative, semantic *logos*) is the specific character which distinguishes the "world" (Fourfold) in its own, self-obstructing essence, as Enframing or Sistence. With this in mind, let us return to Heidegger's essays "The Question Concerning Technology" itself, the scene of which was a meeting of the Bavarian Academy of the Fine Arts in 1953, where Heidegger's lecture followed on the heels of an address by Werner Heisenberg.

4 Causality Displacing the Fourfold

Crucial to Heidegger's Munich lecture is its point of departure in the thesis described above, according to which the essence of technology is nothing technological, which he proceeds to explicate by examining the "instrumental," that is, analyzing the means/end relation defining the instrumental comportment—of *homo faber*, as Hannah Arendt would later call it—and by ranging it within causality. In modern science, as we know, what is constitutive is thought to be the very opposite: i.e., the presumptive elimination of all teleological elements. Heidegger, for his part, claims that the whole sphere of causality remains obscure precisely in that the instrumental (especially as regards technology's finality) is defined in modern terms by "efficient causality" alone as the sole admissible model of causality. Heidegger first refers to the traditional system of four causes (out of which structure modern thought subsequently isolates a single effective cause), raising questions such as: Whence the four causes? And how do they belong together? But then, taking a further step, he even

[14] Humboldt (1961), 30. The import of Heidegger's references to Humboldt, particularly in light of the closing pages of his *On the Way to Language*, has frequently been underestimated. Cf. the author's study, Schmid (1999), 92–98.

declares that ancient thought is ignorant of efficient causality, given that there is not even a Greek word for it (either in Aristotle or elsewhere).[15]

Greek production does not effect an object through subjectivity; as an example, Heidegger demonstrates this Greek character by analyzing the making of a silver chalice, a sacrificial vessel, as it turns out, by a silversmith. (This silversmith may also be read as a critical—if not self-critical—echo to the famous hammer-using artisan of *Being and Time*'s analytic of Dasein.) With regard to Heidegger's example, we may recall that naturally the silver (as *hyle*) and the "aspect" (*eidos*) of "chaliceness" represent material and formal causes. There remains a third that above all is "responsible" (*aition*) for the sacrificial vessel by circumscribing the chalice as belonging within the realm of consecration: the end, *telos*, or final cause, which completes the entity by assigning it the bounds of its sphere—not its purpose. The silversmith, the fourth participant in the responsibility for the finished vessel, is what he is not as efficient cause: "the Aristotelian doctrine," says Heidegger, "neither knows the cause that is named by this term nor uses a Greek word that would correspond to it." What the silversmith does is to deliberate [*überlegen*] and to gather [*versammeln*] the three causes previously mentioned. Deliberation, *Überlegen*, says Heidegger, is in Greek *legein*, *logos*: It is due to this *logos* of the silversmith *that* and *how* those first three modes of *aition* come into appearance and into play.

Three points may strike us in this account of the making of the chalice. First, the denial of an efficient cause (even of a Greek equivalent term), which would, if unconditionally accepted, facilitate a sharp distinction between Greek—namely, in this case, Aristotelian—and modern. However, the texts yield a different impression: for not only does Aristotle know of such a cause, the name that he has for it is exactly "the efficient," understood as the poietic: *to poietikon*.[16] Second, the artisan's doing—*poiein*—is, so to speak, absorbed in the assembling, *legein*; thus it seems that, for Heidegger, sheer "deliberation" brings about the accomplished vessel. In other words, *logos* (the deliberation exhibiting the artisan's *techne*) and *poiesis* become here identical in that *logos* is stripped of its usual meaning "to say" or "to tell," in favor of assembling or "laying," which will turn out to be the more originary sense of *logos*—and *poiesis* as well—that Heidegger had sought. (It could also be observed that *logos* and *poiesis* further coincide with *physis*, nature, with the help of the quotation from Plato directed by Heidegger against the conventional distinction, going back to the Sophists and Aristotle, between natural and cultural or "positive" beings).

Third, the correlate of this latter fusion of *logos/poiesis* is our main interest for the present consideration of "world" (language and Being) in the later Heidegger:

---

[15] This and what follows: Heidegger (1977a), 6ff.

[16] Compare, e.g., *Met.* I, 2, 1013a 31 with *De gen. et corr.* I, 7, 324 b 13 and *De anima* III, 5, 430a 12. Occasionally, as at *Met.* VIII, 6, 1045a 30f., Aristotle unhesitatingly drops all talk of finality to name the efficient cause as solely responsible for any transition from the possible to the actual in the shaping of matter (thereby approaching, once again, the Platonic identification of *physis* with *poiesis* from *Symp.* 205b).

the example of the silversmith's production shows on closer inspection that the play of the four causes is in fact derived as stemming from, and as being a concretization of, the Fourfold. Conversely, the Fourfold constitutes an elaboration of the doctrine of the four causes in the way Heidegger is known to rethink (in terms of the "unthought") loci of ancient tradition in a "more Greek" way. To put this in other terms: Heidegger's idea of the Fourfold is not derived from Hölderlin, as, for example, Reiner Schürmann and others have assumed,[17] but rather from Aristotle. As sky and earth stand for and deepen matter and form, silver and chaliceness, as the *telos* of sacrificial libation leads to the divinities, the region of the mortals then must be the specific site of the *poietikon*, the poetic: in their very act of "assembling," by deliberation: *logos*.[18] So conceived, the fourfold structure becomes concinnous with the equally Aristotelian key thought of the essay, namely the truth-character of technology as *aletheuein*, in using which Heidegger reaches back to his reception of the *Nicomachean Ethics* 30 years earlier.

As the Fourfold constitutes the structure or harmony of the world precisely as refused and silenced by Enframing or Sistence, i.e., by the essence of technology, it is what Heidegger's anamnesis of the Greek inception aims at. In such a retrieval of "the Greeks"—that is: of Aristotle—the un-poetic nature of the essence of technology now accurately echoes the poietic structure of the Aristotelian Fourfold. The poietic doing of the mortals in assembling "things," their *legein*, clearly shows the parallel: just as Enframing is nothing else than self-desisting Fourfold, so *techne*, by now amounting to "Greek" knowing in its entirety (in light of Plato), is essentially obliterated and likewise manifested by the poietic-unpoetic mode of disclosing that is technology's truth.

5 Back to the Pre-Socratics?

There yet remains the riddle of the unpoetic interpretation of language which we seem now in a position to pose more adequately. The further turn to *logos* in Heidegger's reflection *not* as signifying "speech" but something more primordial, leads us one step further back (or ahead) to the pre-Platonic Greeks. It is especially in his essay "Logos (Heraclitus, Fragment B 50)" that Heidegger expounds the allegedly original meaning of *legein* and *logos* as presupposed in the silversmith parable: "laying," or laying-before as letting-lie: this very turn from *speech* to *laying* constitutes the locus where, according to Heidegger, there flashes up the "unthought" essence of language (and "world" alike; that is, the "middle" of the Fourfold as *Sage*). Correspondingly, he comments on what is for us to envision as the unthought in the Greek inception:

[17] See, e.g., Schürmann (1987), 224: "Unfortunately for conceptual clarity, this is where Heidegger's language follows Hölderlin's most closely."

[18] It may be observed that the silver chalice is Aristotle's own example when characterizing the material cause: see *Met.* V, 2, 1013a 25 f. The fact that deliberation, which would expected to be *phronesis*, is shifted to *logos* seems due to the meaning assigned to Parmenides' fr. 7,5 DK.

> had this beginning not safeguarded what has been, i.e., the gathering of what still endures, the Being of beings would not now govern from out of the essence of modern technology. Through technology the entire globe is now embraced and held fast in a kind of Being experienced in Western fashion and represented on the epistemological models of European metaphysics and science.[19]

Metaphysics and science are declared to be based upon the resulting conception of language as tool or organ (*glossa*, "tongue") and as "signifying voice," *phone semantike* (from *semainein*, to mean). By contrast, this flashing up of the primordial unthought essence of language took place in Heraclitus's use of the word *logos*. But this flash was extinguished abruptly so as to obliterate *logos* in the sense of primordial "laying." And hence Heidegger's point is that this "laying" is to be recognized as the originary experience of language: "saying,"*Sage*, which must therefore be thought as the middle of the Fourfold (where it also appears as Fate or Destining, *Geschick*, with an echo of *moira* in Parmenides).

Of the vast field of questions here, we shall only be concerned to address that aspect of *logos* as it relates to Aristotle in transcending him. With the extinction of the flash, *logos* is set on its way to become *ratio*; it will proceed to become, in an ever-renewed application of the form-matter scheme, the human faculty of autonomous reasoning or "logic" as opposed to ("positive") revelation. Meanwhile, it becomes proposition, then concept, ultimately it becomes the word, *verbum*. Thus Heidegger would seem to maintain that *logos*, to the very extent that it took on the meaning of "speech", obscures the more original meaning of laying-out (*lesende Lege*: something like "col-lective layout"). This would be precisely the genesis of the now familiar "unpoetic" interpretation of language, while—with the advent of the "semantic voice"—the unity of World and language in originary *poiesis* falls into oblivion and refusal. Henceforth, in Enframing or Sistence the world speaks only in its concealment. It is important to note that it is this meaning of laying that Heidegger has in mind when he renders logos by *Sage*, saying, as the contrary of speech further to be elaborated as the "ringing of silence." (Another aspect of the saying-laying relation will be mentioned in a moment.) Conversely, *Sage* is not by any means "myth" as some commentators have believed.[20]

In order to measure the enduring presence of Aristotle in all this, while trying at the same time to elucidate the advent of the "semantic voice" as the incisive moment in the history of logos, it may be useful briefly to recall Heraclitus's famous fragment 93 (DK[21]) regarding the diction of Apolline prophesying. It is familiar to all of us, e.g., in Marcovich's translation: "The Lord whose is the oracle in Delphi neither speaks (*legei*) nor conceals, but gives a sign (*semainei*)." Heidegger quotes it repeatedly, since the wording beautifully confirms his main point since *Being and Time*: apophantic "disclosing" (or, "de-claring," with an allusion to Charles Kahn's rendering) as here the sense of *legein* is made evident in opposition to cryptic

[19] Heidegger (1975), 76.

[20] See, e.g., Lacoue-Labarthe (1987), 87; Großmann (1996), 198.

[21] Diels and Kranz (1951).

"concealment."[22] But what about the opposition itself, and what about the *semainein*? Even without intending an overall analysis in our present context, two problems may yet be observed to cohere in this received interpretation (which dates as far back as Plutarch[23]; and, as we recall, Plutarch was himself a Delphic priest): the meaning of *semainein*, on the one hand, together with the meaning of the "neither—nor" opposition on the other, both seeming to center upon the problem of "signifying" (hinting) as the presumptive activity of the oracle. It is to be understood that for commentators from Antiquity, Heraclitus is usually taken to be referring to his own philosophic discourse (*logos*), either metaphorically or by comparing it more or less favorably with the oracle. Thus in the usual understanding of the Delphic way of giving a sign (itself famously ambiguous) is implied something like a scale of transparency between the extremes of total lucidity and total opacity, where *logos*, taken as revealing opposed to concealment, would find its place on the side of lucidity, so that the sign itself comes to stand in the middle: that is, in a chiaroscuro midpoint as a fragile measure between those two extremes.[24] In other words, what we find is Aristotle's conception of the mean (*meson*).

Now, if it were to be accepted that this idea of a moderated mean or middle, between the extremes of concealing and revealing, constitutes but a retrojection of an Aristotelian schema onto the fragment (hereby implying a kind of semantically ambiguous twilight as essential to Pythian sayings), the question would still remain with regard to an earlier meaning of *semainei*. This is not the place to attempt an alternate reading of the fragment according to which the "neither—nor" would refer not to a scale of degrees or *valeurs* of light and darkness but to a qualitative antithesis, in keeping with other occurrences of the neither/nor in Heraclitus. It may be thought, however, that, if anywhere, it is in this Heraclitean saying that something like the "Greek interpretation of language" is to be found and examined as to its poetic or non-poetic character. The crucial point of such a reading would be to emphasize that the lord of Delphi does not declare or "lay open" in the mode of *legein* at all (not even halfway)—not implying as necessary that twilight ambiguity which is a trait of only some of his sayings (for a counter-example here we may recall, in Aeschylus's *Oresteia*, the exactly unambiguous Delphic command that Orestes kill his own mother).[25] With regard to the meaning of *semainein*, "to indicate," it could be argued that its meaning is closer to "instruction" by imperative, giving orders, for instance, indicating where to go for a departing colony.[26] In addition, as

[22] See, for example, Heidegger (1959), 170. Held (1970), 162–206, while emphasizing Heidegger's philological merit in elucidating "the original meaning of the word 'logos'" (204), does not mention fr. 93. Similarly, Bröcker (1965). See Kahn (1979), 43.

[23] See *De Pythiae oraculis*, 21, Mor. 404 HD.

[24] Cf. Marcovich's discussion: "The saying seems to be an image (metaphor); its implication might be the following: 'As Apollo neither speaks out all (100 %) nor conceals all (0 %), but shows forth a part of the truth (50 %), so also Logos inside things is neither inaccessible to human knowledge (0 %) nor self-evident (100 %), but requires an intellectual effort from men,'" etc. Marcovich (1967), 51.

[25] See also Delcourt (1955), 97.

[26] Cf. Detienne (1994), 165ff.; see further Nagy (1996).

semainein is a technical term of mantic and prophetic terminology, to say that the lord of the oracle indicates, *semainei*, would hardly seem for Heraclitus to be a surprising claim but to amount much rather to a tautology. The otherwise inevitable lack of equilibrium (*semainei* must balance *anax*) would point to the previous part of the sentence, i.e., once more to the problematic neither/nor and to the "does not lay open" (*oute legei*). Thus we might be led to improvise a rendering such as, "The ruler who possesses the oracle-chasm at Delphi neither lays open nor conceals but gives orders." If, on principle, the oracle does not "tell" in the way of *logos*, then surely this would encourage enquiring into the Greek interpretation of language beyond *logos* (or, more precisely, beyond the Aristotelian fixations of both *logos* and *semainein*)—all the more so if we recall that the oracles were delivered in verse: in hexameters, like Homer's (unless, with Mallarmé in mind, this were to be put inversely), that is, poetically. It is from this pivotal point of the Greek interpretation of language (i.e., the experience of language and of Being) that the question of *logos* in Greek philosophy in Parmenides and Heraclitus could be reopened. We might expect that it is precisely to the "question concerning technology,"with its identity of Fourfold and Enframing, that such renewed analysis of the limits of *logos* would return: and this would then seem to form a new chapter in the history of the oddly timeless influence of Heraclitus on Hölderlin and Hegel, on Nietzsche and Heidegger. Heraclitus, in his vehement opposition to Homer: after having spoken of the "great Homeric aberration," in a sequel not mentioned by Heidegger, Mallarmé replies to the interlocutor's question, "Before Homer, what?": "Orpheus."

6 Ephesus and the Essence of Technology

In that sense, there is shed more light on the decisive instant when, according to Heidegger, the flashlike appearance of *logos* as saying—i.e., as laying—in Heraclitus was immediately obliterated and obscured so as to set metaphysics on its way: the instant when, through the shift from laying to speech in *logos*, precisely the unpoetic interpretation of language arises, while primordial *techne* and *poiesis* are seen retreating into the unthought, in favor of the incubation of modern technology. That is, exactly when logos came to designate the experience of language to the very extent that it became the occidental *ratio* or calculative reason. This instant is in Aristotle, or as we can further narrow it: in the very opening phrases of *De interpretatione*.[27] What makes the interpretation of language ultimately unpoetic would be the idea of symbols of mental experience as sensual articulation of sentence meaning, in the "semantic voice" (*phone semantike*), where *semainein* first appears as we know it, as signifying. By the same token, *logos* becomes well-ordered, calculative "telling"—it becomes concept, proposition, and at the same time "reason," the thinking faculty of the rational animal, a shift that allegedly dates back to Parmenides (fr. 7,5 DK). At last, on the other hand, *logos* then appears as the Word, once again,

[27] See Heidegger (1971), 97.

after 600 years, in Heraclitus's town of Ephesus on the coast of Asia Minor, in the writings of the fourth Evangelist. All the while, the self-obstruction of the"world" prepares itself, toward its manifestation as *Ge-stell* after 2,300 years.

All of this may then be duly regarded as an exposition of Heidegger's claim in the Heraclitus essay as already cited: "Had this beginning not safeguarded what has been [*das Gewesene*] i.e., the gathering of what still endures, the Being of beings would not now govern from out of the essence of modern technology." Here, the essence of modern technology, the enframing mode of "sisting" and entrapping entities, precisely in its unpoetic character (reduced to *causa efficiens*), is nothing other than the world, the Fourfold, showing itself only in its concealment or refusal, *sub specie contraria*, as Enframing. Or, citing Heidegger once again, it is this essence of modern technology, through which "the entire globe is today transformed and destined into a being which is occidentally conceived and is entrapped within the truth-form of European metaphysics and science." The insight resulting from this anamnesis is not only, first, that the essence of technology is indeed nothing technological but also, second, that it is visible only as seemingly remote in time. It therefore defies any historicist perspective but is emphatically historic, *geschichtlich*, as Heidegger correctly claims. It remains outside the jurisdiction of expert historiography, on pain of confusing the problem with the solution. There is no other way of grasping that direct connection between the height of the technological age and the beginning of metaphysics, i.e., the "Greeks," than what Heidegger calls "thinking." And this will all the more be true to the extent that in light of Heraclitus, as opposed to Aristotle, the Greek experience of language would seem less manifest in Oedipus' struggle for self-determination than in the wisdom of his adversary, Tiresias.

Meanwhile, there is still a corollary to be appended. As we have seen, it is in the totality of aspects concerning the Fourfold no less than the related problem of *logos'* primordial creativity (transcending the "unpoetic" Platonic model of craftsmanship or manufacture)—i.e., in the name of what Heidegger envisioned as originary *techne-poiesis*—that Heidegger turns away from Leibniz and towards Aristotle. He turns to Aristotle in order to depart from him towards the thought of a more primordial, "more Greek" conception of the unity of the four causes in the Fourfold conceived as the "Saying," *die Sage*. Heidegger re-encounters that same Platonism as the innermost character of modernity, if not the essence of technology itself: as anyone can see in today's mediatic reality. It is this constitutive Platonism that Heidegger found embodied, at quite another level, in Heisenberg: symbolically speaking, at the point where Heisenberg himself took up the thought of the four causes, along with other Aristotelian concepts, to articulate the *Zusammenhänge* which he had elaborated 30 years earlier.[28]

On the other hand, an attempt at an even more pointed reflection on language and the "unpoetic," at a greater distance from Aristotle rather than extrapolating what is

[28] See Liesenfeld (1992), 199, n.110, et passim. Subsequent divergences, precisely with regard to Platonism, are mentioned in Pöggeler (1994), 400f.

"more Greek" in rewriting him, would continue the meditation on Greek "basic words," *Grundworte*, by acknowledging above all that they appear, "more primordially," in contexts of poetic composition: which is the case precisely of *logos*, *aletheia*, *semainein*.[29] This would include, and be nourished by, a critical debate, e.g., with the recent book on Pindar by Michael Theunissen, who, coming from a rather un-Heideggerian orientation but nevertheless sharing the historic but non-historicist motivation of presenting a cost-benefit analysis or critical theory to Western rationality, turns to archaic Greek lyric poetry precisely to step out of the tradition pre-given as the discipline of "philosophy" (susceptible of anachronism), in order to grasp, philosophically, the problematic of the experience of time, which would seem to have much in common with the essence of technology.[30]

7 Being and Writing

This would elucidate (such is my concluding observation) further surprising aspects of this Heideggerian Aristotelianism: one of them to be found exactly in the place of the unpoetic interpretation of language, i.e., of Being, where things begin to look somewhat like an everyday evolutionist perspective. In the case of language, *logos*, and of art, *poiesis*, alike, the "Greeks," says Heidegger, dwell in their world without attaining to sufficient concomitant thinking on either.[31] This looks just a bit like conventional thinking about unreflective "primitives" in their *histoire froide*—regarded from modern European perspectives. What is perhaps more crucial is the fact that it also looks like the Husserlian "naive" givenness or "natural" attitude; and we may surmise that this is still an unexpected reflex of the first book of Aristotle's *Metaphysics*. This concerns once again *logos* in what Heidegger claims to be its primordial meaning as "collective laying-out," *lesende Lege*, just as such "laying" as letting-lie represents a remarkable avatar of the Greek *hypokeimenon*, as that which is let, or allowed to, present itself "before": the "underlying" Substance, that is, no less than the metaphysical category *par excellence* since Aristotle. Could that be a coincidence? Concerning the second term, *Lese* (collection, of what lies before), it is hard to escape seeing that it simultaneously refers to the ordinary sense of *lesen*, or *legere*, i.e., reading, in that Heidegger names correlatively, in 1935, the written letters, *grammata*, as the paradigm for the Greek "experience"—not here of *language*, but of *Being*.[32] What could this supposed paradigm have to do with the "unpoetic"? We find therein a final hint at the unity of world and language as revealed by the recourse to the Greeks in the thought of the Fourfold or saying as speaking in its very concealment as Enframing. That is, the essence of technology takes on a

[29] Cf. Boeder (1959); Böhme (1986).

[30] Theunissen (2000).

[31] See, e.g., Heidegger (1975), 77.

[32] See Heidegger (1959), 64.

surprising proximity to the problem of the connection between writing and metaphysics. This would lead to further questions addressed to Heidegger and to the Greeks as well.[33]

References

Aristotle. *De anima*

Aristotle. *De interpretatione*

Aristotle. *Degeneration et corruptione*

Aristotle. *Metaphysica*

Aristotle. *Ethica Nicomachea*

Babich, Babette. 1993. Heidegger on Nietzsche and Technology. *Man and World* 26: 239–260.

Boeder, Heribert. 1959. Der frühgriechische Wortgebrauch von Logos und Aletheia. *Archiv für Begriffsgeschichte* 4: 83–112.

Böhme, Robert. 1986. *Die verkannte Muse. Dichtersprache und geistige Tradition des Parmenides.* Francke: Bern.

Bröcker, Walter. 1965. *Die Geschichte der Philosophie vor Sokrates.* Frankfurt am Main: Klostermann.

Delcourt, Marie. 1955. *L'oracle de Delphes.* Paris: Payot.

Detienne, Marcel. 1994. Manières grecques de commencer. In *Transcrire les mythologies*, ed. Marcel Detienne. Paris: Albin Michel.

Diels, Hermann, and Walther Kranz. 1951. *Die Fragmente der Vorsokratiker griechisch und deutsch.* Sixth Edition. Dublin/Zurich: Weidmannsche.

Großmann, Andreas. 1996. *Spur zum Heiligen. Kunst und Geschichte im Widerstreit zwischen Hegel und Heidegger (Hegel-Studien Beiheft 3).* Bonn: Bouvier.

Heidegger, Martin. 1959. *An Introduction to Metaphysics.* Trans. Ralph Manheim. New Haven: Yale University Press.

Heidegger, Martin. 1967. Hegel und die Griechen. In *Wegmarken.* Frankfurt am Main: Klostermann.

Heidegger, Martin. 1971. *On the Way to Language.* Trans. P.D. Hertz. New York: Harper & Row.

Heidegger, Martin. 1975. *Early greek thinking.* Trans. D.F. Krell and F.A. Capuzzi. New York: Harper and Row.

Heidegger, Martin. 1977a. *The Question Concerning Technology and Other Essays.* Trans. William Lovitt. New York: Harper and Row.

Heidegger, Martin. 1977b. *Vier seminare.* Frankfurt am Main: Klostermann.

Heidegger, Martin. 1984. *Grundfragen der Philosophie: Ausgewählte "Probleme" der "Logik"* Gesamtausgabe (GA) 45, Friedrich-Wilhelm von Hermann. Frankfurt am Main: Klostermann.

Heidegger, Martin. 1994. *Bremer und Freiburger Vorträge (gA79)*, ed. Petra Jaeger. Frankfurt am Main: Klostermann.

Heidegger, Martin. 1996. *The Principle of Reason.* Trans. Reginald Lilly. Bloomington: Indiana University Press.

Held, Klaus. 1970. Der Logos-Gedanke des Heraklit. In *Durchblicke. Martin Heidegger zum 80. Geburtstag*, 162–206. Frankfurt am Main: Klostermann.

Humboldt, Wichelm von. 1967. Latium und Hellas. In *Werke in fünf Bänden*, vol. II, ed. A. Flitner and K. Giel. Darmstadt: Wissenschaftliche Buchgesellschaft.

Jamme, Christoph, and Karsten Harries (eds.). 1992. *Martin Heidegger: Kunst, Politik, Technik.* Munich: Fink.

[33]This essay was originally presented at the 2001 meeting of the Heidegger Circle convened by Babette Babich, Fordham University in New York City, on the 25th anniversary of the question Heidegger offered on April 11th 1976 to the meeting of the Heidegger Circle in Chicago.

Jünger, Friedrich Georg. 2010 [1939]. *Die Perfektion der Technik*, 8th ed. Klostermann: Frankfurt am Main.
Kahn, Charles. 1979. *The art and thought of Heraclitus: A new arrangement and translation of the fragments*. Cambridge: Cambridge University Press.
Kockelmans, Joseph J. 1985. *Heidegger and science*. Washington, DC: University Press of America.
Lacoue-Labarthe, Philippe. 1987. *La fiction du politique*. Paris: Bourgois.
Liesenfeld, Cornelia. 1992. *Philosophische Weltbilder des 20. Jahrhunderts. Eine interdisziplinäre Studie zu Max Planck und Werner Heisenberg*. Würzburg: Königshausen und Neumann.
Macann, Christopher (ed.). 1996. *Critical Heidegger*. London: Routledge.
Marcovich, M. (ed.). 1967. *Heraclitus (Editio maior)*. Venezuela: Los Andes University Press.
Milchman, Alan, and Alan Rosenberg (eds.). 1996. *Martin Heidegger and the Holocaust*. Atlantic Highlands: Humanities Press.
Müller, Severin. 1976. *Vernunft und Technik: Die Dialektik der Erscheinung bei Edmund Husserl*. Freiburg: Alber.
Nagy, Gregory. 1996. *Pindar's Homer: The lyric possession of an epic past*. Baltimore: Johns Hopkins University Press.
Plato. *Symposium*.
Plutarch, *De Pythiae oraculis*.
Pöggeler, Otto. 1994. *Der Denkweg Martin Heideggers*, 30th ed. Pfullinger: Neske.
Rockmore, Tom. 1992. *On Heidegger's Nazism and philosophy*. London: Harvester Wheatsheaf.
Rockmore, Tom, and Joseph Margolis. 1992. *The Heidegger case*. Philadelphia: Temple University Press.
Schmid, Holger. 1999. *Kunst des Hörens*. Cologne and Weimar: Böhlau.
Schmitz, Hermann. 1996. *Husserl und Heidegger*. Bonn: Bouvier.
Schürmann, Reiner. 1987. *Heidegger on being and acting: From principles to anarchy*. Bloomington: Indiana University Press.
Seubold, Günter. 2000. Martin Heideggers Stellungnahme zu Jüngers 'Arbeiter' im Spiegel seiner Technikkritik. In *Titan Technik: Ernst und Friedrich Georg Jünger über das technische Zeitalter*, ed. Strack, Friedrich, 119–132. Würzburg: Königshausen und Neumann.
Theunissen, Michael. 2000. *Pindar: Menschenlos und Wende der Zeit*. Munich: Beck.
Volpi, Franco. 1990. Itinerarium mentis in nihilum. In Ernst Jünger–Martin Heidegger, *Oltre la linea*, ed. 9–45. Milan: Adelphi.
Zimmerman, Michael E. 1990. *Heidegger's confrontation with modernity: Technology, politics, art*. Bloomington: Indiana University Press.

Part III
Philosophical Truth and Hermeneutic Aesthetics

On the Manifold Meaning of Truth in Aristotle

Graeme Nicholson

Abstract When Aristotle treats true and false statements in his logical treatises, he shows that truth and falsity are the pre-supposed, non-discursive grounding for statements themselves. His ethical treatises show that intellectual virtues are constituted by truth. The *Metaphysics* shows that truth in thinking is sustained by the truth of being. All these diverse studies can be connected to one another by way of the Greek term for truth, *aletheia*, as Heidegger has treated it.

It was Heidegger's practice over the years to single out certain words of ancient Greek and assign to them striking and memorable interpretations—we think of his treatment of *physis* and *logos* in the *Introduction to Metaphysics*,[1] where the former is rendered as "emergence" and the latter as "gathering." Equally memorable are his frequent references to *ousia* as "presence" and *alētheia* as "unconcealedness." Sometimes these interpretations were accompanied by etymologizing. But there was always another basis for them as well, a Greek philosophical text whose argument seemed to undergird what Heidegger was saying. So the reading of a particular text was guided in part by etymology, yet, running in the other direction, the reading of the text would add credibility to the etymology, a hermeneutical circle that appeared again and again in Heidegger's many essays on Greek philosophy.

The *Introduction to Metaphysics* had a strong focus on Pre-Socratic thinkers, and earlier in the 1930s Heidegger had devoted courses to Plato. All these writings have been widely studied and interpreted. But in the 1920s, up to 1930, Heidegger devoted many studies to Aristotle that I think have received less attention in the secondary literature. Joseph Kockelmans is one commentator who has devoted attention to this part of Heidegger's work, notably the essay "Being-True as the

[1] Heidegger (1983, 2000). See especially Chapter IV, Sect. 3.

G. Nicholson (✉)
University of Toronto, 350 St. David St. S, Fergus, ON N1M 2L8, Canada
e-mail: graemenicholson@gmail.com

B. Babich and D. Ginev (eds.), *The Multidimensionality of Hermeneutic Phenomenology*, Contributions to Phenomenology 70, DOI 10.1007/978-3-319-01707-5_13,

Basic Determination of Being," which he presented at the Heidegger Conference in 1983 and later published in a book he edited.[2] Kockelmans also calls attention to Heidegger's Aristotle-work frequently in his full-scale interpretation of Heidegger's philosophy.[3]

The topic of truth, or *alētheia*, looms large in Heidegger's Aristotle-commentaries. This is one of the cases that exhibit a circular relationship between textual reading and the investigation of words and etymology. So we are able to put a twofold question to Heidegger's work: does the textual study as such grant support to the etymologizing? And does the study of the word *alētheia* contribute to the philosophical understanding of Aristotle's thought about truth? In particular, there is a question whether Aristotle ever presented a unitary account of truth or whether his scattered comments remain unconnected, an issue that has taxed the minds of many scholars. He has a lot to say about truth in several of the logical writings, especially *de Interpretatione*; he has many comments in different books of the *Metaphysics*; he treats truth in the *Nicomachean Ethics* and in the third book of *de Anima*. When we recognize that these inquiries have different aims, it is hard to avoid the impression that the different treatments invoke quite different ideas of truth. There could be a logical idea of truth, and then a metaphysical idea, an ethical idea and a psychological idea (not to mention even more recondite possibilities). That is why I have adopted the title "On the Manifold Meaning of Truth in Aristotle." I want to explore some of the different references in his text where truth seems to play quite distinct roles; but I am guided by the hypotheses (a) that Heidegger's interpretation of *alētheia* can help us understand each of the distinct contexts, and (b) that it may form a unitary point of reference for them all.

I shall follow the chronology of Heidegger's work in the 1920s, rather than the usual order in which Aristotle's own writings have been arranged.

1 From the Lectures on Ethics: The Many Avenues to Truth

The earliest writings that we have from Heidegger on the topic of truth stem from 1922 when he was working out an interpretation of Aristotle. His research was focussed on the many avenues by which the soul establishes what is true, but of course this must be closely interwoven with an interpretation of what truth itself is. In the late summer of 1922, having just completed a lecture-course on Aristotle offered to students at Freiburg University,[4] Heidegger worked up a prospectus for a book on Aristotle that he sent to several influential philosophy professors.[5] Heidegger indicates what parts of Aristotle's text he proposes to treat, *Nicomachean Ethics*

[2] Joseph J. Kockelmans (1986), 145–160.

[3] Kockelmans (1984).

[4] Heidegger (2005a). Not to be confused with other lecture courses on Aristotle, in *GA* 61, 18 and 22, that will not concern us here.

[5] Heidegger, *Phänomenologische Interpretationen zu Aristotle (Anzeige der hermeneutischen Situation)*, now published as an Appendix to *GA* 62, pp. 343–399.

[N. E.], Book VI, *Metaphysics*, Book I, *Physics*, Books I–V, and *Metaphysics*, Books VII–IX. Later on, this prospectus formed the basis for a much lengthier treatment of those same Aristotle texts in a lecture course that Heidegger gave to the students of Marburg University in 1924–1925 on Plato's *Sophist*,[6] and I shall draw particularly on its treatment of *N. E.*, VI.

In the prospectus, Heidegger is particularly insistent on human finitude. Its first 30 pages highlight such themes as death, care, facticity and fallenness, themes that certainly motivate the question whether we are at all capable of transcendence, of achieving truth. The question is, "What kind of being is being-human such that it is capable of understanding life and being?" (372) Heidegger's question is how truth manifests itself in the midst of such a human life and practice.

It is an ethical and existential inquiry, then, not an inquiry into logic, that motivates Heidegger's original discussion of Aristotle on truth. That already prompts his dissent from a traditional reading, that Aristotle located truth primarily in the judgment, *logos*, and defined it as an agreement or correspondence with its object (*GA* 62, p. 377). The most recent book in English on this topic, by Crivelli,[7] often reaches conclusions similar to Heidegger's, but coming by an utterly different route, from logical studies. The most basic difference is that Crivelli leaves the *N. E.* passages entirely out of the discussion, for they were not based on logic (p. 40). This illustrates the problem that I introduced at the beginning of this paper—the question of a unity to be found among Aristotle's distinct inquiries into truth.

N. E., Book VI, opens with Aristotle's articulation of the human soul. Books II to V dealt with virtues of character, but Book VI treats a topic that is partly foundational for the virtues of character and partly of independent importance: the virtues of thought. On the one hand, virtues of character depend on certain norms of "correct reason," *orthos logos*, and so Aristotle must now examine what qualities of thought are required to define those norms. On the other hand, our power of thought is in some ways a higher function of the soul than moral practice, so that the virtues of thought are valuable in their own right. Practical thinking and decisions aim at what is good for us, but in the thinking that is not enlisted for practice and for decisions "the good and the bad state consists [simply] in being true and false" (1139a 28). The virtues of thought enable the soul to attain this truth (and, as well, of course, the truth that defines the correct norms for moral practice). If we read Book VI in the context of the whole treatise, especially Books I and X, it is clear that truth is an end or *telos* of the soul, one component in the happiness, *eudaimonia*, that constitutes the end of human life, in particular that end which the soul achieves in so far as it is qualified by one of the virtues of thought. And since thought, *dianoia*, has many forms and applications, we shall find several virtues of thought, that is to say, several pathways to truth.

The intellectual virtues (*aretai tēs dianoias*) are five in number: craft, science, practical wisdom, theoretical wisdom and intelligence (*technē, epistēmē, phronēsis, sophia, nous*). Throughout the treatments of them, Aristotle repeatedly shows their links to *alētheia*, truth. The issue of principle that has to be kept in the forefront is

[6] Heidegger (1992, 1997).

[7] Paolo Crivelli (2004).

that science (or knowledge), *epistēmē*, is only one avenue to truth and in no way pre-eminent. Each of the virtues is differentiated according to the things it treats of. But Heidegger takes this further—he aims to show that in each virtue there is a differentiated role for *alētheia*: both in 1922 and 1924, the virtues are divided in view of the truth that constitutes them. Now we come to the core idea. *Alētheia* is referred to in a number of different grammatical forms. There is not only the abstract noun *alētheia* (e.g., 1139a 18, 1139b 12); there is the adjective *alēthēs*, "true" (e.g. 1139a 24, 1140a 10, a 21, b 5, b 21) and, of special interest here, the verb *alētheuō* (1139b 13, b 15, 1141a 3, a 18), which stands at the centre of Heidegger's exposition. The verb certainly signifies an achievement of truth, an access to truth, but it is a challenge to understand and translate the verb. At the opening of Chapter 3 (1139b 15), Aristotle characterizes the group of intellectual virtues as a whole: it is by way of them that the soul *alētheuei* through affirming or denying. Heidegger renders this verb as *erschließen*, "disclose." He translates: "ways in which *das menschliche Dasein als Zu- und Absprechen das Seiende erschließt* (*Soph.*, p. 21): "...how human *Dasein* through affirming or denying discloses that which is." On his interpretation of *alētheuein*, then, there is the disclosing accomplished by us, and then that-which-is that becomes disclosed. Therefore Heidegger says, "Truth [*alētheia*] is a character of what is, insofar as it is encountered, but in the proper sense it is a determination of the being of human *Dasein* itself" (*Soph.*, p. 23).

And this same understanding of the verb is then carried through in the accounts of all the five virtues. We shall try to follow first how Heidegger treats the various modes of *alēthēs*, *alētheia* and *alētheuein* in each of the five virtues, then turn to the question of the general meaning of the term.

Chapter 3 gives Aristotle's account of *epistēmē,* science, differentiated primarily by the domain for which it is competent. As Heidegger reads the text, science is able to disclose that which does not change, which must necessarily be as it is; and so we could reckon geometry and theology as *epistēmai* by virtue of their characteristic objects. The translation *erschließen*, "disclose," treats *alētheuein* as an active and transitive verb. What Heidegger stresses is that such disclosing must range over all times, its object being reliable and constant; it brings to light that which does not change. *Epistēmē* reveals what is absent just as much as what is present, because its insights are valid for all times and places: its object, once disclosed, is *never* concealed from it (p. 32). In the second place, *epistēmē* is that disposition of the soul that is demonstrative [*hexis apodeiktikē*], showing out of the first and highest grounds what must be the case: and this makes possible a disclosing of the object to others. Still, though science may be taught, the practice of research is also a discovering or revealing—the *logos* of science need not be uttered aloud.

The status that Aristotle assigns to *epistēmē* remains of interest even in a modern or post-modern age: we can cite Kockelmans's important paper "On the Problem of Truth in the Sciences."[8] He is able definitely to vindicate truth in the sciences,

[8] Presidential Address delivered before the Eighty-Third Annual Eastern Division Meeting of the American Philosophical Association in Boston, Massachusetts, December 29, 1986.

but this is truth that can be identified only with empirical adequacy. Science cannot claim to embody an absolute, metaphysical idea of truth that would commit us to an ontology of realism. There are of course differences between this view and Aristotle's, but the parallel is that in Aristotle's *Ethics* we see that *epistēmē* and its mode of truth fall short of the achievements of *nous* and *sophia*—to be discussed below.

Technē, craft, is a disposition that functions in the domain of making [*poiētikē*]. But this too is a disclosing. The temporal character of craft is different from that of science, for it works on something that is yet to be. But it is a trained excellence in making, with a *logos* of its own and a *logos alēthēs* at that (1140a 10, a 20). Heidegger's particular emphasis in these pages, 40–47, is to interpret the *logos* that is proper to craft as the grasp of that Idea or Form, *eidos*, that guides the technician e.g., a house-builder, in the process of making. This *eidos* is what *technē* discloses, both projectively, in advance of the work, and in the finished work as well. He refers to a parallel passage in the *Metaphysics*, 1032a 26-b 12:

> …from craft proceed the things of which the form is in the soul of the craftsman…health is the definition in the doctor's soul, i.e., the craft of medicine…the process towards health is called a 'making.' Therefore it follows that in a sense health comes from health, and house from house: that with matter from that without matter; for the medical craft and the building craft are the form of health and of the house…

Practical wisdom, *phronēsis*, is a disposition that is truthful, i.e., revelatory [*hexis alēthē*—1140b 5, b 20–21]. One who acts with reason, *logos*, in all matters central to human life, in what is good and bad, is the guide to all who aspire to moral virtue. What this disposition discloses is not houses and knives, but human life itself (Heidegger, *das Dasein*), life in its central possibilities, especially though deliberation. We tend in everyday life to conceal from ourselves this good and bad through being too much swayed by pleasure and pain—thereby we become opaque to ourselves (Heidegger, 50–52). *Phronēsis* brings transparency to our lives.

We conclude by treating *nous* and *sophia* together (*N. E.* VI, Chapters 6, 7; Heidegger, pp. 57–64). It is by *nous*, says Aristotle, that we *alētheuomen* those highest grounds that science was unable to secure for itself. What is characteristic of it is that it employs no *logos*—it is what Heidegger calls a pure *Vernehmen*, apprehension, but Heidegger thinks Aristotle reserves this properly speaking to God, a point on which one might challenge Heidegger. What is available to some human beings is *sophia*, which reveals to them, *alētheuein*, both what follows from fundamental principles as well as the principles themselves. Here Aristotle is preparing for a difficult theme—to what extent the highest wisdom really is open to human beings. Heidegger will spend a long chapter, pp. 132–188, on that question and the related question whether *phronēsis* or *sophia* can claim the status of the highest wisdom. The topics of *nous* and *sophia* will return in the *Metaphysics*: Books Alpha (I) and Lamda (XII) treat philosophical wisdom, and we have a lengthier study of the role of *nous* in Book Theta (IX) which of course in its fullest extent incorporates *epistēmē* so as to constitute *sophia*. See below.

2 The Word *Alētheuein*: A Common Concept?

If we start from the verb, as it seems Aristotle did, it is hard to imagine any translation of *alētheuein* except "disclose, reveal." As for the general statement that we cited from the opening of Chapter 3 (1139b 15), the typical renderings of this line into English use the noun "truth," and then supply a verb to anchor it: e.g., "…the soul grasps the truth…"[9] It seems that for an English translator, the verbal term is always secondary to the substantive term "truth;" we are hard pressed to find a verb from which the substantive term could then be an offshoot. But "reveal, disclose" does fill that role. Crivelli maintains that no English verb phrase renders the verb adequately.[10] Some writers are forced to invent words for this translation.[11] But this verb is found in a number of other treatises of Aristotle too, e.g., *Metaphysics*, 1011b 28, 1012a 3, a 4, a 6–7, 1051b 15; *de Interpretatione* 17a 3; these texts treat our speaking or saying what is true, in contrast with saying what is false. And since most of the virtues of thought employ *logos*, it might be thought on that basis that our verb does mean "express truth" or "discover truth" where *logos* is the operative agent. But the context of speech is not necessarily involved in the texts we have cited from the *N. E.*—it is the domain of *dianoia*, thought. And in the *Posterior Analytics* 100 b 6, the context is similar to that of the *N. E.*, for it concerns, not speech but the mental powers or thinking states (*dianoian hexeōn*) by which we can engage in *alētheuein*. Even in the *N. E.*, moreover, *nous* does not employ *logos* but engages most pre-eminently in *alētheuein*.

Heidegger treated this word in the Introduction to the *Sophist* lectures. Section 3 gives his etymological rendering of *a-lētheia* as "un-concealment" or "un-coveredness" [*nicht-mehr-verborgen-sein, aufgedecktsein*], p. 16. Thereby, what-is is initially concealed and needs an active intervention to uncover it, which, Heidegger explains, the Greeks thought was generally accomplished through speech, *logos*, p. 17. *Alētheia,* then, is the goal accomplished by an act of *alētheuein*. Heidegger says in the Introduction that he will not translate the verb, but he offers "uncoveringness" [*aufdeckendsein*] as a paraphrase. He summarizes (pp. 17–19): by denying or affirming, we make manifest what-is [*dēloun*] or we let it be seen [*apophansis*] and this achievement of our *logos* is *alētheuein*. But *logos* does not need to be uttered through the tongue; it informs various types of action too, many modes of our life and practice, detailed in *N. E.* VI.

Now, to analyze this further: (a) Heidegger treats *alētheia* as a composite word, a negative word, an alpha-privative added to a root *-lēthē*; (b) it appears as adjective, noun and verb; (c) the roots *-lēthē*, *-lath-* and *lanthanomai* signify "be hidden," "concealment," and "forgetting"; (d) the composite word signified "unconcealment," rendered by a number of different German words, and therefore English

[9] Terence Irwin, Hackett Books. Some other versions: "…the soul possesses truth…" (W. D. Ross, Oxford translation); "…the soul expresses truth…" (Martin Ostwald, Library of Liberal Arts).

[10] *Op.* cit, p. 51, note 24.

[11] E.g., Theodore Kisiel (1993), 250, writes "the soul trues," but that violates English too much.

ones as well, varying from "revealedness" to "unveiling" to "disclosing;" (e) this affords the concept of truth in the Greek language and Greek philosophy; and in conclusion (f) when *alēthēs* is attributed to the soul, to one of the intellectual virtues or to a statement, it means "true," but when attributed to the object it means "unconcealed." So truth arises through tearing off a veil; truth was always preceded by a concealment enshrouding the subject. And most centrally, unconcealment always preserves some relationship to concealment; it does not abolish it utterly; there is some sort of continuity between them, and, in consequence, some continuity between truth and untruth. This is no antiquarian point—Greek philosophy with its concept of truth is to be exemplary for us and our philosophy.

In his phenomenological works, especially *Being and Time*, Heidegger couldshow that there are philosophical grounds for understanding truth as disclosure or uncovering or revealing. The phenomenology can be appraised as it stands independently of Greek scholarship: Heidegger's view is not generated entirely out of his exegesis of the Greek words. But we must still ask whether he is right on the points we itemized above, from (a) to (f). In addition, of course,—and that is the point of this paper—we need to ask what role his understanding of the Greek words plays in his textual exegesis and in his philosophical argument. His work has not stood unchallenged. On (a), Friedländer, in the second edition of his book on Plato[12] said that there was no clear evidence that this was an alpha-privative word; it is likely that "*alēthēs* has nothing to do with *-lēthē*, *-lath-*, *lanth*" (p. 222). Thus we could hardly find "a passage in which the object of the verb could be (let alone must be) the 'unhidden'" (p. 223). Thus Friedländer disposes of points (d), (e) and (f) as well.

But his analysis was refuted by a paper by Heitsch, a Göttingen classicist, that appeared in 1962.[13] In a matter such as this, as he explains, the objective history of word-formation, what is usually called etymology, is not as important as the understanding which the ancient authors (whose texts we possess) had of the relationship of the words they used, how, for example *they* treated the relationship of *alētheia* to *lēthē*, *-lath-*, *lanth*, etc. And from non-philosophers from Homer on down through the tragedians, historians and rhetoricians he assembles about 20 passages showing that *alētheia is* an alpha-privative word: the word-relations in the texts show that the authors placed it in strong contrast to *lēthē* and its cognates. He then adds about a dozen quotations from Plato showing the same thing, e.g. *Apology* 17a "…they almost made me forget who I was [*epelathomēn*]…and yet they hardly uttered a word of truth [*alēthes*]." The grammarians of later antiquity who codified this relationship were only reflecting the word-use that had prevailed for centuries (Heitsch, p. 26); moreover, this interpretation of *alētheia* has been standard in classical scholarship since Johannes Classen in the 1850s (p.24). Other papers, too, have appeared[14] that establish that *alētheia* does mean truth and that it is a privative

[12] Paul Friedländer (1954, 1958), 222–3.

[13] Ernst Heitsch (1962), 24–33.

[14] For instance, Heribert Boeder (1959), 82–112.

construction. When Friedländer came to issue a third edition of his book,[15] he acknowledged (pp. 234–7, 386–7) that Heitsch had established his case; the reproaches against Heidegger were unjustified.

3 The Truth of Statements

One year after the *Sophist* lectures, 1925–1926, Heidegger offered a lecture course, *Logic: The Question of Truth.*[16] In these lectures, Heidegger makes it plain that he is not concerned with syllogistic logic or symbolic logic but rather a set of questions that he calls "philosophical logic," a discipline that would coincide with what Kant and Husserl called "transcendental logic," dealing with fundamental questions concerning the grounds of knowledge, intentionality and of course truth (*GA* 21, pp. 7–9). In earlier studies,[17] he had dealt with the so-called theory of *judgment* (*Urteilstheorie*) but in the main section of this course he uses the terminology of the *statement* (*Aussage*), in addressing the question of truth. How can a statement be true, or also false? He bases his account (pp. 127–195) on Aristotle's explanations of truth in *de Interpretatione*, especially Chapters 1–4 and *Metaphysics*, especially *Gamma* 7, *Epsilon* 4, and *Theta*, 10. The interests of the *Ethics* seem at first to be very far away. Yet we shall see that his interpretations of the Greek words and the Aristotle texts bring the inquiries closer together.

Logic is the study of the *logos*; if we call this provisionally a sentence, we soon see that the particular kind of *logos* that makes a declaration, the *logos apophantikos*, is Aristotle's theme in the treatise: this is the *Aussage*, the statement (*de Int.* 17a 1–4). It has a crucial but intricate relationship to the True and the False. The treatise began by clarifying that a single verb or noun could not be true or false, but only a combination or separation of those two; a mere noun ("goat-stag") may *signify* something, but is capable of *truth or falsity* only where a verb is added, and that could be just the verb for "being," e.g., "is" or "is not." Chapter 4 narrows the discussion down further to the *statement*, employing the criterion of being possibly true or possibly false. A request, an order, may be expressed in a sentence (*logos*) but is not a statement (*apophantikos*) because there is no *alētheuein* or *pseudesthai* in it (as Heidegger makes plain, p. 130, many other kinds of utterance too fall away by

[15] Friedländer (1964). Not yet translated.

[16] Heidegger (1976, in English 2010). Joseph Kockelmans takes his start from this lecture course in his book *On the Truth of Being*; he also treats it in the paper we mentioned that he published in 1986.

[17] Heidegger's researches on logic began with his doctoral dissertation of 1913 on the psychologistic current of the nineteenth century, continued with his habilitation thesis of 1916 on Medieval logic, with the closest engagement with Neo-Kantian and phenomenological logics and epistemologies coming in the years before and during his teaching activity in Freiburg and Marburg. In the years after 1926, we have many treatments in logic in *GA* 24, *GA* 26, *GA* 38 and *GA* 45. An excellent orientation can be found in Mohanty 1988.

this criterion: questions, wishes, and so on). In pp. 127–135, Heidegger derives from Aristotle's text two implications that are of the greatest consequence for logical theory. First of all, it is the *double* possibility, being true *or* being false, that counts for this view of the statement; it is not sufficient to observe that a statement is what is true. Since bivalence (*T* or *F*?) is the essential mark of a statement, it follows that, in logic, truth and falsehood belong under the head of *possibility* (p. 129). That point is also noted by Crivelli (p. 85). The second implication brings a corrective to the common assumption in logical theory that the statement is the home, or location (*Ort*), of truth (and falsehood), i.e., that "True" or "False" are to be *predicated* of statements, or that the statement is a truth-bearer. Aristotle has shown, on the contrary, that (possible) truth and (possible) falsehood are the home, or location, of the statement: *Satz ist nicht der Ort der Wahrheit, sondern Wahrheit der Ort des Satzes* (p. 135). This emerges from the precise words that Aristotle uses at 17a 2–3,[18] which Heidegger understands (p. 129) to mean that it is truth that defines the statement, not vice-versa.[19] Logical theory errs by considering the statement to be a truth-bearer; rather we need to recognize that the true and the false are statement-bearers. This relationship can be comprehended by virtue of the special senses of the Greek words in the text. The verb *alētheuein* means "uncovering, removing the concealment of something" and its contrary *pseudesthai* means "deceiving, covering up" (pp. 131–2). And, moreover, the operative term for the statement, *apophantikos*, *apophansis*, means "letting something be seen" (*aufweisen, sehen-lassen*). We are then able to say that only the discourse that uncovers or covers up achieves a letting-be-seen.[20] We grasp what is apophantic out of the double possibility: true or false, revealing or concealing. Though Crivelli recognizes the point about possibility, he remains with the logical theory that sees statements—or sentences, in his terminology—as truth-bearers (p. 87; not the only truth-bearers, to be sure, for he includes objects, *pragmata*). Heidegger goes further (as he will do in *SZ*, Sec. 33) in deriving from the apophantic character of the statement its further functions as predicating and communicating.

A result of this reading of *de Int.* is that Aristotle has left truth itself (and falsehood) undefined in this treatise. We must look at the texts where he seems to define them, in the *Metaphysics*.

Met. Gamma 7 has always been taken as *locus classicus* on this point. In the Oxford translation, it reads

> To say of what is that it is not, or of what is not that it is, is false, while to say of what is that it is, and of what is not that it is not, is true; so that he who says of anything that it is, or that it is not, will say what is true or what is false (*alētheusei ē pseusetai*)—1011b 26–8.

[18] *Apophantikos ou pas, all' en hō to alētheuein ē pseudesthai hyparchei.*

[19] "Der Satz ist definiert mit Rücksicht auf Wahrheit und nicht umgekehrt,Wahrheit kommt vom Satz her."

[20] "*Aufweisend sehen lassen (Aussage) ist nur das Reden, darin das Entdecken oder Verdecken vorkommt*" 132. Several variants of this point appear 133–5.

Heidegger treats this text on pp. 162–170. There are several critical points he emphasizes. First, that we must reject any relation of copying or picturing [*Abbild*] between the statement and that which is. In part, this is based on his understanding of *legein* (to say) as the apophantic letting-be-seen, the view that already guided his reading of *de Int.* He is particularly intent on warding off any theory that puts an intermediate picture in between the statement and that which is: the statement is *directly* engaged with the thing, no matter how far away it might be. A causal relation such as picturing is out of the question. He claims that the *Abbild* idea arose through a faulty application of *de Int.*, Chapter 1; that chapter dealt causally with perception, vocalic sounds and writing, but it did not seek to explain truth thereby.

Secondly, and this point is connected, he denies that we can assign a "correspondence theory of truth" to this passage of Aristotle. Certainly the statement stands in relation to what-is, but its relation is the active one of *entdecken* or *verdecken*—it reaches out to the thing itself, either to unveil it or to veil it. Here the character of truth and falsity as possibility shows its effect; the statement may be true *or* false. The searchlight can reveal an airplane in the sky, though it doesn't resemble it; the searchlight can also reveal the absence of airplanes. Those two possibilities can appear as prototypes for positive and negative assertions. The searchlight cuts through, and overcomes, the darkness that has concealed airplanes, or, as it may be, that has concealed the absence of airplanes. But the searchlight can miss a plane that is already there shrouded in the darkness, and this is a prototype for false assertion. The action of uncovering cannot be a copying or correspondence because it incorporates as its permanent starting point the cover, the veil; it has in that way a necessary connection to possible untruth.

In the same section, p. 164, Heidegger treats another *locus classicus* from *Met. Epsilon* 4:

> The true judgment affirms where the subject and predicate really are combined, and denies where they are separated, while the false judgment has the opposite of this allocation (1027b 20–22).

The grammatical terms "subject" and "predicate" are not in the Greek text, which speaks of "that which is combined" and "that which lies separated," referring to the state of the things, not of the words. Heidegger calls attention to the circumstance that combination-or-separation (synthesis and diairesis) are found not only in the *logos* but also in the things (*en tois pragmasin*—b 26), so that the truth or falsity of the *logos* (with its combination of terms) is grounded in the state of the things. But the chapter also adds a difficult modification, hard to reconcile:

> …for falsity and truth are not in things but in thought…the combination and the separation are in thought and not…in the things… (1027b 25–31)

This idea has also prompted great controversy regarding the last chapter of Book Theta and so we shall take the point up in that context. Since Chapter 10 of *Theta* introduces another dimension of truth altogether, we shall discuss the *Epsilon* text along with it. Heidegger devotes 12 pages here to *Met. Theta* 10, but 5 years later, in a 1930 summer semester course called *The Essence of Human Freedom: An Introduction to Philosophy*,[21] he gave a much fuller treatment (40 pages), guided by the same questions, so we shall cite the later text.[22]

4 Being as Truth

Sec. 44 of *Being and Time* was devoted to truth, and it began by documenting the ancient connection between the question of being and the theme of truth, so that at certain points in Aristotle the two virtually coincide. Heidegger quotes from the *Metaphysics,* Book *Alpha*, 983b 2 to show that while the "science that we are seeking" is concerned above all with being, (see also *Gamma*, 1003a 21), it can also be said in the very next line that it is the study of truth, *alētheia*—983b 3 (see also 993b 17; 984b 10; 993b 17 and 20—all cited in the first paragraph of Sec.44, quotations from the first two books of the *Metaphysics*). Some of Aristotle's remarks are applied to his predecessors who were investigating the causal powers of fire and earth, for instance (984b 10), or generally what causes and principles operate in the world (988a 18). He repeatedly refers to these as inquiries into *alētheia*. But such inquiries were not focussed on epistemological questions—what the early philosophers sought was not a theory of judgment, as Heidegger underlines (p. 213). *Alētheia* is interchangeable with being or nature in Books *Alpha* and *Alpha Elattōn*, and indeed W. D. Ross regularly renders it "reality" in his Oxford translation of these books. The first two books of the *Metaphysics* speak of wisdom, *sophia* or philosophy, as something sought or aimed for, and the same holds for what this wisdom would attain—the truth is what philosophers have always been seeking. This provides a reason for inquiring into that text of the *Metaphysics* that explores the convergence of being and truth most thoroughly—*Theta*, Chapter 10.

Although Crivelli does not follow Heidegger in understanding truth as unconcealedness, he does recognize that, for Aristotle, truth holds not only for statements and thought, but also for what is (pp. 46–62); as we proceed, we'll review some of what he has to say about the truth of that which is.

The opening of Book *Gamma* and of *Epsilon* pointed us towards a discussion of being, *to on*, and a good part of that inquiry has been accomplished in Books *Zeta, Eta* and *Theta*. They showed how being is realized principally as substance, *ousia*, in relation to the subordinate categories; and then they showed how being is realized supremely as actuality, *energeia*, in relation to potentiality. But then, to the surprise of many commentators, the end of *Theta*, Chapter 10 proposes to discuss being "in the *pre-eminent* sense as the true and the false," *to de kuriōtata on alēthes ē pseudos*, 11051b 1–2. Unlike some commentators,[23] Heidegger accepts the traditional text

[21] Heidegger (1982, 2002).

[22] This is also the main source for the Kockelmans essay, mentioned above, of 1986.

[23] Heidegger himself refers to Jaeger (1948), treats *Theta* 10 on pp. 204–5, though in earlier works he treated it even more critically; Ross (1924), 2 volumes, commentary *ad loc*; and a number of older German scholars. I'll add a remark later on about Crivelli.

(see pp. 82–87). We'll treat some of the controversy about it after an overview of Heidegger's understanding of the whole chapter.

The lines we quoted just above from *Epsilon* 4 (1027b 20–22) are closely echoed here in 1051b 3–5:

> Whoever holds that which is divided to be divided, and that which is combined to be combined, says the truth, and what is false is to reverse this relation that is in the things.

But *Epsilon* 4 seemed to infer from this point that the topic of truth did not pertain to a discussion of being. The present chapter, however, takes a different view: not only does the topic of truth pertain to being—it is the *pre-eminent* sense thereof. Commentators have sometimes found great difficulty in this juxtaposition. But Heidegger undertakes a fundamental distinction (pp. 90–1, 105–6):

(i) What is under discussion in *Epsilon* 4 is the truth of the statement, or truth in thinking, *en dianoiai*.
(ii) But *Theta* 10 is treating *being* as truth, as was proclaimed in the opening lines of Chapter 10 b 1–2. Thus there *is* truth and falsehood in the things, *epi tōn pragmatōn*, b 2.

Heidegger is putting to use his double understanding of *alētheia*, as set out in pp. 87–9. A statement or a thought may engage in *alētheuein*, *entdecken*, disclosing, revealing (*Epsilon* 4). But there is also an *Entdecktheit, Unverborgenheit*, unconcealedness, that is proper to things or beings—*ein Charakter des Seienden selbst*, p. 88 (*Theta* 10). Therefore this chapter brings Book *Theta* to an appropriate conclusion: the theme of actuality that occupies Aristotle in this Book reaches its culmination in the insight that the pre-eminent realization of actuality is in the unconcealedness of what-is. Though Crivelli does not explore the meanings of the word *alētheia*, he is able to reconcile *Epsilon* 4 with *Theta* 10 in a way that is very much like Heidegger's (pp. 62–66).

At 1051b 6–9, Aristotle says:

> You are not pale because we truly think you are, but because you are pale we who say this are telling the truth.

There is a combining of Pale with Face in the human being that deserves to count as truth, one expression of *alēthes on*; this is the grounding for the statement that *alētheuei* (Heidegger, p. 91). The question that Aristotle, according to Heidegger, p. 92, is posing at b 5–6 is: How is it that some being or thing should be true? And Aristotle's answer is that this will depend upon the character of the being of these things.

Aristotle sums up (b 9–15) different kinds of circumstance: some things are always combined (triangle's angles = two right angles), some things never (the diagonal is never commensurable), some things vary in that respect (face may be pale or sunburned). In the first class is being; in the second, not-being; in the third is the accidental. But the main line of Heidegger's interpretation becomes clear from p. 99 onwards: he sees a steady mounting-up in this chapter. Merely accidental things are almost untrue, almost equivalent to not-being (on p. 95, Heidegger quotes a text from *Epsilon* 2 to that effect—1026b 21). The things that are necessarily connected, that cannot be other than they are, are more true than they. The uniting power at work in yielding the nondivision in such a thing is the presence, the *parousia*, of one feature with another. What bestows the power, what is unitive, is the being of beings. So we read, p. 92, that an entity acquires this truth owing to its very being, so that being-true is a character of the being, *einai*, of beings, *ta onta*. And then, at the head of this series stand the things that are incomposite, *asyntheta* b 17–35; they are even more indivisible than the things that exhibit necessary connections. They are utterly simple. This group of things is identified by Heidegger with the principles and grounds of all things [*archai kai aitiai*], citing a few texts from other books of the *Metaphysics* and *de Anima* (pp. 99–104).[24] (He does not connect them, as commentators generally do, with the intelligences that move the heavenly spheres.) Still, the point of this mounting series, ascending to subjects that are utterly simple, is an increasing gain in truth, and equivalently an increasing gain in being. The simplicity of the principle is such that it can only be apprehended simply, what Aristotle here and elsewhere calls a touch, *thigein*, not by a judgment of *dianoia* (in other texts, e.g., *N. E.* VI, Aristotle attributes this grasp to *nous*). The simple has the highest form of truth in that it is not vulnerable to mistakes: you see it or you don't. Correspondingly, such subjects have the highest form of being: pure actuality, *energeia*, without any potentiality (Heidegger maintains, p. 100, that Aristotle infers these subjects' perfection of being from their perfection of truth, not the other way around).

It is apparent that Heidegger's way of reading *Theta* 10 depends on his reading the word *alēthes* at b 23 and b 33 in an ontological rather than logical sense, i.e., attributing it to these subjects as their unconcealedness and not to our *alētheuein* through *nous* (see pp. 104–9). Only that permits him to endorse the statement that stood at the head of the chapter, that being qua truth is the pre-eminent sense of being. On his reading, this unconcealedness is the highest exemplification of the being of beings that *Theta* has taught us to understand as *energeia*, actuality. Ross,[25] on the other hand, reads the text at b 23 as referring to our *apprehending* of the incomposite things through *nous*, while reading b 33, in accord with Heidegger, as referring to the truth of the being of those things. Kockelmans follows Heidegger in both readings (pp. 153–156). To conclude: it might be possible to construe both these occurrences of *alēthes* non-dichotomously, as comprehending both the truth of our *nous* and the unconcealedness of the things. For there is a suggestion that the simplicity and actuality of the things bring about the mode of simple apprehension practiced by *nous*. It seems that Crivelli too (pp. 64–66) is able to assign truth in these passages both to thoughts and to things.

All along, Heidegger has been resisting the readings of Ross and Jaeger, the former excising the word *kuriōtata* from the text, the latter emasculating it so as to mean popular or common. Crivelli (pp. 234–7), following an alternative suggestion

---

[24] *Met.Kappa*, 1059b 35; α, 993 b 28 f.; *de An.* 430a 26, b 6.

[25] *Op. cit.*, II, pp. 276–7.

of Ross and others, wants to connect the *kuriōtata* not with *to on* but with *alēthes*. Thereby, Aristotle would be saying not that truth constitutes the pre-eminent sense of being, but rather that the pre-eminent sense of truth is identified with being. But I believe that this is not the natural, grammatical way to read the line in question. Moreover, little is gained by such a shift. If we have understood that the actuality of a being is expressed in its unconcealedness, it does not matter very much whether a high rank is assigned to the one or to the other.

5 The Aristotelian Equivocals

A pervasive feature of Aristotle's philosophizing is his attention to things which, while they differ in certain ways, have the same name and not by mere accident but because of some common feature—to identify which is a worthwhile undertaking for the philosopher. They are called "things said in different ways" [*pollachōs legomena*], and "equivocals" [*homōnyma*]. The whole of *Metaphysics Delta* is devoted to 30 of them. The most famous of them is being, *to on*, treated in *Delta* 7 in its four senses. As for the character of this equivocity itself, the most famous explanation is in *Gamma* 2:

> Everything which is healthy is related to health, one thing in the sense that it preserves health, another in the sense that it produces it, another in the sense that it is a symptom of health, another because it is capable of it. And that which is medical is relative to the medical art, one thing being called medical because it possesses it, another because it is naturally adapted to it, another because it is a function of the medical art (1003a 34-b 4).

It has emerged that there are different kinds of equivocity, the largest and most important branch being generally known today as "equivocity *pros hen*," or focal meaning, where medicines, foods, exercises and bodies can all be called healthy because each has its peculiar relation to one thing, the state of the organism that is their common focus.[26] Aristotle's doctrine bore very important fruit in the twentieth century owing to the mediation of Franz Brentano's book *On the Several Senses of Being in Aristotle*.[27] It came into the hands of the young Heidegger around 1907, starting him on his way. Brentano was alert to Aristotle's intention to establish substance, *ousia*, as the focal meaning of "being," but Heidegger's later testimonies make it abundantly clear that it was the *question* posed by the many senses of "being" that mattered to him more than the answer.

Aristotle did treat "the false" (*to pseudos*) as such an equivocal in *Delta* 29, a text earlier than *Theta*. He shows three kinds of subjects that can be false: false things, *pragmata*, false accounts, *logoi*, and false persons, *anthrōpoi*. It would not seem that the criteria are exactly the same for the three cases, and we would not expect them to be so if this is an equivocal; nevertheless, it is clear that in all three cases the

[26] See Joseph Owens (1963), Chapter 3.

[27] Brentano (1862, 1975).

contrary is the True. The terms treating the false things resemble some lines of *Theta* 10; some aspects of the false *logoi* recall *de Int*; and there is even a faint echo of the *N. E.* in the section on false persons. So it does seem that there *could* be a mirror image of this text: Truth as an Aristotelian equivocal. Though such a treatment does not exist, the argument of this paper is that such a treatment could be organized around *alētheia* as the focal meaning for logical, ethical, metaphysical and psychological truth.

References

Aristotle. 1941a. *Metaphysics*. Trans. W. D. Ross. Oxford: Oxford University Press.

Aristotle. 1941b. *de Anima*. Trans. J. A. Smith. Oxford: Oxford University Press.

Aristotle. 1925. *Nicomachean Ethics*. Trans. W. D. Ross. Oxford: Oxford University Press.

Aristotle. 1962. *Nicomachean Ethics*. Trans. Martin Ostwald. New York: Library of Liberal Arts.

Aristotle. 1999. *Nicomachean Ethics*. 2nd ed. Trans. Terence Irwin. Indianapolis: Hackett.

Boeder, Heribert. 1959. Der frühgriechische Wortgebrauch von Logos und Alētheia. *Archiv für Begriffsgeschichte* 4: 82–112.

Brentano, Franz. 1862. *Von der mannigfachen Bedeutung des Seienden nach Aristoteles*. Freiburg i. Br: Herder.

Brentano, Franz. 1975. *On the Several Senses of Being in Aristotle*. Trans. R. George. Berkeley: University of California Press.

Crivelli, Paolo. 2004. *Aristotle on truth*. Cambridge: Cambridge University Press.

Friedländer, Paul. 1954. *Platon*, vol. I. Berlin: de Gruyter.

Friedländer, Paul. 1958. *Plato*. Trans. H. Meyerhoff. New York: Harper.

Friedländer, Paul. 1964. *Platon*. Dritte Auflage, Vol. 1. Berlin: de Gruyter.

Heidegger, Martin. 1976. *Logik: Die Frage nach der Wahrheit, GA* 21, ed. Walter Biemel. Frankfurt am Main: Klostermann.

Heidegger, Martin. 1982. *Vom Wesen der menschlichen Freiheit. Einleitung in die Philosophie*. *GA* 31, ed. H. Tietjen. Frankfurt am Main: Klostermann.

Heidegger, Martin. 1983. *Einführung in die Metaphysik*, *Gesamtausgabe* [*GA*], Vol. 40, ed. P. Jaeger. Frankfurt am Main: Klostermann.

Heidegger, Martin. 1992. *Platon: Sophistes*, *GA* 19, ed. Ingeborg Schüssler. Frankfurt am Main: Klostermann.

Heidegger, Martin. 1997. *Plato's Sophist*. Trans. R. Rojcewicz and A. Schuwer. Bloomington: Indiana University Press.

Heidegger, Martin. 2000. *Introduction to Metaphysics*. Trans. G. Fried and R. Polt. New Haven: Yale University Press.

Heidegger, Martin. 2002. *The Essence of Human Freedom: An Introduction to Philosophy*. Trans. Ted Sadler. London: Continuum.

Heidegger, Martin. 2005a. *Phänomenologische Interpretationen Ausgewählter Abhandlungen des Aristoteles zur Ontologie und Logik*, *GA* 62, ed. Günther Neumann. Frankfurt am Main: Klostermann.

Heidegger, Martin. 2005b. *Phänomenologische Interpretationen zu Aristotle (Anzeige der hermeneutischen Situation)*. In *Phänomenologische Interpretationen Ausgewählter Abhandlungen des Aristoteles zur Ontologie und Logik*, ed. Heidegger, 343–399.

Heidegger, Martin. 2010. *Logic: The Question of Truth*. Trans. T. Sheehan. Bloomington: Indiana University Press.

Heitsch, Ernst. 1962. Die nicht-philosophische *alētheia*. *Hermes* 90: 24–33.

Jaeger, Werner. 1948. *Aristotle: Fundamentals of the history of his development*. Oxford: Oxford University Press.

Kisiel, Theodore. 1993. *The genesis of Heidegger's Being and time*. Berkeley: University of California Press.

Kockelmans, Joseph J. 1984. *On the truth of being*. Bloomington: Indiana University Press.

Kockelmans, Joseph J. 1985. On the problem of truth in the sciences. In *Proceedings and addresses of the American Philosophy Association*. Supplement to Volume 61.1. 5–26.

Kockelmans, Joseph J. 1986. *A companion to Martin Heidegger's Being and time*. Washington, DC: Centre for Advanced Research in Phenomenology.

Mohanty, J.N. 1988. Heidegger on logic. *Journal of the History of Philosophy*, reprinted in Martin Heidegger: Critical assessments, Vol. III, ed. C. Macann, 93–120. London: Routledge, 1992.

Owens, Joseph. 1963. *The doctrine of being in the Aristotelian metaphysics*, 2nd ed. Toronto: Pontifical Institute of Mediaeval Studies.

Ross, W.D. 1924. *Aristotle's metaphysics: A revised text with introduction and commentary*. Oxford: Clarendon Press.

The Twofold Character of Truth: Heidegger, Davidson, Tugendhat

Jeff Malpas

In what circle are we moving here? It is the eukukleos alētheiē, the well-rounded non-concealment itself, thought as the clearing

— Joseph Kockelmans, *On the Truth of Being*, 281

Abstract The concept of truth as aletheia, or 'unconcealment,' is one of the founding concepts in Heidegger's thinking. Yet it also appears to be a concept that is as problematic as it is central. Ernst Tugendhat, in particular, famously criticized Heidegger's identification of truth with aletheia in a way that seems to have led Heidegger eventually to abandon that identification. Beginning with Kockelman's own account of the idea of truth as unconcealment, I want to re-examine the questions at issue here, looking particularly at Tugendhat's criticisms, but also drawing on the account of truth to be found in the work of Donald Davidson. My intention will be to show why it remains the case that aletheia has to be understood as indeed a mode of truth; that understanding this involves understanding a certain transcendental-topological structure as pertaining to aletheia, thereby understanding truth as standing in an essential relation to place or topos; and that the fundamental role played by truth as aletheia does not curtail, but itself constitutes the ground for, genuine questioning or critique.

1

The concept of truth as *aletheia*, translated by Kockelmans as *non-concealment*, or as I shall call it *unconcealment*, is one of the founding concepts in Heidegger's thinking. It is a concept present in his early thought as well as in his later. Kockelmans

J. Malpas (✉)
Distinguished Professor, University of Tasmania, Sandy Bay Campus,
Geography-Geology Bldg, Rm 328, Private Bag 78, Hobart, TAS 7001, Australia
e-mail: jeff.malpas@utas.edu.au

B. Babich and D. Ginev (eds.), *The Multidimensionality of Hermeneutic Phenomenology*, Contributions to Phenomenology 70, DOI 10.1007/978-3-319-01707-5_14,

himself refers to Walter Biemel's claim that taken as a whole Heidegger's thinking has a double focus: being and *aletheia*,[1] and the claim is clearly one with which Kockelmans himself is largely in agreement. It is, however, the same idea of truth that appears here that was famously criticised by Ernst Tugendhat[2] in a way that seems eventually to have led Heidegger to abandon the use of 'truth' to refer to *aletheia*.[3] The idea of truth as unconcealment is thus central, but also apparently, problematic. Indeed, in Tugendhat's analysis, it is not merely that Heidegger's characterization of *aletheia* as a mode of truth is without foundation, but that Heidegger's very deployment of the concept is indicative of the limitation that Heidegger's thinking places on the possibility for genuinely critical engagement. Beginning with Kockelman's own account of the idea of truth as unconcealment, I want to re-examine the questions at issue here, looking particularly to the way Tugendhat's criticisms have played been taken up in contemporary discussion, but also drawing, as I have elsewhere,[4] on the account of truth to be found in the work of Donald Davidson. My intention will be to show why it remains the case that *aletheia* has to be understood as indeed a mode of truth; that understanding this involves understanding a certain transcendental-topological structure as pertaining to *aletheia,* thereby understanding truth as standing in an essential relation to place or *topos*[5]; and that the fundamental role played by truth as *aletheia* does not curtail, but itself constitutes the ground for, genuine questioning or critique.

2

In his 'Introduction' to *On the Truth of Being*, Kockelmans sets out an account of Heidegger's thinking of truth as this is developed in both *Being and Time*, from 1927, and in the essay "On the Essence of Truth," from 1930. While these two works both belong to the period of Heidegger's early thinking ("On the Essence of Truth" usually being taken to mark the beginning of the turn to the later work),

[1] Kockelmans (1984), 1. My own claim is that the focus on being and truth are together encompassed by the focus on *place*.

[2] See Tugendhat (1994), 83–97.

[3] See Heidegger (1972), 69—the original essay is in *Zur Sachen des Denkens* (Heidegger, 1969). Although there has been some controversy as to the extent to which Tugendhat's critique was recognized by Heidegger himself (a controversy briefly discussed by Lafont (2000), 116–117), it seems clear that Heidegger was indeed aware of, and responsive to, the issues Tugendhat raises (as indicated by the 1964 letter from Heidegger to Tugendhat cited by Wrathall (2010), 37–38).

[4] See especially Malpas (1991). Unfortunately, I do not discuss the Tugendhat criticism explicitly here, just as Kockelmans does not discuss it explicitly in *On the Truth of Being*. Although, in hindsight, it would have been useful to have taken up the Tugendhat discussion directly in this earlier work, my failure to do so was partly a function of the fact that those criticisms simply do not have the same salience from a Davidsonian perspective as they may appear to have from the Heideggerian—see my discussion in Sect. 5 below.

[5] See Malpas (2006), esp. Chapter Four.

Kockelmans' presentation indicates how they nevertheless provide the basis for the understanding of truth even as it continues into the later thinking. Heidegger's thinking of truth in these earlier works provides, in fact, the essential preliminary to Kockelmans' reflections on the later thinking.

The view of truth that appears in Heidegger, and which Kockelmans delineates with some care, is a view of truth as essentially *twofold*: truth names both truth *as correctness*—the 'adequation' of sentence to thing or of sentence to world—and it names truth *as unconcealment*. The underlying argument here can be put quite simply, and in a way that need not depend exclusively on the language of either *Being and Time* or "On the Essence of Truth." Truth is conventionally understood as correctness. Yet in order for a sentence to stand in the right relation to that which it is about such that the sentence can be said to be 'correct,' not only must the sentence already have picked out something as that about which it speaks, letting it appear as something in relation to which the sentence can be true or false, but both sentence and thing must already stand in a relation of accessibility to one another. Inasmuch as the sentence allows the thing to appear, so a certain capacity for unconcealment is already given in the nature of the sentence—language, one might say, is already disclosive—but the capacity of the sentence to uncover in this way also depends on that mode of unconcealment that allows the uncovering of both sentence and thing. Truth thus names the correctness of the sentence, and it names the original unconcealment that makes such correctness a possibility.

Although much of Tugendhat's presentation of the Heideggerian account mirrors the position just set out, Tugendhat's critique tends to overlook Heidegger's insistence on truth as indeed encompassing both correctness *and* unconcealment. Consequently, one of the responses that can be and has been made to Tugendhat consists in drawing attention to the twofold character of truth that is at issue here.[6] Yet Cristina Lafont and William H. Smith have argued that not only does Heidegger himself not offer any adequate refutation of Tugendhat's critique, but neither has anyone else, and the reason for this, so they claim, lies in a failure to appreciate the nature of Tugendhat's argument—an argument that is not rebutted merely by an assertion of the twofold character of truth. Thus Smith writes that: "no one has yet formulated a successful reply to Tugendhat because the force of his critique is continually misplaced, and therefore the full-force of his objections remains unaddressed."[7] As Lafont and Smith view matters, the real question at issue, a question that remains even if we accept the distinction between correctness and unconcealment, is why unconcealment should itself be understood as a form of truth? Why, for instance, should we not rather treat the concept of truth as just a matter of correctness, and if we are to take unconcealment as the ground for the possibility of truth, treat unconcealment as something other than truth? Thus with regard to unconcealment as it stands in contrast to correctness, Cristina Lafont asks "what justification and what significance does it have that Heidegger chooses 'truth' of all

[6] See for instance, Wrathall (2010), 35.

[7] Smith (2007), 157.

words, to designate this other phenomenon?"[8] The questions put by Lafont and by Smith may be thought to take on a special significance in the light of Heidegger's own apparent change of position on this matter: to what extent, one might ask, does this change of position arise from an inability to provide the justification after which Lafont asks?[9]

The objection that Lafont and Smith restate in Tugendhat's name depends on the idea that unconcealment lacks a feature that is characteristically associated with truth in its normal usage: its normativity. In its ordinary usage, truth is contrasted with falsity, and any claim to truth is always open to critical assessment, and so to being judged as true or false. Even if we use truth to refer to the way in the appearance of something correlates with the nature of the thing (as when one speaks of a 'true' friend as someone who not only presents themselves as a friend, but who actually is one—truth as genuine-ness), still even this usage seems to operate within a framework in which something can fail to be truthful only in virtue of appearing in a way other than it is, and so in a way that depends upon some notion of 'authentic' and inauthentic' appearance that can be normatively construed.[10] Yet no possible failure of truthfulness, and so no possibility of critical assessment, seems to operate with regard to the truth of unconcealment. In fact, this is already indicated by the simple fact that unconcealment is not a form of 'claiming' or asserting (not even in the derivative sense in which an appearance might be seen to carry some sort of assertoric content), but rather provides the ground on which claims or assertions can be made and be assessed.[11]

---

[8] LaFont (2000), 116. Lafont's query echoes the Tugendhat's questioning concerning: "With what right and with what meaning Heidegger chooses the word 'truth' to characterize his metatranscendental reference back [to unconcealment]," (Tugendhat, 1994, 84).

[9] According to Wrathall, not at all—instead, given the way Heidegger's usage deliberately went against conventional ways of thinking, his apparent change of position was "nothing more than a pragmatic response to the refusal to pay attention to his warnings" (Wrathall, 2010, 37). My own reading largely agrees with Wrathall on this point, although, as will be evident below, I see it as a more problematic response than does Wrathall. Having said this, however, it remains the case that he posing of the original question concerning justification is a useful starting point for inquiring into the matters at issue.

[10] Consequently, one cannot adequately respond to Tugendhat by arguing that the notion of truth as correctness represents only one of a range of possible meanings—although truth may be said to have an application outside of the linguistic according to which truth is understood as 'faithfulness', such a sense of truth can itself be construed in terms of the correlation of word with deed, of promise with fulfilment, of semblance with reality, in a way that also lends itself to being understood in terms of something like correctness (especially as connected with correspondence).

[11] In this respect, it seems to me mistaken to attempt to respond to Tugendhat by arguing that there is a properly normative dimension that operates in relation to unconcealment—something that seems to be attempted by Smith, (2007), 174–177, and also, to some extent (although in a very different way), by Daniel Dahlstrom—see Dahlstrom (2001), 419–423. This is an issue to which I shall return, however, in Sect. 5 below, since although unconcealment cannot itself carry any normative element (since it is what makes normativity possible), this does not mean that the *idea* of truth as unconcealment is beyond normative assessment (essentially the point Dahlstrom contests) nor that we cannot critically engage with *particular modes* of unconcealment (the point Smith takes up).

Given that unconcealment is not normatively or critically constrained in this way, the question then arises, not merely *how* it can be understood as a form of truth, but *why* it should be so considered in the first place. Moreover, at this point, the argument can be seen to have an added bite. Unconcealment seems to function in a way that limits critical engagement—the particular mode of unconcealment that is the ground for any specific practice of assertion cannot itself be subjected to critical questioning. As Tugendhat comments: "If truth means unconcealment…then this means that an understanding of world in general is opened up but not that it is put to the test."[12] In this light, Heidegger's position seems to depend, not on a taking of questioning to the most fundamental level, but rather on a radical limiting of questioning: unconcealment appears as a mode of *not questioning*—a dimension into which questioning does not even enter.

3

The problem presented by Tugendhat appears, according to Lafont and Smith, to be clear and straightforward, and yet, in the literature, so they claim, it remains almost entirely unaddressed or even acknowledged. That such an obvious problem could be so completely overlooked or misunderstood ought to prompt some further query, however, and there is, indeed, more to the situation than is apparent in LaFont's, and especially Smith's, presentation. While both are right to point to Tugendhat as asking after that on which Heidegger's identification of unconcealment with truth is based, and right also to point to the way in which what concerns Tugendhat is the lack of any normative dimension in the idea of unconcealment, they go too far in claiming that this has gone entirely unappreciated in earlier discussions or that attempts have not been made to respond to the justificatory demand at issue here. In the case of Kockelmans, in particular, it seems that there is an awareness of the nature of Tugendhat's basic point, as well as an attempt to respond to it.

Kockelmans does not refer to Tugendhat directly, yet not only does he reiterate the twofold character of truth in Heidegger, thereby reiterating the commitment to a notion of truth as correctness (and so to truth as having a normative dimension), but he also argues explicitly that Heidegger's claim that unconcealment is a mode of truth is not an *arbitrary* claim.[13] Kockelmans thus attempts to provide considerations as to *why* unconcealment should indeed be understood as a mode of truth. Consequently, if one is to reject Kockelmans' account, it cannot be on the grounds that Kockelmans ignores the sorts of objections found in Tugendhat, but must instead depend on viewing Kockelmans' responses to those objections as inadequate or unconvincing. How one assesses Kockelmans' position will obviously depend on how one thinks about the concept of truth that is at issue here. It is all too easy, in fact, for the discussion of this matter to slip into a simple confrontation

[12] Tugendhat (1984), 95.

[13] See Kockelmans (1994), 4.

between opposing accounts of truth, rather than taking the form of a genuine engagement regarding the questions at issue. The underlying question here is thus not simply whether unconcealment is a mode of truth, but given a *prima facie* understanding of truth as correctness, whether this is sufficient as a complete account of truth, and whether what Donald Davidson calls 'the structure and content of truth' is exhausted by an approach that focuses on the normativity of truth as this operates in conjunction with the notion of correctness.

On Kockelmans account, there is no question that truth carries an important normative component that operates at the level of particular sentences and is captured in the notion of truth as correctness. Yet the fact that truth carries such normativity with it opens up the further issue as to the ground on which the normative assessment of particular sentences is itself possible. Moreover, if there are reasons for taking the ground for normativity as itself a mode of truth, then that will mean that there is a mode of truth that is not open to normative assessment in the same way as is the mode of truth associated with truth as correctness. Kockelmans claims that there are such reasons, and thus takes truth to refer both to correctness and to that which is the ground for the possibility of correctness, namely, unconcealment.

Kockelmans rehearses the Heideggerian argument for *aletheia* as that which underlies truth as correctness: the correctness of statements is only possible on the basis of a prior comportment towards beings that allows beings to come forth into the open such that things can be stated of them, which statements may then be true or false (the beings themselves providing the measure of such truth or falsity), and this prior comportment is itself based in truth as *aletheia*—as unconcealment. In addition, however, Kockelmans also makes explicit one further claim, concluding that: "if the correctness (truth) of the statement becomes possible only through the openness of the comportment, then that which makes the correctness first possible must also, and with more original right, be taken as the essence of truth."[14] It is this claim that requires further elucidation.

A key element in the argument for the identification of truth with unconcealment, as Kockelmans understands it, is undoubtedly the idea that the inquiry into essence is identical with the inquiry into that which makes possible. Independently of how we view this idea, it certainly has a lengthy and respectable philosophical provenance. Aristotle's inquiries, paradigmatically set out in the *Metaphysics*, into the first principles that underpin the being of things—the inquiry into what is first substance (*prote ousia*)—clearly depend on the idea that what determines the being of a thing (which might be interpreted, in the language Kockelmans employs, as that which makes it possible) is its essence, and there is a sense (although there remains an ambiguity here also) in which the essence of the thing can bear the same name as the thing whose essence it is. Moreover, that the essence of a thing should indeed be called by the same name that belongs to the thing is certainly not an arbitrary suggestion, but one that derives from the idea that the essence of a thing is what that

[14] Ibid., 8.

thing most properly *is*—so the name may be said to designate, first, the essence, and, secondarily, the thing to which the essence belongs. Truth may thus name correctness, but in its primary sense it names that which is the ground for correctness, and it is this that is unconcealment or *aletheia*.

The mere fact that this argument can be reconstructed, and is indeed a type of argument that seems to be assumed, and briefly alluded to, in Kockelmans account shows, at the very least, that it cannot be correct to claim that there is *no* basis, in the existing literature, for the claim that unconcealment is to be identified with truth. Perhaps the argument at issue is too readily assumed, or presented in too schematic a form, but what is surely at issue is not so much whether there is *some* basis for the claim at issue, so much as whether it is an *adequate* basis. What more can be said, then, to defend the adequacy of the position that Kockelmans advances? In the end, what must be done is to show more clearly the way in which the twofold character of truth does indeed follow even from the idea of truth as correctness. It is here that the account of truth found in Davidson proves particularly useful, providing a perspective that, although very different from that to be found in Heidegger (or in Kockelmans), nevertheless moves towards much the same conclusion. Moreover, although there has been discussion of the apparent convergence between the Davidsonian and Heideggerian accounts of truth,[15] the possible relevance of the Davidsonian account to Tugendhat's objection has been largely unexplored. Before we come to Davidson, however, there is still more to be done in order properly to bring to light what is at issue in the twofold structure that truth presents in Heidegger—this is especially so in relation to an aspect of that structure that is clearly present in Heidegger's early thinking, and that is also recognised by Tugendhat, namely, its *transcendental* character.

4

The twofold structure that appears in Heidegger's account of truth is not peculiar only to his treatment of truth alone. It is, in fact, a recurrent structure in his thinking. One can, for instance, discern a very similar structure in Heidegger's discussion of the concept of phenomena in the Introduction to *Being and Time*. There Heidegger distinguishes between two senses of 'phenomenon' writing that:

> …what is designated in the first signification of φαινόμενον ('phenomenon' as that which shows itself) and what is designated in the second ('phenomenon' as semblance) are structurally interconnected. Only when the meaning of something is such that it makes a pretension of showing itself — that is, of being a phenomenon — *can* it show itself *as* something which it is *not*, only then *can* it "merely look like so-and-so." When φαινόμενον signifies

[15] In addition to my own work, see especially Nulty (2006); Wrathall (2010), 40–56; and also Okrent (2011).

"semblance," the primordial signification (the phenomenon as the manifest) is already included as that upon which the second signification is founded.[16]

A similar structure is apparent in a much later discussion of the nature of language. In "The Way to Language," the title of which itself indicates a movement that is at the centre of the essay, Heidegger attempts to find a way to language that nevertheless already finds itself within language. As he writes:

> We are here undertaking something very unusual, which we might paraphrase as follows: we try to speak about speech *qua* speech. That sounds like a formula. It is intended to serve us as a guideline on our way to language. The words: "speak, speech" are used three times in the formula, saying something different each time and yet the Same. It is this underlying Same which, in terms of the oneness that is the distinctive property of language, holds together what is kept separate in the formula. To begin with, though, the formula points to a web of relations in which we ourselves are included. The undertaking of a way to speech is woven into a kind of speaking which intends to uncover speech itself in order to present its as speech and to put it into words in the presentation — which is also evidence that language itself has woven us into the speaking.[17]

The way to language at issue here moves between different senses of speech and speaking, and so different senses of language, that are nevertheless essentially bound together. In uncovering a way to language, which occurs only in and through language, language is illuminated in all of these senses, but the uncovering of that way is an uncovering of the originary phenomenon of language to which we already belong—a phenomenon that Heidegger designates as Saying: "All human language is appropriated in Saying and as such is in the strict sense of the word true language…".[18] In each of these cases—the inquiry into the concept of the phenomenon, the investigation of the way to language, and also the uncovering of the nature of truth—we find a mode of thinking that begins with what is immediately presented ('semblance,' 'speech,' 'correctness') and that looks to elucidate its nature (the conditions of its possibility) by uncovering its essential relatedness within a larger structure ('that which shows itself,' 'Saying,' 'unconcealment'). It is a mode of thinking that can be understood as essentially *hermeneutical* in that it does not rest content with the immediate presentation, but instead looks to uncover the framework of significance (essentially a structure of relatedness) within which that presentation is necessarily situated. There is an essential circularity at work here, since it is only through the immediacy of the presentation that the larger framework becomes at all evident (for the most part it remains withdrawn) at the same time as the presentation is itself dependent on that larger framework—a circularity that, in traditional hermeneutics, is understood in terms of the mutual dependence of whole and part.

The hermeneutical character of the thinking that is evident here is not merely something repeated at different points in Heidegger's thought, but is rather an ubiquitous, one might even say a characteristic, feature of Heidegger's thinking as a

[16] Heidegger (1962), 30.

[17] Heidegger (1971), 112.

[18] Ibid., 133.

whole. Thus, even after Heidegger moves away from any explicit reference to hermeneutics in his work,[19] still the same essentially hermeneutical structure remains.[20] Part of what is so revolutionary about Heidegger's thinking is, indeed, the way in which he brings about a hermeneutic transformation of philosophical inquiry (a transformation that, in its turn, also transforms hermeneutics). Construed as an inquiry into that which is essential—into that which is originary, as well as that which safeguards or preserves—ontological inquiry, in particular, can now be seen to take the form of the uncovering of a twofold structure that encompasses both that which is the initial focus of questioning and that which is brought forth as its proper origin and ground. Moreover, the uncovering of what is essential here is not a matter of the uncovering of some determinate character or entity—not a matter of identifying an *eidos* or *ousia*—but is indeed the uncovering of a structure of relatedness that unifies otherwise multiple, or at least dual, elements, and does so in a way that also preserves their differentiation.

If the structure at issue here is *hermeneutical*, then it is also *transcendental*. In his discussion of Heidegger's idea of truth, Tugendhat refers to the structure of Heidegger's thinking as it moves from the conventional understanding of truth to the idea of truth as unconcealment as involving a "transformed transcendental 'reference back'."[21] The 'transformation' to which Tugendhat refers here is a shift in the idea of the transcendental from Husserl to Heidegger, and although Tugendhat does not himself make this explicit, it is a transformation partly brought about by Heidegger's alignment of the transcendental with the hermeneutical.[22] The fact that the transcendental and hermeneutical might indeed stand in an essential relation to one another is suggested by the presence of an analogous circularity within the transcendental to that which is evident in the hermeneutical—Tugendhat's own talk

[19] See Heidegger (1971), 28–32. The idea of the hermeneutical that emerges here is developed in direct relation to an idea of the twofold, understood in terms of the twofold of presence and what is present, that is also a "simple oneness" (30). What this discussion indicates is indeed the fundamental role of the idea of the twofold in Heidegger's thinking—it does not refer only to the structure of truth nor does the question of truth stand apart from the question of being. The ontological difference is itself one form in which the twofold appears, although to think the ontological difference in terms of the twofold is to think the difference in terms of its essential unity.

[20] The shift away from the hermeneutical, like the shift away from the transcendental that I discuss briefly below, is actually a shift towards the topological. Yet inasmuch as the topological is already at work in the very idea of the hermeneutical as well as in the idea of the transcendental, then, regardless of Heidegger's own terminological preferences, the shift here must be seen as actually a realization of the topological character that belongs to the hermeneutical and the transcendental as such—and also, therefore, as a continuation of the transcendental and the hermeneutical in topological form, which is to say, in the form essential to them.

[21] Tugendhat (1994), 84. Tugendhat also refers to this movement of 'metatranscendental' (see the passage quoted from 'Heidegger's Idea of Truth', 84, in n7 above) as another means to distinguish it from the transcendental as it appears in Husserl. Part of what distinguishes the Heideggerian from the Husserlian notion of the transcendental, although Tugendhat does not make any real use of this idea, is precisely Heidegger's alignment of the transcendental with the hermeneutical.

[22] Heidegger also names the structure at issue here as *phenomenological*, implying an even more significant shift in the conception of phenomenology—something that is evident in Tugendhat's discussion.

of a transcendental 'reference back' might be seen to hint in just this direction (and is perhaps the same circularity that Kockelmans identifies as belonging to truth as unconcealment).[23] The circularity at issue here is itself indicative, however, of the way in which both the transcendental and the hermeneutical already belong, in spite of the various, and often contending, readings and misreadings attached to these notions, within the domain of philosophical *topology* or *topography*.

The topological character of the hermeneutical is perhaps easier to appreciate than is the topological character of the transcendental. The hermeneutical already brings with it, especially in its Heideggerian employment, but also in the Gadamerian, explicit concepts of situatedness and location—even in its mundane forms, hermeneutic inquiry always proceeds on the basis of the concrete engagement of the interpretation with some subject matter as it stands within a larger frame.[24] In comparison with the hermeneutical, the transcendental may appear a more abstract notion, based, not in factical situatedness, but in the relation of condition and conditioned. Moreover, there is an additional difficulty that arises both in the assimilation of the transcendental to the topological and in the use of the transcendental as applying to Heidegger's twofold account of truth in its generality: although Heidegger draws explicitly on the notion of the transcendental in his early work, he explicitly abandons the concept in his later thinking, and this shift is itself associated with a shift towards a more explicitly topological orientation (the transcendental, it appears, gives way to the topological, rather than being an instantiation of it). Since Tugendhat focuses on Heidegger's account of truth primarily as developed in *Being and Time*, in which the transcendental is not put in question, this is not an issue that he is forced to address, but it is a *prima facie* problem for any account—like that developed here—that takes up Heidegger's thinking more broadly. There are thus two issues that need to be further explored: first, what is the idea of the transcendental that Heidegger rejects (and to what extent is it the same idea as is at issue in the hermeneutical structure already delineated above); second, to what extent is the transcendental indeed topological in character (and so to what extent does Heidegegr's topological thinking constitute a continuation, rather than abandonment, of the transcendental as such)? In fact, both these question come down to a question concerning how the transcendental is to be understood—and addressing that question will require that we do not assume too determinate a conception of the transcendental in advance.

No matter what else we say about the idea of the transcendental, the very heart of the concept is a certain way of thinking about the problem of *ground*—it is a grounding that is also a unifying[25]—and it is this that is captured in the commonplace talk of the transcendental as concerned with 'conditions of possibility.' What the

[23] The circularity evident here, both as a feature of the transcendental and the hermeneutical, is explored in Malpas (1997), 1–20.

[24] See Malpas (2010a).

[25] See Malpas (2012b).

transcendental makes possible, on Heidegger's account, however, is specifically *transcendence*—where transcendence is the capacity of Dasein to open up a world in a way that goes beyond any particular thing that may appear within that world.[26] In this respect, transcendence can be understood as equivalent to Dasein's own capacity for disclosedness, and the transcendental as that which grounds such disclosedness. Already this indicates just how closely the structure of the transcendental, at least in Heidegger, is tied to the structure of truth as unconcealment. Yet since it is the relation to transcendence that Heidegger takes to be primary in the idea of the transcendental, so, as his thought develops, and he moves away from the focus on transcendence, he also moves away from the language of the transcendental. The movement away from transcendence, and so away from the transcendental, is also a move away from a focus on human Dasein as the primary locus of truth towards a more direct concentration on the happening of truth as that in relation to which even the human is disclosed.[27]

There is reason to think, however, that Heidegger's particular appropriation of the transcendental as tied to transcendence in this way is mistaken, or, at least, constitutes too narrow an understanding of what is at issue in the concept.[28] On this basis, one might well argue that the idea of the transcendental continues to operate in Heidegger's thinking even after Heidegger has abandoned that particular version of the transcendental that is tied to transcendence. Taken more broadly, and in a way that is also attentive to Kant's, rather than Heidegger's, use of the notion, the transcendental should be understood in terms that connect it with Kant's own geographical or topographic conception of the critical enterprise. Kant's problem is essentially how one can provide a grounding for knowledge or experience that does not appeal to what goes beyond knowledge or experience. Kant's solution, in general terms, is to look to the ground of knowledge or experience in the unity that is given within it and without which it would not be possible[29] (such a way of putting matters clearly echoes the hermeneutic characterisation I set out above). The term 'transcendental' can be used to refer to the ground that is thereby exhibited, to the grounding structure, and to the mode of inquiry by which such a ground is exhibited. In the terms Kant employs, the way such a transcendental grounding proceeds can be taken to be analogous to certain aspects of geometrical or topographic practice: the geometer, for instance, from the measure of a small part of its surface, is able to determine the full extent of the surface of a sphere; the topographical surveyor, by a process of repeated triangulation and traverse, is able to map the entire territory in which she or he is located.[30] The transcendental is indeed a term that describes the inquiry into

[26] See, for instance: Heidegger (1962), 366 and Heidegger (1984), 160–166; see also the discussion in Malpas (2006), 162–171.

[27] See Malpas (2006), Chapter Four, esp. 175–201.

[28] This is an issue taken up at number of places in Malpas (2012b).

[29] See Malpas (1997); see also Malpas (2012b).

[30] See Malpas and Thiel (2011).

a certain place or *topos*, as well as the place thus exhibited, that is *our own* place, and that proceeds *from within* that very place.

Understood in this way, one can see a twofold structure already built into the very character of the transcendental as set out here. The transcendental begins with our being already given over to things; it asks after the ground for that givenness. Since it may be taken to be asking after its own grounds, the twofold character of what is at issue exhibits something of the circularity that is characteristic of the transcendental as well as the hermeneutical.[31] Yet in asking after the ground of our own being-given-over to things in this way, the transcendental does not abandon the givenness at issue, does not attempt to surpass it, but instead remains with it. Invoking the topological character of Kant's own understanding of the transcendental, we can say that the transcendental begins with our being already 'here/there', and what it seeks to uncover is the very place of that here/there, the very place in which we already find ourselves. The twofold is thus evident in the way being here/there is a mode of being-in-place that goes beyond the here/there of our own location—to be here/there is precisely to be opened to a place, and for that place itself to open up.[32] The twofold at issue can thus be said to be identical with the twofold character of place. That twofold character is one explored early on by Aristotle himself, not in the *Metaphysics*, but in his analysis of *topos* in the *Physics*, and also, although in very different terms, in Plato's account of *chora* in the *Timaeus*.[33] In each case, what appears is a structure that combines a movement inwards and outwards (an infolding as well as outfolding), an opening that is also a closing, a relating that is a distinguishing, a limiting that is a freeing up, a withdrawing that is also a coming forth.

In its most basic sense, a sense that underlies any other interpretation of the idea, the transcendental refers us to the inquiry, and the twofold structure, that is named by place or *topos*—an inquiry that, in keeping with the rest of the discussion here, is also hermeneutical (or perhaps one should say that the hermeneutical is essentially topological). Understood in this way, the transcendental can be seen to be closely aligned with that mode of thinking that takes place as the primary focus for philosophy—as closely aligned, that is, with a form of philosophical topography or topology. Such a topology turns out to be present in Heidegger's thought almost from beginning to end (it is what Heidegger calls the 'topology of being'[34]), and is given a particularly clear exemplification in his thinking of truth. Indeed, what Kockelmans refers to as "the *eukukleos alētheiē*, the well-rounded non-concealment itself" is identical with the place, the *topos*, that is the focus of such a topology or

---

[31] See Malpas (1997).

[32] To some extent one might argue that this idea is itself an echo of what is at issue in the idea of transcendence, but it also eschews certain key aspects of transcendence, namely, the move from one element in the direction of another. Here rather than a move *across* or *beyond*, the movement is an opening-up accomplished at the same time as a turning-in.

[33] See my discussion of in Malpas (2012a).

[34] Heidegger (2004), 41.

topography. Even the language Tugendhat employs in his inquiry into Heidegger's idea of truth carries traces of the topological structure at work here—thus Tugendhat speaks of a "depth dimension" as involved in the Heideggerian account as well as talking of what occurs in terms of a "pointing beyond" or a "reference back."[35] The topological character of Heidegger's of thinking is especially evident in his use of the idea of *Lichtung* or 'clearing' to refer to truth as unconcealment. This is no mere 'metaphor,' but a very specific way of referring to the character of place or *topos* as that opening into appearance, into *presence*, that occurs in the midst of the withdrawal into concealment (a withdrawal that is, of course, never complete). Indeed, the very idea of unconcealment itself captures the twofold character of the event that is the clearing: unconcealment is no mere standing in the open (it is not pure transparency), but is instead a dynamic interplay of concealing/revealing. This is itself evident in the characterisation of truth as, indeed, *a-letheia*, un-concealment—the twofold is evident in the privative.

Heidegger's own worries about the idea of the transcendental, and what underlies his eventual rejection of it, are not connected with its topological character, but almost the very opposite—they relate to his particular reading of the transcendental as necessarily implicated with the notion of transcendence. To begin with, transcendence carries with it a problematic tendency towards subjectivism—since transcendence seems to find its ground in human Dasein. In addition, however, transcendence also presupposes a certain separation of Dasein *from* world—a separation itself enshrined in the very idea of Dasein as that in which transcendence finds its ground and world as that towards which transcendence moves. In this latter respect, transcendence can be seen to threaten the very unity that must also be presupposed here (a unity given particularly salient articulation in the topological unity of place). Yet this problematic reading of the transcendental is by no means forced upon us by anything in the notion of the transcendental itself—and Kant's own topographic employment of the idea suggests a very different interpretation, one that is the basis for the discussion above.[36] On this reading, the structure of the transcendental is not to be found in the inquiry into that which underlies transcendence (understood as the move of Dasein in the direction of world), but rather in the inquiry into place or *topos*. Moreover, that inquiry is oriented towards

[35] See especially Tugendhat (1994), 91. Such topological elements are never taken up by Tugendhat, however, and are neither made explicit nor are their implications drawn out. This partly reflects the limitations in Tugendhat's own appreciation in what is at work here, although it is also a function of the fact that Tugendhat's discussion remains so much focussed on the earlier thinking, especially *Being and Time*, in which the topological character of Heidegger's thinking is not yet fully realised. By contrast, if one looks at the account of unconcealment as set out, for instance, in "The Origin of the Work of Art" (or in almost any of the later writings, including Heidegger (1972)—see esp. 65–70), the topological framework of Heidegger's thinking is to the fore: here *aletheia* is clearly understood in terms of *a certain happening of place*.

[36] Which is not to say that Heidegger's reading of the transcendental as tied to transcendence is entirely without foundation, but rather that it is mistaken to see transcendence, in the way Heidegger understands it, as the underlying and determining idea in the structure of the transcendental.

the understanding of a twofold structure that, as twofold, is also therefore essentially unitary; it is a structure that rather than overcome a separation, is the unfolding of an essential relatedness, an originary belonging-together. The twofold character of place is thus quite distinct from the potentially dichotomous separation of Dasein and world that is implied in the idea of transcendence, and that requires a surpassing of one in the direction of the other.[37]

It is precisely because of the transcendental-topological (and also hermeneutic) structure that is at work in Heidegger's twofold account of truth that one cannot prise off truth as correctness from truth as unconcealment. The latter must always be implicated in the former, even though the latter itself tends to withdraw in the face of our concern with truth as correctness. In this respect, Tugendhat's claim that Heidegger's turn towards truth as unconcealment results in the loss of truth as correctness, and so also in the loss of any genuine critical sense,[38] gets things exactly the wrong way round: only by keeping hold of truth as unconcealment, and the twofold structure that it brings with it, can we hold on to truth as correctness. It is for just this reason that Heidegger's later acceptance of Tugendhat's claim that *aletheia* is not the same as truth has to be viewed as problematic, since it threatens to obscure the very twofold unity that is so important here. Although Heidegger clearly did not see this admission as indicating his abandonment of the concept of *aletheia*, but rather as an acceptance of the difficulty that the ordinary understanding of truth as correctness presents for any attempt to think truth differently,[39] still the severing of unconcealment from truth in this way threatens the very structure that is at issue in the idea of *aletheia* itself. *Aletheia*, unconcealment, does not stand apart from truth as correctness, but is, one might say, its 'other side'; there are, in an important sense, not two separated concepts here, but two aspects of a single structure—although a structure that constantly turns a part of itself away from us. This is what it means, in fact, to talk of the *twofold* character of truth: truth as unconcealment is the essence, that is the origin and ground, of truth as correctness.

5

If Lafont and Smith can claim that the nature of Tugendhat's objection to Heidegger's account of truth has been misunderstood and overlooked, then one might equally claim that the real nature of the twofold conception of truth in Heidegger has also been misunderstood and sometimes ignored. Certainly Tugendhat himself seems to have no real sense of the topological dimension in which Heidegger's twofold understanding of truth moves. Indeed, one might view Tugendhat's objection, and the reiteration of

---

[37] The idea of transcendence can itself be seen as based in a misapprehension of what is at issue in the phenomenon of place—but as such, it can also be seen as an attempt to engage, even if mistakenly, with the topology that is at issue here.

[38] See Tugendhat (1994), 94–95.

[39] See Wrathall (2010), 37–38.

that objection, at least in terms of the insistence on the normativity of truth by writers such as Lafont and Smith, as itself constituting a refusal of the very possibility that the inquiry into truth might move in such a direction—a refusal of any transcendental-topological dimension to truth. Although this could well be taken as the real core of Tugendhat's position, one might also take it as the basis for a restatement of Tugendhat's objection: Why should we look to any 'transcendental-topological' structure as necessary at all here? Why should we look to anything beyond truth as correctness? Why should we look for anything as the ground for correctness? One response to such a restated version of Tugendhat's critique would be to rehearse once again exactly the sorts of arguments just considered. Yet there is another, and perhaps more useful, response, that is also available—one that need not draw, at least not initially, on ideas of the hermeneutical, the transcendental or the topological. The source for this response is the work of Donald Davidson. There one finds, as in Heidegger, a conception of truth as also twofold in character: truth belongs to individual sentences, but, on the Davidsonian account, it also inheres in the larger framework within which those sentences are located. What is important about this account from the perspective of Tugendhat's objection, is that it shows the necessity of understanding truth in a way that is not restricted to truth in its normative sense alone—a conclusion arrived at by means of some fairly straightforward considerations concerning the way truth itself operates and the other concepts with which it is implicated. In the larger context of Davidson's thinking as a whole, there is a sense of truth at work that turns out to operate within the same transcendental-topological dimension as can also be discerned in Heidegger, but this is a dimension that is arrived at in the course of Davidson's thinking, rather than one that is assumed from the start.

At first sight, however, far from providing a parallel to the Heideggerian account, the Davidsonian position might be thought to exemplify what Heidegger takes to be the conventional understanding of truth as correctness. A key feature in Davidson's account of truth is that truth is indeed a property of individual sentences, and as it belongs to sentences, so it carries the normative dimension emphasised by Tugendhat.[40] Yet although Davidson does indeed take truth to belong, in its standard usage, to individual sentences, truth also figures, as I noted above, as part of the 'background' against which individual sentences can be true or false, and in this respect, truth not only goes beyond what is given in the individual sentence but it also exceeds what is captured by any notion of correctness. The way in which truth functions here is evident both in Davidson's inquiries into the concept of truth, and in his account of the nature of linguistic understanding and communication.[41]

As developed in the idea of radical interpretation, the possibility of understanding speakers—of interpreting their utterances and actions, and identifying their attitudes—depends on the application of the Davidsonian 'principle of charity'. Charity

[40] Although Davidson does not understand this sense of truth as entailing any substantive notion of truth as correspondence—see n.52 below.

[41] For a brief overall summary of the Davidsonian position see Malpas (2010b); see also Malpas (1991), esp. Chap. 2.

requires that, in interpreting a speaker, one must take their beliefs and utterances to be, for the most part, true (and so as also in agreement with one's own beliefs and utterances). The assumption of overall truth is an assumption that is not only prior to any particular interpretive encounter, but it is not defeasible in the face of any such encounter: that utterances and beliefs are generally true is a requirement if utterances and beliefs are to have content, that is, if they are to be meaningful.[42] In the associated idea of triangulation, a notion that appears in Davidson's more developed thinking, meaning (or, more broadly, content), including the meaning of utterances as well as the content of states of mind, is seen to be dependent on the situation of the speaker within a tripartite structure encompassing self, others, and world.[43] It is not only that we come to understand another speaker's utterances, attitudes and actions through being able to relate aspects of the other's behaviour to our own, as well as to features of the larger environmental situation in which we are jointly located, but that it is only through the relatedness between ourselves, others, and features of the world that utterances, attitudes, and actions take on the meaning (or content) that, in large part, identifies and individuates them.[44] In this sense, it is on the basis of their relatedness within the tripartite structure of self, others, and world, that utterances, attitudes, and actions are constituted *as utterances, attitudes and actions,* and so too, since speakers are in turn constituted *as speakers* by the meanings (the contents) that make up their mental lives, are speakers constituted *as speakers* on the basis of the mutual relatedness that is worked out within the structure of triangulation.[45]

The possibility of truth as a property of sentences, or of individual utterances, arises only on the basis of the conditions that make such sentences and utterances meaningful—that constitute them as sentences or utterances. The conditions that make for the possibility of meaning, and so for the possibility of truth as attaching to individuals sentences, and so as being either truth or false, are the conditions that are identical with the obtaining of the mutual relatedness between self, other, and world within the structure of triangulation. The obtaining of that structure is not a matter of the being true of any particular sentence or sentences, but it is a matter of the being true (and not just being *held true*[46]) of the body of sentences, as a whole and for the most part, and this is because it cannot be the case that the relatedness at issue might fail without an accompanying failure of meaning, which also means a failure in the possibility of individual sentences being true or false—without, in

[42] See, for instance, Davidson (2001a), 153.

[43] See Davidson (2001f), 205–220.

[44] Utterances, attitudes and actions are, on a Davidsonian account, also identified and individuated through their causes and causal effects, but this is not independent of the 'rational' (that is meaningful or contentful) connections that are also at work—see Malpas (1999), 1–30.

[45] See the discussion of this in Malpas (2013).

[46] This is because the distinction between *being* true and *being held* true is itself a distinction that, inasmuch as it is a meaningful or contentful distinction, can only be given meaning in respect of individual sentences—the distinction itself depends on a larger context within which it is embedded.

other words, a failure in normativity. Indeed, without the relatedness that is articulated in triangulation there can be no speakers, no utterances, no attitudes, no actions. When understood specifically in relation to belief and utterance, the obtaining of the relatedness that is worked out within the structure of triangulation is the obtaining of the overall truth of belief and utterance in a way exactly analogous to the requirement that is at issue elsewhere in Davidson in the principle of charity. It is thus that Davidson can write that it "cannot be the case that our general picture of the world and our place in it is mistaken, for it is this picture which informs the rest of our beliefs and makes them intelligible, whether they be true or false."[47]

One way of capturing the point at stake here is by saying that the very possibility of any individual sentence having a truth value (that is being true or false) depends on many other sentences being true (but in a way that does not allow any identification of just which sentences must be true). The symmetry that operates in respect of truth and falsity at the level of individual sentences—a symmetry reflected in the principle of bivalence—does not hold with respect to the larger body of sentences within which individual sentences are always nested. The position described here could be viewed as equivalent to a form of coherentism, since it is similar to the idea according to which any single belief requires connection within a larger system of belief, or as akin to the thesis of linguistic holism, according to which a sentence is only meaningful in the context of a system of sentences, namely, a language.[48] The difference, however, is that neither coherentism nor linguistic holism make explicit the way in which it is indeed *truth* that is implicated here. Meaning does not arise on the basis merely of the interconnection of sentences or beliefs, and so cannot be construed on the basis of some purely 'internal' system of connections, and in a way that stands apart from speakers' engagement in the world (if such a possibility is even conceivable). For there to be meaning is already for meaning to be implicated with the world, and so for it to be also already tied to truth. In this way the possibility of the meaning of any individual sentence or belief mirrors the conditions on which the possibility of the truth or falsity of any individual sentence also rests. Sentences and beliefs are meaningful (have content), and so can be true or false, only inasmuch as they are nested within a larger body of sentences and beliefs that are, for the most part, true—that are already connected with the world.

One might respond to this position by insisting that the truth that is supposed to belong to the larger body of sentences can only be a truth that attaches to sentences individually. So to say that truth inheres in the body of sentences implies that most of the individual sentences that make up that body of sentences must be true. Yet this would be to deny that truth does indeed attach to the *body* of sentences taken together

[47] Davidson (2001f), 214.

[48] To some extent, Davidson himself accepts both such positions, but only to the extent that they are viewed as not concerned only with meaning, but as also encompassing truth—that is, both have to be construed in 'externalist' rather than 'internalist' terms. Yet the characterization of Davidson's position as 'coherentist' misleads more than it illuminates —which is why Davidson later took back his own characterization of his position as a 'coherence theory'. See his 'Afterthoughts (1987)', appended to 'A Coherence Theory of Truth and Knowledge', in Davidson (2001e), 154–155.

(the truth of the body of sentences would directly reduce to the truth of a number of individuals sentences),[49] and would be to reassert a sense of truth that, because it does indeed attach to each sentence individually, and so also entails that each individual sentence might be true or false, opens up the very possibility that has to be ruled out, namely, that the entire body of sentences could, for the most part, be false (if any individual sentence can be false, and if the truth of the entire body of sentences is just a matter of the truth of individual sentences, then the entire body of sentences could be, for the most part, false). The possibility of meaning, and the possibility, therefore, of individual truth or falsity, depends on a sense of truth that is not reducible to the truth of individual sentences, and that is also not normatively constrained. Here, by following through a Davidsonian line of argument, we come up against the real limit of Tugendhat's position, but also to the beginning of Heidegger's.[50]

Although it can be seen to emerge from Davidson's account of the conditions that underpins the possibility of meaning and understanding, the idea that truth cannot be understood in terms of correctness or correspondence tends to be implicit in that account rather than a central theme. Elsewhere, however, Davidson takes up the matter quite directly. This is not surprising, since one of Davidson's distinctive contributions to twentieth-century analytic thought has been the idea that one can take truth as the basic concept in the understanding of meaning.[51] Davidson thus appropriates the formal mechanism of Alfred Tarksi's theory of truth as the template for a formal theory of meaning. Already this should indicate the potential danger in making too radical a distinction between truth and meaning: the concepts are distinct, but as Davidson shows (and as is evident from the discussion above), they are also closely related. Tarski viewed his own truth definition as entailing a correspondence conception of truth. Davidson argues, however, that the Tarskian approach, while it may appear to make use of notions that might be viewed as analogous to correspondence (the idea of 'satisfaction,' as well as the notion of translational equivalence),[52]

[49] A possibility that is ruled out here, since there is no single set of sentences that must be true in order for the body of sentences to be (mostly) true—neither is it the case that the body of sentences makes up a determinate set of sentences nor is it the case that the set of individual sentences that must be true if the body of sentences is to be true is determinate either.

[50] There is also an obvious connection to Wittgenstein here (especially the Wittgenstein [1969]), although on some readings Wittgenstein stands in a closer relation to coherentism than would Davidson or Heidegger.

[51] An idea first set out in Davidson (2001c), 17–42.

[52] In 'True to the Facts', Davidson defends a reading of the Tarskian account as a species of correspondence account, thereby also defending the idea that correspondence captures something important about truth—see Davidson (2001d), 37–54; see also Davidson (2001e), 139–140. The basis for Davidson's original acceptance of correspondence as a core element in the idea of truth is that truth involves "the relation between a statement and something else" (Davidson, 2001d, 38)—a relation, one might say, between words and objects (Davidson, 2001e, 139). This is an admission Davidson later retracts – see especially Davidson (2001a), 154–155. Davidson's retraction is not based, however, on a change of mind about the nature of truth, but rather about whether 'correspondence' is a helpful notion here. Even in "A Coherence Theory of Truth and Knowledge," he acknowledges that his use of the idea of correspondence is neither "straightforward" nor is it "nonmisleading" (Davidson, 2001e, p. 139). In his later comments, he says of the nature of the

does not commit one to an identification of truth with correspondence.[53] Indeed, Davidson devotes considerable attention to showing that correspondence *cannot* be adequate as an elucidation of truth.[54] One reason Davidson gives for this is that there simply is nothing to which true sentences correspond in any interesting or relevant fashion.[55] But one can also say, more generally, that any notion of correspondence must, in any case, always presuppose truth, and so cannot elucidate it. This is because only the right sort of correspondence makes for truth, but saying what counts as the right sort here requires that we be able to specify what truth is in a way that itself constrains the notion of correspondence (in the right way), and this requires a notion of truth that is other than truth as correspondence.[56] The same point applies to any other concept we may use as a means of explaining truth or as a surrogate for it. The fact is that truth is always presupposed in our attempt to explain or give an account of truth.[57] Even a Tarskian truth definition already presupposes that we have a grasp of the way the concept of truth works independently of the Tarskian definition itself, since the Tarskian account depends on the idea of translation into an already understood language (and so a language in which we already have a grasp of truth).[58] Here, once again, truth appears as a concept that

---

mistake that it is "in a way only a misnomer, but terminological infelicities have a way of breeding conceptual confusion…Correspondence theories have always been conceived as providing an explanation or analysis of truth, and this a Tarski-style of truth certainly does not do," Davidson (2001a), 154–155.

[53] See Davidson (2005), 37–42 & 155–156.

[54] There is a sense of correspondence that can be seen to be at work in the notion of correctness—a procedure is correct, for instance, if it matches the rules that govern such a procedure, and a claim is correct if what it claims fits that which the claim is about—but the notion of correspondence at work here is not such as to enable it to be generalised in any useful way, and it certainly cannot serve to provide a genuine explanation or elucidation of the sense of correctness that is a work in relation to truth (see n.52 above). Thus, while there are two senses of truth to be found in Davidson, they are just the sense of truth associated with the truth or falsity of sentences ('correctness')—which is not to be identified with any substantive notion of correspondence any more than it is to be identified with, for instance, coherence, warranted assertibility, or pragmatic udefulness—and the sense of truth that inheres in the larger body of sentences (or better, in the overall involvement in the world as that is expressed in terms of triangulation) against which the truth or falsity of individual sentences is possible.

[55] See Davidson (2005), 41. This is the decisive consideration against correspondence theories and undermines any substantive sense in which truth can be understood as correspondence. It is, however, less relevant to the issues concerning the twofold character of truth.

[56] See Davidson (2001g), 193–194. One might argue that this does not demonstrate that truth cannot be a matter of correspondence, but only that there is no way to elucidate the form of correspondence that belongs to truth. My use of the argument here, however, can be taken as directed only against that weaker claim—the stronger claim is undermined by the Davidsonian point noted above to the effect that there is nothing significant to which true sentences can correspond.

[57] This holds in relation to correctness also: since correctness applies more broadly than just to truth alone, so knowing what sense of correctness is at issue in talk of truth depends on already having a prior sense of truth. Correctness is thus not an elucidation of truth, but merely functions, in the appropriate context, as another way of referring to truth (or to one sense of truth).

[58] Davidson (2001g), 194–195.

constrains discourse, in a way that is additional to the constraint associated with the normative operation of truth in respect of individual sentences, but can never be fully elucidated within such discourse, since it constrains even the normative concept of truth that it operates with respect to individual sentences.

In Davidson, as in Heidegger, truth carries a twofold sense: as 'correctness' and as that on which the possibility of correctness is based. This possibility is understood in Heidegger in terms of the idea of unconcealment, understood, in one form, as the clearing. Correctness thus finds its ground in the prior opening up of the world that first allows for the possibility of action or of assertion. Davidson does not use the same language as Heidegger, and yet the structure that he delineates through the idea of triangulation is also essentially a form of clearing, or opening up. It is fundamentally a structure of *relatedness* that depends on a certain mode of spatiality in which the realisation of meaning, of presence, occurs through the becoming proximate of human beings to themselves, to one another, to other creatures and other things, within the framework of a single world, as that occurs in and through specific places and spaces. While differences in language, sources, and style cannot be ignored, one also cannot afford to allow oneself to be distracted from the points of convergence that may lie beneath. Those points of convergence are especially important in the thinking of truth that is at work in Heidegger and Davidson, since each provides resources to assist in the illumination of the other, and to allow a better understanding of the topology that they both endeavour to explore.

6

The twofold character of truth as it appears in Davidson mirrors the twofold appearance of truth in Heidegger. One might argue that Davidson's account lacks the properly ontological element that is present in Heidegger's, except that the very idea of the twofold that is at work in the concept of truth at issue here brings with it a transformed conception of what ontology might be—a conception that moves the ontological in the direction of the topological, and so in the direction of a structure that does indeed seem to be at work, if through different modes of expression, in both thinkers. A proper understanding of the twofold character of truth as it appears in Davidson and in Heidegger depends on understanding both the transcendental-topological (and also hermeneutical) structure that underpins Heidegger's argument for unconcealment as the proper origin and ground of correctness, and the formal-analytic structure evident in Davidson that reveals the truth of individual assertions as always dependent on truth as it inheres in the larger background to assertion (a sense of truth that stands in the background even of attempts to inquire into or to define truth itself). These two aspects to the twofold conception of truth can both be seen to point towards the character of truth as pertaining both that which is spoken or asserted, and, more fundamentally, to the prior involvement with the world on the basis of which such speaking or asserting is possible.

The idea of truth as other than correctness remains somewhat obscure, but this is because the fact of our being already given over to the world—our being 'in' the world, which is to say, our being always 'placed'—is itself obscure, and must always remain so. Even the attempt to thematize that very 'being-given-over-to,' that very 'being-in,' that very 'placedness,' is to presuppose it. It also remains the case that the idea of truth at issue here cannot be subjected to normative assessment in any direct way—it is not a sense of truth that operates with respect to the idea of a bivalent 'truth value'. Indeed, what Davidson's argument shows, in particular, is that the idea of bivalent truth—normative truth—already implies the idea of a truth that is not bivalent. There is thus a symmetry that operates with respect to truth and falsity at the level of individual assertion, but which does not operate at the level, one might say, of assertion *as a practice*, since the practice of assertion presupposes that assertion is mostly true—and this follows from the close reciprocity that obtains between truth and meaning (a reciprocity that does not hold between truth and falsity).[59] As truth operates within the practice of assertion, so it also operates in the same way as part of the practice of criticism: criticism depends on the same twofold character of truth that distinguishes between that which is the focus of critical assessment and that which enables such critical assessment to proceed. There is thus no failure of critical capacity in being unable to assess as true or false one's original being-placed in relation to things—there is, indeed, no sense to attach to the idea of critical assessment or engagement that could be applied here. Critical assessment, like assertion, is always tied to particular claims or statements, but as soon as we move to attend to such claims or statements, we are no longer dealing with anything that properly belongs to the structure of what Heidegger calls unconcealment. The fact that unconcealment cannot be taken as a direct focus for critical judgment—one cannot say of unconcealment either that it is false (as opposed to true) or that it is not unconcealing (which does not mean that it does not also conceal)—does not mean, however, that the idea of truth as unconcealment stands entirely apart from *any* critical practice.

The idea of truth as unconcealment is not an idea arrived at merely by some sort of unquestioned revelation, but arises out of an original questioning of the possibility of truth as correctness. Indeed, the preceding pages have been concerned with nothing if not the attempt to argue for and to elucidate the idea of truth that is at issue here—they thus operate *within* a framework of criticism, rather than outside of it—and so also within a certain normative practice. One must be clear, however, that this is not a normative practice directed *at* unconcealment as such, but is rather part of the philosophical inquiry *within which* that idea is taken up. One might add here that the idea of truth as unconcealment itself supports and sustains the possibility of any form of critical engagement, and not only in the sense noted above in which it provides the necessary 'background' against which specific criticisms operate. Genuine criticism, genuine questioning, depends on there being a space for thinking

[59] Which reinforces the idea that there are two sense of truth at work here: one in which truth is defined in relation to falsity, and the other in which it is defined as that which makes possible the disjunctive possibly of the true and the false.

in which different possibilities and alternatives can be envisaged, in which connections and disconnections can become evident. It requires something like the very place that is at issue in Heidegger's notion of the clearing—the opening up of truth in the unconcealment of the clearing is thus itself the opening up of the possibility of questioning and of critique.

Aletheia, unconcealment, is one of the terms that Heidegger gives to the original dynamic opening-up of the world—this 'event of truth' is the *Ereignis* that it is also the happening of the Fourfold that is the worlding of world.[60] *Aletheia* does not occur as some strange ethereal event occurring outside of or beyond the concrete world in which we find ourselves—it occurs both in the temporalized-spatialized unfolding of ordinary life and activity, and in the character of the ordinary as part of a larger happening of history as that occurs in and through certain encompassing forms. Heidegger gives a name to that which determines the contemporary happening of truth: *Gestell*, meaning, in English, something like 'framework' or 'enframing'.[61] Gestell, he says, is the essence of the technological mode in which the contemporary world unfolds itself.[62] If this is so, however, then Heidegger's own analysis of *Gestell*, his critique of technological modernity, provides a striking exemplification of the way his twofold conception of truth not only allows for a mode of critical engagement, rather than closing it off, but is centrally oriented towards just such critique—and to a more encompassing and radical mode of critique than is perhaps envisaged in the normativity of truth as captured in the idea of truth as merely "correctness."

It is through the distinction between truth as correctness and truth as unconcealment that we are opened up to the possibility of a form of critique that, even though always expressed in terms of particular claims, can nevertheless engage with and draw attention to the larger framework within which our modes of thinking, and indeed, our very lives, are shaped and oriented. Moreover, part of Heidegger's own argument here, even if not entirely explicit, is surely that the refusal to acknowledge the twofold character of truth—the refusal to allow that there is a larger transcendental-topological determination to thinking, the refusal to recognise the *placed* happening of truth as such—is itself characteristic of the contemporary mode of unconcealment that is *Gestell*. It is thus that Heidegger can say of the world in which we now find ourselves that it is a world that no longer thinks; that no longer holds open a space for genuine questioning; that, one might say, no longer allows for the possibility of critical engagement of a fundamental kind. Kockelmans' draw together some of the various lines of argument at issue here when he writers:

---

[60] I use *Ereignis* here to refer to the event of truth, but *Ereignis* is a difficult term that can also refer to a more fundamental event—a radical turning of and turning back to the originary event of unconcealment. It might be said that *Ereignis* properly means the latter. There is, however, an essential equivocity at work here that cannot and should not be eliminated, and that is common to almost all of Heidegger's key terms (see Malpas, 2006, 12). On the idea that *Gestell* might itself be thought of as a form of *Ereignis*, see Malpas (2006), 288–289.

[61] See Malpas (2006), 280.

[62] See the discussion of this matter in ibid, 288–289.

> When truth became reduced to correctness, man himself became the center and focal point of all beings. And when man began to circle about himself in search of certainty and security, thinking gradually became a pro-posing, positing presentation, and the Being of beings changed into sheer objectivity. All of that prepared the way for the modern era of technicity, concerning which man, thus, has completely lost the truth.[63]

The Heideggerian critique of technology, so central to Heidegger's later thinking, is thus itself closely tied to his twofold idea of truth.

If the idea of truth as unconcealment can in any way be said to set a *limit* to critical inquiry, *a limit to questioning*, it is only in the sense that it functions as its proper *ground*[64]: in understanding the twofold character of truth, we understand how questioning arises only on the basis of our prior being-given over to the world, and so on the basis of a singular opening-up of world in its concreteness, but we also understand how the most fundamental questioning of all must be directed at the very opening into possibility, and so into an inexhaustible multiplicity, that occurs in any and every such opening—even that which belongs to technological modernity, even that which would orient itself only to "correctness."

References

Dahlstrom, Daniel. 2001. *Heidegger's concept of truth*. Cambridge: Cambridge University Press.

Davidson, Donald. 2001a. Afterthoughts. In *Subjective, intersubjective, objective*, 154–158. Oxford: Clarendon Press.

Davidson, Donald. 2001b. Belief and the basis of meaning. In *Inquiries into truth and interpretation*, 2nd ed., 141–154. Oxford: Clarendon Press.

Davidson, Donald. 2001c. Truth and meaning. In *Inquiries into truth and interpretation*, 2nd ed., 17–42. Oxford: Clarendon Press.

Davidson, Donald. 2001d. True to the facts. In *Inquiries into truth and interpretation*, 2nd ed., 141–154. Oxford: Clarendon Press.

Davidson, Donald. 2001e. A coherence theory of truth and knowledge. In *Subjective, intersubjective, objective*, 2nd ed., 139–140. Oxford: Clarendon Press

Davidson, Donald. 2001f. Three varieties of knowledge. In *Subjective, intersubjective, objective*: Philosophical essays volume 3, 205–220. Oxford: Oxford University Press.

Davidson, Donald. 2001g. On the very idea of a conceptual scheme. In *Inquiries into truth and interpretation*, 2nd ed., 183–198. Oxford: Clarendon Press.

Davidson, Donald. 2005. *Truth and predication*. Cambridge, MA: Belknap.

Heidegger, Martin. 1962. *Being and Time*. Trans. John Macquarie, and Edward Robinson. New York: Harper and Row.

Heidegger, Martin. 1969. *Zur Sachen des Denkens*. Tübingen: Max Niemayer.

Heidegger, Martin. 1971. The way to language. In: *On the Way to Language*, ed. Heidegger, M (trans: Peter D. Hertz). New York: Harper and Row.

---

[63] Kockelmans (1984), 233. At the end of the paragraph in which this passage appears, Kockelmans quotes from Egbert Schuurman: "Heidegger pleads for reflective consideration for Being or devotion to Being to show that the origin of technology is a *truth-event*"—Schuurman (1980), 89.

[64] The sense of 'limit' alluded to here is a sense of limit to which Heidegger himself draw attention in a number of places – see Malpas (2012b).

Heidegger, Martin. 1972. The end of philosophy and the task of thinking. In: *On Time and Being*, ed. Heidegger, M (trans: Joan Stambaugh). New York: Harper & Row.

Heidegger, Martin. 1984. *The Metaphysical Foundations of Logic*. Trans. Michael Heim. Bloomington: Indiana University Press.

Heidegger, Martin. 2004. Seminar in Le Thor 1968. In: *Four Seminars*, ed. Heidegger, M (trans: Andrew Mitchell, and François Raffoul). Bloomington: Indiana University Press.

Kockelmans, Joseph J. 1984. *On the truth of being: Reflections on Heidegger's later philosophy*. Bloomington: Indiana University Press.

Lafont, Cristina. 2000. *Heidegger, Language, and World-Disclosure*. Trans. G. Harmon. Cambridge: Cambridge University Press.

Malpas, Jeff. 1991. *Donald Davidson and the mirror of meaning*. Cambridge: Cambridge University Press.

Malpas, Jeff. 1997. The transcendental circle. *Australasian Journal of Philosophy* 75: 1–20.

Malpas, Jeff. 1999. The constitution of the mind: Kant and Davidson on the unity of consciousness. *International Journal of Philosophical Studies* 7: 1–30.

Malpas, Jeff. 2006. *Heidegger's topology*. Cambridge, MA: MIT Press.

Malpas, Jeff. 2010a. The beginning of understanding: Event, place, truth. In *Consequences of hermeneutics*, ed. Jeff Malpas and Santiago Zabala, 261–280. Chicago: Northwestern University Press.

Malpas, Jeff. 2010b. Donald Davidson. *The Stanford encyclopedia of philosophy*. Fall edition, ed. Edward N. Zalta. http://plato.stanford.edu/archives/fall2010/entries/davidson/

Malpas, Jeff. 2012a. The place of space: Geography and philosophical topography. *Planning and Environment C: Space and Society* 30: 226–242.

Malpas, Jeff. 2012b. Ground, unity, and limit. In *Heidegger and the thinking of place*, 73–96. Cambridge, MA: MIT Press.

Malpas, Jeff. 2013. Philosophy, topography, triangulation. In *Triangulation: From an epistemological point of view*, ed. Maria Cristina Amoretti and Gerhard Preyer. Berlin: Ontos Verlag.

Malpas, Jeff, and Karsten Thiel. 2011. Kant's geography of reason. In *Reading Kant's geography*, ed. Stuart Elden and Eduardo Mendieta, 195–214. Albany: SUNY Press.

Nulty, Timothy J. 2006. *Primitive disclosive alethism: Davidson, Heidegger, and the nature of truth*. New York: Peter Lang.

Okrent, Mark. 2011. Davidson, Heidegger, and truth. In *Dialogues with Davidson*, ed. Jeff Malpas. Cambridge, MA: MIT Press.

Schuurman, Egbert. 1980. *Technology and the future*. Toronto: Wedge Publishing.

Smith, William H. 2007. Why Tugendhat's critique of Heidegger's concept of truth remains a critical problem. *Inquiry* 50: 156–179.

Tugendhat, Ernst. 1994. Heidegger's idea of truth. In *Hermeneutics and truth*, ed. Brice R. Wachterhauser. Evanston: Northwestern University Press.

Wittgenstein, Ludwig. 1969. *On Certainty*. Trans. D. Paul and G.E.M. Anscombe. Oxford: Blackwell.

Wrathall, Mark. 2010. *Heidegger and unconcealment: Truth, language, and history*. Cambridge: Cambridge University Press.

What Can Philosophy of Science Learn from Hermeneutics: and What Can Hermeneutics Learn from Philosophy of Science? With an Excursus on Botticelli

Jan Faye

Abstract For a long time hermeneutics and phenomenology were the dominant positions in the philosophy of the humanities. Consequently objects of interpretation and understanding were denied an objective standing. Hence the validity of these constitutive acts of meaning depended on the historical situation of the interpreter and of the object of interpretation. In this paper I deny that this needs to be so. I do agree with the hermeneutic-phenomenological tradition that interpretation plays as significant a role in understanding objects of science as it does in understanding cultural objects. I propose a view of interpretation and understanding that rests on the idea that human cognition is a natural phenomenon. I therefore hold that the science of the humanities is not that different from other empirical sciences as long as we include human intentions as the core object of understanding. Based on these suggestions I conclude that there exists objective understanding in the humanities in the sense that the validity of an interpretation, no more than an explanation in the sciences, needs to depend on the interpreter's historical situation or personal affairs. At the end I use the interpretation of Botticelli's *The Mystical Nativity* as an example.

1 What Is Understanding?

Traditionally hermeneutics and philosophy of science have had a rather antagonistic view of each other. In the hermeneutic tradition going back to Wilhelm Dilthey, it is part of a general account of science and humanities that the notions of explanation and understanding are kept strictly apart. Science, Dilthey famously said, *explains*

J. Faye (✉)
Department of Media, Cognition, and Communication, Section for Philosophy, University of Copenhagen, Karen Blixensvej 4, DK-2300 Copenhagen S, Denmark
e-mail: faye@hum.ku.dk

B. Babich and D. Ginev (eds.), *The Multidimensionality of Hermeneutic Phenomenology*, Contributions to Phenomenology 70, DOI 10.1007/978-3-319-01707-5_15,

the natural world, whereas humanities *understand* human life.[1] This demarcation may also add to the account of why philosophy of science has spilled so much ink on the concept of explanation, but only little on interpretation and understanding. Carl Hempel and his successors regarded understanding in connection with explanation as a psychological notion. It might be specified as a form of subjective expectation.

In recent years, however, philosophers have begun to be interested in questions concerning scientific interpretation and understanding which means that mutual interests between philosophy of science and hermeneutics become apparent. Scientists understand as much as they explain because explanation itself expresses understanding. Some authors, for instance, have argued that scientific understanding can be considered as a skill. If we go back to Michael Polanyi we see that possessing a skill is to have 'tacit knowledge.'[2] Such a suggestion seems reasonable but is not without problems. A skill cannot be ascribed a predicate like true or false, it is a practice that does not necessarily reflect a rule-following procedure. Skills seem always to be functional. A person must have the capacity to do something particular in order for that person to have a certain skill. A person must be able to realize some specific goal. But understanding need not be functional in the sense that it has a practical purpose or leads to a goal. A person may understand a verbal order even though that person is unable to carry it out. A person may understand a joke or a paradox without having other skills than the ability of repeating it. A person may understand something, say, by reading the instructions of a manual but that person may still not know how to carry out the instructions in any practical way. I think that understanding may give rise to skills and that skills are based on understanding. Thus the concept of understanding is just as fundamental as that of skills.

What is 'understanding' then, if it is not a skill? In short, I take understanding to be the organization of beliefs or bodily stored information. It is because of this organization that we have skills. Usually we understand what we believe because a particular belief is connected to other beliefs. It is only when we do not have understanding that we look for an interpretation or an explanation because they may help us to connect a particular belief about some unintelligible phenomenon with other well-established beliefs about other intelligible phenomena.

However, philosophy of science can learn from hermeneutic-phenomenological tradition (as it is presented by Merleau-Ponty) that much of our common experience is acquired as *embodied* cognition. We may consider embodied cognition in contrast to *reflective* cognition. Being a naturalist, I take this to imply that the capacity of embodied cognition is grounded in the human evolution and that a person's embodied knowledge is non-linguistically learned, whereas reflective cognition, of which the sciences and the humanities are parts, is linguistically attained as part of the education of scholars and scientists. So I want to distinguish between two fundamental sorts of understanding. First we have *embodied* understanding of being in the world and practicing a science, and second, based on this

[1] Dilthey (1894), 144.

[2] See, for instance, Polanyi (1966), Chap. 1.

concrete understanding, we have *reflective* understanding that builds on an abstract distinction between subjects and objects, between description and the described, or between representation and the represented.

Thus, embodied understanding is concrete and is a result of non-intentional action and perception, while reflective understanding is abstract and is the result of a purely intentional capacity of thoughts. Embodied understanding is non-interpretive and non-representative experience, whereas reflective understanding is representational thoughts and may be due to an interpretive act of reasoning. In science we meet manifestations of these two kinds of understanding. Often they are intermingled in the actual research processes, and abstract thoughts may sooner or later become concrete experience.

As part of their scientific practice scholars and scientists have concrete, embodied understanding which does not rest on interpretation but which interpretation presupposes or from which it departures. Scholars and scientists' concrete understanding results from their existence in the world (as Heidegger claimed) and from their scientific socialization into the practice of a certain science. This kind of understanding consists of the scientists' lifelong experiences. It does not emerge from a distinction between representations and the represented. In fact it is not possible to uphold such a distinction if one only focuses on embodied understanding.

But scholars and scientists also seek to form new theories or new models in order to explain new phenomena. Before they can get there scholars and scientists have to set up theoretical representations by virtue of interpretation. Here we can separate the abstract representation from what is represented. This is due to the fact that the representation goes behind our immediate perceptual and experimental experience. This abstract form of understanding is neither embodied nor directly based on tacit knowledge. It can be explicitly expressed. Often we arrive at reflective understanding by an appeal to analogous thinking in order to bring in well-known features and structures. What yields reflective understanding is either a fruitful conceptualization of a certain unknown phenomenon in order to bring it within the scope of a conceptual framework or a successful explanation of a phenomenon in terms of already well-understood phenomena. Reflective understanding consists in grasping how pieces of information relate to one another, seeing how they can be connected so they hang together coherently.

I propose that reflective understanding has intrinsic conditions for success, whereas embodied understanding, even viewed internally, involves extrinsic condition for success. In addition, abstract understanding has not only internally accessible criteria but also these are transparent in the sense that it is impossible to understand without understanding that one understands. In contrast to the usual internalist assumption, I hold that it may be possible to possess embodied understanding without knowing that one understands.

Much more can and should be said about the distinction between embodied and reflective cognition. But let me emphasize two things. The scientific practice builds not only on reflective thinking but just as well on tacit understanding in the form of embodied cognition. Moreover, the practice of the sciences or the practice of the humanities is carried out based on already well-established conceptual systems, but if any of these is challenged or developed it involves procedure of interpretations. Scientific discoveries do not necessarily build on interpretation, and the aim of science is not merely prediction and explanation but also the attainment of scientific understanding.

Explanation and interpretation help scholars or scientists to gain reflective understanding, but scholars or scientists rely first and foremost in their daily practice on a reflective understanding, which neither comes from explanation nor interpretation. It consists of what they have already learned and came with education and adopting tradition. Nevertheless, it is also the case that both interpretation and explanation provide scholars and scientists with an abstract and a theoretical form of understanding.

2 The Hermeneutic View of Interpretation

In the hermeneutic tradition it is part of the general understanding of science and humanities that the notions of explanation and interpretation are kept strictly apart. Dilthey basically thought that nature is alienated. It is external to us and given to us only piece by piece via sense experience, while the spiritual life is internal to us and is given in its full continuity. The spiritual lies open to us and can therefore be understood in its particularity. In contrast, science must postulate structures behind observable phenomena together with observable phenomena to be able to bring the latter into a necessary connectedness. Also Dilthey believed that we can only know of other people through a comparison with ourselves. He argued that all understanding in the humanities consists in a reconstruction of another person's mental life based on a perceptible particular like an action, a document, an artwork, or a literary text. The method, by which this is done, Dilthey held, is hermeneutics in Schleiermacher's tradition. So classical hermeneutics associated understanding with meaning and saw interpretation as the method to acquire such an understanding.

In opposition to Schleiermacher and Dilthey, we find Gadamer arguing that understanding does not consist in a reconstruction of the other mind through empathy. The fundamental principle is that we and the other mind, we and the text, always share a horizon of understanding, i.e., a common amount of beliefs, and that any understanding consists in overcoming those divergences that do not immediately fit by virtue of bringing the horizons together.[3] Gadamer believed that understanding and interpretation were impossible to separate because a separation would presuppose setting up two distinct horizons of understanding, the artist/author's and the interpreter's, in opposition to one another. It is, however, impossible to make such a separation since we cannot abandon our own horizon, much less enter another horizon distinct from our own. Our horizon of understanding is always situated in history and therefore becomes historically dependent. Each era

[3] See Gadamer (1960/1993), especially 302–307.

brings its own expectations to the text or the artwork, puts its own questions, and comes up with different answers. There is no objective interpretation of texts or artworks. A text, which is interpreted again and again through centuries, gives rise to different interpretations and validations. I find this view very problematic, but shall refrain from elaborating on this here.[4] Instead, in opposition to Gadamer, I want to emphasize that (1) interpretation should be separated from understanding; (2) many interpretations are explanations of meaning; and (3) that explanations of meaning are objective in the sense that all texts or artworks are produced according to certain intentions of the artist or the author and the object of the interpreter's act of interpretation is the manifestation of these intentions as expressed in the work. The first of these points was addressed in the section above; the other two points will be dealt with in the following sections.

3 Explanation as a Pragmatic-Rhetorical Practise

To begin, let me recall briefly what I take explanation to be.[5] I view explanation as part of a communicative discourse in which the explainer expresses his/her understanding as a response to an explanation-seeking question. In contrast to Hempel's covering law model of explanation, this pragmatic approach denies that the concept of explanation can be characterized solely in semantic or syntactic terms. And contrary to the ontological approach, it denies that explanation is only concerned with ontological categories like causation. The pragmatic approach first and foremost regards explanation to be an appropriate answer to an explanation-seeking question in relation to a particular context. A question is being raised in a situation where the questioner has a cognitive problem because he or she lacks knowledge of some form and now hopes to be informed by an explanatory answer. Therefore, the pragmatic view regards the context of the explanatory discourse, including the explainer's cognitive interest and background beliefs, as what determines the appropriate answer. Pragmatists think that the acceptability of the explanatory product is partly a result of the circumstances under which the explanation is produced. Also, they take scientific explanations to be basically similar to explanations in everyday life. The similarity between different kinds of explanations is found in the discourse of questions and answers that takes place in a context consisting of both factual and cognitive elements. The claim is that we do not understand what an explanation is unless we also take more pragmatic aspects around a communicative situation into consideration. The pragmatic view regards explanation as an agent of change in belief systems.

Thus, the pragmatic-rhetorical approach holds that a response to an explanation-seeking question in science need not follow *valid* deduction from a set of premises, nor does it need to appeal to a causal mechanism; hence, the acceptance of a

[4] In Faye (2012) Chap. 5, I discuss Gadamer's view in greater detail.

[5] For a further presentation, see Faye (2002), Chap. 3, and (2007).

response as an explanation includes lots of contextual elements. It does not pretend to give us more than a *descriptive* account of what the audience will accept as an explanation. Whether an explanation is good or bad, true or false, is not the issue as long as it fits into the general pattern of scientific inquiry. So the insight that can be associated with this pragmatic view of explanation is that scientific inquiry, and thus scientific explanation, is goal-oriented and context-bounded. It is always performed relative to some set of interests and a set of epistemic norms and standards that are context-dependent. Moreover, those norms and standards often vary with change of context without being explicitly acknowledged; thereby leading to controversies about what is an acceptable explanation.

A common objection against any pragmatic theory is that it cannot cope with the widespread wisdom that the understanding one gets from scientific explanations must be objective and invariable. To the extent that this intuition is correct I believe the pragmatic approach can account for it. The pragmatist does not have to deny that scientific explanations are concerned with a mind-independent world against which scientific explanations therefore are measured to find out whether they are true or not. She may be a realist of sorts. But in my opinion the common wisdom has limited value. It is based on a flawed metaphysics that there is *always* one, and only one, correct way of describing the mind-independent world. Our description of the world is dressed in conceptual and theoretical clothing, but the conceptual garb may be renewed from time to time, and norms and standards for evaluating one's beliefs change with respect to the problem in need of an explanation. Such a change of explanation comes not only with historical development over time but also with the context of the problem. The fact is, I believe, that scientific theories may be empirically underdetermined by evidence, which means that the theory one accepts is determined by factors other than mere observations. These other factors are, however, not equally objective, nor do they have an objective ranking. Here personal or shared interests play an important part.

In my opinion, explanation should be understood in the general context of interpersonal communication. Explanation is closely connected with understanding. When we explain things and events to each other, we pass on information about an immense range of different topics. These may cover such things as the structure of the natural world, social tensions, historical events, reasons for our actions, the meaning of words, symbols, literature and art works, or instructions on how to operate a certain piece of machinery. Explaining things and events is thus an appropriate linguistic reaction to what is considered to be an explanation-seeking question by which we distribute information of all kinds to one another.

Faced with the notions of the explanatory act and the explanatory product we must ask ourselves which of them is conceptually prior to the other, or whether they really can be characterised independently of each other. If one is conceptually prior, does it then mean that the secondary sense has to be understood in terms of the primary? A quick glance at the debate shows that most philosophers who defend one of the other approaches, focus entirely on explanation as a propositional outcome.

They never tell us in details how theories, facts, or events possess a capacity of explaining independent of human intentions. Indeed, what they want is to separate objective and subjective features of explanation. They assume that explanation can be completely characterized in terms of formal or ontological categories by abstracting explanation from the pragmatic context in which it takes place. In the right context sentences such as 'The fact that chlorophyll is green explains why plants are green,' 'The decline in interests rates explains the increase in investments,' and 'Maxwell's theory explains that light is electromagnetic radiation' are indeed completely meaningful. But, I surmise, the use of the term 'explanation' is parasitic on the notion of a linguistic discourse that is responsible for binding explanans and explanandum together. The pragmatic theory presupposes that practise is prior to logical status.

Neither facts, nor causes, nor laws explain anything by themselves. There exists no explanatory relationship in nature. Explanation is not an extensional concept, but an intensional one due to the fact that it is meant to confer understanding to the inquirer. Therefore, every explanation is sensitive to our way of describing the facts we seek to explain. The explanatory relation is between utterances or statements. However, this does not mean that the explanatory product is objective in another sense. One might think that theories, propositions, or logical arguments exist as abstract structures that make them publicly accessible. But as such they only have virtual existence. The explanatory product is produced with the intention of bringing forth understanding, and therefore its acceptance depends on the explanatory act of fulfilling this intention. Furthermore, we may say that the ontology of explanation is such that the explanatory product has a concrete and temporal existence as part of a communicative activity. Only as part of a discourse can a response to an explanation-seeking question become accessible for evaluation to other people.

Indeed, we do talk about facts explaining facts; however, this is really an elliptical way of expressing that explanations are concerned with facts, and that we want explanations to be true. We should not blur the distinction between the particular act of explanation and the explanatory force of this action. Nevertheless, what counts as an explanation is not just a question of facts but as much a question of pragmatic communicative strategies. It is fully acceptable to say that facts explain facts as long as we also recognize that in one discursive context, a certain kind of fact is required to provide the requisite understanding, whereas in a different context, a different kind of fact is called for. In many cases it makes good sense to pay attention to the product of the explaining activity whenever we focus on the different *kinds* of explanation. But if we want to understand explanation as such, we must acknowledge that the meaning of the explanatory product is partly determined by the context of the explaining act. It is no surprise that the form and content of explanation offered by different empirical sciences vary according to subject matter. But subject matter only partly determines the manner in which people explain things; other factors include the context of the audience, and the explainer's and the explainee's background knowledge and cognitive interests.

4 Interpretation

Medieval scholars made a distinction between *subtilitasintelligendi* and *subtilitasexplicandi*. These two notions are equally important for understanding the practice of science. Today, however, the modern use of the word "interpretation" seems to cover both senses and thereby blurs an important insight. I shall therefore suggest a distinction between two notions of interpretation: one concerns *construction of meaning*, another relates to *explanation of meaning*. In general, however, I take any interpretation to be a response to a representational question that involves a hypothesis about the connection between the representer and the represented. Originally our representational understanding began as a constructive response and and, later, may be used in an explanatory response.

We ask for an explanation with the hope of gaining understanding, and we make interpretations for similar reasons. It is a common view that interpretation is associated with the understanding of meaning. The objects of interpretation are considered to be intentional objects or objects having intentional properties. Therefore, interpretation is seen as a process that leads us to an understanding of persons, actions, or products of these actions, such as linguistic expressions, texts, painting, sculpture, music, film, dance, plays and social institutions. What we understand is the meaning being expressed by these products and an interpretation is what shows the way to this meaning by means of a hypothesis. So an interpretation is viewed as a response to a question like "What does *X* mean?" This view is in my opinion too simplistic.

First of all, *X* need not be of human origin. Any natural phenomenon can become cognitively meaningful in the right circumstances. Causally produced effects become meaningful if they are designed as data or are considered as evidence of their causes. In addition to being physical phenomena they have gained a cognitive status by being conceived as 'data' or 'evidence.' We convey thereby a certain meaning to them in the sense that we take these phenomena to be capable of informing us about something they are determined to represent. So data and evidence can be objects of interpretation in case someone doesn't understand what they inform us about or are determined to represent.

There is, as already mentioned, a general ambiguity in the way we think of interpretation which seems to have gone unnoticed. Sometimes the object of an interpretation is what is considered to represent something such as data, signs, symbols, and symptoms. The interpretive question is then what these phenomena are evidence of, what they stand for, refer to, or what caused them. But at other times the object of interpretation is types of phenomena which lack an appropriate classification or conceptual representation. Such phenomena may be kinds of entities, relations, sortal properties, etc. No phenomenon by itself points to anything beyond itself. It does not reveal how it must be conceptually understood. In this case the interpretative question is concerned with how they can be made the subject of representation. Finally, there are times where the interpretation question is about whether a particular phenomenon belongs to this or that category.

So what we call "interpretations" are explanations of cognitive meaning if the interpreter addresses questions like "What does *X* stand for?," "What does *X* represent?," "What is *X* evidence of?," etc., and the interpreter intends to answer these interpretation-seeking questions by proposing an appropriate response in form of a hypothesis. I shall therefore suggest that interpretation of this kind, like other types of explanations, can be considered as the interpreter's response to a question which expresses the questioner's lack of understanding.[6] Both question and answer rely on the discursive presupposition that the phenomenon *X* can be understood as intended by somebody to represent something or can be seen as evidence of something which *X* represents. This form of interpretation may also be called *determinative* interpretation since the interpreter determines by explanation the cognitive meaning which a culture, or an individual person in a culture, associates with the phenomenon under consideration. Explanation of meaning is indeed possible only if the interpreted phenomenon *already* carries a meaning independently of the interpreter's act of interpretation because it has already been constructed to have such a cognitive meaning.

However, other acts of interpretation are cases where the interpreter gives meaning to the phenomenon in question. In these cases the focus is on a phenomenon *Y* which the interpreter does not find intelligible as it appears and therefore does not know how to grasp. These situations give rise to questions like "How can *Y* be understood?"and "How is *Y* to be represented?" The interpreter's responses to such questions might be called *investigative* interpretations. What is characteristic of an investigative interpretation is that the act of interpretation makes the phenomenon *Y* intelligible by assigning a certain representational understanding to it. This happens by proposing a new form of classification, conceptualization, or schematization, and then by bringing the resulting beliefs into a coherent connection with the interpreter's belief system. In this case the understanding of *Y* depends on the interpreter's own invention.

I think we need a much broader perspective on 'interpretation' in which it is seen to offer both explanation and conceptual understanding in the humanities and the sciences. As mentioned above, 'interpretation' can be characterized as a means of gaining explanatory understanding of the 'meaning' of a representation. An interpretation in its determinative sense suggests a deliberately formulated hypothesis addressing a representational issue concerning what a representation really represents. The relation between explanation and interpretation is here that the latter is a form of explanation by which one explains *the representational role* of some representation. An explanation of meaning arises in contexts where a phenomenon is considered to represent something else, but where someone has doubts about what the phenomenon really stands for; it may be in connection with the consideration of

[6] Note that the interpreter and the questioner may be one and the same person but may also be two different persons. In the first situation the interpreter eventually answers his own interpretation-seeking question by expressing his own understanding, in the second situation the questioner raise an interpretation seeking question to another person in the hope of being informed by this person's answer.

effects, data, evidence, signs, symbols, texts, pictures, films, statements or actions. Moreover, 'interpretation' in a second non-explanatory form can be seen as a proposal of classification, conceptual representation, or theory formation. Under those circumstances we respond intentionally with an interpreting act only when we do not possess enough information to grasp the puzzling phenomenon in question.

Elsewhere I have argued for a pragmatic notion of interpretation in which interpretation in its determinative form is a context-dependent response to a meaning-seeking question and for applying a pragmatic and rhetorical theory of explanation to interpretation as well.[7] According to this unified theory of explanation and interpretation, the form of interpretation under consideration is a deliberately produced answer to a meaning-seeking question. The result of the interpretive process is certain statements concerning a representational issue, whereas the process is the communicative action that leads to these statements. The way the result turns out depends somehow on the context and therefore, among other things, on the aim and interest of those who do the interpretive work.

My suggestion is that the type of interpretation is determined partly by the interpreter and partly by the object of the interpretation. Indeed, the object plays an important role in the interpreter's selection of the relevant type of interpretation. The interpreter constrains her interpretation in accordance with her grasp of the object by choosing the type of interpretation accordingly. For example, experimental data will give rise to another type of interpretation than a text or a painting. But the interpreter's knowledge of the situation, her goals and interests are also elements in determining the form of interpretation. Thus, the person's background assumptions, beliefs and knowledge of the object influence the hypothesis he generates. This applies not only to the form of hypothesis, but to the content as well. The content of an interpretation is as much context-dependent as its form. But, again, the object of interpretation imposes some constraints on any possible understanding of the content.

A short characterization of the two notions of interpretation I present here would then be that the determinative interpretation signifies a situation in which the interpreter explains the meaning of a certain representational phenomenon, usually an action, an expression, a sign, a symbol, or something similar, by a hypothesis concerning what is represented by this phenomenon. The investigative interpretation, on the other hand, signifies a situation where the interpreter constructs a hypothesis which provides meaning to the phenomenon he wants to have represented in case he does not already grasp the meaning of this phenomenon. I shall also hold, however, that an investigative interpretation may change status and become part of a determinative interpretation whenever the scientific community reaches a common agreement that the conceptual construction of an investigative interpretation should form our general understanding of the phenomenon in question.

Strangely enough, philosophers of science who have been occupied with explanation have shown little interest in characterizing interpretation in spite of the fact that they themselves speak of interpretation. This lack of interest is partly due

---

[7] See Faye (2010, 2011, 2012).

to the fact, I think, that they intuitively assume that these two concepts belong to each side of Reichenbach's famous distinction between the context of discovery and the context of justification. Thus, interpretation has to do with the context of discovery, whereas explanation belongs to the context of justification. This led philosophers like Karl Popper and Carl Hempel to develop the deductive-nomological model of explanation. They simply ignored interpretation as being too much of a psychological notion with its close ties to meaning and understanding; they tacitly seem to have accepted the hermeneutic division between explanation and understanding as important for a characterization of the difference between the natural sciences and the humanities [*Geisteswissenschaften*]. In contrast, hermeneutic philosophers have dealt with understanding and interpretation, but paid no attention to explanation. An important consequence is that rigorousness of the various accounts of explanation is missing with respect to the accounts of interpretation. Explanation was the object of a logical analysis, interpretation involved a subjective synthesis. But if we consider explanation to be an act of intentional communication, there is no reason to uphold the traditional dichotomy between explanation and what I call determinative interpretation. In both cases the explainer or the interpreter supplies some information which is needed for understanding the topic of the explanation/interpretation-seeking question.

Now the challenge we still have to meet is this: scientists often produce causal explanations, and these explanations are taken to be objective because their topics are mind-independent. Causal relationships, which are object of causal explanations, exist regardless of the explainer. But when we turn to the interpretation of art or texts, any explanation of meaning seems not to be controlled (even partially) by similar objective facts. Rather each time a scholar is confronted with an artwork or a text he or she seems to construct the meaning instead of establishing the meaning. However, I shall attempt to prove that scholars working on empirical and scientific grounds both possess and use objective criteria to determine what kind of information is relevant and what isn't in the process of creating a particular interpretation. These criteria are not of the scholars own making but stem from features which the community of scholars believes are controlled by the artist or the author and therefore are considered as meaningful expressions of the artist's or author's intentions.

5 Botticelli's the Mystic Nativity

Among the oeuvre of the famous renaissance painter, Sandro Botticelli, there is one work which for many years has escaped a satisfactory scientific interpretation. Its iconography does not fit the traditional way of depicting similar motifs. The painting is called *The Mystical Nativity* and owned by the National Gallery in London. It shows, as the name suggest, the birth of Jesus Christ. At the center of the painting the Holy family is situated under the roof of a shelter that covers the entrance of a cave where an ox and a donkey appear in the opening. The Virgin Mary, infant Jesus, as well as Joseph, who has fallen asleep, are larger than other figures and their surroundings – as the medieval convention prescribes. A group of men, including wise men, shepherds and a couple of angels, kneel behind the Virgin and Joseph. At the top of the roof sit three angels, the one in the middle holding an open book (the Bible?) and the two others each holding an olive branch. They are dressed in a red, a white, and a green gown signaling faith, hope, and charity. Above them in the sky twelve angels circle under a golden dome, each dressed in similar colors, and each one holding an olive branch and a white ribbon. At the bottom of the painting three more angels embracing three men meet one's eye and around them fivesmitten and self-destructive demons are trying to flee to the underworld. The men hold scrolls proclaiming in Latin: "Peace on Earth to men of goodwill."

The most remarkable feature is the instruction to the viewer which Botticelli has placed at the top of his painting. The Greek inscription translates as: "This picture, at the end of the year 1500, in the troubles of Italy, I, Alessandro, painted in the half time after the time; at the time of the fulfillment of the eleventh of St John, in the Second Woe of the Apocalypse; in the loosing of the devil for three and a half years; then he shall be chained according to the twelfth, and we shall see him [here a word or two is missing] as in this picture."[8] The reference is to Saint John's Book of Revelation where Chap. 11 speaks of the second woe and mentions three-and-a-half years. In Chap. 10, which Botticelli's text does not explicitly mention, it is told that an angel descends from Heaven holding a book in her hand and a voice from Heaven commands Saint John to take it. The central angel sitting on the roof of the stable holding a book supported by two other angels may very well allude to this passage.

Not everything about this painting involves interpretation in order to be understood. Although parts of it are not immediately meaningful for an art historian, there are many other features which are directly understandable even for ordinary people. We can see it as a painting; we see angels, persons, an ox and a donkey without being engaged in any form of interpretation. Everybody speaking a language has the conceptual resources to be aware of these depictions right away. People raised within a Christian culture also immediately recognize the birth of Jesus based on their knowledge of the Gospels' narrative of the birth of Jesus and thousands of other representations of the same story. But ordinary people's pre-existing understanding ends here. The art historians' immediate understanding reaches a bit further. They know already the meaning of the various colors of the various dresses, the iconographic meaning of the olive branches, and the meaning of the enlargement of the Holy Family. By knowing the renaissance context in which *The Mystic Nativity* is painted, art historians can rely on their professional understanding of the iconographic use of colors and symbols at that time. None of these items give rise to any question of meaning, and as long as they do not give rise to such questions, nothing is open for interpretation. It is only if somebody seriously challenges, say, the meaning of the colors that the community may engage interpretive questions.

[8] This translation is due to Hatfield (1995), 98.

But Botticelli's painting is still full of symbols which are not easily understood even by scholars of renaissance art and which provoke them to ask interpretation-seeking questions such as why the painting depicts twelve angels dancing in the sky under a golden dome, why it depicts three angels embracing three olive garlanded men, and why it depicts five demons attempting to escape. What is the meaning of these symbols? Understanding their meaning is not within the scope of well-documented iconography. At this point scholars must formulate certain interpretive hypotheses about their meaning which can be used as a scientific response to that type of questions. Symbols are ambiguous by nature. They are only well-defined in context and the aim of interpretation is to identify this context to determine the meaning.

However, interpretive hypotheses have to be relevant, and the criterion by which scholars can judge that a certain hypothesis is relevant is if the painter can possibly have had this meaning in mind. Besides being relevant interpretive hypotheses must be supported by empirical evidence which comes from historical, cultural and biographical information. The key to the interpretation of this painting seems to be the Greek inscription where Botticelli gives us certain contextual clues. He mentions that Italy faces troubles and two chapters of the Saint John's Revelation. The inscription states that the painting is executed at the time of the fulfillment of the second woe of the Apocalypse after which Satan will be chained as we see in the picture. So any good interpretation must explain this unusual collocation of the birth of Jesus Christ (in the past) and the vision of a world to come (in the future).

The 12th chapter introduces the apocalyptic woman "clothed with the sun, with the moon under her feet, and on her head a garland of twelve stars." She is in labor with a male child. A fiery red seven-headed dragon waits for the birth of the child. However, when born, it is taken up to God's throne while the woman flees into the wilderness for *one thousand two hundred and sixty* days (the same as three and a half years). Exegetic writings often identify the apocalyptic women with Virgin Mary, and she is interpreted as a symbol of the Church. The dragon is taken to be the Devil. This theme also seems to be the object of Botticelli's painting. The second woe will be followed by the realization of the prophecy as Saint John presents in the 12th chapter: Virgin Mary, now having 12 angels circling above her head instead of a wreath of 12 stars, symbolizes the Church, and the Savior's rebirth announces the rebirth of the Church. It is the beginning of a new day that can be seen from the dawn in the background of the picture. The devil will be expelled from the Earth (the five demons beneath) and eternal peace will last forever.

Hatfield mentions three possible interpretations of which this one seems to be the most probable given textual as well as contextual circumstances.[9] But what is the evidence in support of this interpretation? At the end of the fifteenth century an increasing fear accumulated because it was foreshadowed that the end of the world was approaching. The Dominican friar Girolamo Savonarola spoke about it in his stirring speeches, and he urged the people of Florence to give up their luxurious lifestyle, do penance, and show remorse over their undignified and profane life.

[9] Hatfield (1995), 112 ff.

He found many supporters and many of their luxury goods were committed to the flames in a huge bonfire. However, Savonarola was politically too radical, he got many enemies, the pope excommunicated him, and finally he was prosecuted and condemned to death as a heretic and a schismatic, and he was executed together with two of his lieutenants. The execution took place May 28, 1498. There is no doubt that Botticelli was deeply touched by these events.

In his *Compendio di revelatione*, first published in 1495, Savonarola sets forth a vision that had come to him in which he saw an extraordinary heavenly crown. At its base were 12 hearts with 12 banderoles surmounting them and written on these in Latin were the unique mystical qualities or privileges of the Virgin Mary – she is "Mother of her father," "Daughter of her son," "Bride of God," etc.[10] Though much of the writing on the ribbons held by the dancing angels is now invisible to the naked eye, infra-red reflectography has shown that the original words on the angels' ribbons correspond exactly to Savonarola's 12 privileges of the Virgin. In his sermon, preached on Assumption Day 1496, Savonarola went on to explore the 11th and 12th chapters of the Book of Revelation – the precise chapters mentioned in the painting's inscription. He connected the glory of Mary with the imminent coming of the power of Christ on Earth.[11] This theme is exactly the one which Botticelli seems to render by his own constructive interpretation. Thus, it seems very likely that the three men embraced by angels are Savonarola and his two lieutenants who are brought to life again with the rebirth of Jesus Christ.

6 Conclusion

The example illustrates how art history works as an empirical science. Interpretations are subjected to empirical scrutiny and may be rejected as mere speculations if they cannot account for the evidence.

The result is not in accordance with postmodern thinking. There is an 'objective' meaning in the form of the artist's intention and with this in mind it directs scholars in their search for possible interpretive hypotheses. But it may also offend certain hermeneutists. Dilthey was right in so far as the artist's intention is important for a scientific understanding of the work, but he was wrong when he thought that the method to reach such an understanding is empathy into the mind of the artist. It is the artist's intentions as they are expressed in the work which is the object of interpretation, not the artist's psychological moods or motives. Also Gadamer was right to the extent that he denied, in opposition to Dilthey, that the human sciences have their own methods.[12] But in my opinion he was mistaken when he claimed that all understanding is interpretation and when he argued against the objectivity of interpretation. For him, ideas, texts, and works of art are historically bounded and each

[10] Ibid, 94.

[11] Ibid., 96–98.

[12] Gadamer (1960/1993), 7–8.

interpreter is by necessity situated in his or her own time. The historical situation of the interpreter will always influence and become a part of his or her interpretation, and since the historical situation changes over the years, different interpreters standing in different historical situations cannot reach a common ahistorical, hence 'objective', interpretation. I think, however, that the above interpretation of Botticelli's *The Mystic Nativity* proves Gadamer wrong. The most likely interpretation today, of which there is growing consensus among art historians, is probably the interpretation which Botticelli himself would have agreed with, and which contemporary viewers would have recognized seeing the painting. Therefore I don't expect we will see a satisfactory but radically different interpretation in the future.

References

Dilthey, Wilhelm. 1894. *Gesammelte Schriften*, Vol. 5. Leipzig and Berlin, 1914–1936.

Faye, Jan. 2002. *Rethinking science. An introduction to the unity of science*. Aldershort: Ashgate. *An enlarged version of Athenes kammer. En introduktion til videnskabernes enhed (2000)*. Copenhagen: Høst & Søn.

Faye, Jan. 2007. The pragmatic-rhetorical view on explanation. In *Rethinking explanation*, Boston studies in the philosophy of science, vol. 252, ed. J. Persson and P. Ylikoski, 43–68. Dordrecht: Springer.

Faye, Jan. 2010. Interpretation in the natural sciences. In *ESPA epistemology and methodology of science*, ed. M. Suárez, M. Dorato, and M. Rédei, 107–117. Dordrecht: Springer.

Faye, Jan. 2011. Explanation and interpretation in the sciences of man. In *Explanation, prediction and confirmation*, ed. Dennis Dieks, 269–280. Dordrecht: Springer.

Faye, Jan. 2012. *After postmodernism: A naturalistic reconstruction of the humanities*. London: Palgrave Macmillan.

Gardamer, Hans-Georg. 1960/1993. *Wahrheit und Methode*. English edition: *Truth and Method*, 2 edn. New York: Continuum.

Hatfield, Rab. 1995. Botticelli's Mystic Nativity, Savonarola, and the Millennium. *Journal of the Warburg and Courtauld Institutes* 58: 88–114.

Polanyi, Michael. 1966. *The tacit dimension*. London: Routledge.

The Classical Notion of Person and Its Criticism by Modern Philosophy

Enrico Berti

Abstract The paper illustrates the classical notion of person, i. e. the definition of person given by Boethius (fifth to sixth century A.D.) as "an individual substance of a rational nature", showing the derivation of its elements from the philosophy of Aristotle. Afterwards the paper exposes the criticism to this notion formulated by modern and contemporary philosophers (David Hume, Joseph Butler, Alfred Ayer, Derek Parfit). Finally the text shows the reaction to this criticism and the rediscovery of the classical notion of person, or of its Aristotelian elements, by Saul Kripke, David Wiggins, Paul Ricoeur and Martha C. Nussbaum.

1 The Classical Notion of Person

By classical notion we mean the definition of "person" formulated by Boethius (fifth to sixth century A.D.), that is, "an individual substance of a rational nature" [*rationalis naturae individua substantia*, cf. *Contra Eutychen* III 1–6]. This definition possesses the unique characteristic of being theological in origin and of using at the same time purely philosophical categories. The origin of the definition is theological because Boethius introduces it polemically in opposition to the monophysitic heresy of Eutyches, which attributed to Jesus Christ a single nature, the divine one, and against the dualistic heresy of Nestorius, which attributed to him, as well as two natures, also two persons, one divine and one human. Against these positions Boethius defends the Christological dogma of the Council of Chalcedon (451 A.D.), which affirms the "hypostatic union," in a single person [the Greek term *hypostasis* is rendered in Latin as *person*], of two natures, one

E. Berti (✉)
c/o Dipartimento di Filosofia, University of Padova,
piazza Capitaniato 3, I-35139 Padova, Italy
e-mail: enrico.berti@unipd.it

B. Babich and D. Ginev (eds.), *The Multidimensionality of Hermeneutic Phenomenology*, Contributions to Phenomenology 70, DOI 10.1007/978-3-319-01707-5_16,

divine and one human. However, in order to formulate his definition of person, Boethius uses two concepts derived from Aristotle's *Categories*, of which he was the first Latin translator and commentator (together with all of Aristotle's writings on logic, i.e. the collection called *Organon*, which Boethius made known to the Medieval Latin world).

Indeed, the concept of "individual substance" corresponds to what Aristotle in the *Categories* calls "primary substance" [*ousia prôtê*], that is, "what is neither the predicate of a substrate nor inherent in a substrate," because it is itself a substrate. "Substrate" translates the Greek *hypokeimenon*, which literally means "that which lies underneath," which underlies becoming, change, inasmuch as it is its subject, that is, the thing that becomes, the thing that changes and which, in changing, persists during the entire process of change. It might also be translated as "subject" (*subjectum* in Latin is equivalent to the Greek *hypokeimenon*), but modern philosophy has agreed to use this term only for the human subject, while the substrate as intended by Aristotle indicates any subject of becoming, both living and non living. For Aristotle, substrate is that of which universal concepts are predicates, such as species, e.g. "man," and genus, e.g. "animal," and which accidental properties inhere in, e.g. "white" or "grammatical" (i.e., capable of reading and writing). Therefore, as the substrate is not predicated of anything else and is not inherent in anything else, it is "in itself". Since, in order to exist, both the universal and the accidental properties suppose the existence of a substrate on which they may be predicated or in which to inhere, this is termed not only *ousia* (literally "being" in a strong sense, that is, permanent, lasting), which in Latin is translated as *substantia* (literally "what is underneath," like the Greek *hypostasis*), but also "primary" *ousia*, that is, preceding all others. On the contrary, species and genus, which do not exist "in themselves," but only in the substrate, and nevertheless constitute its essence (that is, tell "what it is"), are termed "secondary" *ousia*.

As an example of "primary substance" Aristotle indicates "a certain man," that is Socrates, or Callias, and, more in general "a certain 'this'" [*tode ti*], that is, a determinate individual. Therefore Boethius rightly interprets the Aristotelian concept of "primary substance" as "individual substance." In this case, "individual" does not mean "indivisible" [*atomos* in Greek] but "particular," not universal, because species and genus, that is "secondary substances", are universal. Thus it is not indivisibility which is essential to the Aristotelian concept of primary substance, but individuality, i.e. particularity, the non universality, because the universal, that is the species and genus, is always "in something other," while the primary substance is always "in itself." Individuality, however, is not sufficient to build a primary substance, because there can also be particular or individual properties, for example Socrates' particular whiteness. Thus a primary substance must first and foremost be a substrate, or subject, and must also be individual. This is why Boethius, wanting to say that the person is first of all a primary substance, says that it is an "individual substance."

Even the concept of "nature," used by Boethius to characterise the type of primary substance which the person consists in, derives from Aristotle, where it is expressed by the term *physis*, which alludes to "birth" [the Greek verb *phuô*, in its intransitive meaning, corresponds to the Latin *nascor*, whose participle is *natum*], that is, what a thing is "by birth": e.g. a man is a man because he is born of human parents. In Aristotle "nature," in this sense, is synonymous with "essence," a concept also expressed by the term *ousia*, but with the meaning of "what something is by its own nature," which corresponds to the question "what is it by its own nature?". E.g., if I ask, "what is Socrates?," meaning what is he by nature, that is, by birth, the answer is: "man".

Finally, the term "rational," which Boethius uses to clarify the nature of the person, translates the Greek *logon ekhon*, that is "possessing *logos*." The term *logos*, as is well-known, in Greek certainly means "reason" (Latin *ratio*), but it first and foremost means "word" (Latin *verbum*) and "discourse" (Latin *sermo*, *oratio*). Therefore, Boethius' expression "of a rational nature," contained in the definition of "person," indicates an individual substance that, by its nature, that is, by its essence, possesses *logos*, i.e. speech, language. According to Aristotle, this is what distinguishes man from other animals, what constitutes the specific difference of the species "man" within the genus "animal." Since Boethius' definition applies first of all to divine persons, or to the person of Jesus Christ, the determination of "rational" cannot simply allude to the capability to reason, but must allude more in general to the capability to communicate, to enter into a mutual relationship. Indeed, according to the Trinitarian dogma, formulated by the Council of Nicaea (325 A.D.), the three persons of the Holy Trinity possess the same nature, that is, divine nature, and are distinguished only by the relationship they entertain mutually, that is, because the Son "is generated" by the Father and the Holy Spirit "proceeds" from the Father and from the Son. Already in the *Gospel* of John, the Son is called *Logos*, that is "word" [*verbum*].

Returning once again to speak of the human person, to whom Boethius' definition is applied by analogy with the divine one, we must remark that "substance possessing *logos* by its nature" does not necessarily mean "substance which currently exercises *logos*," but rather also substance that, by nature, possesses the capability of exercising *logos* even when it does not exercise it. Indeed, nature is what Aristotle would call a "primary act," that is the current possession of a body of capabilities, the exercise of which should be called "secondary act" or "activity." Therefore, on the basis of Boethius' definition, a new-born is also a person, even thought he is as yet unable to speak, and so is a human individual affected by aphasia, since he is born of human parents and therefore possesses a rational nature (leaving aside the problem of the human embryo, which would lead to a whole other series of problems, although, in my opinion, what has been said about the new-born can be applied).

2 Criticism of Modern and Contemporary Philosophy

Boethius' definition of person can be considered "classical" because it has remained at the basis of global culture, not only Christian but also Jewish and Muslim, both ancient, medieval and modern, that is of the entire culture which Aristotelian tradition has influenced: indeed, we find it with irrelevant variations in Augustine, John

Damascene, Richard of St Victor, Thomas Aquinas, G. W. Leibniz, Antonio Rosmini, Jacques Maritain and several other thinkers I do not need to mention.[1] However, starting from the seventeenth century the classical notion of person has been jeopardised, not so much because it has been criticised directly, but because the notions on which it is founded, i.e. "substance," "nature" and, more recently, "individual," have been criticised. First of all, the notion of "substance" has been, so to say, over-determined by Descartes and Spinoza, who defined it as "what does not need anything else to exist" or "that which exists by itself," which strictly can only be applied to a divine substance. As a reaction, the notion of substance was criticised by John Locke (1632/1704), who considered it "a complex idea," that is, borne not of direct experience (sensation and reflection), like "simple ideas," but of a combination of several simple ideas, that is, as a construction of the intellect, which does not correspond to any experience. The object of such an idea, that is, the substance strictly speaking, remains for Locke a *substratum obscurum*, that is, something that, so to say, is "underneath" or "behind" the primary or secondary qualities, that can be seen and therefore cannot be seen, cannot be touched, cannot be perceived in any way. With this doctrine we are very far from the Aristotelian notion for which substance is the single individual of whom one has a direct experience, e.g. Socrates. The notion of substance then underwent a further transformation on behalf of George Berkeley (1685/1753), for whom material substances do not exist, inasmuch as existence consists in being perceived (*esse est percipi*), thus the same qualities are nothing but perceptions and the only really existing substance is the percipient subject, that is, the human spirit (besides the divine Spirit).

These transformations led to the explicit criticism of the concept of substance on behalf of David Hume (1711/1776), according to which we do not have a direct experience either of material substances or of spiritual substances (that is, of ourselves as substance), therefore the idea of substance (as indeed also that of cause, which is the object of another memorable criticism by Hume) is only a belief of ours generated by habit, to which we cannot say any independent reality corresponds. For Hume we do not even have experience of ourselves, thus we are not a substance that persists, equipped with its own identity, but only a bundle of impressions that follow one another over time. Personal identity itself, which for Locke was guaranteed at least by memory, that is by conscience, for Hume is not guaranteed by any experience, although this is a problem for him, because in the Appendix to his *Treatise of Human Nature* he declares himself unsatisfied with the doctrine he himself had expounded and admits he has not been able to find a solution.

The Anglican bishop Joseph Butler (1692/1752) and the Scottish philosopher Thomas Reid (1710/1796) reacted to the criticism respectively of Locke and Hume. They referred to the classical notion of substance as the only thing capable of guaranteeing individual identity. But the somewhat narrow notion of experience as formed by individual sensations, or impressions, proper of empiricism, prevented the Aristotelian doctrine from being fully recovered, according to which the true object of experience is the primary substance itself, that is, the

[1] Cf. Berti (1992, 1995).

individual substance perceived in its entirety, with all its properties, including identity and persistence in change.

Even the attempt, made by Immanuel Kant (1724/1804), to give back objective value to the idea of substance (and to that of cause, on which the entire Newtonian mechanics is founded), considering it as an a priori concept, that is, a category of reason, universal and necessary, has not led to an actual recovery of the classical notion, because even Kant continued to admit that we do not have any experience of substance and the perception that we have of ourselves—the "transcendental apperception," or "I think"—is not the experience of a substance but is only the condition of each of our experiences. The idea of "soul" for Kant is an idea of reason, that is, the rational need to unify the psychic phenomena that we know of, which in any case is destined not to be able to be translated into authentic knowledge, for the very lack of an authentic experience of the soul. However, from the practical point of view, Kant has recovered the concept of person as a subject bearing the moral law and thus possessing his own "dignity," i.e., not exchangeability, which distinguishes him from things that are exchangeable and thus only have a "price," and makes him worthy of "respect," worthy of being considered always, in the person proper and in the others, not only as a medium but also as an end.

The concept of "nature," on the contrary, which is still present in Hume, who writes a *Treatise of Human Nature* trying to build a science of this analogous to the one build by Newton for non human nature, is also undermined in the nineteenth century, first by idealist and historicist philosophy and then by evolutionistic anthropology. Fichte, Schelling and Hegel's idealistic philosophy denies the existence of unchangeable essences and, resolving reality in thought, which is a continuous process, dissolves substances, essences and the bodies themselves in moments of a single major process, which is the becoming of the Spirit. However, it is worth noting that Hegel's most important critics, that is, Feuerbach, Marx and Kierkegaard, objected that it is not possible to have a process without a substrate, and conceived this substrate as the individual human subject, just as Aristotle did, explicitly recalling the latter (Marx even went as far as using the Aristotelic term of *hypokeimenon*).[2]

Evolutionistic anthropology, as is well-known, denies the fixed nature of the species and thus the interpretation that has been given of it by positivistic philosophy has gone as far as denying the existence of an unchangeable human nature, which is the same at all stages of evolution and in all the earth's peoples. The concept of "human nature" is thus replaced by the concept of "culture," intended as a differentiated, dynamic reality. However, also for this very reason, we must report a misunderstanding that took place at the beginning of the modern age, when "nature," in particular "human nature," was intended as an unchangeable essence, belonging to a hypothetical "state of nature," that is, to a primitive, pre-political condition of man. This notion, belonging to the so-called "jus-naturalism" (Hobbes, Locke, Rousseau), led to the opposition between "nature" and "culture" or between "nature" and "history," exposing the concept of nature to the criticism of evolutionism and historicism, which have shown that such a "nature" never existed and that the true

[2] Cf. Berti (2004).

nature of man is culture itself, that is, what man makes of himself. But, if we apply such criticisms to the Aristotelian and then to the classical concept of nature, they completely miss the mark, because for Aristotle, as we have seen, the true nature of man is *logos*, that is, speech, therefore political life, "culture." Indeed, man is for Aristotle "an animal who is political by his nature," precisely because of language, and the pre-political condition can belong only to beasts or gods. Besides, Aristotle explicitly states that the true nature of man is the end (*telos*), the achievement, the total fulfilment of human capabilities. Even from the point of view of the modern evolutionistic anthropology I do not think it can be denied that there is a marked difference between the human species and the other animal species, thanks to evolution, and this difference consists precisely in language and culture.

Finally, even the concept of individual, and the connected notion of "personal identity," has been the object of criticism on behalf of contemporary philosophy of empiricist and neopositivist inspiration. Alfred J. Ayer, the greatest representative of neo-positivism in Great Britain, has gone as far as denying the experience that we have of our very thought, declaring that one can never affirm "I think," but can only say "it is thought" or "there is a thought."[3] Derek Parfit, echoing Hume, maintained that the person is nothing but a series of subsequent "selfs" equipped with a collective identity, comparable to what is proper, for example, of a nation, in which individuals change continuously and what persists is only their common quality, that is, the fact of all belonging to the same nation.[4]

3 Reaction to Criticism and the Rediscovery of the Classical Concept of Person

In the Anglo-American philosophy of the second half of the twentieth century, characterised by analytic-linguistic inspiration, that is, by the notion of philosophy as language analysis—not only of scientific language, as was the case in neo-positivism (Russell, the early Wittgenstein, Carnap), but also of ordinary language—we see a progressive rediscovery of the classical notion of person, as an answer to the criticisms of modern and contemporary philosophy of Humean inspiration to the concept of substance and personal identity. To this end we must recall first of all the position of Peter F. Strawson, the continuer of the Oxford and Cambridge School inspired by the late Wittgenstein (Austin, Ryle), who, in the work *Individuals* (1959), tried to describe how the world must be able to explain the way in which we speak of it in ordinary language. By means of this description, which he called "descriptive metaphysics," Strawson showed that the ultimate reference of our language is always made up of particular objects, which are identifiable by means of space-time coordinates and reidentifiable through "sortal" designators (a term

[3] Ayer (1963).

[4] Parfit (1984). For the reaction to criticism and the rediscovery of the classical concept of person, see my article Berti (2006).

derived from Locke to indicate "what sort of" an object it is), that is, of a universal type. Among these particular objects, Strawson remarked, there are some that serve as a reference for the identification of others, which are called by him "basic particulars" or "individuals": they correspond exactly to what Aristotle called "primary substances" and which he indicated as the logical subjects of propositions. Among individuals, Strawson continued, there are some that play an even more basic identificatory role and correspond to original and not further analysable units of physical and psychic facts, which are persons. Persons are thus basic particulars, or individuals, that is "primary substances," with indissolubly united physical and psychic properties.[5] The affinity between this notion and the classical one is evident.

Simultaneously, in the United States he who today is perhaps considered the greatest American philosopher of the twentieth century, that is Willard v. O. Quine, in his work *Word and Object* (1960), maintained that the possibility of referring language to objects, that is, to give meaning to language, requires as a necessary condition the fact of being able to identify objects: indeed, there is no entity without identity.[6] This way, he reproposed the problem of personal identity, denied by Hume and by his most recent continuers. This has given rise to a debate the first document of which was constituted by the seminar on *Identity and Individuation*, which took place at the Institute of Philosophy of New York University during the academic year 1969–70, the proceedings of which were published in a book by the same title edited by Milton K. Munitz.[7] The problem is how it is possible to identify an individual, that is, to distinguish him from others coexisting in space and recognising he has a certain persistence, or identity, over time.

This problem in turn contains various issues, for example what authorises us to affirm the identity of a thing or a person when these change over time? Then there is the issue raised by Leibniz with the so-called "principle of the identity of the indiscernibles": is it true that two individuals who have exactly the same properties, that is, that are indiscernible, are also identical, i.e., are the same individual? Finally, there is a third issue, called forth by the famous essay by G. Frege, *Über Sinn und Bedeutung*, of 1892: how is identity possible between realities that are the object of different descriptions, for instance "morning star" and "evening star?"

A famous solution to this problem was suggested by Saul Kripke in *Naming and Necessity* (1980), according to whom there is identity when two "rigid designators," that is, two signs, that indicate essential properties, have the same referent in all possible worlds. But this supposes, exactly, that there are essences, the object of necessary truths, that is, of necessarily true although not analytic judgements (distinction introduced by Quine), which are first and foremost natural species but can also be classes of artificial objects.[8] The reference to essences naturally calls to mind Aristotle, but this is not essential to Kripke's thesis, which, although criticised and contested,

[5] Strawson (1959).

[6] Quine (1960).

[7] Munitz (ed.) (1971).

[8] Kripke (1980).

is certainly considered an important reference point within the framework of analytic philosophy and thus makes enough sense to be able to be discussed.

David Wiggins is also moving in the same direction as Kripke, but with more explicit references to Aristotle, and in *Sameness and Substance* (1980) he explicitly advocates that, to establish an absolute identity, as is the case in a single individual, it is necessary to resort to the Aristotelian concept of substance. Also for Wiggins natural species are substances and are each characterised by an "activity," that is, life, therefore they are not plain nominal essences in Locke's sense. The same character is possessed, although to a lesser degree, by artificial objects, for which functioning is analogous to activity. Thus, to identify something, it is necessary to say what it is, that is, to subsume it under a predicate that offers for it a principle of continuity or of individuation: this is what predicates indicating a principle of activity or functioning—i.e. the concepts of natural or artificial substances—do.[9]

The debate on identity was finally summarised in the treatise by D. W. Hamlyn on *Metaphysics* (1984), where the author showed that, in order to identity any object, first of all the reference to its space-time coordinates is necessary, then to its "space-time history" and, finally, to the species it belongs to.[10] This can lead to a form of essentialism, which, however—as Putnam noted in *The Meaning of Meaning* (1975)—is indispensable, especially for natural substances, such as water, which has as its essence the fact of being H_2O, whether we know it or not, in all possible worlds.[11]

Within the framework of the problem of the identity of substances, the problem of personal identity was recently taken up again, always in the framework of Anglo-American analytic philosophy. Wiggins suggested an original solution to this problem, indicating as the condition of personal identity not conscience, like Locke, but the continuity of life. Parfit objected to this that the important continuity for the person is not the biological one but the psychological one, which may fail during character mutations,[12] and Williams observed that this notion makes of the human person a simple natural species (it is the accusation of "biologism").[13] These objections of a spiritualistic nature do not take into account the fact that the higher functions of man are strongly conditioned by the biological ones, and that thought itself is a form of life, as proved today by the fact that the *Mind-Body Problem* is no longer addressed by the cognitive sciences by means of information technology or computer science, but especially by recourse to the neurosciences. This emerges clearly from the most recent formulation of "functionalism" by H. Putnam in the book *Words and Life*, where the author goes as far as speaking of a "return to Aristotle after Wittgenstein."[14]

---

[9] Wiggins (1980).

[10] Hamlyn (1984).

[11] Putnam (1975).

[12] Parfit (1973).

[13] Williams (1986).

[14] Putnam (1994).

However, together with the notion of person, analytic philosophy has also recovered the Aristotelian notion of substance. For example, in the *Blackwell Companion to Metaphysics*, the author of the entry "Substance," Peter Simons, illustrated a whole range of possible meanings of this term, affirming the need for a metaphysical perspective in which a single notion of substance can play its role consistently. Indeed, substance can mean: A) being independent, as for Husserl; B) ultimate subject, as for the nominalists Quinton, Price, Quine, Bambrough and Stout, or for the realists Armstrong, Ryle and van Cleve; C) individuating element, as for Strawson and Wiggins; D) what underlies change, as for Mellor, Q. Smith, McMullin, White, Furth and Anscombe; E) fundamental underlying object of reference, as for Campbell, Kim, Loux and Rosenkrantz (I omit further mention of names, although they are present in the text).[15]

Another eloquent example of the topicality of the debate on the substance of analytic philosophy is the article *Substance* by the aforementioned D. Wiggins in the volume *Philosophy. A Guide through the Subject*, edited by A. C. Grayling (1995), of which it constitutes, together with *Causation, Time, Universals* the *Metaphysics* section. Wiggins rightly refers to Aristotle as to the first who focalised the concept of substance and first of all takes into examination the criticisms that Hume addressed to the concept of substance, demonstrating that they start from a prejudicially hostile definition, which oscillates between the "something unknown and invisible" (*Treatise*, I, IV, 4) of Lockian origin, and "that which can exist by itself" (*Treatise*, I, IV, 5) of Cartesian origin.[16] In any case, it has nothing to do with the famous definition of "primary substance" given by Aristotle in the *Categories*, that is, "that which is neither in a subject nor is the predicate of a subject," a definition that can be applied to all those particular concrete realities which can be qualified by other things but do not in themselves qualify other things. Primary substances, which are the basic constituents of the world, are also what survives certain types of change, that is—as Wiggins says with an expression taken from his aforementioned book *Sameness and Substance* (Oxford 1980)—the continuants, characterised by a certain function or activity. In *Metaphysics*—as is well-known—Aristotle further develops the issue, identifying the cause of substantiality in form, intended as principle of activity, of which the latter in living beings fundamentally is life.

The Lockian idea of substance as "a certain je ne sais quoi," that is, something hidden, invisible and thus absurd—observes Wiggins—is the product of the separation of the subject from all of its properties, which has nothing to do with the subject (*hypokeimeon*) which Aristotle speaks of, a perfectly visible reality, which is palpable and possesses quality. The same can be said—I may add—of the Cartesian and Spinozian idea of substance as something that exists in itself, which has nothing to do with the sensible substance that Aristotle speaks of. But Wiggins also criticises some recent misunderstandings of the concept of substance, for example the one that is proper of the constructionalism of David Lewis, while he

---

[15] Simons (1995).

[16] Wiggins (1995).

observes that the Aristotelian idea of substance has been recovered by Strawson and Quine. On the basis of this notion, concludes Wiggins, concrete realities such as animals, human beings and other similar *continuants* are substances, about which one can rather pose the problem of how we can identify them or how they conserve their own identity.

Finally, the thesis inspired by Hume and supported by Parfit, who—echoing Hume—interprets the life of the person as a series of subsequent experiences, comparable to the history of nations, where there is an evident lack of a substantial subject that remains identical at different times, has also been subject to criticism. In particular, Bernard Williams, another exponent of the Oxford School who recently passed away, observed that there must be some kind of link between subsequent "selfs," which should be engendered by change, as proved by the fact that they all fail in the case of the physical death of their "progenitor".[17]

A return to the classical notion of person is not only present in Anglo-American philosophy of analytic inspiration, but also in "continental" philosophy of hermeneutic inspiration. Paul Ricoeur's position is exemplary in this regard. In the article "Meurt le personnalisme, revient la personne," which came out for the first time in the journal that had been the instrument of "personalism," that is, *Esprit*, in 1983, the French philosopher, who had been close to Emmanuel Mounier, founder of this current in the years 1947–1950, and had collaborated with his journal, declares that personalism as a philosophical current is dead because "it was not competitive enough to win the battle of concept," while person returns because "it had been the best candidate to sustain legal, political, economic and social battles" in defence of human rights.[18] I believe that both parts of this diagnosis must be shared, and that for this reason a philosophical foundation of person, more robust than the one previously offered by personalism, must be sought. Besides, Mounier did not consider himself a philosopher and was seeking a philosopher of personalism, after Nazi persecution had parted him from Paul Landsberg, who was the most appropriate to play this role in the *Esprit* group.

The "battle of concept" lost by personalism, although Ricoeur does not say it explicitly, is in my opinion the criticism of the notion of person made by Anglo-American analytic philosophy, which Ricoeur too found himself up against and was able to deal with in his most recent writings. Indeed, we must recognise that not only French personalism but the entire philosophy of Christian inspiration developed in the European continent in the second half of the twentieth century almost completely neglected the comparison with the analytic philosophy tradition, in the conviction that it was too logical, too abstract to say something interesting on the person and on the person's life. Thus not only were the extremist criticisms of a neo-positivist such as Alfred Ayer ignored, so were the much more traditional ones of Derek Parfit.

Ricoeur himself, in his most recent writings, precisely in order to reply to Parfit's objections, tried to solve the problem of personal identity distinguishing identity as

---

[17] Williams (1981).

[18] Ricoeur (1992).

"sameness" (*mêmeté*), on the basis of which each is simply "the same" (*idem, same, gleich*), from identity as "selfhood" (*ipséité*), on the basis of which on the contrary someone is "himself" (*ipse, self, selbst*). The former, in his opinion, supposes the existence of a substance, but it is not important, because it belongs to the sphere, in Heideggerian language, of *Vor-handen* and of *Zu-handen*. The latter is the important one, belonging to the sphere of *Dasein*, that is, of authentic existence. But the latter identity, that is, selfhood, according to Ricoeur is only a "narrative identity," resulting from the effective unity of an entire life, and is ensured by "character," intended as a certain constancy in dispositions, but above all by that loyalty to oneself that one gives proof of by keeping promises. This "loyalty to oneself" (*le maintien de soi*) is, for Ricoeur, the authentic personal identity.[19]

The latter solution may seem insufficient, because it offers a purely ethical, not ontological foundation of the person, which is applicable only to those who are responsible for their own actions, that is, who possess a moral "character," the capability of remaining loyal to themselves, a reliability from the point of view of the others. How could a similar concept of personal identity be valid for someone who is irresponsible, for instance a child, or for someone who is seriously ill, or for a dissociated person? Yet even in these cases there exist rights, such as for example the right to inherit, or the right to property, which suppose a personal identity. If it is true, as Ricoeur himself affirmed, that the person remains the best candidate to sustain the battles in defence of human rights, it is necessary to recur to a concept of person capable of playing this role. Besides, Ricoeur, in the above-mentioned article, had mentioned a similar concept, defining the person as "the support of an attitude," which means the substrate, the substantial subject of the various activities, irreducible to the latter ones. And in his most recent book he points out that the Aristotelian doctrine of potency and of the act does not apply only to human praxis, but indicates "a ground of being, at once potentiality and actuality," which seems to allude to the presence of a substrate as the foundation of acting, equipped with those capabilities that Aristotle indicated with the expression "primary act".[20]

The fact that the person remains, as Ricoeur maintains, the best candidate to sustain the battles in defence of human rights is demonstrated, in my opinion, by the philosophical implications that the formulation of the latter entails. For instance, the right to equality, that is, the right of each to be treated by law in the same way as everyone else, presupposes something that makes all human beings the same, independently of their differences in origin, nationality, social class and culture. Well, this is what the classical notion of person expresses by means of the concept of "nature." Let us then take the right to freedom, freedom of thought, of speech, of press, of religion, of association: it supposes that man, although strongly conditioned by a series of material factors (physical constitution, economic condition, subconscious, education received, etc.) conserves a margin of freedom, that is, of self-determination, of capability of escaping material conditionings, that corresponds to what Boethius called "rational nature." Finally, the right to property, on

[19] Ricoeur (1990).

[20] *Ibidem*, p. 357.

the basis of which the owner of a good conserves its property despite any changes in his life, that is, irregardless of whether he changes civil status, citizenship, religion, etc., presupposes that the owner of said right always remains the same person, that is, is a subject that persists in becoming, which is the same as admitting that he is an individual substance in the sense meant by Boethius.

It is true that not all philosophers recognise human rights as founded, or foundable, on incontrovertible reasons, in fact some believe that they cannot even have an ultimate foundation. However, there is no doubt that they correspond to the way of thinking of the majority of people, i.e. they express "public opinion," as proved by the fact that they have been solemnly proclaimed in universal declarations undersigned by most States, that they are present in many constitutions of democratic States and that even those governments that in actual fact do not respect them are not willing to admit it officially, because they know this would make them unpopular.

Besides, the notion of person that underlies the declarations of human rights has been adopted by some of the philosophers most committed, for instance, to the defence of the rights of women or of people belonging to different cultures than the Western one. I am thinking especially of the case of Martha C. Nussbaum, who, referring to the theory of economist Amartya K. Sen, according to which the most equitable distribution of wealth is the one based on the people's capability of using it, has drawn up an actual list of human capabilities, which outlines an anthropology that is not very distant from the classical notion of person. Besides, M. Nussbaum explicitly echoes the Aristotelian notion of happiness as the full realization of all human capabilities, although she criticises Aristotle for his discrimination of women, slaves and barbarians.[21] All in all, we can say that today, despite the criticisms it has been subjected to by a part of modern philosophy, the classical notion of person proves to be still topical both in the contemporary philosophical debate and in the people's way of thinking.

References

Ayer, A.J. 1963. *The concept of person and other essays*. Oxford: Clarendon.

Berti, E. 1992. Il concetto di persona nella storia del pensiero filosofico. In AA. VV. *Persona e personalismo,* 43–74. Padova, Gregoriana.

Berti, E. 1995. Individuo e persona: la concezione classica. *Studium* 91: 515–528.

Berti, E. 2004. Aristote dans les premières critiques adressées à Hegel par Feuerbach, Marx et Kierkegaard. In *Aristote au XIXe siècle*, ed. D. Thouard, 23–35. Villeneuve d'Ascq: Presses Universitaires du Septentrion.

Berti, E. 2006. How I see philosophy of the 21st century. *International Academy for Philosophy, News and Views* 10: 5–15.

Hamlyn, D.W. 1984. *Metaphysics*. Cambridge: Cambridge University Press.

Kripke, Saul. 1980. *Naming and necessity*. Oxford: Blackwell.

Munitz, Milton K. (ed.). 1971. *Identity and individuation*. New York: New York University Press.

[21] M. C. Nussbaum (2000).

Nussbaum, Martha C. 2000. *Women and human development. The capabilities approach*. Cambridge: Cambridge University Press.
Parfit, Derek. 1973. Later selves and moral principles. In *Philosophy and personal relations*, ed. Alan Montefiore, 149–150. London: Routledge and Kegan Paul.
Parfit, Derek. 1984. *Reasons and persons*. Oxford: Clarendon.
Putnam, Hilary. 1975. *Mind, language and reality, philosophical papers*, vol. 2. Cambridge: Cambridge University Press.
Putnam, Hilary. 1994. *Words and life*. Cambridge: Harvard University Press.
Quine, Willard v. 1960. *Word and object*. Cambridge: M.I.T. University Press.
Ricoeur, Paul. 1990. *Soi-même comme un autre*. Paris: Ed. du Seuil.
Ricoeur, Paul. 1992. *Lectures 2. La contrée des philosophes*. Paris: Ed. du Seuil.
Simons, P. 1995. Substance. In *A companion to metaphysics*, ed. J. Kim and E. Sosa. London: Blackwell.
Strawson, P.F. 1959. *Individuals. An essay in descriptive metaphysics*. Oxford: Oxford University Press.
Wiggins, David. 1980. *Sameness and substance*, 2nd ed. Oxford: Blackwell. 1990.
Wiggins, David. 1995. Substance. In *Philosophy. A guide through the subject*, ed. A.C. Grayling. Oxford: Oxford University Press.
Williams, Bernard. 1981. *Moral luck*. Cambridge: Cambridge University Press.
Williams, Bernard. 1986. Hylomorphism. *Oxford Studies in Ancient Philosophy* 4: 188–199.

Part IV
Hermeneutic Science and First Philosophy, Theology and the Universe

Philosophie des sciences et philosophie première

Pierre Kerszberg

Abstract Depuis son institution à l'époque de Galilée, la science mathématique de la nature a voulu s'affranchir des apparences trompeuses de l'expérience familière. Pourtant la référence à l'expérience familière demeure une exigence de sens que les théories contemporaines ne peuvent pas éviter, même si elles transforment profondément la nature et la portée de cette expérience. Finalement la précompréhension de ce qui est effectivement compris dans les sciences est l'énigme du sens oblitérée par et grâce à leur pouvoir opératoire. Cette énigme impose la tâche philosophique de reprendre l'ancien projet de philosophie première à l'aune de l'horizon d'une *mathesis universalis*. Cet article explore les possibilités d'une épistémologie qui se débarrasse à la fois du fantasme d'une maîtrise absolue de ce qui est et du scepticisme qui suit immanquablement la frustration de ce fantasme. Dans le sillage de Kant, la phénoménologie transcendantale ouvre la voie vers une telle épistémologie. Du fantasme inachevé et inachevable d'une évidence apriorique à propos de ce qui existe effectivement, héritage de la *mathesis universalis*, une telle épistémologie retient des efforts de la science moderne et contemporaine qu'elle commence par inventer des évidences en jouant d'une manière inhabituelle avec les gestes du corps traduits dans les espaces de la pensée.

Les pionniers de la révolution scientifique du dix-septième siècle parlent certainement d'une seule voix lorsqu'ils exigent que la compréhension de la nature doit passer par sa mathématisation. Le contact avec la nature au premier jour n'est pas une source de connaissance fiable, au contraire il est entaché d'erreur et d'illusion par suite de l'intrusion intempestive d'appréciations purement subjectives, et il incombe à une pensée sûre de son chemin, comme peut l'être la

P. Kerszberg (✉)
Department of Philosophy, Université de Toulouse, Toulouse, France
e-mail: kerszber@univ-tlse2.fr

B. Babich and D. Ginev (eds.), *The Multidimensionality of Hermeneutic Phenomenology*, Contributions to Phenomenology 70, DOI 10.1007/978-3-319-01707-5_17,

mathématique, de corriger ces travers. Mais cette voix unanime devient discordante lorsque, au-delà des nouvelles connaissances, il s'agit de justifier la démarche proprement dite.

Il suffit pour s'en rendre compte de considérer deux des auteurs qui ont contribué d'une manière significative à façonner notre image scientifique du monde : Descartes et Newton. On a pu opposer Descartes à Newton comme le monde de l'explication à celui du calcul : l'un fait appel à des atomes accrochés les uns aux autres et entraînés dans des tourbillons qui expliquent les phénomènes suivant les normes de l'étendue géométrique, mais il ne calcule rien ; l'autre se fie au calcul grâce auxquels les phénomènes sont déduits, sans que cette déduction puisse remonter jusqu'à leurs causes ultimes.[1] Pourtant, indépendamment de leurs opinions si éloignées l'une de l'autre, Descartes et Newton s'accordent au moins pour penser qu'au moyen de leurs principes ils font droit à l'expérience commune. Cette fidélité est même décisive quant à la validité de leurs théories – donc rétrospectivement quant à la supériorité d'une théorie sur l'autre. Descartes distingue le mouvement pris selon l'usage commun du mouvement selon la vérité mathématique : le premier est l'action par laquelle un corps se déplace d'un lieu en un autre, mais comme il est impossible de savoir si le lieu est lui-même fixe ou mobile, il faut dépasser cet usage courant et rappeler que là où il y a un lieu, il y a aussi un corps ; par conséquent le déplacement d'un lieu à un autre selon la vérité implique un changement dans la relation du corps en mouvement à d'autres corps qui sont immédiatement en contact avec lui. Newton dit au contraire que le mouvement considéré en relation à son environnement sensible est la notion familière : pour se débarrasser de cette préconception il faut distinguer entre mouvement apparent et mouvement vrai, le mouvement vrai étant justement la translation d'un lieu absolument fixe à un autre. Mais cette distinction opère à partir des termes de temps, d'espace, de lieu et de mouvement dont Newton pense qu'ils sont bien connus de tout le monde, et qu'il ne faut donc pas définir. Faire droit à l'expérience commune (familière) suppose au moins une compréhension commune (partagée) de cette expérience, or les théories physiques de Descartes et de Newton divergent dans la mesure même où ils ne s'entendent pas sur le sens de l'expérience commune.

Il est tentant de conclure que cette dispute est d'un autre âge, et n'a plus raison d'être. Les succès de la physique newtonienne n'ont-ils pas effacé les scrupules sur le sens d'une expérience qui de toute façon n'intervient plus dans la construction théorique ? Ces succès reposent sur des possibilités d'action sur les phénomènes, sur l'efficacité de prédictions qui intéressent principalement le praticien de la science armé de ses instruments de mesure. Lorsque la physique quantique a supplanté la physique newtonienne, la preuve aurait été faite que les physiciens peuvent concevoir et raisonner sans se tromper tout en laissant de côté une référence directe à l'image sensible ou l'intuition physique empruntée à la vie quotidienne. Pour élaborer la mécanique quantique, ils sont partis de caractéristiques pensées au début comme simplement formelles, élaborées à l'aide de mathématiques très abstraites (opérateurs, fonctions d'état, espaces de Hilbert, etc.). Ce faisant, ils les ont

[1] Thom (1983), 12.

transformées en éléments de pensée pleinement conceptuels et théoriques, et ils les ont chargées de contenu physique, par la mise en relation effective et adéquate avec les phénomènes produits dans les expérimentations, et formant par leur moyen leur propre pensée physique de ces phénomènes.

Cela n'empêche pas qu'il reste dans cette théorie des concepts (comme la fonction d'onde) à propos desquels il est impossible de voir à quoi ils correspondent dans la réalité physique. Heisenberg fait valoir que si l'on demande une description de ce qui se passe réellement dans les expériences sur les atomes, les mots « description », « réel », « se passe » ne peuvent justement concerner que les concepts de la vie quotidienne ou de la physique classique. Si le physicien abandonnait cette base, il perdrait le moyen de s'exprimer sans ambiguïté et ne pourrait poursuivre sa recherche scientifique. Comme il ne l'abandonne pas, il décide que la mécanique quantique est inintelligible en dehors de son formalisme mathématique, et que « toute déclaration sur ce qui s'est 'réellement passé' est une déclaration en termes de concepts classiques … incomplète en soi quant au détail des phénomènes atomiques impliqués ».[2] Incomplète, et sans doute incohérente. On aurait des « moments classiques » (aux points d'observation) entrecoupés de « moments quantiques » (entre deux observations), et seuls les premiers seraient détenteurs du sens de la « réalité ». Les deux descriptions ne seraient-elles pas complémentaires l'une de l'autre dans un système complet de la réalité, comme Bohr l'a suggéré ? Fondée sur la perturbation incontrôlable occasionnée par les appareils macroscopiques de mesure sur les objets microphysiques, l'idée de complémentarité n'est pas très convaincante, car pour la vérifier il faudrait disposer d'un moyen d'accéder aux propriétés non perturbées, ce qui est impossible si justement la perturbation est déclarée incontrôlable. D'après le célèbre argument connu sous le nom de paradoxe d'Einstein-Podolsky-Rosen, Einstein répond qu'une condition certes non nécessaire mais néanmoins suffisante pour toute théorie physique (classique ou quantique) est justement que la non-perturbation d'un système est le seul indice du réel : « Si, sans perturber d'aucune manière un système, nous pouvons prédire avec certitude la valeur d'une grandeur physique, alors il existe un élément de la réalité physique correspondant à cette grandeur ». La mise en œuvre expérimentale de cette condition a certes donné tort à Einstein. Mais par son étonnant mélange de réalisme et d'instrumentalisme, c'est la nature même de l'argument qui interpelle. Il se réfère à la réalité physique tout en se limitant à considérer des prédictions quantitatives, c'est-à-dire une représentation des phénomènes conforme aux instruments mis en œuvre pour les observer, et si elle est prise au sens strict cette représentation ne saurait préjuger en rien de l'être qui se tient au cœur des phénomènes. Cherchant à faire la part des choses, Heisenberg dit par ailleurs que « l'emploi des concepts classiques est en définitive une conséquence de la manière générale de penser de l'humanité ».[3] Cela signifie deux choses : d'une part, les concepts classiques seraient devenus les représentants du sens commun, comme si l'effort d'arrachement au sens commun entrepris par les pionniers de la science moderne n'avait servi qu'à le

---

[2] Heisenberg (1961), 187–188.

[3] Ibid., 52.

redéfinir ; d'autre part, la mécanique quantique doit se contenter d'une demi-mesure dans l'arrachement à ces concepts, puisque l'exigence de description entraîne une contradiction intermittente avec l'exigence d'intelligibilité.

L'identification des concepts de la physique classique avec l'ancien sens commun témoigne de la force persistante de l'expérience familière, force d'autant plus insistante qu'elle n'est plus interrogée pour son propre compte. Cela signifie-t-il que la théorie quantique se débarrasse à la fois des concepts classiques et du sens commun ? Dans cette théorie, l'accord entre la théorie et l'expérience va très loin, beaucoup plus loin que les limites de l'expérience ordinaire, puisqu'il atteint parfois jusqu'à plus de dix chiffres significatifs. Mais un tel accord n'a justement plus aucun sens pour l'expérience familière. Les phénomènes microphysiques ne désignent en soi aucune propriété intrinsèque pour les corps auxquels ces phénomènes sont rattachés. Néanmoins, chaque résultat de mesure sur des phénomènes microphysiques succède à un passage dans un appareillage qui impose les contraintes de la familiarité macroscopique sur tout le protocole théorético-expérimental. Il est une occurrence singulière, déterminée par l'irréversibilité des processus qui y trouvent leur aboutissement, et indissolublement rattachée à une histoire expérimentale. Les retrouvailles avec l'ancienne familiarité d'un monde de choses bien définies deviennent ainsi une exigence logique de la théorie, sous la forme particulière d'une *histoire*. On se demande à quelles conditions et dans quelles circonstances particulières on peut rattacher les symboles abstraits de la théorie à l'univers précompris des choses selon un certain déroulement temporel. L'univers du précompris est retrouvé à la fin d'une série d'opérations hautement abstraites, sans se poser la question de savoir ce que le précompris signifie pour commencer et comme s'il allait toujours de soi ! Le fait remarquable est que, à proportion des progrès de la connaissance scientifique depuis l'époque glorieuse de la première révolution scientifique, la mésentente explicite sur l'expérience familière s'est convertie insidieusement en une entente tacite.

En réaction au monde de plus en plus abstrait de la science, la philosophie a développé, depuis le tournant du vingtième siècle, une attention croissante au sens de l'expérience familière. Husserl nous demande d'admettre que le monde abstrait reste malgré tout enraciné dans l'expérience familière considérée pour son propre compte, à savoir dans sa forme systématique qu'est le monde de la vie, même si la connexion entre les deux a été aliénée au point qu'elle est à peine reconnaissable. Heidegger trouve que l'aliénation est devenue désespérément radicale, puisque les sciences élaborent leur propre autofondation sans se soucier d'une fondation encore plus profonde, réservée à la philosophie, seule à même de révéler la compréhension préontologique de l'être dans le domaine régional de l'étant dont s'occupe la science.

À écouter un certain discours épistémologique depuis l'avènement d'une théorie physique aussi extraordinaire que la mécanique quantique, le besoin de comprendre ce qui est précompris ne devrait finalement pas échapper à la théorie elle-même. C'est ainsi que Schrödinger se fait le porte-parole de la mécanique quantique lorsqu'il annonce qu'un premier contact nous lie au monde depuis toujours, un contact que la physique serait en train de retrouver après une longue éclipse. L'éclipse commence avec la révolution scientifique du dix-septième siècle, sans laquelle la nouvelle science n'aurait pourtant jamais vu le jour. Aidée du langage mathématique, la découverte progressive du gigantesque système qui tient et soutient la complexité de la nature ne pouvait avancer qu'au prix de placer le sujet qui effectue cette découverte à distance de ce qu'il découvre. Pour dénouer les fils de cette complexité, le sujet se mettait à son service en se considérant comme un étranger dans le monde qu'il construit – une pure substance pensante face à la substance matérielle et étendue. Schrödinger constate alors avec amertume qu'il est difficile se résoudre à accepter qu'à la suite de la mathématisation de la nature notre conscience soit devenue un point symbolique, une sorte d'assistant pour ce but pratique qu'est la maîtrise de la nature. Pure substance pensante, le sujet vivant s'exclut de la nature dont il est pourtant manifeste qu'il en fait partie, et il en résulte dit-il une « horrible antinomie ».[4] La conscience n'a pas d'espace pour s'étendre et vivre dans un monde qui est pourtant sa création. En réaction à cette horreur, il faut mettre un terme à ce bannissement volontaire et décider que la maléfique discrimination entre le sujet et l'objet a fait son temps. La physique quantique aurait justement accompli un premier pas pour détendre la frontière artificielle qui sépare l'opérateur connaissant de tout ce qu'il y a à connaître. Les moyens dont nous disposons pour observer un objet interfèrent d'une manière irréductible avec cet objet ; entre le sujet et l'objet, l'antinomie qui paraissait si tranchante s'affaiblit pour devenir une discrimination arbitraire. C'est ainsi que la physique quantique entreprend le chemin vers la reconnaissance d'un fait archaïque qui n'aurait jamais dû être oublié : sujet et objet dans la totalité de leurs déterminations respectives ne font qu'un, le monde est donné une seule fois et il n'est pas divisé en monde existant en soi, d'une part, monde perçu et construit par un sujet, d'autre part.

Dans l'interprétation de Schrödinger, tout se passe comme si l'opération qui a réduit le sujet à un point symbolique n'était rien d'autre qu'un moyen simplificateur et temporaire : peut-être nécessaire pour commencer, mais finalement inutile au regard de la tâche à accomplir. Or, même s'il ne s'agit que d'une parenthèse, il est permis de s'interroger sur la raison d'être et la finalité de l'antinomie créée de toutes pièces par Descartes, Newton et leurs successeurs. La cible est facilement identifiable : qui parle d'antinomie évoque la philosophie critique de Kant. Kant n'a-t-il pas expressément désigné par antinomie une configuration de la raison qui se contredit elle-même lorsque, confrontée aux limites de l'expérience possible, elle est obligée de renoncer à son unité ? De tous les philosophes de l'époque moderne, Kant est à la fois le plus proche de cette aspiration à l'unité qui est la marque distinctive de la raison scientifique et le dernier représentant de l'antinomie sujet/objet poussée jusque dans ses derniers retranchements.[5] Le monde existant serait constitué de choses en soi à propos desquelles nous n'aurons cependant jamais aucune connaissance ; mais la théorie quantique est fondée sur les relations d'incertitude, suivant lesquelles les limites de la raison ne sont qu'une traduction des limites constitutives de la nature dans le déploiement total de sa phénoménalité.

---

[4] Schrödinger (1967), 131.

[5] Ibid., 136–137.

Si l'antinomie est peut-être une nécessité pratique pour la vie, dit Schrödinger, elle n'a plus d'intérêt philosophique pour la science. Kant aurait dû se contenter de sa célèbre équation, selon laquelle les conditions de possibilité de l'expérience sont les mêmes que les conditions de possibilité des objets de l'expérience, et abandonner le chemin qui rend ces objets redevables par surcroît de mystérieuses choses en soi. À travers l'exemplarité de Kant, la lecture de Schrödinger condamne la science et la philosophie modernes à une sorte de perversité dans la séparation entre nous et le monde, perversité dont la seule excuse serait qu'elles n'en n'ont pas eu pleinement conscience.

Or, si la physique contemporaine a prétendu venir à bout des différentes figures reconnaissables du conflit de la raison avec elle-même, comme les dimensions finies ou infinies de l'espace-temps, la création ou l'éternité du monde, elle s'est rendue purement et simplement indifférente à une antinomie fondatrice de l'expérience familière du monde. La réalité est si difficile à saisir et semble toujours échapper, parce que tout compte fait, comme Nietzsche l'a signalé avec une pertinence inégalée, personne n'a jamais parlé de réalité en connaissance de cause et personne n'en parlera jamais. Dans la mesure où elles apparaissent comme ceci ou comme cela, les choses sont prêtes depuis toujours à signifier n'importe quoi, sauf ce qu'il en est de leur « réalité ». Aucun savoir n'échappe à la fatalité de ce que Nietzsche appelle « l'apparence au début », qui « finit toujours par devenir essence et *agit* en tant qu'essence ».[6] Ce qui apparaît pour la première fois est tout aussi bien et immédiatement un faux-semblant (*Schein*) de la chose qui est censée apparaître selon sa nature, et malgré tous nos efforts ce faux-semblant collera toujours à la chose. En effet, au lieu de refléter une réalité, l'apparence trahit la manière dont le regard s'est immiscé dans la chose pour la désigner, et cette manière est irréductiblement multiple. Elle habille la chose et l'affuble de toutes sortes de normes : non seulement le nom et l'apparence que le nom donne de désigner la chose même, mais aussi la réputation, la valeur, le poids, la mesure. D'emblée les choses ont été lestées d'un vêtement étranger qui a entaché leur essence d'erreur et d'arbitraire, les faisant apparaître selon une essence qui ne pouvait pas être la leur mais qui était cependant censée être la leur. En raison de cette mise aux normes, la « soi-disant » réalité ne pourra plus jamais être annulée au profit de « la » réalité. Pourtant l'apparence continue d'évoquer une « réalité », et elle compte comme étant justement sa seule marque de reconnaissance. D'où finalement le désir de trouver un moyen de supprimer ce qui ne fait que passer pour essentiel au profit de la pure essence. Désir qui fait preuve de délire : pour saisir ce que l'étranger aurait d'étranger vis-à-vis du réel, il faudrait pouvoir le comparer avec ce réel dans sa nudité totale, ce qui est impossible car un monde mis à nu est un monde où personne ne vit. Pour que les choses se livrent dans leur absolue réalité, il faudrait paradoxalement qu'elles n'apparaissent pas du tout. Einstein est tombé dans ce piège : dans son effort pour préserver contre Bohr la séparation classique entre l'instrument de mesure et les résultats d'une mesure, il s'appuie sur un critère indépendant de la réalité, qu'il énonce de la façon suivante : « tout élément de la réalité physique doit avoir sa

---

[6] Nietzsche (1982), 96.

contrepartie dans la réalité physique ». Même s'il admet qu'il ne s'agit que d'une condition suffisante de la réalité, et non une condition nécessaire, Einstein se place pour ainsi dire dans la réalité sans médiation, pour voir ensuite si la médiation introduite par la théorie lui est fidèle.

Nietzsche insiste : comme personne ne peut se satisfaire de penser les choses en les désignant, l'impact du nom sur la chose prend du temps. C'est seulement à la longue, comme fruit d'une tradition, que le nom finit par passer pour le corps de la chose. La motivation de la science moderne a été de rompre avec les habitudes héritées depuis la nuit des temps, pour corriger et surmonter une fois pour toutes une longue suite d'erreurs à propos du choix des normes. Parmi toutes les normes identifiables et les significations possibles des choses, il a semblé que leur représentation mathématique allait soulager l'apparence au début des charges qui s'étaient indûment accumulées sur elle. Connaître, c'est se trouver en terrain de connaissance, donc reconnaître qu'il y avait quelque chose de connaissable en plus de ce qui passait inaperçu à force d'aller de soi : évaluer les normes les unes par rapport aux autres, et les mettre en ordre. Or, comme le souligne Nietzsche, la mathématique se prête parfaitement à ce jeu précisément parce que ses règles s'utilisent si facilement comme des recettes qui dévoilent la raison de ce qui semble aller de soi. Par exemple reconnaître notre table de multiplication dans le comportement des choses. Les charges résiduelles de l'apparence au début s'en trouvent si allégées que l'opération semble donner raison au vieil adage de Galilée : parce qu'elle est écrite en langage mathématique, la nature *est* mathématique au plus profond de son être. À la persistance d'un savoir qui va de soi dans tout savoir, aussi neuf soit-il, la science depuis la révolution scientifique du dix-septième siècle répond en accordant à un certain type de savoir un privilège qui ne pose pas problème. C'est ainsi que, malgré l'abîme fantastique creusé par le chemin semé de ruptures qui les sépare, l'arithmétique la plus élémentaire a fini par léguer aux algorithmes de la mécanique quantique l'impression de fouiller le même terrain de connaissance qui a été ouvert pour la première fois.

Délestée de tout ce qui ne concerne pas directement l'être supposé, l'apparence au début dans la pleine richesse du premier contact avec la nature n'est plus un objet pour la science ; elle est tout au plus une sorte d'aspiration nostalgique dans un rêve éveillé, comme en témoignent les spéculations d'un Schrödinger. Mais si l'apparence au début agit toujours comme essence, pour ainsi dire en sous-main, qu'est-ce que cela signifie pour la pensée scientifique de ne plus s'en préoccuper, ne pas même se donner les moyens de s'en préoccuper, sinon en retenant son aspect mathématique ? Qu'est-ce qu'une première connaissance, à propos de laquelle il est impossible de s'entendre avec les moyens de la connaissance qui dépendent pourtant d'elle ?

La pensée de la première connaissance renvoie à une discipline éminente que toutes les disciplines scientifiques supposent alors qu'elle ne les implique pas. Une discipline éminente de ce genre a été identifiée bien avant la naissance de la science moderne : depuis Aristote elle est la philosophie première. Ce qui ressort de la nature au premier contact dans toute sa plénitude, que la physique recueille sans pouvoir en rendre compte, Aristote l'appelle : l'être ; certes la physique aussi, écrit-il, est une sagesse, mais elle n'est pas première, car la nature est seulement un genre déterminé de l'être.[7] Contre les sciences qui en font un amalgame, la philosophie première distingue la nature de l'être. Aristote attaque ses prédécesseurs immédiats, les physiologues ioniens, qui au contraire faisaient comme nous le faisons nous-mêmes aujourd'hui depuis la révolution scientifique du dix-septième siècle, à savoir qu'ils identifiaient la nature et l'être dans le monde unique et total qu'est l'univers. Dans la conception naturaliste de la nature, tout élément de la nature s'explique pour les Anciens comme pour nous par un autre élément de la nature qui en est la cause ou l'effet ; les Modernes que nous sommes ont seulement ajouté un outil privilégié pour réaliser cette opération : l'outil mathématique. La sagesse des premiers physiciens, objecte Aristote, est pourtant seconde si la nature n'est elle-même qu'un genre de l'être. Depuis l'avènement de la science moderne la sagesse naturaliste est redevenue première, et Aristote nous paraît donc à bien des égards complètement dépassé, ne fût-ce que parce qu'il justifie la sagesse au sens le plus large comme simple désir de savoir, en réponse à l'expérience du plaisir que nous éprouvons dans nos sensations, non par le souci qui est désormais le nôtre d'une maîtrise de la nature. Toute science selon Aristote ne sera qu'un grand déballage de tout ce que comporte la sensation pour commencer. Un déploiement qui s'effectue au moyen d'idées, sans que des Idées s'en détachent pour constituer un monde propre de la science au détriment de la sensation – comme le monde des Idées mathématiques dans le platonisme mathématique de la science moderne.

La philosophie première selon Aristote n'est plus la physique, mais elle n'en est pas moins une science. En effet, comme la nature, l'être est lui aussi une totalité, ce qui ne s'aperçoit que lorsqu'il est considéré *strictement en tant qu'être*. Cet Etre est ce qui est commun à toutes les choses, de sorte que l'être en tant qu'être se dit de tous les êtres en tant qu'êtres. L'ensemble de tous les êtres comporte aussi bien des êtres de pensée que des êtres naturels : le nombre, la ligne, le feu. Toutes les choses qui participent de l'Etre, le philosophe doit les connaître dans leur essence et dans leurs attributs, ce qui n'est possible que parce que l'être en tant qu'être possède lui aussi des attributs propres. Mais comme son objet est ce qui est premier, la science de l'être en tant qu'être est une science d'un genre particulier.[8] De ce qui est premier tout le reste dépend pour ainsi dire à sens unique, puisque les autres connaissances supposent la connaissance de ce qui est premier alors que celle-ci ne les implique pas. Pourquoi au juste appeler science la philosophie première, puisque l'être dont elle s'occupe va au-delà du savoir de type démonstratif qu'est la science ? La réponse est une échappatoire grandiose : la philosophie première est la science que l'on cherche, celle qui est toujours recherchée en toute science alors même que toute science cherche pour trouver. Cette recherche incessante ne l'épuisera jamais. Or, il suffit que cette science soit recherchée pour qu'en elle s'affirme une puissance causale supérieure à la physique : la première découverte qui affleure à même le mouvement de recherche vers ce qui est premier, c'est la cause formelle, qui n'a plus rien à voir avec une cause de type matériel (physique).

[7] Aristote, *Métaphysique*, 1005b.

[8] Ibid., 1003b.

Ce qu'Aristote vient de formuler est à la fois décisif et insatisfaisant, en attente d'une nouvelle décision que la science moderne viendra justifier rétrospectivement. Décisif, en ce sens que l'explication de ce qui est sollicite la forme, ou essence, qui est cause de l'être matériel dont s'occupe la physique ; la forme répond à la question de savoir quelles sont les causes les plus hautes de la nature, qui sont aussi plus hautes que la nature. Insatisfaisant, parce que cette décision ne va pas de soi : elle implique à la fois un saut au-delà de la nature et un retour vers la nature. Par exemple, la forme est une substance immobile, et pourtant en tant que premier moteur elle est la cause des substances en mouvement dont s'occupe la physique. Aristote rassemble en un seul mot cette double notion de la cause la plus haute qui s'abaisse pourtant dans la source d'où elle a jailli : la cause est aussi principe. D'une part les multiples sciences connues et pratiquées s'organisent suivant les genres d'être dont elles s'occupent en particulier, et chaque science constitue le genre qui pose son propre principe dont l'existence est impossible à démontrer ; par suite, chaque science doit observer une règle d'incommunicabilité des genres, qui interdit dans la démonstration de passer d'un genre à l'autre (par exemple l'arithmétique pose l'unité comme son principe, et la géométrie la grandeur). D'autre part, la philosophie première est universelle parce que comme science elle n'a pas son propre genre ; mais elle ne devient science qu'au prix d'un saut au-delà de la sensation, dont on est en droit de se demander s'il conserve le « ce qu'est » d'où il provient. Il aura suffi à la science moderne de nier ce pouvoir exorbitant de la sensation pour restituer au monde des Idées une place qui lui revient en priorité dans l'ordre du savoir.

Or, ce refus plane déjà comme une potentialité plus ou moins cachée dans la conception aristotélicienne de la science. La cause formelle, Aristote la voit d'abord à l'œuvre en mathématique où le nombre est forme parce qu'il est agencement d'un et de multiple. Dès le début, la philosophie première est ainsi empreinte de la forme mathématique comme accès privilégié à la pensée de ce qui est premier. Aristote voit un rapprochement possible entre la philosophie première et une mathématique générale, dont l'idée est empruntée à la théorie générale des proportions d'Eudoxe. La science moderne a tiré parti de cette indication pour démocratiser la forme, et rapprocher la nature de l'être. Si tout dépend de la forme, cela n'est concevable désormais selon la tendance démocratique que si en retour elle implique ce qui dépend d'elle. La géométrie analytique de Descartes est au cœur d'une mathématique générale où un genre supérieur est posé, tel qu'il enveloppe les diverses branches de cette science sous un même principe. Descartes appelle *mathesis universalis* la science « qui explique tout ce qu'il est possible de rechercher touchant l'ordre et la mesure, sans assignation à quelque matière particulière que ce soit ».[9] En suivant jusqu'au bout la voie mathématique pour déchiffrer l'essence de ce qui est premier, la science universelle permet de fonder les divers genres de sciences sur un même principe général. Dans la philosophie première ainsi élevée au rang de science universelle, ce qui est recherché n'est plus astreint à basculer dans la forme au détriment de la matière, au contraire l'explication de qui est recherché porte sur

[9] Descartes (1963), 98.

la matière en général qui est la forme de n'importe quelle matière particulière. Cette forme se décline dans la mathématique pure des formes spatio-temporelles applicables aux corps matériels par le procédé de la mesure. Par exemple, l'immobilité n'est plus la cause/principe du mouvement des corps : repos et mouvement sont plutôt deux états de la matière en général que la mesure distingue, sans que l'un soit ontologiquement plus éminent que l'autre.

Par suite de ce recouvrement de la mathématique générale avec la matière en général, la philosophie première devient la science mathématique de la nature matérielle. Ce qui est recherché en toute science se confond-il alors avec ce que la science recherche effectivement ? Les sciences mathématiques de la nature héritent de l'ancienne philosophie première le souci d'une forme universelle qui, sans dépasser la nature pour basculer dans l'être, dépasse néanmoins la nature en tant qu'elle est spécifiable en choses naturelles déterminées. C'est ainsi que Kant distingue le concept d'une nature en général de toute théorie particulière des corps physiques. Il se place d'abord sur le terrain de la science en affirmant que « dans toute théorie particulière de la nature, on ne peut trouver de science *à proprement parler (eigentlich)* que dans l'exacte mesure où il peut s'y trouver *de la mathématique* ».[10] La mathématique fournit le fondement de la partie empirique dans la science de la nature, et cela seul doit suffire à faire de cette science une science authentique, c'est-à-dire autre chose qu'un art systématique ou une théorie purement expérimentale comme pouvaient l'être (au temps de Kant) la chimie ou la psychologie. Mais une science fondée sur une partie d'elle-même est-elle fondée authentiquement ? Prenant le relais de la science, la philosophie pure recherche ce qui constitue le concept d'une nature dans toute son universalité. Or, ajoute Kant, si ce concept universel est pensable, il est possible qu'il le soit sans mathématiques. Le sens universel de nature ne sera jamais fixé par les mathématiques, et pourtant il sera toujours recherché dans la science qui théorise la nature au moyen des mathématiques. Un espace de pure pensée reste ainsi ouvert au cœur même de l'activité théorique qui porte la pensée vers le réel concret en le construisant selon les normes mathématiques.

Lorsque la *mathesis universalis* forme le projet de réunir la mathématique et le réel concret dans un espace abstrait, elle imagine que la pure pensée est épuisée dans cette forme abstraite. Selon une tendance qui n'a fait que s'accentuer depuis l'époque de la première fondation et des succès de la science mathématique de la nature, elle n'aperçoit pas qu'elle confond ainsi la nature en général avec toutes ses manifestations dans des choses déterminées. Une loi de la nature est du domaine de la pure pensée, et pourtant elle n'a de sens que vis-à-vis de phénomènes déterminés : comment cela est-il possible ? Considérons un physicien placé devant un phénomène quelconque. Ce phénomène est par lui-même privé de signification : cette signification devra donc être inventée. Particulièrement significatif est l'exemple de la loi de Torricelli qui relie l'élévation de l'eau dans les corps de pompe et la pression atmosphérique. Il est connu depuis l'Antiquité que l'eau s'élève dans les corps de pompe jusqu'à une hauteur déterminée. Mais le sens de

[10] Kant (1980), 367.

ce fait ne réside pas du tout dans le fait même. Torricelli n'a pas rapproché cette expérience de celle de la pression atmosphérique : au contraire il a supposé la pression atmosphérique parce qu'elle était nécessaire à l'explication ; l'existence d'un phénomène correspondant au concept n'a été prouvée que plus tard par Pascal. Donc Torricelli n'a pas découvert sa loi, il l'a inventée dans un exercice de pure pensée face au sens que doit avoir la nature en général. L'expression contenue dans la loi de la nature n'est pas moins inventée que l'expression « la nature a horreur du vide » si la relation à l'expérience est différente. Comme l'a remarqué Husserl en digne successeur de Kant, quand bien même la mathématique est applicable au réel concret dans certaines limites, « nous ne possédons pas l'ombre d'une évidence apriorique à l'égard de ce qui existe effectivement dans la nature ».[11] Tout ce que nous pouvons en connaître exige une induction à partir des faits d'expérience ; même une induction conforme aux opérations logiques ne donnera jamais qu'une évidence rétrospective, et elle se heurte par principe au mur d'un fait nouveau susceptible de révoquer complètement des conclusions provisoires. La mathématique pure des formes spatio-temporelles appliquées au réel concret est fondée sur une évidence apriorique, qui à la fois court-circuite le progrès de l'expérience et dicte la forme que doit épouser ce progrès. Elle s'adresse à la nature concrète de tel ou tel phénomène comme si une nature en général transparaissait en lui, et elle est donc une manière de glaner une connaissance absolue de tout ce qui existe effectivement dans et pour un entendement divin. Pour nous, êtres finis et mortels, la seule différence avec Dieu est que la connaissance de toutes les formes dans une Forme absolument universelle de l'Espace-Temps-Matière prend du temps, elle exige un effort intellectuel pour les déployer selon un déroulement systématique ; nous pensons que la systématicité de l'effort est précisément garante de la validité rétrospective de certaines évidences qui en ressortent comme des évidences ultimes, de sorte que l'évidence apriorique pour commencer se trouve finalement justifiée par son usage. D'où l'idée que science et système s'impliquent l'un l'autre d'une manière si intime qu'il n'y a de connaissance propre que systématique. Depuis Descartes et Galilée tout progrès dans la connaissance scientifique consiste à unifier sous l'égide de la Forme des domaines du savoir a priori disparates. Mais comme en témoigne le progrès effectif des sciences jusqu'à aujourd'hui, l'illusion rétrospective de l'évidence ultime se répercute dans l'irréductibilité de certains faits systématiquement réfractaires à l'unification. Ceux-ci ne sont derechef que des stimuli pour avancer systématiquement vers l'unification soi-disant finale. Quand des résultats expérimentaux se montrent rebelles à une interprétation évidente dans les termes d'une théorie existante, cette résistance est considérée comme un indice de la manifestation de la nature « en soi », qui appelle une nouvelle théorie.

Peut-on envisager une épistémologie qui se débarrasse à la fois du fantasme d'une maîtrise absolue de ce qui est et du scepticisme qui suit immanquablement la frustration de ce fantasme ? La phénoménologie ouvre la voie vers une telle épistémologie. Du fantasme inachevé et inachevable d'une évidence apriorique à propos

[11] Husserl (1976), 64.

de ce qui existe effectivement, héritage de la *mathesis universalis*, une telle épistémologie retient des efforts de la science moderne qu'elle commence par inventer des évidences en jouant avec les sensations d'une manière inhabituelle. Les sensations, dont Aristote disait qu'elles sont le commencement de la science, s'y épanouissent en gestes.

Galilée nous demande de nous enfermer dans la cabine d'un bateau, avec tout un attirail incongru de choses qui n'y ont pas leur place « naturelle » : des papillons et des mouches, des poissons dans un aquarium. Tant que le bateau vogue paisiblement sur une mer calme, la vie à l'intérieur de ce milieu très artificiel est tout aussi naturelle que si les papillons volaient libres dans l'air, et si les poissons nageaient tranquillement dans la mer. Par rapport au rivage le mouvement est quelque chose, par rapport aux parois de la cabine il n'est rien. Donc un mouvement partagé par des corps dans les limites bien définies d'un système de référence est sans effet, et en bonne orthodoxie aristotélicienne ce qui n'a pas d'effet n'a pas non plus de cause. En son essence le mouvement est « comme nul ». Les sensations éprouvées par procuration dans l'étrange cabine viennent de produire le principe de relativité. Il reste alors à formuler mathématiquement les lois du mouvement des corps en partant de cette situation où les lois ne sont pas affectées par des circonstances adventices, tout comme le voyageur dans la cabine du bateau n'était pas affecté par le mouvement du bateau.

Le tout jeune Einstein a jeté les bases de la théorie de la relativité restreinte lorsqu'il s'est imaginé à la poursuite d'un rayon lumineux : si j'ai la même vitesse que lui, il n'y a plus rien à voir (un champ électromagnétique stationnaire n'est plus un champ électromagnétique) ni à penser (les équations de Maxwell ne sont plus valables). Ce mouvement ne se laisse pas annuler. Pour continuer à voir et à penser, il faut redonner à l'observateur la capacité de faire l'expérience de ses propres mouvements, c'est-à-dire ses sensations propres de mouvement, ce qui se traduit dans la théorie par l'abandon du temps absolu.[12] Le vertige de la sensation dans l'expérience en pensée réactive d'une manière étonnante le plaisir que nous éprouvons dans la sensation, en constatant que notre désir de savoir est tiré dans une direction inhabituelle. Il instaure une pratique symbolique en amont du formalisme pour produire des idéalités : celles-ci ne sont ni abstraites des sensations ordinaires ni posées dans un monde séparé. Elles flottent pour ainsi dire dans la seule conscience du physicien, et ne valent que pour cette conscience.[13]

Au-delà du rôle paradigmatique joué par ces gestes fondateurs dans la science depuis Galilée, la philosophie phénoménologique de la connaissance scientifique considère que le vide ouvert par la désorientation dans le monde concret est la caractéristique essentielle de toute science qui construit des idéalités représentatives de ce monde. Quel est le sens de ce vide avant qu'il ne soit recouvert par l'évidence

[12] Einstein (1949), 53.

[13] Voir Châtelet (1993), 33–36. L'histoire des sciences apprend à voir les gestes fondateurs dans deux registres de la pensée : non seulement l'expérience en pensée, mais aussi le diagramme. Il est essentiel au diagramme qu'il soit parcouru de pointillés tels que seule la libre imagination puisse les intuitionner.

toute prête qui se révélera rétrospectivement ? À la conscience vide du physicien répond la prise de conscience radicale du phénoménologue, qui invite à considérer la conscience vide comme un cas particulier d'une légalité universelle.[14]

On la voit fort bien à l'œuvre dans la *mathesis universalis* élaborée par Descartes, où la science est délibérément restreinte aux idées claires et distinctes dont la transparence pour l'esprit est comparable à une sorte de figuration du vide. Les figures géométriques incarnent ces idées, de telle sorte que l'essence du réel est elle-même géométrique. Le moment inventif des idéalités se dédouble en deux phases qui différencient nettement l'idéal du réel : une phase conventionnelle et arbitraire, suivie d'un ajustement à la réalité.[15] D'abord l'invention d'un schème conventionnel. Ainsi la couleur, dans les termes de Descartes, est tout ce qu'on voudra, précisément parce qu'on ne sait pas pour commencer ce qu'elle est. Néanmoins il est impossible de nier qu'elle est étendue, donc qu'elle a une figure. Descartes veut faire droit à ce fait sans préjuger à la légère de l'essence de la couleur, et c'est pourquoi il met de l'ordre dans la diversité des couleurs en leur attribuant ce qui a seulement la nature d'une figure, par exemple des lignes parallèles pour le blanc, des carreaux pour le bleu, etc. Toutes les différences de nuances dans les choses sensibles colorées peuvent en principe être mises en correspondance avec ces symboles, car a priori il y a une diversité infinie de figures prêtes à représenter des nuances aussi fines que possible. Reste à savoir comment l'espace représentant arbitraire de la réalité de la couleur en est aussi un représentant adéquat. La solution de ce problème passe par l'invention d'une idéalité dont l'essence spatiale a la particularité de s'effacer au moment de passer dans la représentation spatiale d'une réalité tout autre. Cette idéalité est la *dimension*.[16] Notion spatiale, la dimension signifie longueur. Cette notion permet de reconstituer la réalité spatiale, par composition de trois dimensions. Mais elle se prête aussi à la reconstitution des autres réalités, comme la pesanteur ou la vitesse. Pour y arriver il suffit de voir que la dimension n'est pas seulement la longueur mais tout élément analogue à la longueur. Le mode de composition des dimensions n'est alors qu'un aspect du mode général de composition des grandeurs. Des phénomènes comme la pesanteur ou la vitesse sont des dimensions des choses pesées ou des choses en mouvement, c'est-à-dire des représentations spatiales de ces phénomènes qui ne dépendent plus de l'essence spatiale de la dimension.

Dans tout ce raisonnement, le souci de Descartes est de ne rien ajouter arbitrairement aux choses et de cantonner les idéalités au travail du seul esprit aux prises avec lui-même. En identifiant pour commencer une couleur avec une étendue figurée, Descartes raisonne au moyen d'une hypothèse qui n'a pas d'autre prétention que d'être commode. Il trouve ensuite un moyen pour étendre cette représentation commode à toutes les réalités possibles. La chose réelle se fond dans la méthode de représentation pour se confondre finalement avec elle. Autrement dit,

---

[14] Husserl, *Logique formelle et logique transcendantale*, §16a.

[15] Descartes, *op.cit.*, 137.

[16] Ibid., 177–178.

la clarté de l'outil de pensée qu'est la figure constitue une préfiguration de sens déjà organisée à l'avance, et le réel se tient prêt à remplir cette préfiguration avec sa matérialité propre.

Mais quand bien même elle fonctionne en recouvrant l'objet par la méthode de sa construction symbolique, dans cette science véritable faite d'idées claires et distinctes Descartes ne calcule rien. Considérant le calcul en tant que tel, Leibniz revendique au contraire les droits de la connaissance symbolique comme une connaissance *aveugle*. Les opérations sur les symboles nécessitent de longues chaînes de raisonnement, si longues que la signification soi-disant claire et distincte des symboles et combinaisons de symboles peut et doit être mise en parenthèses. En effet, en utilisant des signes, en algèbre « et même presque en toutes choses », il nous faut omettre « de préciser dans notre conscience leur conception explicite, sachant, ou croyant que nous l'avons en notre pouvoir ».[17] Selon une tendance qui n'a fait que s'accentuer depuis lors dans tous les compartiments de la science mathématique de la nature, l'aveuglement au sens des outils de pensée est devenue la condition de leur efficacité.

Fruit d'un oubli volontaire, l'aveuglement de la méthode mathématique convoque pour ainsi dire malgré lui des questions ontologiques auxquelles la science ne pourra jamais répondre. Elle y reste pourtant sensible, comme si elle ne pouvait pas éviter de se rendre conforme à un appel qui la dépasse.

Aucun domaine de la physique contemporaine n'échappe à ce qui a été pensé en elle pour la première fois, implicitement ou explicitement, au tournant du vingtième siècle. Dans les notes autobiographiques qu'il a rédigées vers la fin de sa vie, Einstein a accordé à Minkowski le mérite d'avoir réalisé le plus profond bouleversement dans les relations entre la physique et la mathématique pour répondre à cette attente. D'après Minkowski, une nouvelle notion de « monde » émerge de la théorie de la relativité restreinte, dans la mesure où, comme il l'écrivait en conclusion de sa célèbre conférence de 1908 sur l'espace et le temps, « l'espace par lui-même et le temps par lui-même sont condamnés à ne devenir que des ombres, et seule une sorte d'union des deux sera préservée comme réalité indépendante ». Espace et temps étaient justement les deux concepts fondamentaux séparés dans une loi comme la loi de la chute des corps de Galilée. À cette réalité répond ce que Minkowski appelle un « postulat de monde » corrélatif de la substitution de la notion de causalité par celle de symétrie. Ainsi, Lorentz a d'abord découvert et formulé dans sa théorie des électrons le premier principe de symétrie important ; il s'agissait d'une propriété mathématique des équations de Maxwell qui elles-mêmes étaient fondées sur les lois expérimentales de l'électromagnétisme. Mais alors que chez Lorentz la symétrie était une découverte pour ainsi dire secondaire, Minkowski part de l'invariance mathématique au sens de Lorentz et pose l'exigence que les équations du champ aient une forme particulière (connue sous le nom de covariance) en rapport avec cette invariance.

En raison de la priorité de la représentation mathématique formelle, l'image de la nature dans la physique contemporaine n'est-elle pas une manière élégante

[17] Leibniz (1972), 152.

d'évacuer toute la problématique de l'être de l'horizon de la science ? C'est une conséquence à laquelle Einstein aura tenté de résister jusqu'au bout. En 1950, vers le terme de sa vie, ne nous demandait-il pas de réfléchir en profondeur à cet énoncé apparemment trivial : « Les équations de Maxwell impliquent le groupe de Lorentz, mais le groupe de Lorentz n'implique pas les équations de Maxwell ».[18]

Forts de cette conclusion banale mais fondamentale, revenons une dernière fois à la théorie de la relativité restreinte. Héritiers d'un mode de penser classique, les tenants de la conception électromagnétique du monde pensaient qu'il existe une classe privilégiée de systèmes d'inertie, les systèmes au repos dans l'éther ; ils pensaient aussi que les phénomènes relativistes, tels que la contraction des longueurs, exigent une explication détaillée en termes de forces électromagnétiques, en tout cas une explication dynamique. La théorie de la relativité restreinte fut le premier geste qui dit au contraire que ces forces sont des artefacts de l'état de mouvement relatif des systèmes de référence utilisés pour la représentation. Autrement dit, elle « explique » les phénomènes relativistes en se dispensant de les expliquer autrement que comme des conséquences simples de la géométrie de l'espace-temps. Cela revient à tourner la flèche de l'explication à l'envers. Que les lois de la nature pour les phénomènes physiques non gravitationnels soient invariantes par rapport au groupe de Lorentz, cela serait une sorte de fait brut qui ne demande pas d'explication – on en tiendra pour preuve que l'espace-temps de Minkowski est justement le cadre propre à représenter ces phénomènes. En substituant l'espace-temps à l'espace et au temps, il s'agit de reculer le moment où la nature commence à se manifester telle qu'elle est en soi, pour ne pas s'en laisser conter et l'affronter de face à armes égales. Il se pourrait donc que le principe de l'invariance de Lorentz ne nécessite aucune explication, mais la question reste entière de savoir si la géométrie de Minkowski ne l'explique pas malgré tout. L'explication devrait en appeler à l'espace-temps conçu comme une entité existant indépendamment de tout le reste, de sorte que la structure métrique de l'espace-temps contraint les lois de la nature à revêtir une certaine forme.[19] Cette « explication » ne fait qu'exaspérer le renversement de la flèche de l'explication au lieu de la retourner dans le bon sens : comment des barres rigides et des horloges censées fonctionner « normalement » pourraient-elles jamais savoir dans quel espace-temps ils sont plongés ? Le mystère des forces est reporté sur une mystérieuse intention de la nature qui s'exprimerait dans la forme de ses lois.

Depuis le début Einstein cherchait pour sa part une « théorie fondamentale » qui servirait de canevas pour l'intelligibilité future de la nature. Comment les mathématiques pures viendraient-elles à la rencontre du monde concret dont s'occupe la physique ? D'un côté, il s'agit d'introduire une dose d'a priori dans la nature, en transformant un fait comme l'invariance de la vitesse de la lumière en loi ; d'un autre côté, si des propositions logiques comme celles de la géométrie sont vides vis-à-vis du réel, il s'agit de leur fournir une dose d'empiricité en reconduisant le système des axiomes à une science des relations mutuelles entre des corps

[18] Einstein (1950), 15.

[19] Voir Brown (2005), 143.

considérés comme pratiquement rigides. Mais les échecs répétés pour réaliser une théorie fondamentale de ce genre, tout comme les échecs essuyés par d'autres tentatives en vue d'un tel but, démontrent finalement que la physique n'a pas les moyens d'expliquer l'explication comme tendance réductrice à l'unification. Une vision totalement unifiée du monde échappe à une justification interne à la physique, qui pour sa part pourra et devra toujours se contenter d'aménager tant bien que mal ses reconstructions fragmentées du réel.

La décision de Leibniz d'utiliser les mathématiques pour les rendre aveugles au sens est la décision capitale sur laquelle repose aujourd'hui l'intelligibilité du monde naturel. Husserl a justement désigné par « crise » l'état de chose qui découle de cette décision, et il a tenté de lui substituer un remède sous la forme d'une autre décision, qui consiste à élaborer l'ontologie d'un monde de la vie sous-jacent à toute prise de position théorique. Ce monde de la vie contiendrait toutes les possibilités de penser sans lesquelles aucun sens ne pourrait advenir, y compris le sens des formations de pensée oubliées dans l'aventure du sens. Heidegger en a tiré toutes les conséquences. Le monde naturel qui se déploie depuis Galilée, rappelle-t-il, obéit à une position fondamentale de l'être : la position mathématique. Cette position est un mode de l'apprendre : l'apprendre mathématique est un apprendre à connaître ce que l'on a déjà, « un prendre suprêmement remarquable, un prendre dans lequel celui qui prend ne prend que ce qu'au fond il a déjà »,[20] de sorte que ce qu'il a déjà d'une manière indéterminée se prête à l'effort d'un marquage par des déterminations possibles. Sollicité par les choses, le prendre mathématique n'a pas d'égard pour elles, contrairement au prendre ordinaire qui veut et obtient une connaissance totale de la chose qui l'intéresse.[21] Le prendre mathématique fait retour sur le fonds indéterminé d'où la connaissance s'extrait et s'élève à partir d'elle-même. Certes, lorsqu'il entre au contact des choses dont nous faisons l'expérience, le prendre mathématique devient mathématique au sens déterminé de nombre ou de figure. Mais en tant que mise en œuvre de l'apprentissage au sens le plus général, Heidegger voit dans le prendre mathématique dont le sens a été déterminé en discipline mathématique depuis Galilée la marque d'une précompréhension de l'être qui n'arrive jamais à son terme. La position mathématique fondamentale ne persiste à nous renseigner sur les propriétés de la nature que pour autant que le fonds indéterminé de la mathesis galiléenne *peut et doit rester indéterminé*. C'est dire que la science n'est pas une connaissance authentique tournée vers l'essence des choses, peut-être n'est-elle même pas une connaissance du tout, mais une pratique tournée vers l'exploitation sans relâche d'un indéterminé qui, du fait de son retrait, se prête à toutes les déterminations imaginables.

Toutefois, la liberté inouïe de l'imagination scientifique se prête à son tour à une reprise critique de la question de savoir ce que signifie habiter un monde avant de le comprendre. Tandis que Descartes considérait qu'il appartient au physicien de décider parmi les essences librement inventées celles qui ont un fondement réel, Husserl considère que « toutes les sciences concevables conformes au réel et au

[20] Heidegger (1971), 85.

[21] Ortega y Gasset (1970), 223.

possible sont des formes transcendantales qui se dessinent (*vorgezeichnet*) conformément à l'essence et qui se dessinent comme devant être réalisées dans une libre activité ».[22] C'est pourquoi il a pu emboîter le pas au projet de la *mathesis universalis* et soutenir que la phénoménologie est tout de même une science, et même la philosophie première par excellence pour notre époque, dans la mesure où les formes transcendantales qu'elle met en œuvre donnent vie à une manière d'habiter l'espace du monde, constituer des environnements vivables à partir de configurations qui caractérisent schématiquement une situation environnementale. L'environnement peut désigner tout aussi bien le milieu où nous vivons naturellement que le milieu manipulé pour en faire un pôle d'intérêt purement théorique. En tant que schèmes, les premiers gestes de la pensée rejoignent ainsi la condition de toute vie, ce qui rapproche l'homme de l'animal tout autant qu'il l'en éloigne radicalement.[23] Alors que l'animal est capté dans les schèmes de son comportement, et que le sens des schèmes est tout entier contenu en eux sans aller aussi loin que la saisie de la forme complexe de l'objet naturel, les schèmes proprement humains se libèrent de la capture pour s'enchaîner dans d'authentiques perceptions, où ils trouvent un sens que les schèmes primaires n'ont pas du tout. Il s'agit pourtant dans les deux cas des mêmes schèmes.

References

Aristotle. 1974. *La Métaphysique*. Trad. J.Tricot. Paris: J.Vrin.

Brown, Harvey. 2005. *Physical relativity: Space-time structure from a dynamical perspective*. Oxford: Oxford University Press.

Châtelet, Gilles. 1993. *Les Enjeux du mobile : Mathématiques, physique, philosophie*. Paris: Le Seuil.

Descartes, René. 1963. *Règles pour la direction de l'esprit*. F. Alquié (éd.) *Œuvres philosophiques*. I. Paris: Garnier.

Einstein, Albert. 1949. Autobiographical notes. In *Albert Einstein: Philosopher-scientist*, ed. P.A. Schilpp. La Salle: Open Court.

Einstein, Albert. 1950. On the generalized theory of gravitation. *Scientific American* 182(4): 13–17.

Heidegger, Martin. 1971. *Qu'est-ce qu'une chose?* J. Taminiaux et J. Reboul (trad.). Paris: Gallimard.

Heisenberg, Werner. 1961. *Physique et philosophie*. J. Hadamard (trad.). Paris: Albin Michel.

Husserl, Edmund. 1957. *Logique Formelle et Logique Transcendentale*. S. Bachelard (trad.). Paris: Presses Universitaires de France.

Husserl, Edmund. 1976. *La Crise des sciences européennes et la phénoménologie transcendantale*. G. Granel (trad.). Paris: Gallimard.

Kant, Emmanuel. 1980. *Premiers principes métaphysiques de la science de la nature*. F. de Gandt. (trad.) *Œuvres philosophiques*. I. Paris: Gallimard.

Leibniz, G.W. 1972. Méditations sur la connaissance, la vérité et les idées. In *Œuvres*, ed. Prenant, L. Paris: Aubier-Montaigne.

[22] Husserl (1957), 362.

[23] Voir Lorenz, *Les fondements de l'éthologie* (1984), 190.

Nietzsche, Friedrich. 1982. *Le Gai savoir*. Pierre Klossowski (trad.). Paris: Gallimard.
Ortega y Gasset, José. 1970. *L'évolution de la théorie déductive*. J.P. Borel (trad.). Paris: Gallimard.
Schrödinger, Erwin. 1967. *Mind and matter*. Cambridge: Cambridge University Press.
Thom, René. 1983. *Paraboles et catastrophes*. Paris: Flammarion.

A Re-Reading of Heidegger's "Phenomenology and Theology"

Adriaan T. Peperzak

Abstract This close reading of Heidegger's "Phenomenology and Theology" is motivated by critical questions concerning his basic statements about the presence and absence of certain relations between faith and philosophy.

Shortly after the publication of *Being and Time* in 1927, Martin Heidegger presented a paper on the task of theology and its relations with philosophy in Tübingen (on March 9, 1927) and in Marburg (on February 14, 1928). The title of his paper was programmatic: *Phenomenology and Theology*, but the text was not published until 1969. In a short preface to its long delayed publication, Heidegger suggests that his early paper might still be useful for rethinking not only the questionability [*Fragwürdigkeit*] of the Christian character [*Christlichkeit*] of Christianity and its *theology*, but also the questionability of *philosophy*.[1] During his life, Heidegger did not publish any other extensive discussion on the relations between philosophy and theology; but at the occasion of a theological conversation on "the problem of a non-objectifying thinking and speaking in today's theology," held in April 1964 at Drew University in Madison (USA), Heidegger sent the participants a letter, dated March 11, 1964, with remarks or hints [*Hinweise*] about the topic of that conference, and this letter was added to the 1969 publication of his 1927 text.[2] As far as I know, most or all of his other observations on theology, as published before his death, are in the

[1] Heidegger (1969), 45–78.

[2] In 1969 the German text, accompanied by a French translation, was published in the *Archives de Philosophie* 32 (1969), 356ff, and the letter of 1964 has there received the title «*Some hints concerning the main perspectives for the theological conversation about "The problem of a non-objectifying thinking and speaking in contemporary theology"*». Both texts are reprinted in *Wegmarken*, GA 9, 47–67 and 68–77. To the latter version Heidegger added a short preface

A.T. Peperzak (✉)
Arthur J. Schmitt Chair of Philosophy, Loyola University Chicago,
Lake Shore Campus, 1032 W. Sheridan Rd., Chicago 60626, IL, USA
e-mail: apeperz@luc.edu

B. Babich and D. Ginev (eds.), *The Multidimensionality of Hermeneutic Phenomenology*, Contributions to Phenomenology 70, DOI 10.1007/978-3-319-01707-5_18,

form of brief passages, most often in a polemical context. Many, but not all of them, fit into the framework of his early essay, which I would like to reread and annotate here in memory of my dear friend Joe Kockelmans, for whom Heidegger's struggling with his Catholic past has been a profoundly significant drama.[3] I doubt whether Heidegger ever published a direct or indirect *retractatio* of the basic thoughts expressed in his essay of 1927, although Hans-Georg Gadamer has assured me that the letter he added in 1964 to it, ought to be read as such. Toward the end of the present rereading I will come back to the question of whether that letter indeed can be read as a correction or partial withdrawal, and if so, what this then would mean for the validity of *Phenomenology and Theology.*

Heidegger's own preface (45–46) to its later publication situates his early paper in a tradition of writings that reaches from Overbeck's pamphlet *Über die Christlichkeit unserer heutigen Theologie* [On the Christianness of Today's Theology, 1873],[4] which opposes Christian faith to all forms of *Bildung* and *Wissenschaft*, and the fierce attack on David Strauß's "philistine" theology found in Nietzsche's first of his *Untimely Meditations* (1873), all the way to Heidegger's own courses of the 1930s on "Nietzsche and European nihilism" and his essay on "Nietzsche's word 'God is dead'," published in *Holzwege* (1950).

1 Heidegger on Theology and Philosophy

In *Phenomenology and Theology*, Heidegger defends the thesis that philosophy does not need theology, whereas theology *can* use, but does not *need* philosophy, except for some of theology's formal aspects, insofar as it claims to be a *Wissenschaft* or "science."[5] Christian *faith* [*Glaube*] however, on which a Christian is totally dependent, is and remains "the mortal enemy [*Todfeind*] of *philosophy*" (66), because the existenz*ielle* roots of faith and philosophy are radically different. To appreciate this—for many, and especially for Catholic, readers shocking or exaggerated—statement, we must understand that Heidegger's essay does not present faith as an ensemble of theoretical beliefs, but instead as a most radical and fundamental "form" or "mode of existence" (an *Existenzform* or *Existenzweise*), i.e., as a characteristic *existenz**ielle*** possibility of performing one's *Dasein* (and not as an

(45–46) and an "Appendix" (78). I will quote from this edition. An English translation of the text is available (1976), 5–21.

[3] Cf. Kockelmans (1973), 85–108, on Heidegger's "Phenomenology and Theology," to which the present essay is my belated response, too late—alas!—for further discussions with Joe.

[4] The second edition (1903) of Franz Overbeck's *Über die Christlichkeit unserer heutigen Theologie* was enlarged by an Introduction and an Epilogue. I used the photographic reprint of this second edition, as published by the Wissenschaftliche Buchgesellschaft (1989).

[5] Heidegger (1969), 61 (Further citations are given parenthetically in the text itself). As we will see, the latter statement must be qualified. Is it really possible to isolate the formal elements of theology from all its claims about its content?

existenziale structure of *Dasein* as such). Intellectuals who try to combine a faith-ruled way of life with a serious commitment to the philosophical mode (or form) of existence, so different from a faith-inspired basic stance, lead a contradictory and false existence. If they position themselves as philosophers, while already being truly committed to Christian faith, they hide or ignore their own most radical commitment, because this does not allow for the radical and all-risking kind of "free" and "purely rational" [*rein rationale,* 63] questioning that is proper to philosophy. Their "philosophy" would in fact be no more than a game ("as if" one were a real philosopher), or else their faith would be a lie. The contradiction between a faith-bound existence and philosophy's free and total responsibility for the way an engaged philosopher performs his own *Dasein* excludes any synthesis between the two. Christian philosophy is a "wooden iron" (66); and theology can neither be a form of phenomenology, if "phenomenology," as Heidegger thinks, is synonymous with philosophy (67), nor can a theologian at the same time be member of any particular brand or school of philosophy (66).

Heidegger begins his paper with a very succinct summary of what he calls "the vulgar conception of the relation between theology and philosophy" (47). According to this widespread conception, philosophy and theology are two competing world-visions [*Weltanschauungen*] regarding one and the same reality: human life in the world. Whereas theology appeals to faith in revelation for announcing its interpretation of the human world, philosophy, being "the faith-free interpretation of world and life" (47), turns to reason [*Vernunft*] alone. Instead of either immediately discussing this vulgar view or proposing a phenomenologically more accurate contrast between theology and philosophy, Heidegger states that he will treat their relation as a relation between two sciences [*Wissenschaften*].[6] He further restricts his scope by discarding any inductive or empiricist definition of philosophy's or theology's "scientific" character that would be based on their factual performances. Instead he presents "an *ideal* construction of the *ideas* of both sciences" (47, my emphasis), i.e., a meta-scientific outline of the different types of scientificity that should characterize philosophy and theology, if they indeed do realize the essence or the "idea" of science as such. Only thus, Heidegger insists, a fundamental [*grundsätzlich*] insight in their relation can be won (47).

Noticing that such a "construction" presupposes a clear concept of "science," Heidegger presents us with a formal definition: *Wissenschaft* is "the founding [*begründende*] disclosure, exposure, or unveiling [*Enthüllung*] of an in-itself-enclosed [*in sichgeschlossen*] domain of beings [*das Seiende*] or being [*das Sein*] for the sake of [its or their] unveiledness or disclosedness" (48). After reminding the reader of the distinction between beings (entities) and the kind of being [*Sein*] that characterizes them as belonging to different groups according to each group's specific mode of

[6] The meaning of the word *Wissenschaft*, which Heidegger uses constantly in this essay, is closer to the medieval *scientia* and the ancient *epistēmē* than to the modern Anglo-Saxon meaning of "science." Among the academic disciplines, theology and philosophy are here still called *Wissenschaften* by Heidegger, but later, like in his letter of 1964 (74), he opposes their formal character quite strongly to that of the natural sciences and mathematics.

being [*Seinsart*], Heidegger states that each particular science has its own mode of approach and knowledge according to the specific mode of being that determines the domain to which those beings belong (48). Then, focusing on philosophy and theology as two sciences that have their own objects, problems, and characteristics, he rehearses the distinction between the *ontic* and the *ontological* perspective, as it was set out in the first sections of *Being and Time*. As the investigation of "*being itself*," philosophy is the only "*ontological*" science, whereas all the positive sciences presuppose a "*positum*" of their own, i.e., a specific type of beings [*onta*] that distinguishes them from other beings and thus makes it possible to gather and thematize them as the collective *Gegenstand* [or object] of a particular science. As non-ontological—and thus non-philosophical—thematizations of beings that have a particular character, the latter sciences are *ontic sciences*. Their themes or "objects" [*Gegenstände*] have already become familiar and to a certain extent manifest—though not yet in a scientific way—before they are investigated scientifically, but they still need a rigorous conceptualization and an examination of their relations to the ontological findings of philosophy. Not only for philosophy and theology but also for all other sciences, this means that their objects are known in a provisional—perhaps even hardly self-conscious—way before they are studied explicitly.

Against the theoretical background sketched above, Heidegger repeats his disagreement with the vulgar conception of the difference between philosophy and theology: as an ontic science, theology is closer to chemistry or mathematics than to philosophy, because theology does not have any insight into the truth of *being itself*, a topic that—as we already have heard—is reserved for philosophy.[7] He concludes then by stating the following thesis about the scientific status of theology: "*Theology is a positive science and as such it is therefore absolutely different from philosophy*" (p. 49).

How does Heidegger delineate the *positum*, the specific being that is studied and the mode of existence that is invested in theology? Warning us that he will not focus on any other theology than the *Christian* one (49)—in fact, he also abstains from saying anything about other forms of religious or irreligious faiths, including all Greek, Roman, Asian, or Germanic ones—Heidegger answers that the *positum* of (Christian) theology lies in "*Christlichkeit*" as such. Through an expression that figures in the title of Overbeck's book of 1873, to which Heidegger refers (see above), he thus affirms that the object [*Gegenstand*] of theology is "Christianness" (52). But what makes (Christian) theology—and all that is attached to it, such as Christian beliefs, opinions, behavior, morality, prayer, and contemplation—characteristically Christian? The answer lies in the particular form or mode of (human)

[7] The statement that theology is not concerned with "ontological" questions implies that it cannot compete with philosophy, because it does not and cannot have any competence for pursuing questions with regard to Heidegger's main philosophical concern, called "being" [*Sein*]: "being as such" or "being itself." Below I will suggest that Heidegger's discarding of 2000 years of ontology from philosophical and biblical theology is quite dogmatic and unjustified, especially if it is not preceded by a thorough discussion of late Greek and medieval theology of creation and providence, which cannot be simply dismissed as containing nothing else than "metaphysical speculation."

existing that is expressed in them or, in other words, in the "ontic" and *existenzielle* reality that can be summarized by the word *faith* [*Glaube*].

Heidegger's description of faith's distinctive character is not very eloquent, but he does give some hints. As appropriate response to the reality (the *Gegenstand*, object, or theme) in which Christians believe, faith cannot be a product or effect of *Dasein's* own free initiative.[8] Instead, it is presented by faith as the believed (–in): *das Geglaubte,* of which the believer believes that it has been or is being revealed (52).

Before we follow Heidegger's description of the object [*Gegenstand*] of Christian faith, it is necessary that we have a preliminary, although not yet scientifically justified, idea of that object [*das Geglaubte*]. What Christians believe *in*, the "object" of their faith, is neither a list of articles, statements, or dogmas, nor even a coherent ensemble of propositions that compose one whole. Faith is much more fundamental than a creed, because it is the decisive mode of existence [the *Existenzform* or *Existenzweise*] that expresses itself in all the theoretical and practical tenets and facets that constitute the Christianness of a Christian life. Faith does not coincide with believing *that* certain statements or doctrines or convictions are true (e.g., I state as an objective truth *that* God exists, *that* Christ has revealed God's compassion, etc.). "Believing-*in*" is closer to the total self-delivery and self-abandonment through which true believers entrust themselves to the revealed reality in which they put their faith. Heidegger repeatedly emphasizes the radicality of the faithful's devotion to which an authentically Christian existence through its faith, confession, dedication, and activities testify. Faith (or believing-*in*) is the fundamental and encompassing position that turns a Christian into a wholly different direction than, for example, the mode of existence that orients a non-believing philosopher.[9] "Faith" is the name for a basic and unifying existen*zielle* turn and orientation, and only secondarily a question of convictions that can be spelled out in catechisms, sermons, or theology (52–56).

After Heidegger's general remarks about our prescientific acquaintance with certain beings that a positive science then can investigate as its *positum*, the attentive reader hopes that Heidegger first will describe the prescientific familiarity of Christians with their faith and its object, in order to thereafter give some clarifications about the way in which theology, as a science, can transform that naïve form of understanding into a scientifically ascertained interpretation and a conceptually[10] accurate knowledge. It is not easy, however, to follow the articulation of his text on this point—not only because of its selective character but also (1) because it emphasizes some aspects of the Christian *creed* while silencing others, which are at least

[8] Cf., 52: "not generated by *Dasein* and *not* developed (or brought to maturity) through it in complete liberty" ["*nicht aus dem Dasein und nicht durch es aus freien Stücken gezeitigt*"].

[9] Faith implies "a turnaround [or conversion, *Umgestelltwerden*] of someone's existence in and through God's faithfully assumed mercy" (53).

[10] Throughout "Phänomenologie und Theologie" Heidegger frequently uses the words "*Begreifen*," "*Begriffe*," and "*begrifflich*" to indicate scientific knowledge as distinct from the prescientific acquaintance with phenomena. See, for example, 50, 54, 56, 57, 60, 63, and 65.

as central and fundamental if not more fundamental than the ones he mentions, and (2) because Heidegger's description itself already privileges certain *theologies* over others. Many commentators have pointed to the influence of Luther, Karl Barth, and Rudolf Bultmann on Heidegger's quite un-Catholic presentation of theology in his 1927 essay, but his silence about (or suppression of?) certain theologoumena that are essential of the entire Christian tradition from its very beginnings demands an explanation of its own. Whatever the truth about Heidegger's changes of mind around 1920 may be, his essay of 1927 answers only the question of what he, in 1927–1928 deems central for the core of Christian faith—and consequently also for a theology that is inspired by this faith.

Implicitly appealing to Luther's plea for theology as a *theologia crucis*, Heidegger summarizes the Christian message through the formula "Christ, the crucified God" (52), but he does not offer much help for a faithful understanding of this ambiguous expression, which hardly makes sense if it is isolated from the Christian faith in God's incarnation.[11] What individual Christians and their community [*Gemeinde*] in faith embrace, the object of their faith, is, according to Heidegger, not primarily God, but instead the crucifixion of Christ, which, as understood by faith, has the character of a sacrifice [*Opfer*, 52]. Faith and theology thus focus on a unique fact of history (51–54) whose revelatory meaning only faith can recognize. For the faithful this fact is a liberating occurrence or event [*Geschehen* or *Geschehnis*] that, as revelation, cannot be "known" as such and adhered to except through faith alone. Since this event has also a very special mode of historicity, we cannot be informed about it by world historical or local reports. Only faith itself is open to "revelatory history" [*Offenbarungsgeschehen,* 52–54]. Revelation is neither a news update about the past nor a piece of scientific research about the historical evolution of humankind, but it challenges the believers and makes them participate in the sacrificial crucifixion of Christ, which reveals at the same time the *sin* [*Sünde*] of their own *forgetfulness* about God and their having been forgiven by God's *mercy* (52–53). To live *coram Deo* as a sinner who has become crucified with Christ and thus reconciled with God, changes the believer's destiny [*Geschick*]. The history [*das Geschehen*] into which faith grants the Christians participation, is not an episode of the general history [*Geschichte*], but instead a "specific form of enabling" or aptitude [*Geschi**ck**lichkeit*] that changes their destiny (53–54). For faith in Christ is a revolution [*Umgestelltwerden der Existenz*]. It converts the basic position [*die Existenzweise*] of the faithless into a form of participatory living with and "in" Christ before God. Notwithstanding the very special character of Christian historicity, Heidegger calls theology a *historical* science; but since faith causes a radical conversion [*Umstellung*] of the sinner's position [*Einstellung*] thanks to his being crucified with Christ (53), Heidegger stresses that Christians belong to an eschatological history by being born again. Through faith, one is reborn as a "new creation"

[11] The German phrases on 52: "*der gekreuzigt Gott,*" "*das so durch Christus bestimmte Verhältnis des Glaubens zum Kreuz*" (the relation to the cross that thus is determined by Christ), and "*die Kreuzigung*" (the crucifixion) do not by themselves clarify whether the one who is crucified is God or only a kind of god or pseudogod in the line of Zeus, Prometheus, Apollo or Ares.

[*Neue Schöpfung*, 63]. "Faith = rebirth" [*Glaube = Wiedergeburt,* 53]. Instead of being dominated by sin, the Christian has become "a servant [*Knecht*] of God" (53, 62). Through God's mercy the entirety of his *Dasein* can no longer be separated from the cross of Christ (53–63).

It is clear that Heidegger's résumé of Christian faith does not pretend to offer a complete summary of its main tenets but instead prioritizes its *existenzielle* specificity: faith is a mode of existing, an *Existenzform*, in which sin and rebirth are central concerns. It is almost as if Heidegger wants to show us the quintessence of Christian faith by showing the reader a crucifix, surrounded by a couple of symbolic words (such as *sin, cross, sacrifice, pardon, conversion, service,* and *new creation*) which would remind expert theologians of their biblical roots and 1,900 years of discussion among Catholic, Orthodox, Lutheran, Calvinian, and other commentators. Because Heidegger himself does not even try to justify his selection of *credita* and their roots, the reader remains puzzled by the very partial outline of the "*positum*" that, according to Heidegger's view, distinguishes a pre-theological but already believing acquaintance with Christian faith from any other "ontic" basis for departure in theology. That Heidegger, after his early conservative-Catholic period, has become a fervent reader of Luther, might explain his emphasis on *sin, sacrifice, cross, forgiving*, and *new creation*, but it does not explain his quite blatant silence about *creation, incarnation, trinity, resurrection, eternal life, prayer, agapē* and—above all—*Godself as not being one of many beings and incomparable to any "other" being*. All his emphases remain enigmatic if they are not explained in their coherence with the other—at least equally fundamental but silenced—tenets of Christian faith. For example, what does the sacrificial character of the Cross mean, if, in a profoundly anti-Paulinian way, Christ's *resurrection* is not even mentioned at all? How should we understand the very ambiguous, perhaps even improper, expression of "the crucified God" (or "god"), if it is not accompanied by a reminder of the equally silenced (or perhaps repressed?) *incarnation* of God's "Word" or "Son" (52)? And what about a Christianness in which prayer and *agapē* are not even mentioned as primarily and most radically constitutive of a faithful Christian existence? Is *agapē* between God and humanity, between humans mutually, and from singular humans to God not the one and all-encompassing "new command," through which the "new creation" (63) is distinct from a God-forgotten life in sin?

A particularly surprising aspect of Heidegger's hints regarding the object of theology lies in his often repeated emphasis on faith as concentrated on itself, in contrast with the parsimony of his indications about the ways in which God—and not (wo)man—constitutes the primary, fundamental, central, absolute, ultimate and originary source and meaning of the believer's existence.

Heidegger defends his position by pointing out that the philosophical study and all positive but non-theological studies of religion approach Christianity from a perspective that is neither rooted in nor guided by faith, while Heidegger's distinction between theology and other "positive" sciences, such as, for example, zoology (59), can be justified by the infinite difference between the *Seinsart* of their objects, and the *Seinsart* of God. Apparently he presupposes that God's being is a particular kind of being instead of "being" beyond all species or varieties of being. But why does

Heidegger restrict the "object" of theology to human faith—a faith in the crucified Christ, whom he interprets as a historical occurrence—without any attempt to, at least somewhat, clarify the relation (or coincidence) of this "fact" with the omnipresence of the Absolute and Incomparable Origin or "Father," who, according to Christian faith, is not only responsible for all kinds of being in all their created meanings, but, as God, also one with the man Jesus.

It remains unclear to me how Heidegger could combine his almost exclusive concentration on human faith without any serious attempt at a phenomenological description of the characteristically Christian attitudes, practices and motivations that immediately refer to God. Christian faith cannot be restricted to a central conviction ("I believe that…"), because it conditions and orients all the dimensions of the true believers' individual and common[12] existence (54). All the expressions Heidegger uses in these pages focus on human sin and rebirth, but no echo is heard of the overwhelming presence of God in the biblical and postbiblical liturgical, devotional, or mystical documentation about three millennia of sacred history and the abundant literature about piety as seeking of and walking with God, described in many styles by holy and not so holy, learned and hardly learned but faithful men and women, some of whom (also) used a considerable amount of philosophical and theological expertise to illuminate their wealth of religious experience.

To what extent it is possible to write a phenomenological evocation of God's hidden presence (more hidden than that of "being as such" or "being itself") is of course a thorny question, but Heidegger, who taught us how to adjust our descriptions to the characteristic phenomenality of characteristic experiences, could have helped us, if he had taken Gregory of Nyssa, Augustine, Eckhart, or Juan de la Cruz and many others as seriously as Parmenides, Heraclitus, or Hölderlin. However, his choice for philosophy—which, according to his own words, excludes him from faith—does not seem to have prepared him well for a phenomenology of the Christian experience. As far as biblical exegesis is concerned, he does not even try to hint at it and his exclusion of the first testament from the Bible does not forbode a faithful understanding of the "New Testament."[13]

Toward the end of his explanation of theology's *positum*, Heidegger quotes, by way of summary, a word of Luther: "Faith is self-imprisonment [*das Sichgefangengeben*] in things we do not see,"[14] and once more he emphasizes that faith is not just a special kind of knowledge about a salvific history, but instead a particular way of *participating* in the occurrence of a very special kind of history, of which theology itself is a part. Theology is oriented, guided, motivated, and encompassed

[12] Heidegger mentions the Christian *Gemeinde* [religious community] a few times, for instance on 52 and 56, but he silences all questions regarding the universality and internal diversity of the Church.

[13] On 57, Heidegger restricts the Bible, out of which all Christian theology thinks, to the New Testament alone. The first part of the Bible and the meaning of Israel are not even mentioned (Does he *ever* refer to them?).

[14] The passive moment of *Sichgefangengeben* (53) can be strengthened by translating the expression as: "to allow (or let) oneself (to) be imprisoned."

by or "imprisoned" in the ongoing history of being crucified with Christ (53–54). To what extent this "prison" of faith and theology has consequences for the scientific character of theology has to be seen; but in any case Heidegger's framing of theology (*if* it *is* a science) emphasizes its radical difference from all other sciences.

Regarding the ontological monopoly of philosophy, Heidegger's position is clear: do not turn to theology for questions or answers about being and its why, its meaning, or its distinction from beings (including the "highest"ones) and all that cannot be called *a being* because it lacks or *surpasses* all beings. Whether Heidegger's expression "speculative knowledge of God" (60) is pointing at a honorable part of philosophy is not clearly affirmed or denied, but in any case theologians should not expect any contribution to their science (if it is one) from that side. Insofar as theology is seen as a "science of God's action [*das Handeln Gottes*] with regard to human's faithful action" (58–59), it focuses on God's creative, providential, redeeming, and re-creating activity.[15] It should then not remain enclosed within the borders of the basic and all-encompassing (but in Heidegger's view *ontologically* completely irrelevant) faith from which it emerges. If theology indeed *is* a science, it should borrow from philosophy, because only philosophy possesses the highest authority about the conditions of scientificity. Heidegger's statements according to which theology, despite its special character, is subordinate to philosophy (27), because all ontic sciences are rooted in ontology, are not easy to combine with his sharp distinction between theology and philosophy; but he is very explicit (27, 31), at least about the *formal* demands that philosophy, as a meta-scientific methodology, imposes on theology, if this discipline indeed must be understood as a science.

Right from the beginning of his essay, Heidegger separates theology not only from philosophy, but also from all other positive sciences, despite the ontic character it shares with them. The reason lies in its Christian character, which determines theology as an "ingredient of faith" (54), whereas all other sciences, like philosophy, operate in a "free" [*freie*] and *purely rational* [*rein rational*, 65–66] way. As a faith-obeying interpretation of faith at the service of faith itself, theology does not coincide with a special part or specialty within some faith-free and merely rational or "profane" (58) science that seems open to all possible kinds of faith or religion. Consequently, theology must not only avoid any mixture with philosophy of religion (59–60), but also refrain from confiding in or borrowing from other kinds of religious study, such as the history or the psychology of religions, and even from a general world history with particular attention to the Christian religion (51–52, 59–60). Because all non-theological sciences, according to Heidegger's "ideal construction of their idea" (47, 53), discard faith from their essential presuppositions, just as philosophy does, theologians are completely on their own, i.e., they are left alone to their faith (and not to any form of universal evidence or shared but faith-free convictions).

---

[15]These activities of God are presupposed in a "new creation" [*neue Schöpfung*, 63], if this expression is meant seriously, but neither creation nor the entire pre-Christian history are mentioned.

By sharply distinguishing Christian theology not only from philosophy and all other sciences, including profane history and religious studies insofar as these are not entirely dominated by Christian faith, Heidegger avoids a host of difficult questions regarding both theology and philosophy. For example, the following: (1) Do the faiths of other religions express themselves in theologies of their own, which have a similar scientific status and relation to philosophy? Could being and ontology unite humanity by showing that a "free and purely rational" but faith-free understanding of Dasein precedes or accompanies (or contradicts) *all* faiths? (2) To what extent is it reasonable to defend that faith—or, at least, some kind of radical and pre-rational commitment, which analogically might be called a "fundamental trust" or a "profound affinity"—not only belongs to the *existenzielle*, but *also to the existenziale, conditions of any mode of human existence?* (3) Does not each epochal—e.g., modern or postmodern—meta-scientific idea of science imply an un-provable, but hardly escapable, trust or commitment or "faith" that precedes and orients all human existence before it can appeal to boast about a "free" form of "pure rationality"?

These questions can be multiplied, but Heidegger—perhaps wisely or strategically—confines his discussion to the relations between (1) a particular (rather Lutheran than phenomenological) interpretation of Christianness, (2) his own rather early version of philosophy, and (3) a meta-scientific definition of science as such. Within a horizon of 2,000 years of faithful writing and reflection (of which he disregards the first 1,400 years that separate Paul from Luther), he confines theology to a faith-bound commentary on some fragments of the fundamental documentation about one century: some selected texts of the *New Testament* (57). The question of whether the first (or—according to a Christian tradition—"old") Testament and the very long tradition of its Christian explanations can, should, or ought to play a role in theological assimilations of the Christian kerygma, he leaves to those who are disappointed about his casual mention of their biblical basis (57). Heidegger's delineation of (Christian) theology isolates its domain not only from any invasion of "profane" (58), pagan, "free," "purely-rational," faith-less, or philosophical questions and curiosities, but also from its Jewish past about which he shows neither interest nor acquaintance.

If all ontological questions are reserved for the "free" and exclusively "rational" realm of philosophy, however, how then exactly does a theological understanding of *creation* and *providence*, *historicity*, *incarnation*, *resurrection*, and the all-encompasing bond between God and humanity relate to philosophical questions about the amazing givenness of the world and humanity? The question of whether one and the same person can at the same time be a Christian and a philosopher—i.e., how one can *live* and practice a union of two *existenzielle* modi—remains very unclear. Heidegger seems to deny the possibility of such a union, but how then can he write so confidently about faith versus rational freedom, philosophy as the only superscience, and theology as an unfree, and especially philosophy-free (perhaps also non-rational?), discipline?

To liberate theology from profane or pagan interference might be seen as the creation of a refuge for believers and theologians, but also as a quarantine or exile.

By relegating theology to the "prison" of its faith, the philosopher liberates himself to think freely without being directed, oriented, or limited, and reminded of his sins by theology or faith. Is not this the illusion of a completely autarkic existence many philosophers longed for since the beginning of modernity? Or has Heidegger his own, but perhaps no longer Christian, faith?

Heidegger's sketch of faith and its content is part of an introduction to his thesis about the radical difference and *existenzielle* incompatibility between philosophy, on the one hand, and theology as the science of, from, and for faith, on the other. Before reacting to Heidegger's view of the relation that keeps theology and philosophy apart, however, we must acquire a better understanding of how he determines the scientific character of *philosophy*. In contrast to all other sciences, philosophy is not focused on a specific kind or cluster of beings, as specified by their particular modes of being [*Seinsart*], but instead on (the being of) being as such and in itself. In the phase of his development reached in 1927, Heidegger identifies philosophy as the one and only *ontological* science whose task concerns not only the exploration of "being" as such, but also the ways in which being itself conditions and connects the various *modes* of being that are proper to the various dimensions studied by the "positive" or "*ontic*" sciences (62–63). The title of *Phenomenology and Theology* limits that task by confining it to an examination of the relation between philosophy and theology.[16] To what extent does ontology then still concern theology? If theology is completely dependent on, surrounded by, and immersed in faith, what could still remain for any collaboration with or dependence on philosophy and vice-versa?

Heidegger's conception of philosophy is somewhat clearer than his characterization of theology. As produced on the basis of a free, non-Christian experience [*Erfahrung*, 63 and 67] that uses "purely rational" means [*rein rationale*] of knowing, philosophy emerges from a peculiar mode of existence, which is called "free" and "posited on its own" [*rein auf sichgestellt,* 63–66]. As completely independent of any kind of *doxa* or faith, philosophy is rooted in a mode of existence that has nothing to do with Christianity. Being a free form of questioning [*freies Fragen*, 65] in search for a "transcendental ontology" (67), philosophy emerges from a non-Christian but "free self-appropriation of one's entire Dasein" (66). It is supported by its own free experience, which gives access to the existenzi*ale* mode and structure of being. With regard to all positive sciences, philosophy thus has the direction of their ultimate, purely ontological, foundation (65–66).

Heidegger wants to state without ambiguity that philosophy does not need to pay attention to (Christian) faith and that the existenzi*elle* attitude of faith, as servitude to God and sacrificial participation in Christ's crucifixion, is narrow and too different from the "free self-appropriation" of the philosopher's own "*Dasein* in its entirety" to justify any cooperation between philosophy and faith in a theology that would constitute some kind of mediation [*Vermittlung*, 66] or synthesis. The core of faith (its particular mode of existence) remains the deadly enemy [*Todfeind,* 66] of

[16]That Heidegger sees "phenomenology" as another name of philosophy is clear not only from the title, but also for example, from 67.

any truly philosophical existence. However, Heidegger also maintains that the fundamental *opposition* [*Gegensatz*] between philosophy and theology, insofar as both are *sciences*, does not exclude the *possibility* that theologians borrow from philosophy. A "Christian philosophy" is a "wooden iron" (66), but insofar as theology claims to be an authentic science, it must follow the formal and *existenziale* findings of philosophy, whereas philosophy, as a non-Christian ontology free from faith, *cannot* borrow anything from faith or theology.

The definition of philosophy as an entirely self-directed mode of thought that emerges from the philosopher's "free self-appropriation of his own entire *Dasein*" is surprising; but who would dare to accuse the writer of *Being and Time* that he naively repeats the modern slogans about philosophy's autonomy or to even suggest that a thinker could claim full responsibility for his own being born in a particular epoch of the European history and his entire education in a particular language by German parents and schools in the aftermath of a neo-Kantian and post-Hegelian epoch of philosophy? Would Heidegger in 1927–1928 really think that anyone's thinking can be entirely self-directed, free from all those unchosen influences that no one can either prove or re-create after the fact? His own essay is an eloquent example of being inspired by a *particular* trend in theology (and a revolt against his own Catholic past), and his later work is incomprehensible without awareness of his glaring dependence on Aristotle, Kant, Nietzsche, Hölderlin, and some others.

What exactly does Heidegger mean, when he writes that "all the fundamental concepts of theology "have (…) in themselves an (…) *ontologically* determining pre-Christian, and consequently a *purely rationally understandable* content" [my emphasis] and that they necessarily imply [*bergen*] in themselves "*that* understanding of being [*Seinsverständnis*], which human *Dasein* as such and out of itself [*von sich aus*] has, insofar as it exists *überhaupt*" [and thus not yet insofar as it follows an existentially specified mode]? (63) Does he want to evoke here the traditional distinction between the abstract dimension of a "natural" or generally human philosophy, on the one hand, and its historical concretizations, on the other? Does he presuppose that a philosophy is only then authentically "free," "merely rational," and fully responsible, if it gets rid of all historical forms of trust or faith in particular *doxai*, *mores*, literary, theoretical and affective traditions? Could he really suggest that the modern bible of human Reason and Freedom should be written by philosophers without any faith? Hegel called the main condition for such a philosophy "faith in Reason" [*Glaube an die Vernunft*]. But already *Sein und Zeit*—and certainly Heidegger's later work—point into a different direction. Could any postmodern thinker really be so naïve as to deem a profoundly presuppositionless philosophy possible at all?

Heidegger's characterization of philosophical freedom as "free self-appropriation of the entire Dasein" (66) places us before a puzzle. If it is not naïve, it seems to imply a rather shocking arrogance. Several modern philosophers have accustomed us to the posture of an ideal thinker who, as a kind of miniature god, offers us from an Archimedean nowhere his interpretation of the universe; but we are reluctant to rank Heidegger among them, because we are grateful for his lessons in phenomenology and especially for his later focusing on the receptive and "listening" aspects of thinking [*Denken*] as thanking [*Danken*] and appropriate responding [*Entsprechen*]. Should we read his early text on *Phenomenology and Theology* as already implying—or at least as not denying—that a free and responsible manner of thinking may include and integrate the historical person who, thanks to an epochal and personal destiny, has become a well-educated and well-read thinker? But if that is a more accurate sense of freedom—a certain form of *Gelassenheit* perhaps?—the question of philosophy's presuppositions and the measure of its freedom and rationality are still not quite answered. With regard to the relation between Christian theology and philosophy, the main problem we would then like to see resolved is whether any philosophy is possible that not only emerges from pre-philosophical, pre-scientific, and insufficiently tested conceptions of the most important phenomena, but also is rooted in a pervasive and profound form of trust or faith concerning more superficial, but not irrelevant, questions, guesses, suspicions, and procedures. In 1927–1928, Heidegger's procedure avoids not only all confrontations with explicitly—*and implicitly—religious* kinds of non-Christian faith, which avoidance permits him to ignore the hypothesis according to which no real philosophy is completely faithless. But if no philosophy can avoid at least some connections with some basic faith or trust (or superstition), the postulates supporting Heidegger's essay must be revised.

Much of Heidegger's later work might be diagnosed as based on a specific kind of non-Christian faith (a faith in "the truth of Being" as "opening"and "granting" a certain belief in a kind of "space" where finite gods and human mortals meet). If he agreed that philosophy, in its own way, emerges from a *fiducial* mode of existence in search for understanding, a comparison of philosophy with theology and of their—perhaps different—roots, as he understood them, would be easier than a comparison between a "wooden iron" (66) and a work of illusory "free" rationality or "rational" autarchy.[17]

The last page (67) of Heidegger's essay can perhaps be read as an indirect hint about softening the sharp opposition between philosophy and theology that was formulated in the preceding pages. Without mentioning theology, the text focuses on the general problem of a researcher in one of the positive sciences who examines the basic concepts and connections of his own science. When such an examination is so thorough that the researcher must reconsider the traditional conception of his central *positum* or theme, including its characteristic mode of being, in order to replace it with a better understanding of its ontological foundation, then this researcher needs also to be a philosopher or to consult with one in order to readjust the basic concepts of his science through interdisciplinary cooperation. Heidegger concedes that there are no "fixed rules" for such coordination, but he points at the

---

[17] Heidegger does not refer to the traditional formula of "*fides quaerens intellectum*." Probably he would hear it as a justification of the (neo-)scholastic mode of mobilizing reason for the conquest of insights into (parts of) the patrimonium of Christian faith. With regard to a metaphilosophical interpretation of philosophy as search for understanding of an explicitly or implicitly religious faith or trust, see Peperzak, (2013), Chaps. 7 and 8.

importance of an "instinct for the topic at stake" and of "scientific tact."[18] He does not return once more to the special status of theology as a unique science founded on faith, but instead refers theologians to philosophy's pre- and non-Christian analyses of *Dasein's* existential structures, insofar as these are presupposed in theological commentaries on the Christian faith. Insofar as theology confides in itself or borrows from philosophical analyses, theologians must remain attentive to philosophy and accept the needed corrections from that side (64–65). And here Heidegger comes much closer to the great tradition from Origen and the Capadoceans to Cusanus, Malebranche, Blondel and some other twentieth century philosophy than he suggested before. He uses even the word *"aufgehoben"* in the sense of *"hinaufgehoben"* (63, 67) when he writes that faith can take certain outcomes of philosophical analysis under its wings [*"in Verfügung nehmen"*] in order to "lift them up" to the level of theological interpretation and faith-based praxis (63, 67).

Heidegger also gives a particular example of cooperation between philosophy and theology by confronting faith's disclosure and reflective awareness of sin [*Sünde*] with his own philosophical analysis of debt [*Schuld*], given in *Being and Time* §§ 54–60. While suggesting that an authentic theology of sin cannot be satisfied with a merely philosophical phenomenology of debt, he underlines that no theology should contradict correct philosophical analyses of debt. Without discussing or accepting either Heidegger's presuppositions expressed in his own analyses of debt and deficiency as an *existenzial* moment of *Dasein* as such, or his opinion that "sin" is an exclusively Christian concept or phenomenon, I would like to ask instead what exactly the difference might be between a theological integration of good analyses and acceptance of corrections that are offered by a free, rational, and phenomenologically accurate ontology, on the one hand, and theology as a "faith in search of self-understanding" that integrates philosophically adequate analyses, like, e.g., those offered by Plato, Aristotle, Plotinus, or the Stoics, or, for that matter, by Heidegger himself.

Heidegger might be right in stating that the fundamental stance and motivation of non-believing philosophers is radically different from that of a faith-inspired existence (although the evidence is not immediately obvious), but what seems highly incredible to me—and difficult to combine with Heidegger's experience of the historical and epochal determination of all philosophies—is his claim that theology is a special kind of exclusively *ontic* and, more specifically, *historical* (or historial, or rather "destinal," *geschickliche*) science, while philosophy is a wholly "free" thinking that is not at all oriented, turned, motivated, inspired, or biased by any specific—Christian, Jewish, or otherwise religious, pre-Christian, anti-Christian, post-Christian, quasi-Christian, Nietzschean, Aristotelian, Parmenidean, polytheically Hölderlinian, or other, but in any case pre- or beyond-"philosophical"—trust or *faith*. As a *philosopher*, I find it very difficult to believe that

[18] Ibid., 67. Giancarlo Tarantino, whom I want to thank here for his assistance in producing this chapter, reminded me of a possible link between these expressions and Heidegger's explanation of Aristotle's analysis of *phronēsis* in book VI of his *Nicomachean Ethics*. See Heidegger's course on Plato's *Sophist*, GA 19, 48–57 and 138–165.

the relatively young Heidegger of 1927, just as most neo-Thomists of a 100 years ago, believed in the possibility of clearly separating the reality of a generally-human, merely "natural," completely unbiased, unhistorical and un-epochal intellect, together with its "merely rational" and entirely "free" insights, on the one hand, from the factual *doxa* and *ethos* that define the established convictions and mores in which even the leading thinkers of a certain period feel at home, on the other. I rather believe that a basic trust or faith is unavoidable and effective in all who dare to think "on their own." And let's not forget that philosophers too have their own adherence to variations of *das* (elitarian) *Man*.

In tandem with his strong opposition between theology and philosophy, Heidegger emphasizes the thematic (faith-given) independence, but not the "freedom" of theology. Since theology, according to Heidegger's analysis, is and "has meaning and right only as" an "ingredient" of faith (53), the question of its scientific character, including the character of its conceptuality, remains problematic (60). This problematic character is stressed when Heidegger declares and partially italicizes without commentary that "*theology itself is primarily* [not entirely?] *based on* [*begründet durch*] *faith*, even if its statements and argumentations originate from formally free operations of reason" (61).

One cannot conclude that the relations between theology and philosophy have been satisfactorily solved by Heidegger's recourse to the old—and controversial—distinctions between *form* and *content* and between *dependence on faith* and *rational independence from any trust or other authority*. On the one hand, theology borrows from philosophy, when it accepts philosophical corrections, whereas, on the other hand, philosophy declares that theology is not able to think freely and must accept corrections of its analyses where philosophy deems these necessary. In order to be a science, theology must follow philosophy's indications about scientificity and the status of *ontic* research, while leaving the truly radical and most fundamental—*ontological*—questions concerning being and its ultimate meaning to philosophy.

If Heidegger is right, good philosophers cannot be theologians—they cannot even be interested in theology—and good theologians cannot be authentic philosophers. Consequently he cannot recognize Augustine, Anselm, Bonaventura, Aquinas, Scotus, Cusanus, and innumerable others as theologians who were also philosophers. According to his criteria, however, even Plato, Aristotle, Plotinus, and Proclus were inauthentic philosophers, because they were very much involved in *speculative study of God* [*speculative Gotteslehre*, 59–60], just as Descartes, Leibniz, Spinoza, Fichte, Kant, Hegel, and Schelling were. Heidegger avoids evoking any of the post-Socratic Greek or pre-modern European philosophers in order to discuss with them the relations between philosophy and theology, but he cannot, of course, deny that at least some of them are still revered by many of his colleagues who consider them to be outstanding theologians and philosophers at the same time. As for his own past or present colleagues in theology, such as Augustine, Thomas, Scotus, Cusanus, Newman, Barth, and Bultmann, Heidegger cannot recognize them as also relevant for philosophical studies. Some of them have indeed been able to reflect so thoroughly on the foundation of their own ("positive") science that they have become capable of discussing philosophical questions with professors of philosophy. They might even have learned some parts of the philosophical tradition, but if they mix it with their theology, they betray their own discipline and bastardize philosophy. A certain, purely formal, cooperation of philosophers with theologians seems however possible, as we have seen, even for Heidegger. Though far from being well-informed about post-Platonic and medieval authors, Heidegger still knew more double-sided thinkers than many of his colleagues in philosophy. If he had been more interested in a benevolent or simply accurate and less-selective reading of, e.g., Plato, Plotinus, Bonaventura, Aquinas, Scotus, or Cusanus, he would probably have been more nuanced and convincing about their handling of the relations between faith and reason, theology, ontology and phenomenology, revelation and philosophy. Perhaps, he would then also have refrained from dogmatically declaring that philosophy must exclude any idea of God. Such theologians, whose philosophical or "speculative" sophistication was clearly not inferior to that of Hegel's logic, might have impressed him, if he had participated in their discussions about the possibility of approaching God through the stammering language of human discussions about such (quasi-)concepts like—for example—*esse ipsum*, being itself, or *infinity*. Indeed, some of those starling philosophers who were also theologians have shown that divine infinity has very little or nothing in common with either a highest being or the being that is common to all, and that, consequently, all gods who claim to be God, are idols, whereas only the infinity of God makes free.

2 The Letter of 1964

In his letter of 1964, which he added to the publication of *Phenomenology and Theology* in 1969, Heidegger maintains that the task of theology lies in a clarification [*Erörterung*] of the Christian faith, including its meaning and its claims; but he refrains again from sketching a theological methodology. Instead, he focuses on a question that was implied in the topic of the Madison conference to which he was invited in 1964. From that topic, "the problem of a non-objectifying thinking and speaking in today's theology," Heidegger isolates the question of whether objectification is essential and unavoidable in science as such (and not only in, for example, the natural sciences). As we might expect, his answer is a clear "No," and this might be seen as part of the *retractatio* that Gadamer mentioned.

In the paper of 1927 the word "*Vergegenständlichung*" [objectification] was used rather often in order to indicate the transition from the prescientific acquaintance with some kind of being to its scientific positing, which makes it the *Gegenstand* (that which confronts it) of a specific science. In the letter of 1964, however, Heidegger not only states that *Objektivierung* and *Vergegenständlichung*[19]

[19] In the original essay, "Phänomenologie und Theologie," Heidegger uses *Gegenstand* and *Vergegenständlichung* instead of the words *Objekt* and *Objektivierung*. I neglect his distinction here, because it is not immediately relevant for the present discussion.

are *not* essential for thinking or speaking; he also gives a few hints about his new understanding of thinking and speaking as forms of intuitive acceptance [*Erblicken* and *Hinnehmen*] and responsive obedience to what, in its own way of being, comes forward to a receptive "listener." Both thinking and speaking are now forms of appropriate responding [***Ent****sprechen*, 72–73]; they allow the phenomena to show and tell in their own way what the thinker should think and say about them (74).

At the end of his letter (77), Heidegger has only one advice for theologians: Discover, in your concentration on the essence of Christian faith, "what theology should think and how it should speak." One of the questions that are implicitly answered in this advice is whether theology still can be a science at all. And Heidegger already suggests an answer in adding that theology "presumably is not allowed to [*darf nicht*] be a science." If this suspicion is true, then Gadamer's remark about Heidegger's own rejection of his early view on theology is certainly justified. However, if theology is *not* a science, then much of what the text of 1927 said about theology and its relation to other sciences, including philosophy (if this is a science), must be forgotten or denied. What was said about the special character of theology as confined to faith alone might still be valid, but nothing new is forwarded about the difference between faith and theology (which is no longer considered a science) or about the meaning of theology for the faithful, the university, or the general culture.

While arguing, in the beginning of his letter, against the identification of speaking as a form of objectification, Heidegger offers the examples of "*Trost zusprechen*" [to console] and "*ansprechen*" [to address, call, speak to] as modes of speaking that certainly cannot be described or understood as objectifying activities. What the later Heidegger has written about speaking as responding [*Entsprechen*] is here also present in the background, but the most current form of everyday speaking to and with other speakers as *interlocutory* event is hardly ever thematized by Heidegger. Where has he offered a phenomenology that highlights the addressing of one speaker by another, the mutual exchange in a conversation or dialogue, the interplay between teachers and students, the reactions evoked by a prophetic address, or a simple prayer to God? Regrettably, Heidegger has shown little interest in the fact that all forms of speaking are directed to someone, either as response or as provocation, or rather as both at the same time. Like many other philosophers, he seems to have been more interested in poems and *soliloquia* by authors who did not wait for human responses before they wrote their next poetic or noetic work.

In his later essays Heidegger has shown how a master of phenomenology tries hard to approach various phenomena not in an objectifying but each time in a carefully appropriate way—for example, by describing a thing, a goblet, a landscape, a bridge, a space, or an event. We owe him thanks for those lessons in phenomenological observation and accuracy. What neither he nor most of his philosophical predecessors have often shown, however, is how *persons* stand, walk, look, speak, think, gesticulate, dance, fight, enjoy life, invite, serve, love, discuss, or sadly feel alone. What he, insofar as I know, has hardly ever done—alas!—is to retrieve phenomenologically the very abundant literature of first hand testimonies about human encounters with the one and only God, who, despite the densest hiddenness, is experienced by the adoring faithful as being present around and beyond but also within them.

Both addressing and responding—conversation with other humans and encounters in prayer with God—are constitutive of Christian faith and theology, if it is true that the one and only "new" and all-encompassing "commandment" identifies *agapē* as loving God while at the same time loving every and all human persons. Consequently, an adequate understanding of theology does not seem possible without a phenomenologically correct analysis of faith as being drawn by and moving up (i.e., as praying) to the lovable God, in whose love one participates by loving one's neighbors. Faith is not primarily a believing that… or a belief in dogmas but rather a believing *in* and a reaching out of the whole person *toward* the hidden Love that moves the universe. This truth cannot be unveiled from the standard perspective of traditional philosophy, because this presupposes the standpoint of a solitary thinker (ego, I), who, as a supreme mental being, dominates the all of all beings thanks to the panoramic view that such a position at the top allows for. That our mind from *such* a point of view cannot discover God is inevitable, however, because God is *neither* a being that can appear as one among or above other beings within the all-encompassing (or "metaphysical") horizon of all beings together, because such a horizon makes *all* gathered beings finite, *nor* as the beingness that is proper to all beings. Even "Being itself" seems incapable of granting us a contact with God, as long as we understand "Being" as the absolute and ultimate condition, *Lichtung*, "space," or horizon of an all-encompassing vision. Plato has tried to think and name the unique non-being humanity is driven to, by referring to it as "beyond-being" because it is *radically incomparable* to all essences [*ousiai, Wesen, Seiendes*]; and later (Greek, Latin, and Germanic) thinkers have introduced the word "infinite" to point at its ungraspable (and therefore utterly dark) unicity.

Are both ruptures with the panoramic overview—interlocution as well as infinity—necessary for philosophy, at least in the form of experimental hypo-theses? Must philosophy overcome its obsessive fascination with being and "metaphysics" by reaching beyond them to point at the ultimate condition(s) of all being and "Being itself"? Or can we continue our philosophical search for truth, like Hegel and many of his predecessors, in the direction of a highest or most originary X that explains the coherence and the gathering of *ta panta* as the full unfolding of the One [*to Hen*] that is also *to Pan* (the universe)? Must philosophers begin by *apriori* excluding both the God of the Biblical tradition and the philosophical Infinite of "speculative" philosophy from their own domain, where only finite gods *and other idols* would have the right to dwell? Or should philosophers, like Plato and his offspring, discover that only faith or trust in a *beyond* supports their own daring to live and think and speak responsibly? Would it perhaps even become again possible to experience a kind of affinity between the best of theology and the best of philosophy? Only, of course, if truly reasonable freedom and non-superstitious faith are no longer perceived as contrary.

3 Looking Back

Throughout his entire essay of 1927, Heidegger has treated theology as a science. On its first page he even declares that one of the main differences between his own approach of the relation between philosophy and theology and the "vulgar," non-scientific but *weltanschauliche,* approach lies in the fact that he will treat both as sciences, and that their difference therefore must be determined through "scientific argumentation"—not only by way of persuasion and conviction (47). A few pages later (49) however, a doubt emerges: immediately after declaring that theology is a *positive science*, Heidegger asks, rather unexpectedly, "whether theology is a science at all" (49). He recognizes that this question "is the most central question," but adds that an answer to it must be delayed because the "*idea*" of theology must first be determined more thoroughly. The following pages explain the concept of a "positive science" (50–51) and then give a description of theology according to its idea. Theology emerges from faith, but, as a science, it must make transparent [*durchsichtig*] what faith is, believes, and does, and how it profoundly marks the believer's existence. Theology is motivated by faith and its purpose is the concrete development and strengthening of that faith (51–54).

Then Heidegger confronts this idea of theology with the definition of a science in general. If science is "a freely performed, conceptually disclosing objectification [*Vergegenständlichung*]" of its specific topic, its application to theology, as a science so utterly determined by faith, implies an answer to the following consideration. "If faith essentially [*von Hause aus*] would resist a conceptual explanation [*eine begriffliche Auslegung*], then theology would be a quite inadequate way of understanding [*Erfassen*] its object [faith]. It would then miss an essential moment, without which it from the outset never could become a science" (54).

Without giving the reader an answer, the text passes on to other aspects of the central question (see p. 49); but a warning has been given. During the entire essay the task of theology is seen in the "*conceptual* explanation" of faith as guided by faith itself. In the just quoted passage, the expression is presented as constitutive of authentic science. Does it reinforce the doubt that was hinted at on p. 49? Or does Heidegger take here "conceptual" [*begrifflich*], together with *Begriffe* and *Begreifen* in a more rigorous sense than in the many sentences where he usesthese words in a somewhat loose way, more or less similar to "understandable," "rational," "reasonable," or "clear"? (e.g., 60)

On page 60, Heidegger points to the "peculiar [*eigentümliche*] conceptuality" of theology, which he illustrates by underlining its characteristic mode of access to its "object" (faith) and the specific evidence by which theological theses are supported. But, without further elucidation, he only concludes from that peculiarity that theology should neither borrow from other (positive) sciences, nor allow them to interfere in its own mode(s) of demonstration and conceptual rigor. "Theology's own conceptuality can only grow out of itself" (60).

Maintaining that (1) "theology itself is founded primarily in faith," Heidegger also maintains that (2) its statements [*Aussagen*] and demonstrative procedures

[*Beweisgänge*] are generated by formally free actions of reason [*Vernunft*, 61]. Despite the peculiarity of theology's object, character, motivation, purpose, evidence, conceptuality, and methodical procedures, and despite its "imprisonment" within the circle of faith-in-faith, of which it is an ingredient (54), theology is a science—not quite similar to other sciences, but still respectable because of the extent of *freedom* and *rationality* it nevertheless shares with all other sciences, including even philosophy.[20]

Heidegger's letter of 1964, written almost 40 years later, disagrees with his early text by cautiously suggesting that theology "probably cannot (or is not allowed to) be a science at all" (77). In a much earlier letter, however, written on August 8, 1928, half a year after presenting *Phenomenology and Theology* in Marburg, he confides to Elisabeth Blochmann: "I am personally convinced that theology is *not* a science—but today I am not yet capable to *really show* this"[21] And after indicating that the relation between philosophy and theology would demand a more fundamental approach of both than he could offer in his talk of 1928, he adds: "My work in Marburg was also always consciously double-sided—helpful and quite disquieting—and I freed more than one person from theology…"[22]

The word "freed" [*befreit*] or "liberated" in the second quote is particularly interesting; it rhymes with the conviction, shared by Heidegger in 1927–1928, that theology is characterized by the servitude of faith, whereas science, including philosophy, is a "free" and "purely rational" manner of taking full responsibility for one's own thought.

However, if, already before August 1928, Heidegger gave up the postulates on which his replacement of the "vulgar" conception of theology in relation to philosophy was based, then the entire argumentation of his 1927 essay crumbles. If theology is *neither a Weltanschauung nor a science*, what is it then? Just a faith imprisoned in a horizon that isolates it from philosophy and the ontic sciences, but also forbids it to interfere in them or compete with them? Perhaps Heidegger would answer that, in any case, theology has other roots, and therefore also a radically other style of intelligibility than philosophy.

---

[20] Heidegger does not explain the implicit distinction between the form (*formale* structures and procedures) and the content [*Gegenstand*] which are united in each science, nor does he analyze the peculiar way in which they (as formally free and rational action regarding a neither free nor rational faith) constitute one scientific and at the same time faithful, but not schizoid and still intelligent mode of existence.

[21] Cf. Martin Heidegger and Elisabeth Blochmann (1990), 25.

[22] Cf. 26. The same letter of August 8, 1928 clearly shows that *Phenomenology and Theology* represents only a phase of Heidegger's struggle with the problem on which it focuses. Especially the following statements would prompt a further development (or even a radical overhaul, especially with regard to philosophy) of the printed essay: "It belongs to the essence of human *Dasein* that it, insofar as it exists, philosophizes. To be human already *means* to philosophize […]" (25). And "Religion is a fundamental possibility [*eine Grundmöglichkeit*] of human existence, although of a completely other kind than philosophy. ***This [philosophy], in turn, has its [own] faith [Glauben]*** [my emphasis, A.P.]—which is the freedom [*Freiheit*] of *Dasein* itself, which, of course [*ja*], becomes existent only in *being* free [*im **Freisein***]" (25).

But what about the opposition between faith and reason [*Vernunft*], in which the vulgar view takes refuge for an explication of the formal features of theology and philosophy? Doesn't Heidegger's contrast between the *faithful* "conceptuality" of theology on the one hand, and the "purely rational," *faithless*, "free" and "self-directed" operations of philosophy, on the other, come close to the "vulgar" (though typically modern) contrast between *faith* and *reason*? Like Hegel, Heidegger seems to agree with the modern (and, to a certain extent, also postmodern) conviction that the freedom of reason is the "deadly enemy" of a faith that tries to understand itself.

And what about the exile of theology, as an ontic discipline, from the domain of ontology? The answer to this question depends on the way in which one responds to the long history of Western thought, in which creation has been thought as indivisibly related to the most amazing of all wonders: that all beings *are* and that the "fact" of this "arc" is the most striking of all wonderful "facts" for those who are in awe of it. Especially the question of God—but *not as an entity* (which cannot be God)—is then a criterium for depth in thinking, but that question has already been exiled by too many (true or false?) thinkers from genuine philosophy.

References

Aristotle, *Nicomachean Ethics*.

Heidegger, Martin. 1969. Phänomenologie und Theologie. In *Wegmarken*, GA 9, ed. Friedrich Wilhelm von Hermann, 45–78. Frankfurt am Main: Klostermann.

Heidegger, Martin. 1976. Phenomenology and Theology. In *The Piety of Thinking*. Trans. James Hart, and John Maraldo, 5–21. Bloomington: Indiana University Press.

Heidegger, Martin. 1992. *Platon: Sophistes*. GA 19, ed. Ingeborg Schußler. Frankfurt am Main: Vittorio Klostermann.

Heidegger, Martin. 2003. *Plato's Sophist*. Trans. Richard Rojcewicz and Andre Schuwer. Bloomington: Indiana University Press.

Heidegger, Martin and Elisabeth, Blochmann. 1990. *Briefwechsel 1918–1969,* 2nd ed, ed. Joachim W. Storck. Marbach am Neckar: Deutsche Schillergesellschaft.

Kockelmans, Joseph. 1973. Heidegger and theology. *The Southern Journal of Philosophy* 4(3): 85–108.

Overbeck, Franz. 1903. *Über die Christlichkeit unserer heutigen Theologie*. Leipzig: Naumann.

Overbeck, Franz. 1989. *Über die Christlichkeit unserer heutigen Theologie*. Darmstadt: Wissenschaftliche Buchgesellschaft.

Peperzak, Adriaan T. 2013. *Trust. Who or what might support us?* New York: Fordham University Press.

The Remainders of Faith: On Karl Löwith's Conception of Secularization

Rodolphe Gasché

Abstract The essay explores Karl Löwith's notion of secularization arguing that this notion presupposes a conception of faith found only in the religions of the Book. In addition, it is shown that his analyses of history whether eschatological or progressive are carried out against the background of the Greek experience of the physical cosmos characterized by cyclical time.

In its most compressed form, Karl Löwith's thesis in his landmark investigation of the theological underpinnings of modern historical consciousness asserts that the problem of history has a "supramundane origin," to use Dieter Henrich's incisive expression.[1] This thesis, according to which the modern conception of history as an open-ended process of progress is a worldly reflection of the Christian history of salvation, has been the subject of many controversies and fierce criticism. In particular, Löwith's use of the category of secularization in order to conceptualize the transition from the Christian view on history—with its idea of providence and eschatological endtime—to worldly history has been contested. Indeed, notwithstanding all the caveats and qualifications that Löwith broaches in the "Conclusion" to his study—particularly the distinction "between a historical source and its possible consequences," which serves him to explain how Christianity itself could

This essay was originally published in *Divinatio. Studia Culturologica Series*, Sofia: MSHS. Vol. 28 (2008): 27–50.

[1] Henrich (1967), 459.

R. Gasché (✉)
SUNY Distinguished Professor & Eugenio Donato Professor of Comparative Literature
Department of Comparative Literature, State University of New York (SUNY),
639 Clemens Hall, Buffalo, NY 14260, USA
e-mail: gasche@buffalo.edu

B. Babich and D. Ginev (eds.), *The Multidimensionality of Hermeneutic Phenomenology*, Contributions to Phenomenology 70, DOI 10.1007/978-3-319-01707-5_19,

produce an anti-Christian "illegitimate child"[2]—the general thesis of *Meaning in History* states that modern historical consciousness, with its conception of history as progressive, originates in Hebrew and Christian thought, and is but a secularized version of the Judeo-Christian "eschatological outlook toward a future fulfilment" (196). As Hans Blumenberg has convincingly shown, the historical notion of secularization as a category of illegitimacy only accomplishes this task of accounting for the discontinuity between the world of Christianity and the modern world by presupposing an identity of substance between both. Construed as a secularization of sacred history, the worldly conception of history characteristic of the modern world, despite its seemingly radical alienation from the Christian history of salvation, entertains an intimate relation of continuity with theological history. Secularization, Blumenberg holds, is a category of continuity. Yet, in his 1934 book on *Nietzsches Philosophie der ewigen Wiederkehr des Gleichen*, where the first rudiments of his later theory of secularization are worked out, Löwith argues, in a discussion of Nietzsche's conception of nihilism, that what remains after the decay of Christian faith is also much more than just a remainder of faith. He writes that "this seeming *remainder* is in fact [*im Grunde*] the kernel" of Christian faith.[3] In other words, the replacement of Christianity by the secular view of world-history as a history of progress does not only fully remain within the framework set by Christianity, but is even the actualization, if not the highest potentiation of its very kernel—the kernel of its constituting faith. And indeed, on more than one occasion in *Meaning in History*, Löwith speaks of the faith in progress and of the religion of progress. The category of secularization is thus not only a category of continuity as Blumenberg has shown, and as such unable to account for discontinuities in history, in short, structurally incapable of recognizing autonomous developments, novelties, and innovations. Rather, it is above all, I will hold hereafter, a category demonstrating that it itself has raised the kernel of Christian faith to its ultimate potentiality.[4] As a consequence, Blumenberg's argument that as a category of continuity, the category of secularization remains a Christian conception—a conception intent on extending and salvaging theological history itself—and thus essentially is 'the *ultimate theologumenon*," would be even more compelling.[5] But if, at the same time, secularization also represents a potentiation of the very essence of Christian faith, then what follows from this for the very conception of this category?

The history sketched out in *Meaning in History*, that of how modern historical consciousness derives from the Christian salvation story, not only shows that this history has its unifying substance in a prime concern with the future, but also that it is a history which does not evidence any true epochal breaks. Thus, the passage from sacred history to secular history is only a seeming discontinuity, and what

[2] Löwith (1949). All citations in the text refer to this edition. Whereas the English original of *Meaning in History* refers to a "natural child" (112), the translation into German speaks more appropriately of an "illegitimate child." Löwith (2004), 123.

[3] Löwith (1956b), 53.

[4] Blumenberg (1983).

[5] Blumenberg (1964), 265.

distinguishes modern historical consciousness from the Christian eschatological view of history is ultimately only a question of degree. If, as Löwith contends at the very beginning of his work, the "philosophy of history is [...] *entirely* dependent on theology of history, in particular on the theological concept of history as a history of fulfilment and salvation" (1; emphasis mine), then it is clear from the start that the modern consciousness of history does not represent a radical rupture with a previous epoch. By contrast, this profoundly continuous, that is, linear, history predicated on a concern with the future stands in a sharp contrast to the Greek understanding of history and the world. It is not only in Löwith's work posterior to *Meaning in History*, as Blumenberg contends, that one can see that his concept of secularization becomes really intelligible only when thought from, or against, the backdrop of the divide between the pagan world with its cyclical view of nature and history on the one hand and the Jewish-Christian futuristic understanding of history on the other. That the divide between the Christian and the modern secular world is only secondary at best when compared to the epochal break between paganism and Jewish-Christian faith is obvious already from Löwith's 1949 work. Undoubtedly, in his later writings this point is made in a much blunter fashion as when, for example, he writes that "the transition from Greek culture to Christianity is a decisive and deciding break compared to the development from the Christian tradition to the modern attitude of consciousness which is only a secularization of the Christian *saeculum*."[6] Indeed, all that occurs in secularization is a rendering worldly of the world that Christianity had rendered un-wordly [*Verweltlichung der entweltlichten Welt*] through having tied it to the salvation plan, thus establishing, paradoxically, the world's mere worldliness in the first place. Löwith remarks: "Secularization remains one of the Christian *saeculum*."[7] Blumenberg, no doubt, echoes Löwith's statement about the relative significance of the divide between the Christian and the modern world when he observes: "The secularization of Christianity that produces modernity becomes for Löwith a comparatively unimportant differentiation as soon as he turns his attention to the unique and epochal break that in one stroke decided in favor of both the Middle Ages and the modern age: the turning away from the pagan cosmos of antiquity, with its cyclical structure of security, to the one-time temporal action of the biblical/Christian type."[8] The break with pagan cosmology is the only epochal break that truly counts. This one break puts all other breaks into perspective. In light of this unique break all the other breaks, in particular the secular break away from Christianity and hence also modern self-affirmation carried with it, are only of relative significance. If Löwith can so easily dismiss modernity's novelty, i.e., its claim to independence from Christianity, and depict secularization as entirely dependent on what it seeks to overcome, it is because of the significance he attributes to the one fateful break, that of Judeo-Christianity with the pagan world. Accordingly, this explicit valorization of the pagan natural world and its cyclical history for understanding the notion of secularization in Löwith will interest

[6] Löwith (1960), 254.

[7] Ibid., 237.

[8] Blumenberg (1983), 28.

us in the following. Löwith's whole analysis of secularization and its concomitant maxim of history as progress as an exclusively Christian conception, and, by extension, his analysis of modernity's attempts at self-emancipation and self-affirmation as still indebted to Christianity, is predicated on this valorization.

In my view, Blumenberg's criticism of Löwith's conception of secularization is irrefutable: this conception does not do justice to modernity's accomplishments. At the same time, by establishing the intra-Christian credentials of the conception of secularization, and, therein, its restricted nature as a historical category, another facet of his concept of secularization comes into view, one which, by highlighting the continuity between Christianity and modern secularization, puts the assumption that history is continuous into its place. Furthermore, by following up on the problematic of the difference between the pagan and the Christian world, a difference that also includes the modern world, I also hope to bring to light an aspect of Löwith's understanding of secularization that to my knowledge has received little attention, one that significantly restricts the concept's interpretive power to those cultures and civilizations founded on the religious movements that broke away from the pagan world of the Greeks, that is, the religions of the Book.

First, however, an additional issue needs to be addressed, one that inevitably poses itself when reading *Meaning in History*. This is the question of whether, by resorting to the category of secularization, Löwith can escape subscribing to the kind of understanding of history that he sets out to criticize, namely that of a future-oriented conception of historical time. By locating the origin of the modern understanding of history as an open-ended progress in the secularization of the biblical conception of an end-time—a conception that itself thus comes to an end in modernity, which in turn, according to Löwith's diagnosis, faces its own end in the present in which all faith in progress has vanished—the temporal pattern associated with the category of secularization seems also to subtend Löwith's own historical presentation of this development. Helmut Kuhn—one of the first to critically respond to Löwith's work—noted already in his 1949 book review of *Meaning in History* that the author "seems to embrace the very type of theory which he combats. He, too, constructs a philosophical history directed towards an *eschaton*. In a peculiarly inverted manner he is among the believers in progress–the goal being the undoing of the things done."[9] Kuhn refers here in particular to Löwith's "inversion of the customary way of historical presentation" in *Meaning in History* through "developing the historical succession of the interpretations of history regressively, starting from modern times and going back to their beginning" (2). Indeed, in his book, which opens with a chapter on Jacob Burckhardt's view of history, a view that refrains from any theological or metaphysical frame of reference, Löwith traces the theological implications of the philosophy of history backwards from Burckhardt through Marx and Hegel, Voltaire and Vico, Joachim and Augustine, to name a few, and concludes his study with a chapter on the Bible. This approach is justified "on three grounds: didactic, methodical, and substantial" (2). But, before briefly discussing these three reasons, a word about the addressee of Löwith's book is warranted. It is the modern reader, more precisely the

[9] Kuhn (1949), 825.

post World War II generation, which, having witnessed the atrocities committed in Europe, "is just awakening from the secular dream of progress which replaced the faith in providence but which has not yet reached Burckhardt's resolute renunciation" (2). In other words, it is a reader who, similar to Löwith, finds himself "more or less at the end of the rope," at the end, that is, of the secular belief in progress and, in the same breadth, of the belief in Christian providence, of which this philosophy of history is but an antireligious perversion [*Verkehrung*] (3, 192). However, the addressee of Löwith's work, although disillusioned with the modern faith in progress, is not aware of its theological foundation. By choosing to invert his historical presentation of the origination of modern historical consciousness, Löwith thus works toward awakening the present generation not only from the dream of a history of progress which has reaped nothing but disaster, but from that of the history of salvation as well. Although vaguely reminiscent of the Husserlian conception of *Abbau* as the dismantling of the historical sedimentations that cover over the originary evidences of the life-world, and of Heidegger's notion of *Destruktion* of the history of metaphysics in order to reach back to its covered up unthought so as to be able to begin anew, Löwith's regress from the present understanding of history to its origin in the biblical view of history does not aim at bringing the hidden religious presuppositions of the conception of history to light in order to reactivate these covered up presuppositions in some more originary way, but, on the contrary, to awaken his addressee from the dream of history altogether. This is the first didactic reason that Löwith offers to justify his inverted historical presentation of the supposedly linear sequence that is the origination and development of the philosophy of history from the secularization of the eschatological conception of Jewish and Christian faith in a fulfillment of time—a linear progress that has come already to a provisional end in the suspension of both conceptions of history in the work of Burckhardt. It is more expedient, Löwith argues, to start with what is familiar to the modern mind, namely that conception of history as progress from which the new generation is just awakening, before turning to "the unfamiliar thought of former generations" to which this conception is indebted. He writes: "It is easier to understand the former belief in providence through a critical analysis of the theological implications of the still existing belief in secular progress than it would be to understand belief in progress through an analysis of providence" (2).

The methodical ground for inverting the sequence of his historical presentation is the following: since "history is moving forward, leaving behind the historical foundations of the more recent and contemporary elaborations [...] historical consciousness cannot but start with itself, though its aim is to know the thought of other times and of other men, different from our times and ourselves" (2). However one goes about interpreting history, one is always only "reading the book of history backward from the last to the first page" (2). The second reason for his inverted presentation is, thus, merely a formalization of what actually happens, consciously or not, even in the customary way of historical presentation. It is a reason based on the claim that "history is moving forward," and that it leaves behind the historical foundation of the present. Yet if history is held to move forward, then it is essentially oriented toward the future whether or not this future is thought in terms of the *eschaton* of religious and theological history or of the open-ended progress that secular historicism advocates. By justifying his approach to history through invoking what factually obtains in historical consciousness, and by holding that history "moves forward," this second rationale for presenting history in an inverted way fully subscribes, it would seem, to that kind of history that, in going back to its beginning in theological history, Löwith seeks to undo.

So what about the substantial [*sachliche*] reflection that demands the inverted sequence of the presentation in question? Löwith explains: "The methodical regress from the modern secular interpretations of history to their ancient religious pattern is, last but not least, substantially justified by the realization that we find ourselves more or less at the end of the modern rope" (3). In the subsequent German publication of the book, the reason for proceeding this way rests on the diagnosis that currently we find ourselves at "the end of modern historical thought," that is, at the end of a conception of history predicated on open-ended progress.[10] This substantial diagnosis that today modern historical consciousness has (more or less) come to an end subtends, in fact, the two previous reasons for proceeding to write history backwards. It explains the existence of a new generation for which modern historical consciousness has lost all credibility, but one which also needs to be awakened to the rootedness of the secular dream of progress in the theological foundation of faith in providence so that it may resolutely renounce the Judeo-Christian conception of history as well. The conclusion that we find ourselves "at the end of the modern rope" calls as well for the formalization of what occurs in all turns to the past. When an end has been reached and history no longer moves forward, then it is necessary to revert backwards to the stages of the development of historical consciousness, to its foundation in Judeo-Christian faith in order to be able to sanction this end and make it irreversible. Only by returning back to the beginning of the modern compound in the "ancient religious pattern" through a step-by-step "analytical reduction" can the full meaning of that beginning that led to the whole process be grasped and a definite deathblow be dealt to what already has come to an end. Furthermore, this is the condition as well for overcoming the uncertainty and "suspense" that characterizes the contemporary world.

In his above mentioned review article of *Meaning in History*, Kuhn devotes a number of insightful remarks to the reversed temporal sequence of Löwith's historical presentation and points out that, indeed, this reversal, "far from being a mere expository device, is vividly expressive of the writer's thesis," namely that modern philosophy of history is a secularization of the history of salvation.[11] But, I think that Löwith's decision to present the history of this development in a reverse fashion accomplishes something more. As Kuhn rightly points out, "everything in history cries out against a regressive report" because it puts into question the elemental and "minimal faith—faith in the directed continuity of the forward move of events in time." By inverting the sequence of history, Löwith not only keeps the reader "from chapter through chapter [...] in expectation of the past to come," but he undermines,

[10] Löwith (2004), 13.

[11] Kuhn (1949), 823.

I would hold, the very elemental faith in question.[12] This is the faith, or expectation, that history moves forward. As we have seen, the assumption that history moves forward allows Löwith to justify his regressive approach to history insofar as, through this forward movement, the present's antecedents become covered over, but this same approach also appeared to underwrite the very concept of history that he sought to undo. Now, however, it becomes clear that this very method also undermines the faith in history's movement to begin with. If the modern conception of historical progress is a secularization of eschatological history, if, furthermore, every reflection on history within this process needs to be retraced to a previous position, then not only is there no progress, but on the contrary, history stagnates. Although future-oriented, secularization implies that nothing (new) happens. In addition to being a category of continuity, secularization may thus also be a category that, from within history, keeps history from moving forward. It is a category that draws one's attention to a past that inhibits the present from being something new, a category, that is, of the inner inertia of a conception of time that is essentially geared toward the future. If modernity is understood in terms of secularization, then it is from the start an aborted project.

However, before following through with this line of thought, let me first ask what, according to Löwith, is at stake in history and historical consciousness. The question is all the more warranted since modern historical consciousness, which itself is the secular offshoot of theological history, has given rise to what Löwith describes as "the modern overemphasis on secular history as *the* scene of man's destiny" (192). Compared to the Christian conception of secular history which considers the whole of worldly history as merely an interim or detour within the scheme of providence which is scheduled to be ultimately overturned at the end of times, modern historical consciousness overrates, overestimates worldly history. Löwith writes:

> The modern overemphasis on secular history as *the* scene of man's destiny is a product of our alienation from the natural theology of antiquity and from the supernatural theology of Christianity. It is foreign to wisdom and faith. Classical antiquity believed that human nature and history imitate the nature of the cosmos; the Old Testament teaches that man is created in the image of God; and the Christian teaching is focused on the imitation of Christ. According to the New Testament view, the advent of Christ is not a particular, though outstanding, fact within the continuity of secular history but the unique event that shattered once and for all the whole frame of history by breaking into its natural course, which is a course of sin and death. The importance of secular history decreases in direct proportion to the intensity of man's concern with God and himself. (192–93)

Even though modern historical consciousness has its roots in Christianity, Christian sacred history does not valorize worldly history and, in this, its view resembles that of antiquity. In light of the eternal laws of the cosmos, history for the ancients, is in some way of as little importance as is worldly history from a Christian perspective, a history that is interrupted and transcended by the one single event—the advent of Jesus Christ. However, this is also where the similarities between the historical views of classical cosmology and Christian theology end. Even though Christianity makes worldly history into a mere interim in the plan of salvation, it is

[12] Kuhn (1949), 822–23.

not therefore less important. On the contrary, as Löwith remarks: "This 'interim,' i.e., the whole of history, is neither an empty period in which nothing happens nor a busy period in which everything may happen, but the decisive time of probation and final discrimination between the wheat and the tares" (184). In characterizing worldly history as an interim, that is, in linking it to the promise of redemption and salvation, history is rendered intelligible and made meaningful in itself. Accordingly, Löwith reports that "man's sin and God's saving purpose—they alone require and justify history as such, and historical time. Without original sin and final redemption the historical interim would be unnecessary and unintelligible" (183–84). In the same way, to suggest that, with the second coming of Christ, worldly history would come to an end, is to make history highly significant in the first place. In short, in Christianity, worldly history, always experienced as a mere interim or necessary detour within the history of salvation and hence as limited in a way similar to classical cosmology's view of history which "restrained the experience of history and prevented its growing into indefinite dimensions [*dass sie masslos wurde*]," has nonetheless endowed history itself with a specific meaning (193). Although limited and restrained, this very meaningfulness of worldly history serves as the ground for explaining how modern historical consciousness can be construed as the secularization of the history of salvation to begin with and how within modernity history, could assume a disproportionate, if not measureless or even hubristic, importance.

By making worldly history meaningful even in such a limited fashion, Christianity brought about the fateful break with classical cosmology, from which modern historical consciousness is thus doubly alienated. This quest for meaning—of the meaning of worldly history and existence—which according to Löwith, arises in Judaism, also sets Christianity and its later secularized forms radically apart from the classical world. Beginning with his early work on *Nietzsches Philosophie der ewigen Wiederkehr des Gleichen*, in which Nietzsche is described as the philosopher who attempted to recover the lost world of the Greek cosmos, Löwith has consistently linked the question of the meaning of history (and the problem of the value of existence) to the Christian interpretation of existence. Nietzsche, he contends in *Wissen, Glaube und Skepsis*, is "the sole modern philosopher, who radically sought to overcome the question of [...] meaning and purpose," and can in this sense be said to be Greek.[13] Compared to the visible order of the cosmos, history does not show any order of its own. Historical or political events do not have the power to interrupt the cyclical movement of the cosmos that itself regulates the order of human affairs. The ancients did not ask what the meaning or purpose of history was as such; they did not endow it with a meaning separate and independent from that of the natural order of things. Furthermore, in comparison with the cosmological order, human history was seen to be rather insignificant. As the realm of the contingent, history was understood as political history and thus as an object not worth the attention of the philosopher whose eyes were turned to the necessary, immutable, and eternal laws of the visible cosmos, but only of statesmen and historians concerned with retelling and learning from past events. Löwith writes:

[13] Löwith (1956a), 80.

> The Greeks were deeply impressed by the everlasting order and beauty of the visible world, but it never occurred to any Greek thinker to connect in his mind this well-ordered eternal *cosmos* to the transitory *pragmata* of human history into a 'world-history' [...] The classical philosophers did not make history their subject matter and they did not inquire into the meaning of history because, as philosophers, they dwelt on the one and all which has its existence from nature and is for ever and so they left the continuously changing fate of history to the political historians.[14]

So, what triggers the emergence of the question of the meaning and purpose of history? At the beginning of *Meaning in History* Löwith notes that "the basic experience of evil and suffering" that comes with historical action is "the outstanding element [...] out of which an interpretation of history could arise at all" (3).[15] He adds: "The interpretation of history is, in the last analysis, an attempt to understand the meaning of history as the meaning of suffering by historical action. The Christian meaning of history, in particular, consists in the most paradoxical fact that the cross, this sign of deepest ignominy, could conquer the world of conquerors by opposing it. In our times crosses have been borne silently by millions of people; and if anything warrants the thought that the meaning of history has to be understood in a Christian sense, it is such boundless suffering" (3). As Löwith's analysis of the Prometheus myth in "Das Verhängnis des Fortschritts" demonstrates, for the Greeks evil and suffering are indeed intrinsically tied to historical action, but only as the inevitable dark side of all historical accomplishments.[16] There is no historical great deed that does not also have evil consequences. What, then, must have happened for suffering and evil to become such an issue in biblical and post-Christian thinking? Why does the evil and suffering that accompanies historical action demand an interpretation of history and set off a quest for its meaning? Is it not because, with Christianity, history is experienced as oriented toward the future, and, therefore, expectations are necessarily bound up with history? According to Löwith, "there are only deceptions, where there are expectations." [17] Only because history is no longer, as in the pagan world, only concerned with what occurred in the past for the sake of political edification, but now is thought to move forward and toward something, does the evil and pain associated with it become an issue and the question of its meaning, as a question of what is in it for the human being, becomes pertinent. For evil and suffering to become an issue at all, is it not, first, because history as such has become a concern and becomes endowed with the sense of a promise? Understood as holding something in wait for the human being, all of the evil and suffering associated with history lets itself be interpreted as the price to be paid for

---

[14] Löwith (1969), 47–48.

[15] See also Löwith (2004), 13.

[16] Löwith writes: "It is true that Prometheus frees the human being thanks to the gift stolen from the gods, but he does not redeem them; on the contrary, he is chained and punished by Zeus ... The Greeks have atoned in the cult of Prometheus for the theft of the fire of the heavens by way of the myth of the chained Prometheus, because they profoundly sensed that this theft provided the human being with a power that needed the most powerful chains so as not to bring about the ruin of man." Löwith (1964).

[17] Löwith, *Weltgeschichte und Heilsgeschehen*, p. 14.

an ultimate delivery from history, either in the outer-worldly Kingdom of God or in an enlightened secular world of infinite progress. Only because history, beginning with Judaism and Christianity, is conceived as a directed process, can there also be the "modern illusion" of history "as a progressive evolution which solves the problem of evil by way of elimination" (3).

But the inquiry into the meaning of history from biblical times to the post-Christian philosophy of history has another side still to be accounted for. For history to become a topical issue, it must first be experienced as being meaningless in light of the promise of redemption from everything worldly. But such experience is possible only in view of, and with respect to, a meaning that, since it is not manifest in history itself, must be of the order of the hidden. Only with an idea of a meaning of history in place, but one that is not revealed by history itself because it is nothing historical, can history appear as the realm of the meaningless and thus set off the quest for its hidden meaning. According to Löwith, "there would be no search for the meaning of history if its meaning were manifest in historical events. It is the very absence of meaning in the events themselves that motivates the quest. Conversely, it is only within a pre-established horizon of ultimate meaning, however hidden it may be, that actual history seems to be meaningless. The horizon has been established by history, for it is Hebrew and Christian thinking that brought this colossal question [*masslose Frage*] into existence. To ask earnestly the question of the ultimate meaning of history takes one's breath away; it transports us into a vacuum which only hope and faith can fill" (4). Compared to the ancients' much more modest speculations, the question of the ultimate meaning of the contingent realm of history is a "colossal question," more precisely, an immoderate question, one that does not know its limits. From a Greek perspective it would be a hubristic question that transgresses the limits of the knowable. No cognitive answer to this question is conceivable. As a consequence, this question "transports us into a vacuum which only hope and faith can fill."[18] In other words, only hope and faith could possibly provide a response to the question regarding the ultimate meaning of history as a whole that arises with Judeo-Christian thought. It is a question that can only be asked if faith in a history of salvation and hope for redemption have already turned worldly history into the saeculum and have made history worldly to begin with.

As Löwith pointed out:

> the ancients were more moderate in their speculations. They did not presume to make sense of the world or to discover its ultimate meaning. They were impressed by the visible order and beauty of the cosmos, and the cosmic laws of growth and decay was also a pattern for their understanding of history. According to the Greek view of life and the world, everything moves in recurrences, like the eternal recurrence of sunrise and sunset, of summer and winter, of generation and corruption. This view was satisfactory to them because it is a rational and natural understanding of the universe, combining a recognition of temporal changes with periodic regularity, constancy, and immutability. The immutable, as visible in the fixed order of the heavenly bodies, had a higher interest and value to them than any progressive and radical change. (4)

[18] Löwith (2004), 14.

Unlike the Hebrews and the Christians, the ancients were not so presumptuous as to claim a right to fathom what cannot be known because it is not manifest in the course of history itself, that is, the ultimate meaning of history. They focused on the order of the visible—the cyclical order of the visible manifest in the rational organization of the movements of the heavenly bodies, the cyclical changes of the seasons, and life and death. No other order distinguished the world of human affairs. Limiting themselves to what they could see, the ancients rather than speculating about what cannot be known, also held history to be circular, a cyclical recurrence of growth and decay, rather than moving toward an *eschaton*. Löwith concludes: "In this intellectual climate dominated by the rationality of the natural cosmos, there was no room for the universal [*weltgeschichtliche*] significance of a unique, incomparable historic event [such as the advent of Christ] [...] They were primarily concerned with the *logos* of the cosmos, not with the *Lord* and the *meaning* of history" (4).[19] In any case, what should be clear at this point is that it is this quest for the ultimate meaning of history, and, in particular, the question of faith (as opposed to knowledge) implied therein, that distinguishes the epoch-making advent of Christianity, which is epoch-making in that it breaks with the wisdom of the ancients. By implication, it should also be clear that modern historical consciousness, as a secularized eschatological history, must rest on a secularized conception of faith, namely faith in history. But if this is so, does one then not have to consider the possibility that the concept of secularization is intrinsically tied to faith and thus that secularization is conceivable only where it is preceded by faith? In the following I wish to argue that, for Löwith, secularization not only presupposes faith, and in particular Christian faith, but also that one can only meaningfully speak of secularization where there is a faith in the strict sense that can be rendered worldly in the first place. Despite Löwith's own failure to consistently clearly distinguish between faith and belief in *Meaning in History*, these two concepts are entirely distinct.[20]

[19] Since by juxtaposing the ancients' exclusive concern with the immutable and rational order of the cosmos to the Christian and post-Christian concern with history which is a function of an article of faith, namely the singular advent of Jesus Christ, Löwith intends to dismiss history as an infatuation, it should be kept in mind that the singularity of the interrupting advent of the first coming of Christ, although unique in that it is also the only event worth its name known by Christianity, has set the stage for the thinking of the event and singularity.

[20] In the 1949 English original, Löwith speaks somewhat indiscriminately of "Christian faith," "revelation and faith," "faith in providence,"but also of "the belief in providence," "the belief in salvation," "the belief in reason and progress," and so forth (1–2). It is therefore not unimportant to recall that in German only one word—*Glaube*—covers the religious and epistemic meanings expressed by the English terms faith and belief respectively. It is not clear whether Löwith has been aware of the semantic difference between the two words—between an adherence to a religious dogma based on a *credo*, and an adherence to a judgment of existential import which although impeccable (because it does not contain any internal contradiction) does nevertheless not allow for proof. In any case as the examples given seem to suggest, Löwith did not rigorously distinguish between the two English terms. Furthermore, the German translation by Hermann Kesting, revised by Löwith himself, blurs whatever distinction there may have been in the English original between faith and belief, by translating both by *Glaube*, or *Glauben*. For the distinction in German between *Glaube* and *Glauben*, see the entry by Büttgen (2004).

Belief is based on a logically flawless judgment, even if the latter cannot be scientifically proved. Faith [*Glaube*, *foi*], by contrast, is an unconditional confidence or trust in someone or something. It is not the merely pragmatic confidence in the reliability of this or that thing, even if it cannot be rationally supported. Faith is, rather, a steadfast trust in something despite the absence of any evidence for it. In the context of Löwith's work, faith refers primarily, if not exclusively to Christian faith. But, all differences aside, the way faith is determined here makes it not the sole property of Christianity, but extends it to the other two religions of the Book. Nonetheless, Löwith highlights the fact that for the Christian believer faith is not "the unquestionable possession of a constantly available certainty of faith." In *Wissen, Glaube und Skepsis* he writes: "Faith, as it is expressed in Hiob and Paulus, as well as in Augustine, Luther, Pascal and Kierkegaard, is an unconditional confidence that it has been difficult to achieve and that can also easily be lost again."[21] However, it is not enough to say that the concept of faith which is presupposed by all talk about secularization is primarily Christian faith. Rather, it must be noted that faith is a function of Christianity insofar as it is a form of monotheism. As will become clear, faith is intimately tied to a monotheistic conception of God, and it is only in the context of monotheism that it makes sense to speak of secularization. This is, then, also the reason why the concept of secularization can, in principle, be extended to the other two religions of the Book, but only to them. Within the context of non-monotheistic religions all talk of secularization is meaningless.

Anyway, in the crucial chapter in *Meaning in History* devoted to Voltaire, who was the first to have coined the expression of a "philosophy of history," Löwith argues that the modern conception of a universal history [*Weltgeschichte*] "directed toward one single end and unifying, at least potentially, the whole course of events was not created by Voltaire," whose conception of worldly history still lacks a central meaning that imparts a uniform orientation to all histories. Rather, this directed universal history first arises through "Jewish messianism and Christian eschatology, on the basis of an exclusive monotheism" (111). In the German translation of the 1949 book, Löwith adds that "it is only the One biblical God who universally orients and centers history."[22] The faith in One God, then, is the presupposition for being able to conceive of history as the conflict between the will of God and that of men, a history whose impact is the salvation of God's sinful creature. "One single theme: "God's call and man's response to it" allows for a transformation of the whole of history into an interim, that is, a time of probation, by which worldly history becomes intelligible and justified as such, while at the same time being programed to come to an end at the end of time (183–84). In the "Conclusion" to *Meaning in History*, Löwith remarks that "the problem of history as a whole is unanswerable within its own perspective. Historical processes as such do not bear the least evidence of a comprehensive and ultimate meaning. History as such has no outcome. There has never been and never will be an immanent solution of the problem of history, for man's historical experience is one of steady failure" (191).

[21] Löwith (1956a), 26.

[22] Löwith (2004), 122.

Needless to say, history is experienced as a steady failure only if it is measured against expectations, anticipations, and hopes in the first place. When Löwith holds that "Christianity, too, as a historical *world* religion, is a complete failure,"Christian faith remains the standard in relation to which the immanent meaninglessness of history is asserted. But what Löwith also suggests here is that, since no immanent meaning can be discovered in the actual course of history, any meaning to be attributed to it must come to history from the outside. And, it his here that faith comes in, as it offers at last an answer to such a demand for an ultimate meaning that unifies all historical events and directs them toward one definite outcome. Within Christianity, the answer to this demand is the second coming of Christ at the end of times which will confirm the belief that all worldly history will only have been a detour in the history of salvation.

If history is to be meaningful as an interim in the promised salvation, then such meaning cannot simply have its source outside of history. Rather, this source must be something that, from the perspective of historical experience, is impossible, namely, in the very words of the Scriptures, a *scandalon*, "an offence," i.e., an occasion of disbelief. As Joachim Gnilka notes, in the New Testament, the scandalon is above all one that God imposes on human beings, in that His ways of salvation occur in a manner that does not correspond to human expectations and representations. Faith is based on, and defined by, this very offence that the incarnation and the resurrection of Christ represent to the human mind.[23] Broadly speaking, the *scandalon* on which the Christian act of rendering history intelligible as a history of salvation rests is the advent of Christ. Löwith writes:

> The Christian claim that the whole and only meaning of history before and after Christ rests on the historical appearance of Jesus Christ is a claim so strange, stupendous, and radical that it could not and cannot but contradict and upset the normal historical consciousness of ancient and modern times [...] The possibility of a Christian interpretation of history rests neither on the recognition of spiritual values nor on that of Jesus as a world-historical individual; for many such individuals have had a world-wide effect and more than one claimed to be a savior. The Christian interpretation of history stands or falls with the acceptance of Jesus as Christ, i.e., with the doctrine of Incarnation. (184)

If, for a classical mind such as the second century Platonic philosopher Celsus, "the Christian claim is ridiculously pretentious," it is "because it endows an insignificant group of Jews and Christians with cosmic relevance. To a modern mind like Voltaire it is equally ridiculous because it exempts a particular history of salvation and revelation from the profane and general history of civilization. Both Celsus and Voltaire realize the *scandalon* of a history of salvation" (184). More precisely, the *scandalon* consists in believing that one supposedly historical event can, by interrupting history, render history as a whole intelligible and meaningful, even if realizing this amounts to positing that worldly history is meaningless. The faith that

[23] From the offences that God imposes on the faithful, one must distinguish "the offences [that, according to Mattheus 18, 6] will come." These are offences to the faithful by the incredulous, admonitions that is to Christ's followers to remain steadfast in the faith. Gnilka (1973), Vol. I, 111–115.

subtends theological history is "the faith in an actual event which has come to pass," that is, "the faith in an accomplished fact," namely, the faith that the advent of Jesus Christ is the advent of fulfilled time which only needs to be completed and that, with the second coming of Christ, history can be done away with once and for all (188, 189). For the pagans, the beginning of history is "a decisive political event (e.g., the foundation of Rome or a new revolutionary beginning) as the lasting foundation of the following happenings" (182). Therefore, for the pagan, the biblical view of history in light of which all mundane history is ultimately meaningless, even though it also originates in a beginning—the creation of the world—and unfolds in view of an *eschaton*, that is, the still outstanding central event at the end of time as the coming of the Messia, must appear as incredulous. Even more so, Christian time-reckoning, which begins with a central and decisive event that has already occurred—the *perfectum praesens* of the advent of Jesus Christ who, by having died for humankind, has already begun the process of salvation—and from which "the years of the history B.C. continuously decrease while the years A.D. increase toward an end-time," must, of needs, be judged aberrant (182). And, indeed, for the pagan mind, the Christian hope for and expectation of the accomplished fact of Jesus Christ which is required to make sense of history is simply not an option.

As Löwith emphasizes, faith is the unconditional credo in a central event such as the creation of the world or, especially, the advent of Jesus Christ, through which time is reckoned forward as well as backward. Thus, "the New Testament concept of faith did not exist in Greek thought."[24] Nor did the ancients know the distinction between intra-worldly knowledge and trans-worldly faith, a distinction which, according to Löwith, is, strictly speaking, "an inner-Christian affair" in that it presupposes a beforehand relation of philosophical knowledge to faith, a relation which is simply not present in ancient thought.[25] By contrast, the Greeks distinguished between two forms of knowledge—true knowledge [*episteme*] and opinion [*doxa*]. Although *doxa* can be translated as belief, and in German even as *Glauben*, it is done so simply in the sense of believing something to be the case. Löwith writes: "Held against the standard of *episteme* as true knowledge *doxa* is not faith in the New Testament sense of *pistis*, but belief in the sense of merely holding something to be true."[26] *Pistis*, rather than meaning genuine faith, is, for the ancient Greeks, only a "subordinate form of *doxa*."[27] Does it not follow from this that the only fundamental epochal break which Löwith recognizes, the one between the world of the ancients and the Judeo-Christian world, is predicated on

---

[24] Löwith (1956a), 14.

[25] Ibid., 11. In a passing remark in the *Critique of Practical Reason*, Kant observes that the Greeks thought that the principle of the good was sufficient to establish the principle of morals, and that it seemed to them that they did not need the postulate of the existence of God as a further condition of its possibility. It follows from this that faith, and even "pure *rational belief* [*Vernunftglaube*]" has no place in the world of the ancients (Kant 1996, 241). The question whether they believed in their gods, is a wrong question. Rather they *knew* of their existence.

[26] Löwith (1956a), 13.

[27] Ibid., 14.

the difference between knowledge in all of its forms on the one hand, and faith in something which like a fact can be held to be true on the other? From his remarks in *Wissen, Glauben und Skepsis* regarding the relation of philosophical knowledge to faith and, in particular, from his elaborations on theolology and philosophical thought in St. Augustine, for whom faith holds precedence over all cognitive insight, it is clear that, for Löwith, philosophy must already have accepted the priority of faith over knowledge to even be able to broach the question regarding the relation of faith to knowledge in the first place.[28]

Now, as we have seen, for Löwith, modern philosophy of history with "its faith in the absolute relevance of what is the most relative, namely history," is the result of "a philosophical secularization of Christian faith," of faith in providence and the coming of the Kingdom of God.[29] According to Löwith, "true, modern historical consciousness has discarded the Christian faith in a central event of absolute relevance, yet it maintains its logical antecedents and consequences, viz., the past as preparation and the future as consummation, thus reducing the history of salvation to the impersonal teleology of a progressive evolution in which every present stage is the fulfilment of past preparations. Transformed into a secular theory of progress, the scheme of the history of salvation could seem to be natural and demonstrable" (186). Undoubtedly, modern historical consciousness has abandoned the faith in history that is predicated on one unique event—the "single once-for-all which happened once-upon-a-time" (186). If Christian faith rests on, and is defined by, the assumption of such a unique event of fulfilled time in which past and future converge, it would seem that the philosophy of history has indeed transcended faith altogether. However, as Löwith argues, modern historical consciousness may well have broken with faith as a faith in a history of salvation, but only insofar as it replaces—or, "reduces"—biblical and Christian futurism to the impersonal teleology of a progressive evolution. For, fundamentally, the notion of history that modern historical consciousness opposes to that of faith does not break with faith's formal structure. This formal structure of the history of salvation that survives in modern time-consciousness is not limited to "the articulation of all historical time into past, present, and future," but is above all the formal scheme involved in Christianity's valorization of one unique now—the advent of Jesus Christ—which constitutes the core of faith, the scheme most properly taken up in what Löwith terms "the secularization [*Verweltlichung*] of Christian faith."[30] For the modern experience of qualitative historical time there is no now-point that would be neutral, insignificant, or indifferent. In fact, by viewing each present now as the opening of the horizon of a past and a future fulfilment, modern historical consciousness has not only generalized the Christian *kairos*, it has indeed fully realized the very kernel of faith. Secular historical consciousness, rather than being a radical break with Christian faith, is thus not only in full continuity with the latter, it is, for Löwith, the very realization or accomplishment of it, more precisely, the completion of the

[28] Ibid., 18–21.

[29] Löwith (1960), 170, 174.

[30] Ibid., 169.

essence of faith itself. It is therefore that Löwith can, with Benedetto Croce, qualify the modern faith in history as a faith in progress and as the last and ultimate religion.[31] In a historical consciousness that is a secularization of Christian faith, the formal schema characteristic of the unique historical advent of Jesus Christ has been extended to all present nows. The modern faith in history is therefore the last, and ultimate, religion.

However, if what remains of Christianity in modern historical consciousness is the very kernel of Christian faith itself, and if secularization consists above all in universalizing faith's formal schemes, then the implication is twofold: first, that faith in history is, indeed, the last religion—that is, a religion understood as constituted by faith in the sense we have seen, and, second, that not only does faith here achieve its completion, but also comes to its end. Recall the new generation which, according to Löwith, has become disenchanted with the modern faith in progress and to whom he dedicates his analyses of the biblical and Christian underpinnings of the philosophy of history in a effort to consolidate the defeat of the eschatological and teleological tradition. In order to glean what, according to Löwith, may come after the failure of both Christianity and modernity, it is necessary to return one more time to the epoch-making break between antiquity and Christianity, that is, to the break between knowledge and faith. Löwith writes: "Faith in history is the result of our alienation from the natural cosmo-theology of the ancient world and from the supra-natural theology of Christianity both of which provide a frame for history as well as a non-historical horizon of experience and understanding." If it is true, indeed, that faith in history could arise only as a result of the dissolution of these two pre-modern conceptions and also, then, with historical existence's loss of "a determinate place in the whole of nature [*des von Natur aus Seienden*]," that is, in therefore having "become completely independent and confined within its own temporality," then it is true as well that the "abstraction" from "the whole natural world, from the physical cosmos" which is presupposed by the modern assumption that "history is *the* dimension of human existence," characterizes already the break of Christianity with the pagan world.[32] No doubt, if "the experience of history was still bound, ordered, and limited *cosmologically* by Greek thought through the order and the logos of the physical cosmos, and *theologically* by Christian faith through the order of creation and God's will," then the biblical history of creation and the theological limitation of worldly history could only give rise to what Löwith qualifies as "the illusion of modern historical consciousness" because it had already dismissed the natural world to begin with.[33] Only an "essential incongruity [*prinzipielles Missverhältnis*] of the individual to the world in general" can explain "the singular incongruity to the historical world" characteristic of modernity, an incongruity that originates already in Christianity's transformation of the human historical world into an interim in God's salvation plan. A "no man's land," as it were, the "wholly other, natural, and physical world, which precedes all world- and human history,"

[31] Ibid., 160.

[32] Ibid., 160.

[33] Ibid., 159, 155.

that is, all religious, public, and political being-together, is the world that had to become denaturalized for a history of salvation, and, subsequently, a secularized conception of history, to become possible.[34] Löwith remarks that the biblical faith in creation implies "that the whole visible world, the totality of what is, the human being included, is not *by nature* [*von Natur aus*] there. The critical function of the doctrine of creation consists in denaturalizing *physis* and the *cosmos*, and to render absurd the discovery of nature in its naturalness [...] A *physis* arising from itself as well as a *cosmos* that originates in itself, or has been formed from chaos—this first and final theme of all natural philosophy—becomes annulled right away by the belief in a creation."[35]

Löwith's concern with the living physical cosmos goes at least as far back as his 1935 study on *Nietzsches Philosophie der ewigen Wiederkehr des Gleichen*, where this issue is addressed for the first time.[36] It has been pointed out that the physical cosmos, as it has been sketched out by Löwith, "remains, in spite of frequent invocations of originary experience and sensible immediacy," strangely empty.[37] Yet, were one to take into account that Löwith's natural world is a form of life-world, albeit distinct from Husserl's total horizon of the world and of Heidegger's world-project, it would be possible to construe this notion in a much more substantial manner, even though Löwith himself acknowledges that, while the natural world is the greatest and the richest, it is at the same time "as empty as a frame without a picture."[38] As he submits, "the word *cosmos* corresponds to the peculiar Greek experience of the world, but who could just like that claim that we do not live anymore in a *cosmos*."[39] If *cosmos* means the "wonderfully organized and surprisingly reasonable natural world,"—Löwith points to the discoveries of modern biology to support his claim—or "the omnipresent physical world within which world-history is something minute," in short, "a cosmos that exists by nature [*von Natur aus*] and is organized in a lively fashion," then it is difficult to hold that we no longer live in the *cosmos*.[40] To characterize the visible world in which we live, and which, as such, encloses all of human history, as a *cosmos* does not exclude the expansion of the immediately visible order of this form of life-world by means of, say, prosthetic devices. But to qualify the omnipresent natural world as *cosmos* is also to evoke the law of its temporality as distinct from the eschatological time-conception of the bible and Christian theology, as well as from the open-ended future-oriented conception of progress of secular modernity. To determine the all-present physical world as *cosmos* is to suggest that the law of the world is, as the ancients held it, a cyclical law, or, in Nietzschean terms, the law of eternal recurrence of the same.

[34] Löwith (1956a), 60.

[35] Ibid., 68.

[36] See also Hosoya (1967), 163.

[37] Hosoya (1967), 168.

[38] See in particular, Löwith (1960), 228, 239.

[39] Löwith (1956a), 76.

[40] Ibid.; Löwith (1956b), 109; Löwith (1960), 240.

To conclude, I would like to return to the question left in abeyance that of whether or not Löwith in fact embraces the kind of futuristic history, whether eschatological or progressive, which he combats in his work. It has been argued that, although *Meaning in History* calls "for a renewal of the model of cyclical time characteristic of ancient Greek thought patterned on natural phenomena," this idea has been advanced "more as a portent than as a definite philosophical program," and that it, therefore, also "remained quite vague."[41] Furthermore, Löwith's critique of Nietzsche's "questionable," because ambiguous and contradictory, doctrine of the eternal return in his 1935 study on Nietzsche could be mentioned as further proof that the idea of a return to a cyclical time conception is not of any real concern to Löwith. And, does he not even declare in *Meaning in History* that "a return to such views as had satisfied the ancients," that is, "to the goalless, cyclical conception of the Greeks regarding the course of history," is no longer possible (111)?[42] But, that statement is made in the context of a discussion of the weight of the Christian belief in a universal history directed toward one single end and unifying, at least potentially, the whole course of events, a belief that weighs in on all the Enlightenment attempts to break away from it. However, as we have seen, times have changed! Not only has the new generation for whom Löwith is writing become disenchanted with the idea of progress, but, by seeking to reveal the theological underpinnings of this secular conception of history, Löwith aims at the same time at dismantling the formal structures characteristic of the eschatological futurism of Judaism and Christianity. For Löwith, it is no longer a question of replacing [*ersetzen*] the major theological tenets regarding the Jewish and Christian view of history by a secular one because the religious views on history are no longer "the established horizon" (111). Let me also note that Löwith's intransigent critique of Nietzsche's conception of the eternal return of the same in his early study takes place after all in the name of the originary Greek conception of cyclical time. Nietzsche's interpretation of the eternal return of the same is termed "un-Greek," and is said to amount to a fateful modernization—because indebted to the Jewish-Christian tradition—of the idea of the eternal recurrence of the same.[43] Furthermore, the opposition between cyclical history and eschatological and teleological history, including modern history as an open-ended progress, is crucial to *Meaning in History*. Take, for example, the following statement: "It is not by chance that the religion of progress did not emerge and develop in antiquity, with its veneration for the past and the ever present. It is Jewish-Christian futurism which opened the future as the dynamic horizon of all modern striving and thinking. Within a cyclic *Weltanschauung* and order of the universe, where every movement of advance is, at the same time, a movement of return, there is no place for progress [...] The whole significance of progress depends on 'looking forward,'" that is, on the assumption that the future brings something new (111). It is in light of this cyclical time conception of the ancients that Löwith consistently evaluates both the Jewish and Christian infatuation with an end of time, as well as that of the moderns with

[41] Barash (1998), 75.

[42] Löwith (2004), 122.

[43] Löwith (1956b), 125–26.

unrelenting progress. As Löwith's qualification of this faith in progress as immodest and inordinate demonstrates unambiguously, it is the Greek conception of time that guides his analyses from the start. But the cyclical conception of time is operative in *Meaning in History* in still another way. Indeed, as should have become obvious, Löwith's diagnosis that currently the secular conception of history is coming to an end, though not in order to make place for a revival of its Jewish-Christian underpinnings, but, rather, to end both the Jewish and Christian's ways of understanding history (given that the secular understanding of history represents, in fact, the full realization of the essence of both conceptions of history), is not an indication of a subscription to a linear end-time-oriented history. Instead, it is already a cyclical conception of time that animates Löwith's inverted account of the development of the genesis of modern historical consciousness. Rather than indicative of a linear and continuous conception of history, the category of secularization in Löwith is subservient to a cyclical time conception. Furthermore, such a conception of cyclical time is, it seems, even necessary to be able to account for the continuous history from biblical and Christian conceptions of history to those of modernity. One more time, Löwith's profound dedication to Greek thought is demonstrated when, echoing Aristotle's suggestion in *Physics IV* that a punctual now as both a beginning and an end is required for time to be a continuous succession, while, at the same time, also transforming such synthesis of time to imply the circularity of time, he writes: "To be theoretically consequent [...], the trust in continuity would have to come back to the classical theory of *circular* movement; for only on the basis of a circular, endless movement, without beginning and end, is continuity really demonstrable. But how can one imagine history as a continuous process within a *linear* progression, without presupposing a discontinuing *terminus a quo* and *ad quem*, i.e., a beginning and an end?" (207).[44] By recounting the continuous history of the historicized consciousness from its origin in Jewish-Christian eschatological conceptions to its secularization in modernity in an inverse way, Löwith has forcefully made a case for history's steady failure. He has demonstrated, in a gesture that is Stoic indeed, that "the world is still as it was in the time of Alaric," that "only our means of oppression and destruction (as well as of reconstruction) are considerably improved and are adorned with hypocrisy" (191).

References

Aristotle. 1985. In *The complete works of Aristotle*, ed. J. Barnes. Princeton: Princeton University Press.

Barash, Jeffrey Andrew. 1998. The sense of history: On the political implications of Karl Löwith's concept of secularization. *History and Theory* 37(1): 69–82.

Blumenberg, Hans. 1964. 'Säkularisation'. Kritik einer Kategorie historischer Illegitimität. In *Die Philosophie und die Frage nach dem Fortschritt*, ed. H. Kuhn and F. Wiedmann. 240–265. Munich: Anton Pustet.

[44] Aristotle, *Physics*, Book IV (218a-222a), in (1985), Vol. 1, 370–376.

Blumenberg, Hans. 1983. *The Legitimacy of the Modern Age*. Trans. R. M. Wallace. Cambridge, MA: MIT Press.

Büttgen, Philippe. 2004. Glaube. In *Vocabulaire européen des philosophies*, ed. B. Cassin, 507–508. Paris: Seuil.

Gnilka, Joachim. 1973. Aergernis. In *Handbuch theologischer Grundbegriffe*. Munich: Deutscher Taschenbuch Verlag.

Henrich, Dieter. 1967. Sceptico sereno. In *Natur und Geschichte. Karl Löwith zum 70. Geburtstag*, ed. H. Braun and M. Riedel. Stuttgart: Kohlhammer.

Hosoya, Sadao.1967. Zwischen Natur und Geschichte. Eine unzulängliche Bemerkung zu K. Löwith. In *Natur und Geschichte. Karl Löwith zum 70. Geburtstag*, ed. Braun, H. and M. Riedel. Suttgart: Kohlhammer.

Kant, Immanuel. 1996. In *Practical philosophy*, ed. M.J. Gregor. Cambridge: Cambridge University Press.

Kuhn, Helmut. 1949. Book review of *Meaning in history. The theological implications of the philosophy of history*. *The Journal of Philosophy* 46(25) (Dec 8): 825.

Löwith, Karl. 1949. *Meaning in history. The theological implications of the philosophy of history*. Chicago: Chicago University Press.

Löwith, Karl. 1956a. *Wissen, Glaube und Skepsis*. Göttingen: Vandenhoeck & Ruprecht.

Löwith, Karl. 1956b. *Nietzsches Philosophie der ewigen Wiederkehr des Gleichen*. Frankfurt am Main: Kohlhammer.

Löwith, Karl. 1960. *Gesammelte Abhandlungen. Zur Kritik der geschichtlichen Existenz*. Stuttgart: Kohlhammer.

Löwith, Karl. 1964. Das Verhängnis des Fortschritts. In *Die Philosophie und die Frage nach dem Fortschritt*, ed. H. Kuhn and F. Wiedmann, 27–28. Munich: Anton Pustet.

Löwith, Karl. 1969. *Permanence and change. Lectures on the philosophy of history*. Capetown: Haum.

Löwith, Karl. 2004. *Weltgeschichte und Heilsgeschehen. Die theologischen Voraussetzungen der Geschichtsphilosophie*. Stuttgart: J. B. Metzler.

The Hermeneutics of God, the Universe, and Everything

Simon Glynn

Abstract Hermeneutic interpretation entered modern thought as a means of clarifying and resolving apparent incoherencies and contradictions within the scriptures, its potential for determining the meanings of legal, classical, and other texts being soon recognized, and even extended to the discernment of the meanings of plays, paintings, and other artistic and cultural artifacts and performances. And while some argued such meanings were to be ascertained by interpreting them within the contexts in which they appeared, others maintaining that artists' or authors' intentions were ultimately authoritative, were forced to concede that these too could only be interpretively derived, often in similar manner. Moreover conflicting interpretations suggest that the concepts which shape our "perceptions" of such matters are relative, while Gestalt psychologists and Ames and his school empirically demonstrated that even our most basic empirical perceptions are interpretations shaped by our pre-conceptions; an insight which clearly undermines the objectivistic pretensions of the natural sciences. The paper concludes, along with Heidegger, that hermeneutic interpretation is central to all epistemological understanding, as indeed it is to our very existence or being as humans.

1 Epistemological Hermeneutics: A Brief Outline

Hermeneutics, like Hermes, the winged messenger of the gods for whom it is arguably named,[1] is concerned with the interpretive derivation of meaning and correct understanding. Heidegger traces hermeneutic practice back, via Plato, to

[1] The term has its roots in the Greek term *hermeneuein* meaning to interpret.

S. Glynn (✉)
Florida Atlantic University, 777 Glades Road, Boca Raton, FL 33431, USA
e-mail: glynn@fau.edu

B. Babich and D. Ginev (eds.), *The Multidimensionality of Hermeneutic Phenomenology*, Contributions to Phenomenology 70, DOI 10.1007/978-3-319-01707-5_20,

Parmenides and Heraclitus,[2] but it is with the Reformation that the widespread practice of hermeneutics is more generally associated. And while certain Protestants claimed that the intended, and therefore supposedly absolutely authoritative, meanings of the scriptures, which they took to be the "inerrant word of God," had been divinely revealed to them, those souls who felt themselves to be less well connected generally proceed in the opposite direction, attempting to approach an understand of what God meant through an understanding of the scriptures. And against the Fundamentalist claim that a supposedly literal meaning was *immediately* evident, and thus needed no interpretation, many, made wary of such claims by the apparent incoherences, not to say contradictions, exhibited by and between certain scriptures, argued that it was only by interpreting particular scriptures, and the individual passages constitutive of them, in relation to the whole, that they could be properly understood. Catholics on the other hand, generally anxious to maintain the authority of the church, not to mention the power accruing to its role as an indispensable intermediary between the laity and the Deity, argued against interpreting the scriptures solely in terms of other scriptures, insisting that a proper understanding of them could only be derived from the *extrinsic* context or perspective offered by ecclesiastical tradition.

Now it was clearly only a short step from employing hermeneutic interpretation for the purposes of understanding scriptural texts, to its deployment for the purposes of understanding texts, such as works of classical antiquity, more generally, with disputes similar to those just outlined between the would-be fundamentalists and hermeneuticists regarding scriptural exegesis besetting attempts to understand legal texts, and to implement the law. And while it fell to Schleiermacher to articulate general hermeneutic principles which hopefully could be applied to all attempts to understand the full range of these various, scriptural, legal or classical texts, it was predictably suggested that such general principles of interpretation might also be extended to attempts to understand the meaning or significance of literary works more generally, as well as cultural artifacts and performances, such as paintings, plays and the like.

Thus the Romantics, who—in eliding the distinction between divine creation *ex nihilo*, and human creativity, which in contrast simply consists in the structuring or restructuring of form, whether material or immaterial—regarded the artist or author as god like, consequently insisted that so far from the meaning of a work being derivable entirely *intrinsically*, which is to say wholly upon the basis of relations between elements entirely internal to it, the artist's or author's *intentions* were definitive. However unlike those divines who claimed direct revelation as the source of their insight into The Creator's intentions, cultural critics could not always, or arguably ever, claim direct access to authorial and artistic *intentions,* which could perhaps best be understood in terms of the context from which they arose, and from which, accordingly, they were to be hermeneutically or interpretively derived.

Enter Dilthey who, opposing those (Positivists and neo-Positivists etc.) who suggested that the human and social sciences adopt the supposedly objective epistemologies,

[2] See for instance Heidegger (1971, 1975, 1976a, Sect. 44, 256–273).

quantitative methodologies and causal explanations that had seemingly served the natural science so well, argued that such an approach, developed for the study of physical objects and their properties, relations and interactions, was singularly inappropriate for the study of those of human subjects, which should be understood hermeneutically. Thus insisting that while "Nature we explain; man (on the other hand) we understand,"[3] Dilthey held that, so far from being *causally explicable,* as were the behavior and interactions of material objects, human behavior etc. was the result of *free choices* which could be *understood* in terms of the interpretively derived meaning or significance of the situations from which they arose. And if human behavior and relations could be understood hermeneutically, so too, Dilthey suggested, could socio-cultural systems and institutions. As he therefore further insisted:

> A rigorous hermeneutic of social organizations is needed in addition to single textual works.... Hermeneutics is possible here because between a people and a state, between believers and a church, between scientific life and a university there stands a relation in which a general outlook and unitary form of life find a structural coherence in which they express themselves. There is here the relation of parts to the whole, in which the parts receive meaning from the whole, and the whole receives sense from the parts; these categories of interpretation have their correlate in the structural coherence of the organization, by which it realizes its goal teleologically.[4]

Thus just as when we look up a word in a dictionary its meaning is determined by its relation to others, which are determined in turn by their relations to still others, and eventually by their relation to the original individual (this being the famous hermeneutic circle, or at least its epistemological aspect),[5] and just as Saussure has shown that the meaning of an individual linguistic utterance (*parole*) is similarly dependent upon its relation to the whole of the language in which it is uttered (*le langue*)[6] which such utterances collectively constitute, Dilthey here suggests that the meaning or significance of particular human and/or socio-cultural, relations and institutions, may similarly be understood on the basis of their relations to the socio-cultural whole which they collectively constitute. Like linguistic, artistic and other forms of human expression then, human actions, cultural systems and social organizations also have meaning or significance which hermeneutics can help us understand.[7]

---

[3] Dilthey (1957), 144 as quoted by Schrag (1980), 101. My addition in parentheses.

[4] Dilthey, quoted in Polkinghorne (1983), 221, quoting Earmarth (1978), 303.

[5] As we shall see later, Heidegger argues that *epistemological* hermeneutics, concerned with meanings of the sort with which we are here concerned, is merely an aspect of an *ontological hermeneutics* which is concerned to understand the significance of being, of what it means to be, which is in its turn to be differentiated from a *hermeneutic ontology,* for which, irrespective of the (epistemological) recognition thereof, the pursuit of meaning or significance is itself a mode of being.

[6] de Saussure (1959), 114.

[7] For a more detailed discussion and references concerning the aforementioned development of Hermeneutics see Polkinghorne (1983), 215ff.

However while Kant argued that human understanding, and indeed experiences also, were mediated by *universal categorial structures*, conflicting interpretations of the scriptures, as well as of legal and literary texts and works of art suggest that it is historical, and indeed socio-culturally etc. *relative conceptions* and/or *preconceptions*, rather than universal or transcendental categories, that have the last word in this regard.[8]

Furthermore any understanding an inquiring scientist may have, of the meaning or significance the actor attaches to his/her own *behavior*, is of course a (second order) understanding of an understanding, or Double Hermeneutic as it has been called, as so too is the scientist's understanding of the actor's understanding of his/her *intentions* or *motives*, and *experiences* etc. While focusing for a moment upon experience, it would seem to follow from the neo-Kantian insight above that, as the cognitive experiments of the *Gestalt* psychologists and of Ames and his school empirically demonstrate, the *sensible* is inextricably intertwined with the *intelligible,* and that consequently even our most basic empirical *perceptions* or *observations* are relative to our *conceptions* and/or *preconceptions.* In which case insofar as the natural sciences claim to be grounded upon just such empirical perceptions and observations, they too would seem to be substantially hermeneutic as Heidegger suggests.[9]

Having thus briefly delineated the range of, and some of the debates surrounding, hermeneutic epistemology, let us now examine some of the more important issues and possibilities arising therefrom.

2 Scripture

Returning firstly then to the question of scriptural exegesis, fundamentalists claim that the intentions of the authors are revealed in the very texts they are subsequently taken to illuminate. However, reflecting upon their understanding of "Thou Shalt Not Kill" for example, which many currently take as prohibiting abortion, but as entirely consistent with the death penalty and a couple of wars, it is evident that all that is in fact reveled are the presuppositions or assumptions, not to say prejudices, of the interpreter. As Heidegger confirms: "If one is engaged in—textual Interpretation, one likes to appeal [*Bruft*] to what "stands there," than one finds that what "stands there" in the first instance is nothing other than the obvious undiscussed assumptions [*Vormeinung*] (i.e. theories or pre-conceptions) of the person who does the interpreting."[10]

---

[8] See Bauman (1978), 9.

[9] On this issue, see for example, Heidegger (1976b) and Martin Heidegger (1962, Sect. 32, 188–195).

[10] Martin Heidegger (1962), 192. My addition in provided parentheses.

Clearly then, rooted in his or her socio-historical-cultural etc. context (let us call it "B") the interpreter can neither truly step into that of the authors of antiquity in an attempt to interpret them on their own ground (let us call it "A") nor can s/he step outside his/her context and establish a transcendent perspective from which to evaluate the degree of correspondence between the initial, authorial, intention in context "A," and his/her (the interpreter's) subsequent interpretation in context "B." Rather s/he can only attempt to understand the intentions of the author in socio-historical-cultural context "A" as they appeared through the text as perceived in socio-historical-cultural context "B." Moreover, insofar as the authors of the scriptures are dead, the only clues as to the supposed *correspondence* of the interpreter's interpretation to the author's original intention are the *intrinsic* coherence of the interpretation, and its *extrinsic* coherence with other related texts, or the whole of the available scriptures (always assuming *their* coherence of course),[11] perhaps along with the interpretation's coherence with what is more generally thought to be understood of the socio-cultural etc. context of the text's initial production. Not forgetting of course that whether the interpreter attempts to understand related texts and contexts directly, or understands them as interpreted, and thus mediated to him/her by others in the scholarly community, the interpreter's understanding of all of this is, in the final analysis, mediated by his/her own conceptions and preconceptions, often drawn in larger part from his/her socio-historico-cultural circumstances, which such understanding therefore reflects.

Thus, Gadamer informs us that:

> Every age has to understand a transmitted text in its own way, for the text is part of the whole of the tradition in which the age takes an objective interest and in which it seeks to understand itself. The real meaning of a text as it speaks to the interpreter does not depend upon the contingencies of the author and whom he originally wrote for… it is always partly determined also by the historical situation of the interpreter and hence by the totality of the objective course of history.[12]

Hence "…understanding …in the human sciences is essentially historical…in that a text is understood only if it is understood in a different way every time."[13]

However what is perhaps not widely recognized is that while some, in what we may loosely refer to as the Hellenistic or Greek tradition (dominated by the notion of *Logos* as absolute foundation or basis for a teleological pursuit of a definitive or literal truth) regard the concomitant demise of a rigid absolutism as lamentable, others, in what we may loosely refer to as the Hebraic or Jewish tradition (which while insisting that Talmudic interpretation be constrained by its need to be intrinsically and contextually coherent, nevertheless embrace its interpretation and reinterpretation) regard

[11] An enormous assumption that many would argue is clearly unwarranted. Thus while, to give but one of many examples, the Old Testament insists "An eye for an eye and a tooth for a tooth" the New Testament of course councils that we "Turn the other cheek" while the attempt to overcome this apparent contradiction by limiting each to its own Testament provides no solution to a religion whose "holy book" The Bible, consists of both!

[12] Gadamer (1975), 263.

[13] Gadamer (1975), 275–276.

the fact that such contextually delineated relativism facilitates an understanding that changes with the historically changing circumstances as laudable, facilitating as it does the propriety of the text relative to the ever changing situation, in or to which it is often to be applied.[14] Yet no matter which of these *attitudes* individuals may adopt, the *fact* remains that, absent a transcendental (or quasi-divine) perspective or insight, we cannot know what the author's intentions were.

3 Law

And as with the scriptures, so too, of course, with regard to the law. Thus taking the US Constitution as an example, the authors are long since deceased, and so here again, even the aspiration, much less the claim, to have divined the "original intent" of the authors of this document, is problematic for the same reasons as was the claim to have divined the original intent of the authors of the scriptures.

Nor does the fact that the authors of the Constitution left other documents behind, and that, in comparison to the authors of the scriptures, we have a much more *extensive* and *recent* record of the doings of the Constitutional authors, make any *essential* difference. For the reading of the Constitution within the context of other documents and historical circumstances is essentially not terribly different from reading parts of the scriptures within the context of the whole. What we are doing in all such cases is looking for provisionally coherent interpretations, and seeking to hone them in such a manner as to avoid conflict with the progressively revealed context, while eliminating those interpretations that recalcitrantly remain incoherent. Thus so far from changing the *essential* features of the hermeneutic process, other texts and an historical record simply act as filters which, in restricting interpretive latitude or indeterminacy, merely increase its *degree* of rigor; the *coherence* of an *interpretation,* no matter how constrained, while arguably *necessary*, nevertheless always remaining *insufficient,* to definitively establish its *correspondence* to the animating intention underlying the text, which could only be ascertained by an illusive transcendental reflection.

Moreover as with scriptural interpretation, while those would-be absolutists who, unlike "strict constructionists," do not simply deny the intranscendable relativism of legal interpretation, regard it as a lamentable fact, non-absolutists, usually of a less authoritarian or more liberal disposition, regard the fact that the meaning of the (therefore "living") Constitution or legal code is dependent upon its being interpreted from the perspective of the historical situation to which it is to be applied as insuring its continuing propriety.

[14] On this distinction see, for instance, Mathew Arnold, *Culture and Anarchy*, as quoted by Derrida (1978), 79.

4 The Arts

And just as scriptures and legal texts are subject to hermeneutic interpretation, so too of course are works of the literary imagination, not to mention painting, sculpture and the performing arts etc. And although limited space precludes a detailed examination of each, a paradigmatic example of an artistic work, *viz.* the much analyzed Mona Lisa, will prove most instructive and helpful in enabling us to identify hermeneutic principles which may often be applied, *mutatis mutandis,* to the arts in general, and not infrequently beyond.

Now many have interpreted the Mona Lisa's supposedly "enigmatic" smile as indicative of Leonardo's, possibly unconscious, intention to evoke the lost love of his mother; an intention which—as with those scriptural and legal hermeneuticists who claim to have derived this "animating spirit" of the scripture or law from the very text or letter whose interpretation it is to inform—some simply claim to have derived from the very picture whose meaning they take this intention to inform. Moreover even where access to Leonardo's family history etc. may provide evidence of his early separation from his mother, only if this was interpreted or seen as being significant enough to the older Leonardo as to inform his intention, conscious or otherwise, would it provide significant support for the initial interpretation. Furthermore, the very seeing of the smile as enigmatic, not of course to mention the seeing an oil paint on a two dimensional canvas as being a picture at all, much less a picture of a three dimensional young woman, and one who is smiling at that, may all be understood as grounded upon interpretive acts.

Additionally, even if we assume for the moment that Leonardo indeed intended, *consciously or otherwise*, to evoke his mother's lost love, which intention we assume to have been *unambiguous*, this might still fail to illuminate the full range of meanings or significance that the work had *even for or to Leonardo himself*, who may have had *multiple* intentions. Moreover, just as the context of his family history may inform *an*, or perhaps even *the*, understanding of Leonardo's *intentions*, there seems to be no *a priori* reason why other *contexts*, such as that *within which the work was received*, should not *equally*, or perhaps we should say *more fully*, inform the work's meaning or significance; *and this regardless of whether or not Leonardo could even have* been aware of such contexts, much less *intended such meanings or significance*.

That is to say that while, as Gadamer claims "…the discovery of the true meaning of a text or a work of art is never finished; it is in fact an infinite process"[15] this is not merely because there may be many *intentions underlying the work*, which could be *understood from within the constantly changing context or contexts of the interpreter or interpreters*, but also because, moving beyond a concern with *intentions,* there seems to be absolutely no good reason why, *entirely independent of authorial intentions* (and as with the meanings of scriptural and legal texts) the *constantly changing historical, (not to mention social, economic or political* etc.*)*

[15] Gadamer (1975), 265.

context(s) which the work exists in and may therefore contribute to, and within which it is received, should not also inform its meaning or significance.

For instance, the status of the Mona Lisa as *the* emblematic icon of the Renaissance can be seen as central to the work's significance, irrespective of the fact that even had Leonardo been aware of participating in *a* renaissance of some sort, he is extremely unlikely to have anticipated, much less intended the works iconic status. And even if he did both, given that the Renaissance had, at the time of Leonardo's execution of the painting, yet to run its full course, and he did not know how long this would take, he still could not have entirely understood the significance of such emblematic status anyhow; the more so in light of the fact that the way in which the Renaissance itself is understood varies between cultures and epochs etc. Nor indeed could he have known that several centuries later Marcel Duchamps would draw a moustache and goatee on a reproduction and retitle it L.H.O.O.Q., which is a pun in that when these letters are pronounced in French they sound like "Elle a chaud au cul," or, in translation, "She has a hot ass." Nor could he have anticipated that in consequence Duchamp would later be able to refer to a subsequent reproduction of the standard Mona Lisa, *sans* moustache and goatee, in somewhat *risqué* fashion, as "L.H.O.O.Q. Shaved." Neither could Leonardo have envisioned the iconic significance of his work as embodied in Andy Warhol's silk screened re-colored acrylic multiple of the same image, which he re-titled "Mona," thereby ambiguously both invoking the title of a blues song, and providing a homonym for a term signifying someone given to vocalizing sexual pleasure, to which the song title may or may not have been intended to apply.

Regardless of authorial intentions then, the full meaning or significance of a text or artifact, and indeed, as we shall see, human and social experiences and actions also, is dependent, at least in part, upon *the context of origin*, its/their *accumulating history,* and the changing *contexts of reception*. Clearly then, as Gadamer has duly noted "….not occasionally only but always, the text goes beyond its author. That is why understanding is not merely a reproductive but a productive attitude as well."[16]

Bauman, affirming this last point, insists that "All meaning results from interpretation; it is something to be constructed, not discovered."[17] In which case it should not surprise us to learn that it is not only history, family or otherwise, that is capable of providing the context in terms of which works and actions are to be interpreted. Leonardo's historical separation from his mother may, for instance, be ontogenetically generalized, and the Mona Lisa interpreted, in psychoanalytic terms, as the product of an Oedipus Complex, while even more generally, a Freudian may interpret all artistic production as having its origin in the sublimation of repressed libido, whereas a Marxist might understand the Italian Renaissance in general, and portraiture of this type in particular, from an economic perspective, in the context of maritime trade which resulted in the rise of bourgeois patrons who commissioned non-religious works.

[16] Gadamer (1975), 264. See also 263.

[17] Bauman (1978), 181.

5 History

Moreover not only does the interpretation of the arts, or indeed the scriptures, or the law, not necessarily have to be in historical terms (Marx famously interpreted the law as "the mystification of class conflict" for example) but neither does the interpretation of history itself.

For instance, Marx's understanding of history in terms of the movement from "Eden" (characterized by no private ownership) through Slavery (where Labor was privately owned) Feudalism (where Land was privately owned) Capitalism (where Capital was privately owned) to Socialism (where there was public ownership of the means of production) and finally to Communism (where, to come full circle, there was again to be no ownership) is of course the concomitant of an *economic-interpretation*. Indeed for a "vulgar," or reductive, Marxist, all history may be *understood* and *explained* in economic terms. Thus the rise of Fascism may be understood as the result of large war reparations extracted from Germany at the end of WWI, which while plausibly interpretable as being *intended* to insure peace by blocking the reconstruction of a powerful, militaristic, German State, escaped the "authors" (of these reparations) intentions in that, by handicapping economic recovery these reparations served to delegitimize the ruling Junker elite. This in turn made them ever more reliant upon the lower middle class dominated National Socialists and their policing to maintain power and order, thereby facilitated the rise of the Nazis, and German militarism. While, moving from an *economic* to a *psychological* interpretation, the rise of the Nazis and Hitler's power may be understood as deriving from a population which felt that its insecurity and self-doubt—resulting from their defeat in WWI and exacerbated by the economic collapse of the late 1920s and after—could be assuaged by embracing an authoritarian father figure or *Fuhrer* who, in return for unquestioning acceptance of his authority, would insure order, security and empowerment to his "family," which is to say to Arian members of the "Fatherland." From a vulgar Freudian perspective, all wars may be interpreted in terms of Thanatos, or the death instinct, or perhaps, in terms of Eros; the old men sending the younger men off to war so that they would have unobstructed access to those young women left behind, and, if victorious, access to those of the defeated nation also.

6 Human and Social Sciences

Now it will be noted that such interpretations of history, in this case from economic and psychological perspectives, also subsumed social and political elements or dimensions within their prevue, and just as shifts in political power (the rise of Nazism) may be interpreted, and thus understood, in economic terms (large war reparations) or indeed in psychological terms (insecurity) psychological insecurity for example may in its turn be interpreted, and thus understood, in economic (debt burden and the concomitant lack of economic empowerment) or even political (the loss of a war) terms etc.

Furthermore not only may psychological *states* be thus understandable in the sense of their *origination* being explainable, in economic and/or political terms, but psychological *terms* may be understood in the sense of their *significance* of *meanings* being explicateable, and concomitantly their deployment being explainable, in similar terms.

Take the term "Psychosis," for instance. Observing a friend talking to her/him self, who I initially take to be "thinking out loud" as we might normally say, I notice, upon drawing closed, that what s/he is in fact saying, together with the patterns of speech and pauses, are more consistent with the act of conversing with an interlocutor, which, despite the fact that I can see none present, s/he claims to be doing.

Worried about my friend's psychological state I call another friend, who upon arrival confirms my observation as to the absence of our friend's supposed interlocutor. A psychiatrist is called, and upon being told by our friend that the interlocutor, who s/he claims is standing in plain sight, is male, about 5 ft 10 in. tall, and has long, light brown hair, comes to the conclusion that our friend has lost contact with reality, and as such is psychotic!

Asked further about the supposed interlocutor, our friend volunteers that usually s/he converses with him about once a week, when s/he comes to visit him—in what happens to be a church—and that his name is Jesus! Moreover I, my other friend and the psychiatrist observe that we are in the midst of an evangelical congregation, all of whom claim to be literally in the presence of Christ. And while we will perhaps interpret, and thus understand, their experiences, and even the congregants themselves, as psychotic, they may interpret our failure to see Christ as indicative of our being heretical. Clearly then, as Schütz has put it: "When I become aware of the segment of your lived experience, I arrange what I see within my own meaning context. But meanwhile you have arranged it in yours. Thus I am always interpreting your lived experience from my own standpoint."[18]

Yet there is another, or further, hermeneutic dimension to this situation. For, perhaps, by virtue of the fact that the congregants are in their own Church, to which, unlike some "cults," they might readily grant public access, and therefore are perceived neither as a threat to society, nor even as disruptive, and perhaps because such behavior is not uncommon within the broader society, and perhaps even, in part, because they may be economically affluent and politically well connected, we might refrain from calling or even thinking them psychotic, as we might if none of this were so! The significance of the collective or social, not to say economic and political, context within which the meanings and deployment of psychological terms or concepts are negotiated could not be more evident.

And as with the understanding and deployment of psychological terms or concepts, so too of course with that of political, economic, cultural, and indeed all, terms or concepts, as can readily be demonstrated by a couple of examples drawn from US politics. For instance, having refused to allow the Vietnamese elections

[18] Schütz (1972), 106.

(scheduled for 1956) to go ahead because Eisenhower had been advised that the "Communists" or Vietnamese Nationalists, under the leadership of Ho Chi Minh would win, the US government actually claimed that in the war it subsequently waged against Ho's Viet Minh it was fighting *for democracy* which, despite that term normally referring to particular system or form of *political* governance, was redefined in contradistinction to communism, which is normally understood either as a system or form of *economy* characterized by the absence of private ownership, or as a system or form of equalitarian *socio-cultural* relations. And in similar vein many neo-liberal economists and their neoconservative allies continue to insist that a "free market"—unregulated even by those democratically elected governments which might wish to insure their constituents' freedom from exploitation, starvation and pollution etc., by enacting minimum wage and environmental legislation—is a necessary, and even a sufficient, condition of a liberal democracy. Or again, former US President George W. Bush's insistence that "we do not torture" which has, in light of the facts, something of an Alice in Wonderland quality. And although Humpty Dumpty's claim that "When I use a word, it means just what I choose it to mean…" this only holds, of course, if one is not necessarily concerned to communicate with, and be understood by, others. Otherwise the definition or delineation and deployment of words and the (signified) concepts associated with them—in terms of which we interpret and come to "understand" human experiences, personal behavior, and socio-cultural and political etc. interactions and institutions etc.—are subject to the linguistic community's "understanding" of their meanings; negotiated meanings which inevitably reflects the political, economic, social and cultural interests and power dynamics at play within that community.

And just as, like linguistic contexts as narrowly understood, economic, political, social and cultural contexts also define and delineate the meaning or significance of linguistic terms and concepts within linguistic communities, so too the very meanings or significance of the *experiences, behavior, interactions* and *institutions* which these concepts and terms help us understand and communicate, are also dependent upon all kinds of contexts.

Beginning then with experience, let us examine the case of Marnie (from the Alfred Hitchcock film of the same name) who experiences extreme anxiety when perceiving the color red. Psychoanalytic regression reveals that Marnie's mother, a prostitute, had worn red shoes and a red dress when going out to attract clients whom she brought back home with her. One night, when 6 years old, Marnie had observed her mother struggling with a man she had brought back to their apartment, whom she killed by bloodily splitting his skull with a blow to the head with a fire iron. Marnie's experience of anxiety when perceiving the color red was then *interpreted*, and thus *understood* and *explained*, in terms of the *context* of the repressed memory of a painful event, which it threatened to revive via association. Thus affirming Schütz and Luckmann's claim, that "Living experiences first become meaningful when they are explicated…in respect to their position in a reference schema,"[19] we can see how access to a wider *context*, provided by her repressed

[19] Schütz and Luckmann (1974), 16.

past, of which Marnie was not consciously aware, (not to mention an understanding of psychoanalytic theory) provided a "reference schema" which enabled the inquirer to understand the meaning or significance of her experience better than she had understood it herself. While in contrast to the case of Leonardo's understanding of the Mona Lisa then, it is not the inaccessibility of *future* contexts, but rather the inaccessibility of *past* contexts which is in play here.

And like *experiences*, and works of *art*, etc. etc., human *actions* also derive their meaning or significance, at least in part, from the contexts within which they play out or are *received*, their meaning or significance consequently evolving to the point were they too may escape their "author's" intentions.

Take, for example, a foreign invasion. Perhaps it was initially intended to protect a nation against terrorism, either by preempting an attack, or by setting an example of the retribution that might follow an attack, and/or by destroying or otherwise incapacitating any future potential threat etc. However let us imagine that this invasion in fact demonstrates the relative impotency of the invader, refutes previously widely held beliefs regarding the invader's moral rectitude, economically impoverishes and therefore diminishes the invader's real power, and causes widespread civilian casualties, which fuel anti-invader sentiment and recruitment to the resistance, thereby increasing the invader's vulnerability to terrorist attacks. In such a case then (isomorphically with Leonardo's painting) the significance of the act clearly transcends the actor's intentions, even perhaps to the point that its consequences could not even have been imagined, much less fully comprehended, by the actor.

Nevertheless actor's *intentions remain important*, for irrespective of any light they may throw upon the meanings of actions, *an understanding of the intentions or reasons motivating human and social actions may facilitate the prediction of future acts*. For instance, while an invasion may be interpreted as motivated by a desire for security, it might also be interpreted in terms of the desire to topple a despot, to spread democracy, or as a prelude to a resource grab, while an understanding of which intentions or group of intentions were primary motivators would clearly be useful in attempting to predict the future behavior of the invader.[20]

Now although, unlike the case with the scriptures and the US Constitution, the perpetrators of the acts in question may well be available to the inquirer, and therefore apparently able to confirm or refute the inquirer's interpretation and concomitant understanding of the actors' intention, this is far from unproblematic, for several reasons. To begin with (i) actors may not *remember* their intentions at all or fully or (ii) actors' intentions may nevertheless have been to some degree *ambiguous* even to the actors themselves, and may in any event (iii) (as in the previously alluded to case of a war motivated by desire of those older males instigating the war to

[20] This is not, of course, to imply that the motivating intention, and the sort of actions following from it, would necessarily remain the same over time, and regardless of circumstance. Indeed were the motive a resource grab, sufficient success, and/or the diminishing importance of the resource, brought about by technological change for instance, might well lead to the prediction that the hitherto belligerent invader will cease being so.

remove those young males who mostly participate in it, in order to have unobstructed access to the young women they leave behind) be unconscious. Moreover, (iv) actors may seek to *mislead* the inquirer as to their intentions, while even when this is not the case (v) *actors' communications* of their understanding of the intentions underlying their acts may be unintentionally cryptic, ambiguous or misleading, while (vi) even if it is not, *if this communication is to be understood by the inquirer, it will be no less in need of interpretation than the act or behavior it is supposed to illuminate.*

Thus even when the inquirer or social scientist turns to actors to verify his/her *interpretation* and concomitant *understanding* of the actors' intentions, as Schütz has observed: "...the concepts formed by the social scientist are constructs of the constructs formed in the commonsense thinking by actors on the social scene... constructs formed at the second level."[21]

So far from obviating the need for interpretation, or providing assurance as to its veridicality then, the inquirer's direct access to or communication with actors involves a second order or Double Hermeneutic, by which, if the inquirer is to attempt to *understand* the actors' intentions s/he must attempt to *interpret* the actors' *interpretations* and concomitant *understanding* of their own intentions.

Clearly then, as Bauman has observed: "There is no essential difference ... between the sense actors make of their actions and the meaning assigned to this action by an...external observer for that matter; all of them are equally in a basically similar process of meaning-construction-through-interpretation."[22]

Thus even actors' immediate accounts of their intentions at best provide contextualization which is not necessarily any more transparent and less in need of interpretation than the subsidiary writings and histories of the long dead Constitutional authors for example. While in any event, the coherence of the inquirer's *interpretation* of actors' or an actor's intentions with the available evidence, although perhaps *necessary* to insure its *correspondence* to the actors' or actor's actual intentions, is in no way *sufficient* to guarantee it.

Moreover, in view of the previously mentioned difficulties actors may have, even when not intentionally misleading the inquirer, in *remembering*, *disambiguating* (even to themselves), and/or being fully *conscious* of their intentions, it should be apparent that (as with the attempt to understand a subject's *experiences* better than s/he does him/herself) it may be possible, by adopting a *broader or different context or perspective*, as Ricoeur would have it, to "...understand an author/actor" and thus her/his *intentions* and *actions* etc. "better than s/he understood (them) her/himself."[23]

And as with human experiences and actions, so too with cultures. That is to say that while an anthropologist who belongs to a culture will tend to shares the conceptions, preconceptions and presuppositions prevalent therein, and will consequently understand that culture as it understands itself, if on the other hand s/he is an

---

[21] Schütz (1962), 246. See also 243.

[22] Bauman (1978), 181.

[23] Quoted by Ricoeur (1981), 151. My addition in parentheses.

outsider, this implies a critical distance, which, by enabling her/him to escape certain of the preconceptions and presuppositions endemic to it, may facilitate his/her understand it from a different, and in some cases a better, or more illuminating, perspective or context, than it understands itself.

Take for example the story told about the two space probes, sent by the Martians and the Venusians respectively to earth. The first, arriving on a Los Angeles freeway, proceeds to send imagery to Mission Control Mars, where the Martians conclude that earthlings are 3 or 4 ft high, 10 or 11 ft long, go around at about 50–70 miles per hour, and at night their eyes light up! The second, arriving in a Hollywood cemetery, proceeds to probe the earth, and observing that a number of the otherwise biodegrading corpses still have well preserved breast implants the Venusians conclude that these are the corpses of deceased members of a fertility cult! And although our initial reaction might be to dismiss both interpretations as equally incoherent with our understanding of things, further consideration may lead us to conclude that while the first is evidently so, the second may have something to it, perhaps offering an insight into our own culture which has hitherto eluded us, and thereby effecting our cultural self-understanding!

Clearly then, in the first case, *participation* in a culture enables us to reject what obviously appear to be spurious interpretations, while in the second, it is precisely the Venusians' *detachment* from the culture that facilitates what we may come to regard as genuine insight. Synthesizing these apparently contradictory requirements, that in order to facilitate our hermeneutic understanding we should both become familiar with, yet retain critical distance from, the "object" of our study, Plessner has pointed out, "Understanding is not the identification of the self with others, so that distance is illuminated; it is becoming familiar at a distance."[24]

On this view then, while the inquirer, be s/he an anthropologist, psychologist, sociologist or cultural critic etc., should attempt to engage with, and adopt the perspective of, those whose cultures, experiences, acts and/or artifacts, etc., s/he wishes to study, nevertheless s/he should also attempt to remain or become something of an estranged outsider, thus enabling her/him to retain or adopt an *alternative*, maybe *broader*, hopefully more *complete perspective*, and consequently an even more compelling and perhaps better, understanding than that of those who are more directly involved. It is after all surely precisely to the degree to which, upon returning from genuinely[25] foreign travel for instance, we feel ourselves to have become estranged from our own culture, and thus to have become (albeit temporarily) strangers in our own land, that we regard ourselves as having gained a better understanding than previously of our own culture and everyday existence.

[24] Helmuth Plessner (1978), p. 39.

[25] I use his qualification to distinguish "Travelers" from Tourists, who sojourning in Europe, South America, Africa or Asia, typically journey by air or with coach touring parties, from one Holiday Inn, Hilton or Ritz Carlton resort to another, and, like visitors to Disney's Epcot "World Showcase," make occasional forays from these hotels into "alien" cultures.

Not that it is always necessary to be, or come from, outside a culture in order to gain such a perspective; a certain reflective detachment may be sufficient as suggested by an example Plessner draws from another sphere:

> ...the estranged vision of the artist fulfils an indispensible condition of all genuine understanding. It lifts what is invisible in human relations because it is familiar, the counterfoil which puts the familiar into perspective as foreground and background and makes it comprehensible...[26]

Similarly then with the entry into a culture by Simmel's "Stranger"[27] for instance, which results in those in the culture attempting to adopt what they *imagine* to be the perspective of the Other in order to see the culture as they imagine the stranger must see it; an act of reflection which provides a sufficiently different perspective upon the culture to enhance the participants' understanding of it.[28] While like foreign travel, or the presence of strangers, so too the study of foreign cultures may also promote the adoption of alternative conceptions, preconceptions and presuppositions, thereby similarly affording an alternative, reflectively critical, perspective upon our own, to the point where Ricoeur is driven to conclude that:

> It is ... the growth of his [sic] own understanding of himself that he pursues through his understanding of the other. Every hermeneutics is thus, explicitly or implicitly, self understanding by means of understanding others.[29]

Nor should this surprise us, for it is, of course, the implicit proto-structuralist rule of the hermeneutic circle that, as with the meanings of words in a dictionary, each individual, as well as each shared world-view, human intention, action, experience or artifact etc. is to be understood in terms of their/its relations to the others, to which in turn—and here we come full hermeneutic circle—the same applies.

7 Critical Reflection and the Conflict of Interpretations

Now we have seen that regardless of whether one claims to understand others' writings, artistic works, experiences, intentions or behavior etc., better than the author/creator/actor etc. does her/himself, or whether it is a revision and/or enhancement of one's own understanding that is claimed, all such claims are predicated upon the capacity to gain critical distance from, or to reflect critically upon, the relevant phenomena and/or context, as well as upon whatever preconceptions, presumptions and prejudices may have mediated the understanding of them. This then has lead critics of hermeneutics and champions of neo-positivistic epistemologies to suggest that the adoption of a critical perspective can enable one to escape

---

[26] Plessner (1978), 31.

[27] See Simmel (1970).

[28] See for example Koepping (1981).

[29] Ricoeur (1981), 17.

preconceptions, presuppositions and prejudices etc. altogether, and concomitantly to attain an absolutely objective or transcendental perspective.

However it should be noted that from all that we have seen so far it seems evident that, as the example of the Martian misunderstanding of their experiences as being of Earthlings suggests, so far from critical distance or, as we shall see, critical reflection, facilitating such a *transcendental* perspective or "*View from Nowhere,*" it merely offers an alternative *perspective* or point of view, which although it *transcends* the particular pre-reflective or uncritical view which it may call into question, comes replete with its own preconceptions and presuppositions, which may or may not be superior, in the sense of offering a more coherent and/or extensive understanding.

Take for example an individual who is firmly convinced that everyone is out to get him/her. This individual may well see even others' apparently innocent behavior as evidence of their attempt to disguise this fact, to the point of regarding others' attempts to reassure him/her as a ploy to put him/her off guard. A psychoanalyst on the other hand, having some critical distance from the person's experience, may interpret and thus see her/his reports and behavior as paranoid, and her/his denial of this as evidence of repression (and the denial of repression as further evidence of it). However, as William Burrows famously insisted "Just because you are "paranoid" doesn't necessarily mean that they are not all out to get you!" Perhaps supposed "paranoiacs" are the only people who really know what is going on. Similarly, absent a truly *transcendental* perspective, how are we to adjudicate the methodological dispute between, say, a vulgar Marxist, who may understand and explain the form of all socio-cultural behavior and institutions, and the fact of all political change, in terms of *deterministic* economic or *material* factors, and a Neo-Hegelian who perceives everywhere in the same phenomena evidence of spirit or *ideas* and *free choice*, and who therefore regards the Marxist, as s/he will in turn be regarded by the Marxist, as a victim of "false consciousness."[30] And how are we to settle the dispute between Freud and Marx regarding the significance of religion? For unlike Freud who saw religious belief as an "illusion" or quasi-psychotic fantasy primarily aimed at overcoming the *fear of death*, Marx on the other hand saw it as "the opium of the (oppressed) masses" intended to ameliorate the *suffering of life* resulting from economic exploitation and political inequality (for instance by the Christian promise that "the meek will inherit the earth," and "the first will be last and the last will be first").

Gadamer's assertion made apropos coherent historical perspectives that "It is by no means settled (and never can be settled) that any particular perspective …is the right one"[31] would seem to hold in all these examples, thus suggesting that it might indeed apply to the full range of the human and social sciences.

---

[30] See my note 10. While I am aware that Marxists normally apply this term to those who, they claim, do not understand what is in their best interests economically, there is no reason why it should not be more widely understood as indicative of delusions of many sorts.

[31] Gadamer (1975), 484. Gadamer's parentheses.

Although forced to acknowledge that none of the examples we have examined so far immediately evidence a presuppositionless or objective perspective, the Positivists will surely object that this in no way implies that there is none, before going on to insist that it is most certainly possible to adjudicate between such intrinsically coherent yet extrinsically conflicting interpretations as those outlined above by appealing to the facts. That, to take the last cited example for instance, if the ending of oppressive exploitation and gross inequality results in the demise of religion, then Marx is correct, while if it does not then Freud would seem to be.

However although the Marxist may regard the contrast between the virulence of fundamentalist religions and religious beliefs amongst the poor with the comparative secularity of much of affluent Western Europe as confirming her/his view, a Freudian might respond by noting that in the US, which is generally more affluent still, religion continues to be much more significant than in Western Europe. To this a Marxist might reply by pointing out that it was the Puritans and other religious zealots followed by the "poor huddled masses" who initially founded and predominantly populated the US, and then insisting upon the "birth marks" interpretation of history (according to which every society will inevitably bear the birth marks of the society from which it emerges) thereby interpreting the continuing significance of religion in the US as a vestigial consequence of past economic hardship soon to fall victim to economic affluence.

What the Positivists appeal to the facts fail to appreciate then is that access to the facts is always mediated by the very preconceptions and presuppositions that they are supposed to verity,[32] and that consequently each individual or group will tend to *conceive* of the "facts" in such a manner as to have them fit their own preconceptions, presuppositions, or *theories* as we might call them. And when confronted with "facts" which prove absolutely recalcitrance to being perceived/conceived in a manner coherent with our theoretical preconceptions, we are often as likely (as we shall presently detail) to ignore them as to abandon our preconceptions, presuppositions or *prejudices* as we might then call them. While even if we indeed abandon our theories or preconceptions etc. in face of apparently recalcitrant facts, it should be noted that there may remain any number of alternative conceptual frameworks or theories in terms of which they may be *coherently* interpreted. Therefore while such *coherence* may be *necessary* to insure the veridicality of our perceptions and understanding, it nevertheless remains *insufficient*.

And while if and insofar as such *intrinsically* coherent alternative interpretations do not conflict, or are not *extrinsically* incoherent with each other, they may possibly be seen as *complementing* or as *supplementing* each other (each offering a different perspective upon, or nuance or aspect to our understanding of, the phenomenon in question) if they are *conflicting*, which is to say mutually exclusive of each other, Gadamer's previous assertion seems to be correct, there being no way, as things stand, of determining which particular perspective is the right one.

Thus although it may, for instance, be possible to see the previously outlined Marxist and Psychoanalytic interpretations of the rise of Nazism as *complementing*

---

[32] See my note 10 above.

each other (in the sense that each can be seen as elucidating the phenomena from a different perspective) and it may be possible to see Marx's and Freud's different interpretations of religion as *supplementing* each other, (in the sense that religion may perform the function of alleviating anxiety in relation *both* to economic hardship and death) the same cannot be said of the just mentioned vulgar Marxist and neo-Hegelian interpretation of social change, or the psychoanalyst's and the "paranoiac's" interpretation of the latter's experiences. And it is by no means clear, given that the perception, by each, of the "facts" is mediated by their own theoretical preconceptions, how we might adjudicate between them, save on the basis of our own preconceptions or prejudices, or perhaps normatively upon, or at least upon *our understanding of,* those of most others.

In recognition of the efficacy of such arguments even that neo-Kantian champion of the Enlightenment—and thus of the Hellenistic side of the Hebraic/Hellenistic debate—Jurgen Habermas, is forced to concede that "…it is always illusory to suppose an autonomy, free of presuppositions, in which knowing first grasps reality theoretically…" Nevertheless, Habermas continues "…the mind can always reflect back upon the interest structure … this is reserved to self-reflection." And "If the latter cannot cancel out interest it can to a certain extent make up for it." [33] That is to say that, recognizing that, as a *contingent* matter of fact, as these and any number of other examples demonstrate, we are certainly *often* unable to free ourselves of the preconceptions and presuppositions which therefore continue to mediate our perceptions and understanding, Habermas none the less remains unwilling to concede that this is always and entirely so, and thus to abandon the Enlightenment pretension altogether.

To this Gadamer's emphatic response is that "The customary enlightenment formula according to which the process of demagnification of the world leads necessarily from mythos to logos seems to me to be a modern prejudice,"[34] or myth as we might perhaps say. Indeed the *concept* of "en-*light*-enment" itself may be understood as a metaphorical notion derived from the Platonic *myth* of escape from the cave of "*dark*" prejudices and presuppositions characteristic of *mythos*, and emergence into the clear "*light*" of *logos.* While its *substantive* application, marked by an attempt to rid oneself of, or escape from, presuppositions and prejudices, is itself based upon a presupposition or prejudice; *viz* that an unprejudiced view, even if possible, is superior to a perspective which embodies presuppositions and prejudices; a presupposition or prejudice which, as such, negates what is supposedly affirmed and thereby undermines itself! As Gadamer's succinctly formulates this self-contradiction at the heart of the Enlightenment the "…one prejudice of the enlightenment that is essential to it: the fundamental prejudice of the enlightenment is the prejudice against prejudice itself…".[35]

Nor should this critique surprise us unduly, for as Godel formally demonstrated, no system can be self-axiomatizing or justifying; any attempt to be so always necessarily ending either in circularity or infinite regress. Thus not even reason—the

[33] Habermas (1978), 313–314. See also 315.

[34] Gadamer (1976), 51.

[35] Gadamer (1975), 239–240.

logos which is the Enlightenment's defining *ethos*—can save us,[36] all such justifications therefore being, as the "Habermas/Gadamer debate" demonstrated[37] and as the Postmodernists insist, ultimately non-rational.[38]

In "light," or perhaps we might better say in view?, of all this it should then be "clear" that, as Gadamer insists "Prejudices are the biases of our openness to the world."[39] While the absence of unprejudiced or unmediated access to the facts thereby asserted would seem to present a direct challenge to the supposed objectivity of the natural, no less than of the human and social sciences!

8 The Natural Sciences: The Hermeneutics of Empiricism

Having had their attempted assault upon the human and social sciences decisively repelled the die hard Positivists now then find themselves in the unenviable position of attempting to retrench around and defend the natural sciences against the onslaught of a counter attack. Forced to acknowledge the role, if not indeed the prevalence, of non-rational prejudices and presuppositions, and subjective interpretations, at least in the human and social sciences and the wider field of inquiry onto which they open, the Positivists nevertheless continue to maintain that the natural sciences at any rate are based upon presuppositionless observation of the empirically given facts, and that they thereby achieve a disinterested or value-free objectivity which they are forced to concede may be absent elsewhere.

However in the first place, as Max Scheler has observed, "To conceive of the world as value-free is a task which men (sic) set themselves on account of a value: the vital value of mastery and power over things."[40] So far from being disinterestedly objective, enlightenment science was, from its inception, animated and guided by human interests. Indeed as that "founding father" Bacon, recognized, "…human knowledge and human power meet in one …truth and utility are here the very same thing,"[41] our conception of knowledge being then, as it still is, pragmatic. Nietzsche explains:

> The *compulsion* to form concepts, genera, forms, ends, and laws ("*one world of identical causes*") should not be understood as though we were capable through them of ascertaining the *true* world, but rather the compulsion to adapt to ourselves a world in which *existence* is made possible. Thereby we create a world that is calculable, simplified, understandable, etc., for us.[42]

---

[36] See Barnes (1974), 5. For a fuller discussion of the Postmodern critique of reason see Glynn (1991).

[37] See How (1980).

[38] See Gadamer (1975), 245, and (1976), 51, and Jacques Derrida (1974), 11. For a fuller discussion see Glynn (2005), 59–76. Note that so far from being synonymous with the irrational, and thus simply the logical negation of rational, the "non-rational" is nether.

[39] Gadamer (1976), 9.

[40] Scheler (1960), 122, FN 2, as quoted by Leiss in (1974), 109.

[41] Bacon as quoted in Rozack (1973), 149.

[42] For but one example see Nietzsche (1970) 3:526.

And as with knowledge, so too with reason also. Thus Nietzsche sees "…logic and the categories of reason as means to the adaption of the world to ends of utility (that is 'in principle,' for a useful *falsification*)."[43]

And as with this reflective *conception* of the epistemology of natural science, so too with the *perception* of the "empirical facts," upon which such science is supposedly based. That is to say that, so far from being disinterestedly objective, the empirical perceptions or experiences of the supposed empirical "facts" of the natural sciences reflect the often pragmatic preconceptions and presumptions, not to say prejudices, of the observing subject no less than we have seen do those of the human and social sciences. As Heidegger, building upon the Kantian insight that the sensible is inextricably intertwined with the intelligible—which implies that while *perception and conception* may be *analytically distinguishable* they nevertheless remain *ontologically inseparable*–affirms:

> The greatness and superiority of the natural sciences during the sixteenth and seventeenth centuries rests in the fact that all the scientists were philosophers. They understood that there are no mere facts but that a fact is only what it is in the light of the fundamental conception… [44]

Beginning with experience itself then, so far from experiencing what William James notably characterized as a "…blooming, buzzing, confusion" of *incoherent* impressions, rather, as Husserl noted, in everyday experience "I do not see color sensations, but colored things, I do not hear sensations of sound, but the song a women is singing etc.,"[45] that "…in immediate givenness, one finds anything but color data, tone data, other "sense" data …. instead …. I see a tree which is green; I hear the rustling of its leaves, I smell its blossoms etc."[46] Maurice Merleau-Ponty affirms, "Pure sensation … corresponds to nothing in our experience…," "…there is no experience of sensation…"[47]

However although many take this, in and of itself, as evidence of the conceptual mediation of our experiences, if and insofar as all of our perceptions really were conceptually mediated interpretations, then it seems that we could not know, in the sense of be certain of, this. For insofar as we experience what we take to be objects and events rather than undifferentiated sensations, then while these experiences *might* be taken to *imply* the mediation of pure sensations by conceptions, equally they could be taken as unmediated reflections, and therefore evidence of the existence, of just such objects and events.

But although our everyday experiences of objects and events provide insufficient grounds for concluding that these perceptions are the product of interpretation, there are other grounds for doing so. Thus drawing upon *Gestalt* experiments in

---

[43] Friedrich Nietzsche (1970) 3:726, as quoted by Habermas (1978), 297. Nietzsche's addition in parentheses.

[44] Heidegger (1976b), 247–248.

[45] Husserl (1913), 374 quoted by Lubbe (1978), 108.

[46] Husserl (1970), 233.

[47] Merleau-Ponty (1962), 3 & 7.

cognition, Thomas Kuhn notes that "The duck-rabbit shows that two men (sic) with the same retinal impressions can see different things; the inverting lenses show that two men with different retinal impressions can see the same thing."[48] While if someone who only perceives, say, the "rabbit," to begin with, is told that the picture may also be seen as a "duck," s/he will then often be able to see the "duck." Furthermore "Ames and his school have shown that when a ball set against a featureless background is silently and rapidly inflated (by an air hose obscured from the observer by the ball itself) it is seen as if it retains its size and was coming nearer," for the reason, as Polanyi explains, that on the basis of most of our past experiences we "…construct() a universal interpretive framework that assumes the ubiquitous existence of objects, retaining their size and shape…"[49] However the experimenter then demonstrates to the observer that what s/he previously interpreted as an *apparent* increase in size due to the ball's *approaching*, was in fact a *real* increase in size due to its *inflation*. Consequently, when the now fully inflated ball (suspended by thin, and therefore invisible, wires) is slowly propelled, at uniform speed and with a linear trajectory, directly towards the observer, who has adopted the inflation "framework" or conception, s/he interprets the now *apparent* increase in size of the ball due to its *approaching*, as a *real* increase in size due to its continued *inflation!*

We can then see that certain conflicts in even our most basic experiences or perceptions reveal them to be mediated by our conceptions or preconceptions regardless of whether or not there is in fact an objective reality existing beyond the realm of experiences or appearances. Thus while Kant, who also denies the existence of (catagorially) unmediated experiences, nevertheless insisted that there *was* a noumenal realm of objects or "Things-In-Themselves," existing independently of the empirically given, or phenomenal, realm of experiences, clearly there can, by definition, be no *direct* empirical evidence of such an objective, experience or appearance transcending, reality. As Husserl affirms, while: "The empiricist talk of the natural scientists often, if not for the most part, gives the impression that the natural sciences are based on the experience of objective nature … The objective is precisely never experienceable as itself."[50]

Indeed: "The contrast between the subjectivity of the life-world and the "objective," the "true" world, lies in the fact that the latter is a theoretical-logical substruction, the substruction of something that is in principle not perceivable, in principle not experienceable…"[51]

That is to say that, as Hume noted, so far from observing objects or "things-in-themselves" we actually experience *bundles* of *constantly changing, interrupted and different perceptions.* It being precisely the *intrinsic continuities, resemblances and similarities* exhibited by or between them that lead us to *interpret* these *changing, interrupted and different perceptions* as appearances of relatively *unchanging, or self-identical and continuously existing* "things-in-themselves";

[48] Kuhn (1970), 126–127.

[49] Polanyi (1958), 96.

[50] Husserl (1970), 128–129.

[51] Husserl (1970), 127.

objects which, with breathtaking (hermeneutic) circularity, we take to *explain* or be the *cause* of, the very experiences of the *continuities, resemblances and similarities*, from which we infer their existence.[52] Clearly then such empirically unverifiable "things-in-themselves" or objects are nothing more nor less than *interpretively* derived theoretical constructs or *theories*, which, regardless of their ontological status, enable us to *explain*, and make *law like predictions regarding the future nature or properties of, our perception*. For instance Einstein informs us that:

> ...the formation of the word, and hence the concept "ball," is a kind of thought economy enabling the child to combine very complicated sense impressions in a simple way.... Mach also thinks ... the formulation of scientific theories ...takes place in a similar way. We try to order the phenomena to reduce them to simple form, until we can describe (and explain and predict we might add) what may be a large number of them with the aid of a few simple concepts.[53]

That is to say that, as with supposed objects, whose hypothetically conceived existence we interpretively infer from the properties and relations displayed by and between the very conceptually mediated perceptions and conceptions they are subsequently taken to explain, so too *mutatis mutandis,* our scientific theories are interpretively inferred from the very properties of and relations displayed by and between such "objects" which we subsequently take them to explain and predict. An interpretation of an interpretation, or Double Hermeneutic; not, as *per* the human and social sciences, of an observer's understanding of an actor's understanding of his/her intentions or experiences etc., but an *explicit* (scientific) theoretical interpretation and consequent understanding of the properties of and relations between "objects," our experiences of which have already been *implicitly* mediated by or interpreted in terms of the pre-conceptions informing even our most basic perceptions. A theoretical interpretation, or understanding, of the "facts" arising from the interpretive preconceptions informs our perceptions supposedly thereof.

Thus turning to the derivation of such scientific theories, observing, for instance that an unsupported ink stand, apple, musket ball and cannon ball all fall down (or close with the earth) I explain this in terms of the theory of gravity, which like the "objects" supposedly influenced thereby, is inexperienceable in and of itself, being derived from the very phenomena which it is, again with breathtaking (hermeneutic) circularity, taken to explain. As C.S. Pierce has so succinctly put it, "...the force of gravity is the cause of the ink stand and other objects falling-although the force of gravity will consist merely in the fact that the ink stand and other objects will fall."[54] And not unlike supposed objects or "things-in-themselves," which enable us to explain and predict our experiences and the relations between them, the theory of gravity enables us to both explain and predict the behavior and relations displayed by and between such objects; *viz* that under similar circumstances all objects will fall.

---

[52] See Hume (1967), 204–208, 211–212 & 215.

[53] Albert Einstein quoted in Heisenberg (1971), 64–65. My additions in brackets.

[54] Pierce (1931–5), Vol. VII, 344.

But some will surely object that so far from being an inexperienceable "occult force"[55] (as members of the Royal Society disparagingly dubbed it) gravity can be and in fact is directly experienced, in aching outstretched arms for instance. However this merely demonstrates the degree to which our *theoretical assumptions or preconceptions* pervade our *perceptions* or experiences themselves, for just as Newtonians will *interpret,* and thus come to understand, such experiences as being experiences of gravity, Aristotelians—believing that, by their very nature, entities removed from their "natural" place on earth strive to return there—will *interpret* the same experiences, as well as the motions of unsupported objects, as earthly bodies striving to reach their earthly homes. As Michael Polanyi tells us "...within two different conceptual frameworks the same range of experiences takes the shape of different facts and different evidence."[56]

And as with gravity, or indeed gravitons also, so too with anti-matter, curved space, or even such mundane "entities" as electrons, atoms and molecules, which, as never having been experienced "in themselves" are therefore clearly reified hypotheses, which similarly owe their supposed existence to the *conceptions* and *preconceptions or theoretical assumptions* underlying our *interpretation* of the very phenomena—such as the behavior of material bodies, tracks across cloud or bubble chambers, lines across photographic emulsions, etc.—which we derive them from, and which, again, to come full hermeneutic circle, we take them to *explain.*

Given such dependence of our "*perception*" or "*understanding*" of the putative "*facts*" upon such *implicit preconceptions* etc. it is perhaps unsurprising that, faced with a conflict between the *explicit theoretical hypotheses* characteristic of science and "empirical observations" of "*facts*" which appear to refute them, so far from abandoning or adjusting the theories so that they fit the facts, on the contrary, scientists often *reinterpret* the "facts" so as to make them fit their theories, while facts that prove recalcitrant are often simply ignored.

For instance, having formed the hypothesis or theoretical preconception that all heavenly bodies move in circular orbit, Galileo, upon looking through his telescope, observed that comets and asteroids, not to mention planets, moved in elliptical orbits. So far from this leading him to abandon his theory he simply interpreted his observations as illusions resulting from imperfections in the telescope's lenses. Or take the famous experiment by which Galileo supposedly proved the theory of uniform acceleration. If the musket ball and cannonball had been dropped simultaneously from Pisa's leaning tower, differences in the air resistance they would have encountered would have meant that the musket ball, being smaller, would have hit the ground first. Nevertheless as Einstein, in accord with all we saw previously regarding our most basic perceptions, affirms "In reality it is the theory that decides what we can observe,"[57] in which case, either Galileo, already committed to the theory of uniform acceleration, would have observed them as

[55] See Rozack (1973), 362ff.

[56] Polanyi, *Personal Knowledge*, 167.

[57] Albert Einstein quoted in Heisenberg (1971), 63.

hitting the ground together, or, if he did not, he would not have "observed" this fact in the sense of considering it noteworthy, and would simply have ignored it. Furthermore, so far from us understanding this incident as a refutation of his theory by the facts, in light of what we have seen we can now re-interpret it as a clash between his explicit scientific theory (of uniform acceleration) and whatever other theoretical preconception was mediating his perception. In other words, if, as Popper, in accordance with what we have seen previously, affirms, "...observations ...are always *interpretations ...in the light of theories,*"[58] then quite clearly, as Imre Lakatos insists "...clashes between theories and factual propositions are not "falsifications" but merely inconsistencies"[59] between the *explicitly deployed* theoretical hypotheses of the sciences, and the often *implicitly employed* interpretive preconceptions informing our supposedly immediate perceptions of the "facts;" clashes between the two aspects of the Double Hermeneutic of the natural sciences.

In light of this we can now see that, so far from the consistency or coherence of the empirically observed "facts" and our scientific theories being a consequence of the theoretical conceptions of science being brought into line with the facts, it is, on the contrary, often the result of our "perceptions" of the "facts," being brought into line with the theoretical conceptions of science. Furthermore, while inconsistencies or incoherences between the scientific theories and the "fact" may indicate that one or the other must be rejected or revised, on the other hand their consistency or *coherence* is never *sufficient* to insure their veridicality, and can do no more than provide a degree of verisimilitude. Moreover the complete consistency of Popper's notion of falsification [60] with this ramification of a hermeneutic understanding of the natural sciences serves as yet further evidence, if such evidence were needed, that in addition to its already well established applications, hermeneutic interpretation is central to the natural sciences also.

9 Ontological Hermeneutics

Now if not only the scriptures, but the classics, and the arts and the humanities in general, not to mention the human, and social sciences, and, as we now see, even the natural sciences also are all hermeneutic, then one might further speculate, as Heidegger goes on to claim, that all human inquiry and understanding is hermeneutic. Indeed arguing that all inquiry, including that into existence itself, is grounded upon reflective human consciousness, Heidegger insists that "...man (sic) should be understood, within the question of being (which is to say from an

[58] Popper (1959), 107, Fn. 3. My addition in brackets.

[59] Lakatos (1970), 99.

[60] That Universal theories can never be definetively verified but gain greater credability with the failure of diverse observations to falsify them.

ontological perspective) as *the* site which being requires in order to disclose itself. Man is the site of the Openness of the there"[61]; the (reflective or human) being or *entity* which Being or *existence* in general, requires in order to reflect upon, and thus understand, itself. Further, Heidegger continues "...to work out the question of Being" or understand what it means to Be "... is an entity's mode of *Being*... this entity we shall denote by the term "*Dasein*"."[62] "*Understanding of Being is a defining characteristic of Dasein's Being.*"[63]

It is then ultimately within or upon this most general of all, or *ontological*, hermeneutics, the quest to understand the meaning of existence or Being, that *epistemological* hermeneutics, the quest to understand the totality of particular entities or modes of being, is subsumed or grounded. As Heidegger affirms:

> Philosophy is Universal Phenomenological Ontology and takes its departure from the hermeneutics of *Dasein*, which as an analytic of *existence*, has made fast the guiding line for all philosophical inquiry at the point where it *arises* and to which it *returns*.[64]

In which case finally, as Ricoeur concludes:

> ...the properly *epistemological* concerns of Hermeneutics...are subordinated to ontological preconceptions...*understanding* ceases to appear as a simple *mode of knowing* in order to become *a way of being*.[65]

References

Barnes, Barry. 1974. *Scientific Knowledge and Sociological Theory*. London: Routledge and Kegan Paul.

Bauman, Zygmunt. 1978. *Hermeneutics and the Social Sciences*. London: Hutchinson.

Derrida, Jacques. 1974. "White Mythology: Metaphor in the Text of Philosophy." *New Literary History* 6(1): 5–74.

Derrida, Jacque. 1978. *Writing and Difference*. Trans. Alan, Bass. Chicago: University of Chicago Press.

Dilthey, Willhelm. 1957. Die Geistige Welt. In *Gessammelte Schriften*, vol. 5. Stuttgart: Teubner.

Earmarth, Michael. 1978. *Wilhelm Dilthey: The Critique of Historical Reason*. Chicago: Chicago University Press.

Gadamer, Hans-Georg. 1975. *Truth and Method*, 2nd ed. London: Sheed and Ward.

Gadamer, Hans-Georg. 1976. *Philosophical Hermeneutics*. ed. D. Linge. Berkeley: University of California Press.

Glynn, Simon. 1991. "The Deconstruction of Reason." *Man and World: An International Journal of Epistemology* 24(3): 311–320.

Glynn, Simon. 2005. "The Mythos Logos Deconstructed." In *Dialogue and Universalism* XV(3/4): 59–76.

[61] Heidegger (1959), 205. My addition in parentheses.

[62] Heidegger (1962), 27. My addition in parentheses.

[63] Heidegger (1962), 32. See also 27.

[64] Heidegger (1962), 62.

[65] Ricoeur (1981), 44.

Habermas, Jurgen. 1978. *Knowledge and Human Interests* (2nd ed.). Trans. J. Shapiro. London: Heinemann.
Heidegger, Martin. 1959. *Introduction to Metaphysics*. Trans. R. Manheim. New Haven: Yale University Press.
Heidegger, Martin. 1962. *Being and Time*. Trans. J. Macquarrie, and E. Robinson. New York: Harper and Row.
Heidegger, Martin. 1971. "*Plato's Doctrine of Truth*." Trans. J. Barlow. *Philosophy in the 20th Century* (Vol. 3), eds. W. Barrett, and H. Atkins, 251–270. New York: Random House.
Heidegger, Martin. 1975. *Moira* (Parmenides VIII, 3, 4–41) & *Aletheia* (Heraclitus Fragment B. 16). *Early Greek Thinking*. Trans. David, Krell and F. Capuzzi. 79–101 & 102–123. New York: Harper and Row.
Heidegger, Martin. 1976a. "On The Essence of Truth." In Martin Heidegger: Basic Writings, ed. D. Krell, 112–141. New York: Harper and Row.
Heidegger, Martin. 1976b. "What is a Thing" reprinted in part as Modern Science, Metaphysics and Mathematics. In Martin Heidegger: Basic Writings, ed. Krell David, 243–282. New York: Harper and Row.
Heisenberg, Werner. 1971. *Physics and Beyond*. Trans. A. Honeruns, ed. R. Rashon. New York: Harper and Row.
How, Alan. 1980. "Dialogue as Productive Limitation in Social Theory: The Habermas-Gadamer debate." In *The Journal of the British Society for Phenomenology* II(2): 131–143.
Hume, David. 1967. *A Treatise on Human Nature*. Oxford: Clarendon.
Husserl, Edmund. 1913. *Logische Untersuchungen* (revised ed.). Halle: M. Niemeyer. Vol. II, Part I.
Husserl, Edmund. 1970. *The Crisis of European Science and Transcendental Phenomenology.* Trans. D. Carr. Evanston: Northwestern University Press.
Koepping, Klaus-Peter. 1981. "Leib und Leiben, Sprache und Spiel, Schweigen und Scham." In *Der Wissenshaftler und des Irrationale*, ed. Duerr Hans-Peter, 296–329. Frankfurt: Syndikat.
Kuhn, Thomas. 1970. *The Structure of Scientific Revolutions*, 2nd ed. Chicago: University of Chicago Press.
Lakatos, Imre. 1970. "The Methodology of Scientific Research Programs." In *Criticism and the Growth of Knowledge*, ed. I. Lakatos and A. Musgrove, 91–196. Cambridge: Cambridge University Press.
Leiss, William. 1974. "*The Domination of Nature*." Boston: Beacon.
Lubbe, Hermann 1978. "Positivism and Phenomenology: Mach and Husserl." In *Phenomenology and sociology*, ed. Luckmann Thomas, 90–118. Harmondsworth: Penguin.
Merleau-Ponty, Maurice. 1962. *The Phenomenology of Perception.* Trans. C. Smith. London: Routledge and Kegan Paul.
Nietzsche, Friedrich. 1970. *Werke*, 2nd ed. Munich: Hansar.
Pierce Charles, S. 1931–5. Lectures on Pragmatism. In *Collected papers*, ed. C. Hartshorne and P. Weiss, 13–121. Cambridge, MA: Harvard University Press.
Plessner, Helmuth. 1978. With Different Eyes. In *Phenomenology and Sociology*, ed. Thomas Luckmann, 27–41. Harmondsworth: Penguin.
Polanyi, Michael. 1958. *Personal Knowledge*. London: Routledge and Kegan Paul.
Polkinghorne, Donald. 1983. *Methodology for the Human Sciences*. Albany: SUNY Press.
Popper, Karl. 1959. *The Logic of Scientific Discovery*. London: Hutchinson.
Ricoeur, Paul. 1981. *Hermeneutics and the Human Sciences*. ed. J.P. Thompson. Cambridge: Cambridge University Press.
Rozack, Theodore. 1973. *Where the Wasteland Ends*. London: Faber and Faber.
Saussure, Ferdinand de. 1959. *Course in General Linguistics*. Trans. Wade, Baskin. New York: NewYork Philosophical Library.
Scheler, Max. 1960. *Weissesformen. Gesammelte Werke, Band 8*, 2nd ed. Bern: Verlag.

Schrag, Calvin. 1980. *Radical Reflection*. Indiana: Perdue University Press.
Schütz, Alfred. 1962. Concept and Theory formation in the Social Sciences. In *Collected papers*, 1st ed. The Hague: Nijhoff.
Schütz, Alfred. 1972. *The Phenomenology of the Social World.* Trans. G. Walsh, and F. Lennert. London: Heinemann.
Schütz, Alfred, and Luckmann, Thomas. 1974. *The Structure of the Life-World.* Trans. R. Zaner, and H. Engelenhardt Jr. London: Heinemann.
Simmel, Georg. 1970. *Grundfragen der Soziologie*. Berlin: de Gruyter.

Notes on Contributors

Babette Babich is Professor of Philosophy at Fordham University in New York. She is author of seven books, including *The Hallelujah Effect: Philosophical Reflections on Music, Performance Practice and Technology* (2013), *La fin de la pensée? Philosophie analytique contre philosophie continentale* (2012), and *Nietzsche's Philosophy of Science* (German 2010; Italian 1996, English 1994). Her work has been translated into Chinese, French, German, Hungarian, Italian, Spanish, and Slovenian. Editor or co-editor of eight book collections, she is founding (1994) and Executive Editor of *New Nietzsche Studies*. Her interests include continental philosophy of science and technology, musical aesthetics, animal rights/ethics, ancient Greek bronzes, the academic politics of the analytic-continental divide, and distinguishing science from scientism.

Enrico Berti (born in Valeggio sul Mincio, Italy, 1935) is emeritus professor of the University of Padua and Doctor *honoris causa* of the National and Capodistrian University of Athens. He was professor of philosophy in the universities of Perugia, Padua, Geneva, Bruxelles and Lugano. He is member of the Accademia Nazionale dei Lincei (Rome), of the International Academy for Philosophy (Yerevan, Armenia) and of the Pontifical Academy of Sciences (Vatican City), and president of the International Institute of Philosophy (Paris). His scientific interests are directed to ancient and contemporary philosophy, especially Aristotle's thought and its survival today, so in logic as in metaphysics, anthropology, ethics and politics. He is author of many publications (books and articles), some of which have been translated in English, French, Polish, Spanish and Portuguese.

Robert P. Crease is a philosopher and historian of science. He writes a monthly column, "Critical Point," for the international physics magazine *Physics World*. He is Chairman of the Department of Philosophy at Stony Brook University, where he has taught since 1987. He is a Fellow of the American Physical Society and the Institute of Physics. He has written, co-written, translated, or edited over a dozen books. His most recent books include *World in the Balance: The Historic Quest for a Universal System of Measurement* (2011), *The Great Equations: Breakthroughs in*

B. Babich and D. Ginev (eds.), *The Multidimensionality of Hermeneutic Phenomenology*, Contributions to Phenomenology 70, DOI 10.1007/978-3-319-01707-5,

Science from Pythagoras to Heisenberg (2009); *J. Robert Oppenheimer: A Life*, by Abraham Pais, with Robert P. Crease (2006); and *The Prism and the Pendulum: The Ten Most Beautiful Experiments in Science* (2003).

Jan Faye, Section for Philosophy, University of Copenhagen, Denmark. He has written extensively on the philosophy of the natural sciences and the humanities in both English and Danish. Among the English books published are *TheReality of the Future* (1989), *Niels Bohr: His Heritage and Legacy* (1991), *Rethinking Science* (2002), and *After Postmodernism* (2012).

Robert Frodeman holds a doctorate in philosophy and a master of science in geology and is Professor of Philosophy at the University of North Texas, where he specializes in environmental philosophy, the philosophy of science and technology policy, and the philosophy of interdisciplinarity. He served as a consultant for the US Geological Survey for 8 years and was the 2001–2002 Hennebach Professor of the Humanities at the Colorado School of Mines, and ESRC Fellow at Lancaster University in England (2005). In addition to more than 80 published articles, Frodeman is author/editor of nine books, including *Geo-Logic: Breaking Ground between Philosophy and the Earth Sciences* (2003), *Encyclopedia of Environmental Ethics and Philosophy* (2008), *Oxford Handbook of Interdisciplinarity* (2010) and *Sustainable Knowledge* (2013).

Rodolphe Gasché is SUNY Distinguished Professor & Eugenio Donato Professor of Comparative Literature at the State University of New York at Buffalo. His books include *Die hybrideWissenschaft* (1973); *System und Metaphorik in der Philosophie von Georges Bataille* (1978); *The Tain of the Mirror: Derrida and the Philosophy of Reflection* (1986); *Inventions of Difference: On Jacques Derrida (Harvard, 1994); The Wild Card of Reading: On Paul de Man* (1998); *Of Minimal Things: Studies on the Notion of Relation* (1999); *The Idea of Form: Rethinking Kant's Aesthetic* (2003); *Views and Interviews. On "Deconstruction" in America* (2006); *The Honor of Thinking: Critique, Theory, Philosophy* (2007); *Europe, or The Infinite Task. A Study of a Philosophical Concept* (2009); *Un Arte Muy Fragile. Sobre la Retorica de Aristoteles* (2010; *The Stelliferous Fold. Toward a Virtual Law of Literature's Self-Formation* (2011); *Georges Bataille: Phenomenology and Phantasmatology* (2012). He has just finished another booklength study entitled: *Geophilosophy: On Gilles Deleuze and Félix Guattari's What is Philosophy?*

Dimitri Ginev is founder and editor-in-chief of the journal *Studia Culturologica*. Among his books are: *Entre Anthroplogie et Hermeneutique* (2003), *Das hermeneutische Projekt Georg Mischs* (2011), and *Practices and Possibilities* (2013). His main research interest is the history of modern hermeneutics.

Simon Glynn received his Ph.D. from The University of Manchester in England and is Professor of Philosophy at Florida Atlantic University. His research interests are in Contemporary Continental Epistemology (from Phenomenology and Hermeneutics, to Structuralism, and Post-structural Deconstruction) and its application to the natural, human and social sciences, and cultural analysis and interpretation, as well as in philosophy of technology. In addition to numerous articles and chapters in journals and books, he is contributing editor of book collections including *Sartre: An Investigation of Some Major Themes* (1987); *European Philosophy and the Human and Social Sciences* (1987) and contributing co-editor of *Continental and Postmodern Perspectives in the Philosophy of Science* (1995). He is currently working on a book applying Contemporary Continental Epistemology to the methodology of the sciences.

Patrick Aidan Heelan is William A. Gaston Professsor of Philosophy emeritus at Georgetown University and a Jesuit priest. He has a doctorate in theoretical physics specializing in geophysics and high energy physics, having studied with Erwin Schrödinger (Dublin Institute for Advanced Studies) and Eugene Wigner (post-doc Princeton) and a doctorate in philosophy (at the University of Louvain in Belgium) specializing in the study of Heisenberg's physical physical from the perspective of Husserl, Merleau-Ponty, and Heidegger and in active dialogue with Heisenberg himself which may be found in his books *Quantum mechanics and Objectivity* (1965), *Space-Perception and the Philosophy of Science* (2nd edition: 1989), and *The Observable* (2013). In addition to his interest in questions relating to science and religion, Heelan has published extensively on the philosophy of the quantum theory, contextual logic, the hermeneutics of theory and experiment, and the task-dependent geometry structure of human spatial vision in everyday life and in pictorial art and architecture.

Pierre Kerszberg is Professor of Philosophy at the University of Toulouse since 2000. He was previously Researcher at the National Foundation for Scientific Research (Belgium), Senior Lecturer at the University of Sydney (Australia), and Professor at the Pennsylvania State University (USA). His main interests lie in the history of physics and cosmology, Kant, and phenomenology. His book publications include *The Invented Universe* (1989), *Critique and Totality* (1997), *Kant et la Nature* (1999), *L'Ombre de la Nature* (2009), *La Science dans le Monde de la Vie* (2012). Future research projects include the philosophy of music.

Theodore Kisiel is Distinguished Research Professor emeritus of philosophy at Northern Illinois University, is translator of Heidegger's Summer 1925 course, *History of the Concept of Time: Prolegomena* (1992) as well as author of *The Genesis of Heidegger's Being and Time* (1995), *Heidegger's Way of Thought: Critical and Interpretive Signposts* (2002), in addition to co-editing, with Joseph J. Kockelmans, *Phenomenology and the Natural Sciences: Essays and Translations* (1970). He is also author of numerous articles on Heidegger and hermeneutic phenomenology.

Giovanni Leghissa is Research Professor of Epistemology of the Humanities at the Department of Philosophy, University of Torino, Italy. He was Visiting Professor at the Institut für Philosophie at the University of Vienna, Austria, and at the Hochschule für Gestaltung, Karlsruhe, Germany. His work focuses on phenomenology, continental philosophy, psychoanalysis, gender and cultural studies, comparative philosophy. He has authored five books: *L'evidenza impossibile. Saggio* *sull' immaginazione in Husserl* (1999); *Il dio mortale. I potesi sulla religiosità moderna* (2004); *Il gioco dell' identità. Differenza, alterità, rappresentazione* (2005), *Incorporare l'antico. Filologia classica e invenzione della modernità* (2007). *Neoliberalismo. Un' introduzione critica* (2012). He has edited six collective volumes and special issues of journals as well as Italian editions of Husserl, Derrida, Blumenberg, Hall, de Certeau, Overbeck, Tempels. He is member of the editorial board of the journal of philosophy *aut aut*.

Lin Ma is associate professor of philosophy at Renmin University of China (Beijing). She is the author of *Heidegger on East–west Dialogue: Anticipating the Event* (2008), and has published essays on comparative studies of Levinas and Daoism. Currently she is co-organizing a special issue "Re-Discovering Heidegger and Chinese Philosophy" for the *Journal of Chinese Philosophy*.

Jeff Malpas is Distinguished Professor at the University of Tasmania and Visiting Distinguished Professor at Latrobe University. He is the author or editor of 21 books, and has published over 100 scholarly articles on topics in philosophy, art, architecture, and geography. His work is grounded in post-Kantian thought, especially the hermeneutical and phenomenological traditions, as well as in analytic philosophy of language and mind, and draws on the thinking of a diverse range of thinkers including, most notably, Albert Camus, Donald Davidson, Martin Heidegger, and Hans-Georg Gadamer. He is currently working on topics including the ethics of place, the failing character of governance, the materiality of memory, the topological character of hermeneutics, the place of art, and the relation between place, boundary, and surface.

Graeme Nicholson, Professor Emeritus of Philosophy at the University of Toronto, has been an active member of the Heidegger Circle and Society for Phenomenology and Existential Philosopophy; co-editor of the series *Contemporary Studies in Philosophy and the Human Sciences* at Humanities (later Humanity) Press; author of some books in that series and, more recently, of *Plato's Phaedrus: The Philosophy of Love* (1999); and *Justifying Our Existence: An Essay in Applied Phenomenology* (2009).

Adriaan T. Peperzak is Arthur J. Schmitt Professor of Loyola University in Chicago. His research in the history of philosophy has focused on Hegel—in six books and numerous articles—as well as Emmanuel Levinas—two books and three edited collections. In particular, he is author, most recently, of *Modern Freedom* (2001), *Thinking* (2006) and *Trust* (2013). He has also published extensively on Plato, Aristotle, Bonaventura, Descartes, Heidegger, and Ricoeur, and on thematic questions in ethics, social and political philosophy, metaphilosophy, and philosophy of religion.

Nicholas Rescher (Ph.D., Princeton, 1951) is Distinguished University Professor of Philosophy at be University of Pittsburgh where he has also served as Chairman of the Philosophy Department and a Director (and currently Chairman) of the Center for Philosophy of Science. In a productive research career extending over six decades he has more than 100 books to this credit. Thirteen books about Rescher's philosophy have been published in five languages. He has served as a President of the American Philosophical Association, of the American Catholic Philosophy Association, of the American G. W. Leibniz Society, of the C. S. Peirce Society, and of the American Metaphysical Society as well as Secretary General of the International Union of History and Philosophy of Sciences. Rescher has been elected to membership in the American Academy of Arts and Sciences, the Academia Europea, the Royal Society of Canada, and the Royal Asiatic Society of Great Britain. He has been awarded the Alexander von Humboldt prize for Humanistic Scholarship in 1984, the Belgian Prix Mercier in 2005, and the Aquinas Medal of the American Catholic Philosophical Association in 2007. In 2011 he was awarded the premier cross of the Order of Merit (Bundesdienstkreuz Erster Klasse) of the Federal Republic of Germany.

Gregor Schiemann is Professor for History and Philosophy of science at the Department of Philosophy at the Bergische Universität in Wuppertal. He holds a diploma in physics and a PhD in Philosophy. His areas of specialization are: history and philosophy of physics, and the concept of nature. He is coeditor of the *Journal for General Philosophy of Science*. Recent publications among other things are: *Hermann von Helmholtz's Mechanism: The Loss of Certainty* (2008); *Werner Heisenberg* (2008); and, as editor (with Michael Heidelberger) *The Significance of the Hypothetical in the Natural Sciences* (2009) (with Alfred Nordmann and Hans Radder) *Science Transformed? Debating Claims of an Epochal Break* (2011).

Holger Schmid teaches philosophy at the University of Lille III, France. He has published *Nietzsches Gedanke der tragischen Erkenntnis* (1984) and *Kunst des Hörens: Orte und Grenzen philosophischer Spracherfahrung* (1999) as well as articles published in German, French, and English bearing on subjects ancient and modern, with special attention to problems of language and art.

Michael Stöltzner is associate professor of philosophy at the University of South Carolina. He has studied physics and philosophy at Tübingen, Trieste, Vienna, and Bielefeld and held positions at the Universities of Salzburg, Bielefeld and Wuppertal. His main areas of research are philosophy and history of physics and applied mathematics, core principles of mathematical physics; history of logical empiricism; the development of formal teleology; and the philosophy of applied science, in particular the role of models and ceteris paribus laws.

Jaap van Brakel is professor emeritus at the Higher Institute of Philosophy of the University of Leuven. His current research interest concerns the necessary preconditions of comparative and intercultural philosophy.

Index

B. Babich and D. Ginev (eds.), *The Multidimensionality of Hermeneutic Phenomenology*, Contributions to Phenomenology 70, DOI 10.1007/978-3-319-01707-5,

CPSIA information can be obtained at www.ICGtesting.com
Printed in the USA
LVOW01*1244020214

371959LV00003B/397/P

9 783319 017068